AN INTRODUCTION TO FLUID DYNAMICS • **Principles of Analysis and Design**

AN INTRODUCTION TO FLUID DYNAMICS

PRINCIPLES OF ANALYSIS AND DESIGN

Stanley Middleman
University of California, San Diego

John Wiley & Sons, Inc.
New York • Chichester • Weinheim • Brisbane • Singapore • Toronto

ACQUISITIONS EDITOR Wayne Anderson
MARKETING MANAGER Harper Mooy
SENIOR PRODUCTION MANAGER Lucille Buonocore
SENIOR PRODUCTION EDITOR Monique Calello
COVER DESIGNER Carol C. Grobe
ILLUSTRATION COORDINATOR Jaime Perea
ILLUSTRATION Thunder Graphics, Inc.
MANUFACTURING MANAGER Monique Calello

This book was set in 9.5/12 Times Roman by Bi-Comp, Inc. .

Recognizing the importance of preserving what has been written, it is a policy of John Wiley & Sons, Inc. to have books of enduring value published in the United States printed on acid-free paper, and we exert our best efforts to that end.

The paper in this book was manufactured by a mill whose forest management programs include sustained yield harvesting of its timberlands. Sustained yield harvesting principles ensure that the number of trees cut each year does not exceed the amount of new growth.

To order books or for customer service please, call 1(800)-CALL-WILEY (225-5945).

Library of Congress Cataloging-in-Publication Data
Middleman, Stanley.
    An introduction to fluid dynamics : principles of analysis and
design / Stanley Middleman.
      p. cm.
    Includes bibliographical references and index.
    ISBN 0-471-18209-5 (cloth : alk. paper)
    1. Fluid dynamics—Mathematical models. I. Title.
TA357.M525 1998
620.1'06—dc21
                                                                  97-9530
                                                                     CIP

10 9 8 7 6 5 4 3

This book is dedicated to

Yocheved bat Miriam

Ani L'Dodi

Preface

This textbook is the outgrowth of 36 years of teaching this material to students in chemical engineering, mechanical engineering, bioengineering, and applied mechanics. As such, it represents my ideas regarding how the material should be taught to engineering students. Primary is my belief that the development of a mathematical model is central to the analysis and design of an engineering system or process. Hence this text, like my classroom approach to this material, is oriented toward teaching students how to develop mathematical representations (models) of physical phenomena. The key elements in model development involve assumptions about the physics, the application of basic physical principles, the exploration of the implications of the resulting model, and the evaluation of the degree to which the model mimics reality. This latter point—evaluation—is critical. It requires that the model builder have some a priori sense of what is required of the model, in terms of both the specific phenomena to be described and the desired degree of correspondence of the predictions to the observations. As a consequence, a great deal of effort has been put forth to provide many examples of experimental data against which the results of modeling exercises can be compared. Students need to see that simple models often are quite accurate, and that when a model fails to yield the required accuracy, the assumptions of the model need to be explored and reconsidered.

Another goal of this text is to expose students to the wide range of technologies to which they may be asked to apply their skills. Hence, few of the problems illustrated in this text, or raised in the Problem sections, are sterile mathematical analyses. Where possible, the examples presented are motivated by real engineering applications. Many of the problems are derived from my years of experience as a consultant to companies whose businesses cover a broad spectrum of engineering technologies. As a teacher, I have found that students are more motivated by problems having a basis in commercial technology than by those whose orientation is primarily mathematical or analytical.

The text assumes that the reader/student has been exposed to a basic course in ordinary differential equations. My observation is that most students learn their mathematics in an environment that is separated from application. In addition, many students enroll in the required calculus courses early in their academic program, whereas a course in fluid dynamics calling for the solution of ordinary differential equations may come more than a year after the completion of the math sequence. This is a serious handicap to both the student and the instructor, and remedial instruction may be required. Since, however, space does not permit any significant review of mathematical concepts here, instructors must evaluate the needs of their students and proceed accordingly.

This textbook provides more than sufficient material to support a one-semester course in fluid dynamics; for curricula that present fluid dynamics in a single quarter, there is an abundance of teaching material. The book is deliberately overwritten. The instructor may choose to illustrate certain concepts with different sections of the text from one year to the next. Indeed, there are so many exercises in the Problem sections that it should not be necessary to assign the same problems in successive years. Upon reading the Solutions Manual carefully, the instructor will find a selection of problems, the modeling of which can be used to extend the classroom presentation beyond the confines of the text proper. Many of these problems are worked out in sufficient detail, and with accompanying commentary, to provide a basis for use by teaching assistants as material in discussion sections.

Stanley Middleman

ACKNOWLEDGMENTS

I am indebted to many generations of students and colleagues whose questions and criticisms have improved the text. Over the years, many undergraduate students have worked in my laboratory carrying out experiments aimed at illustrating models presented in the classroom. Such examples are typically cited as "unpublished data" throughout the text. A number of colleagues reviewed the manuscript prior to my final revisions. I make special acknowledgment of my gratitude to Professor Richard Calabrese of the University of Maryland, College Park, for his gift of a line-by-line critique of the text. The final form of this book is considerably improved in clarity because of his input. The inconsistencies, obscurities, and out-and-out errors that remain are all mine to correct in subsequent editions, and I hope that my attention will be drawn to these imperfections. I have taught fluid dynamics from many textbooks over many semesters, incorporating material from these books into many generations of course notes. Over the years the origin of much of this material has been forgotten. Readers are encouraged to bring my attention to any examples of material in the text or in the Problem sections that warrant acknowledgment and attribution of the original source.

S. M.

Contents

Chapter 1

What Is Fluid Dynamics?

This introductory chapter gives some examples of the variety of problems we deal with in this textbook. Fluid dynamics is a very broad field, and the range of interesting and important problems is great. Thus it is essential to develop the habit of thinking and speculating about the *physics* of any problem in fluid dynamics. This involves two main issues: How does the system respond if operating conditions change, and what variables and parameters play an important role in the observed physics? These considerations lead us naturally to a discussion of dimensional analysis, a topic we take up again in more detail in Chapter 5.

Fluid dynamics is a branch of mechanics, or physics, that seeks to describe or explain the nature of physical phenomena that involve the flow of liquids and/or gases. It will be necessary to understand and describe in quantitative (mathematical) terms the manner in which forces are transmitted through a fluid, and by (or on) a fluid that is acting on (or is acted upon by) another fluid or solid at its boundary. We will learn to come to terms with the precise physical meaning and mathematical representation of such ideas as velocity, stress, pressure, momentum, and energy. We will see how statements of fundamental principles such as conservation of mass, momentum, and energy lead to equations that govern or describe the interrelationships of velocity, stress, pressure, and so on. We will learn how to predict the variation of velocity and pressure with position within a fluid, and as a function of time.

> *One of our primary goals will be to produce mathematical models that permit us to understand, describe, and design engineering systems and processes that involve the mechanics and dynamics of fluids.*

This is a textbook in *engineering analysis*. We will often be more concerned with obtaining a mathematical model that has engineering utility than with developing an analysis that is aesthetically and logically satisfying. We will not use illogical methods to gain our goals. But we will often sacrifice precision to obtain a quick estimate of the behavior of a system of interest to us. Quick approximate estimates based on sound physical principles and good engineering judgment are of enormous value to the engineer. One of the primary goals of this textbook is to teach this skill.

There are two central ideas that I want to emphasize in this book. **One is that the principal goal of an exercise in mathematical modeling is to produce testable predictions.**

While we would not waste time developing models based on very flimsy premises, we must be bold in exploring the consequences of a set of *rational* premises, even though we recognize that these premises represent what seems to be a very poor approximation—an oversimplification of the phenomenon or process of interest to us.

> *A testable—albeit poor—model can be an important stepping-stone in the path toward improving our understanding of the world about us.*

The second important idea is that a primary characteristic of the engineer is interest in the *design* of a device or a process. We develop mathematical models as tools for design.

The design process can exist without analysis; that is, it can be a trial-and-error process in which we ultimately produce a solution to an engineering problem by utilizing intuition and experience. The history of engineering development offers many such examples. But design can also be a process *directed* by application of sound engineering principles. This is the kind of design we describe in this book, not because it is better than the trial-and-error kind, but because it is teachable.

1.1   THINKING ABOUT FLUID DYNAMICS: SOME TYPICAL PROBLEMS

Let's begin our study of fluid dynamics by describing a sampling of interesting and important problems that call for a knowledge of the flow field. We will discuss a few of these examples from the perspective of "parameterization." That is, we want to know what parameters determine the behavior of the flow system. The parameters we can choose include the physical properties of the fluids, the length scales and shape factors that characterize the geometry of the boundaries of the flow, and the variables that we can control or impose on the system, such as pressures and flowrates, or the motion of portions of the boundaries.

1.1.1   Drop Breakup in a Stirred Tank

Suppose we place a liquid (e.g., water) in a baffled tank (T) and agitate it with a high speed rotating impeller (I) as in Fig. 1.1.1. A small amount of an immiscible oil is then added to the water in the tank. We wish to be able to predict the size of the oil droplets formed in this system. Here's a specific "thought experiment." We place some volume fraction (denoted by ϕ) of a light oil in the water-filled tank. The oil is assumed to be

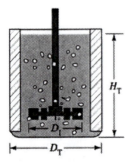

Figure 1.1.1 Oil droplets in an agitated tank of water.

Increasing stirring speed ⟶

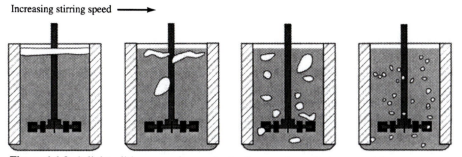

Figure 1.1.2 A light oil becomes increasingly dispersed as stirring speed increases.

immiscible in water. If we do nothing else, the oil will float on top of the water as a distinct film or layer.

What do we expect to observe as we increase the rotational speed N of the impeller? Probably we will see something similar to what is suggested in the sketches of Fig. 1.1.2. Our expectation (our intuition) is that the oil will be broken up into blobs and drops, and that the drop size will get smaller as the agitation speed increases.

Now we turn to the idea of "parameterization." What physical properties do we expect to be important in determining the dynamics of this system? Table 1.1.1 presents a reasonable list.

An important practical feature of this system is the size of the oil drops formed as a consequence of the agitation. If we make careful observations we will find that the dynamics of this complex flow create not a *single* drop size, but a range or distribution of sizes. The goal (the end result) of an analysis of this problem would be a mathematical model for the drop size distribution function, and the *mean* drop size, as a function of relevant physical properties and operating and design parameters. It is no simple task to develop such a model.

Table 1.1.1 Important Physical Properties for the System Illustrated in Fig. 1.1.1

Parameter	Symbol	Dimensions[a,b]
Parameters that are physical properties		
Viscosity of phase 1 (water)	μ_1	m/Lt
Viscosity of phase 2 (oil)	μ_2	m/Lt
Densities of each phase	ρ_1, ρ_2	m/L^3
Interfacial tension between the phases	σ_{12}	m/t^2
What *operating* parameters do we expect to be important?		
Rotational speed	N	t^{-1}
Water temperature	T	T
What *design* parameters would affect the system?		
Geometrical parameters	H_T, D_T, D_I	L
Is the volume fraction ϕ expected to matter?		
If so, we add	ϕ	None

[a]The following conventional symbols are used: L, length; t, time; m, mass; T, temperature.
[b]It may not be intuitively obvious that the dimensions of viscosity and interfacial tension are as stated here. Accept these as fact for the moment. We define the terms explicitly, later.

Suppose, instead of a mathematical model, we have some measurements (observations) of the mean drop size, call it $\overline{D}$, obtained through experiments in agitated tanks of various sizes, at different stirring rates N, and with different immiscible liquids in water. Suppose that all our stirred tanks are geometrically similar, though of different sizes. By "geometrical similarity" we mean that if we had photographs of all the tanks, we could not tell one from another except by size. In other words, the tanks all have the same shape. Hence each tank would have the same values of *ratios* of characteristic lengths, such as H_T/D_I and D_T/D_I, and we could specify or distinguish each tank by giving the value of one of the length scales, such as D_I.

When data are obtained over a wide range of parameters, it is common to seek a form of correlation of the data, usually in graphical form, in terms of *dimensionless variables*. Thus we might ask the question: If we had data for the mean drop size $\overline{D}$, over a range of design and operating parameters, how would we plot the data? We would soon observe that the mean drop size *decreases* as the stirring speed increases. Hence we might choose to plot a dimensionless mean drop size as a function of some dimensionless stirring speed. If we then decided to use the stirring speed as the *operating parameter* for controlling the drop size, we would have another reason for plotting the data in this form. Since the drop size depends on stirring speed, a more formal way of saying this is that the *dependent* variable is drop size and the *independent* variable is stirring speed, and we commonly plot the results of an experiment in the form of the dependence of the dependent variable on the controlled or independent variable. By convention, we put the dependent variable on the y axis and the independent variable on the x axis.

But drop size and stirring speed have dimensions,[1] and we have elected to plot a *dimensionless* mean drop size as a function of some dimensionless stirring speed. We could make all the drop size data dimensionless by dividing each measured value of $\overline{D}$ by the impeller diameter in the tank in which that measurement was taken. How do we choose a dimensionless stirring speed, V? Whatever the choice, we start with the obvious idea that V is proportional to the rotational speed N. Keep in mind that N is a *rotational, not a linear*, speed. It has dimensions of inverse time, not length per time—the dimensions of *linear* velocity. We can make a linear velocity of it by taking the product of N and the diameter of the impeller. That product is proportional to the tip speed of the impeller, in dimensions of L/t. Since we will settle for something *proportional* to velocity, we choose ND_I as the *characteristic linear velocity* of each experiment.

If we do all experiments with water as the continuous phase, the parameters μ_1 and ρ_1 are not changed. In such experiments, if the viscosity μ_2 of the dispersed phase (the drop phase) is not too high, we would observe that the mean drop sizes do not depend on μ_2 or ρ_2. Hence we are faced with the task of defining a dimensionless speed using just the parameters ND_I, D_I itself, and σ_{12}. If we examine the dimensions of these parameters, we find that this is impossible. (You need to know that the dimensions of interfacial tension σ_{12} are m/t^2, and none of the parameters listed two sentences before has *mass* as part of its dimensions.) Thus we need to include a parameter with mass in its dimensions, and we choose the density of the continuous phase, ρ_1. (But see Problem 1.10.)

---

[1] This is a good place to distinguish *units* from dimensions. A unit is the system in which we measure a dimension. Length is a dimension. The *meter* is but one of many *units* we could choose to measure length.

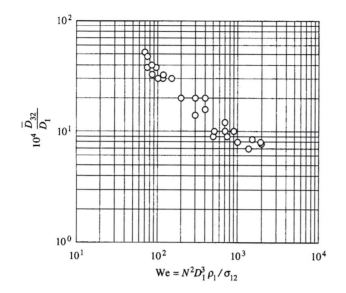

Figure 1.1.3 Data on the mean drop size in an agitated oil/water dispersion. Chen and Middleman, *AIChE J*, **13**, 989 (1967).

Now if we play around with this "short list" of parameters, we can convince ourselves that the following grouping of these parameters is dimensionless:

$$\frac{N^2 D_I^3 \rho_1}{\sigma_{12}}$$

Some experimental data are presented in Fig. 1.1.3, plotted as a dimensionless mean drop size versus the dimensionless group just defined above.[2] Within the experimental accuracy of these data, we conclude that the data are correlated exclusively by this choice of parameters for plotting. By "exclusive correlation" we imply that within experimental uncertainty, the data define a *single* curve independent of liquid properties and tank size. (See the original reference for the range of these parameters tested.)

The data in Fig. 1.1.3 were obtained with a variety of low viscosity organic liquids as the "oil" phase, using various tank and impeller diameters. The horizontal axis, called the Weber number (We), is not an arbitrary choice of a combination of parameters. We will find later that the Weber number arises naturally in certain fluid dynamics problems. We will also learn that dimensionless variables play an important role in mathematical modeling, and in the presentation of experimental tests of mathematical models. The measured drop size plotted in Fig. 1.1.3 is the average drop diameter. It is a specific average, defined as

$$\bar{D}_{32} = \frac{\sum_{i=1}^{N} n_i D_i^3}{\sum_{i=1}^{N} n_i D_i^2} \tag{1.1.1}$$

---

[2] Note how the y axis is labeled. In this book, $10^4 Y$ means that the value you read from the graph is 10^4 times greater than the actual value. Hence, in Fig. 1.1.3, at We = 100, $\bar{D}_{32}/D_I$ has the value 30 × 10^{-4} = 0.003. That is, we *divide* the value read from the graph by 10^4, we don't multiply it by 10^4. Another way of saying this is that we multiply $\bar{D}_{32}/D_I$ by 10^4, so we plot the values of $10^4 D_{32}/D_I$.

We assume here that the measured distribution is grouped into *discrete* drop size classes (like a histogram) such that in the size range $D_i \pm \delta$ we have counted n_i such drops. Our choice of the width of the separation between one drop size D_i and the next, D_{i+1}, is δ.

It is possible to predict just such a correlation as is observed in Fig. 1.1.3 by using some basic principles of fluid dynamics. (See Chen and Middleman, cited in the figure legend.) The preceding brief discussion suggests that if we give some thought to the physics of the problem, prior to carrying out a formal analysis based on fluid dynamic principles and prior to designing an experiment, we often can draw some tentative conclusions that help us to design an efficient experimental program and to interpret our experimental observations. (In this regard, see Problem 1.11.)

Let's look now at another interesting problem.

1.1.2 Removal of an Oil Film from a Surface

Surfaces are often contaminated by an attached or adherent layer of an undesired material. The contaminant could be in particulate form (e.g., dust) or in the form of an undesired liquid that wets the surface. Surface cleanliness is very important in industries such as semiconductor manufacturing, and in biotechnology. The problem of removing thin viscous films from surfaces arises in the area of decontamination of surfaces that have been exposed to toxic spills. Many examples can be cited.

Suppose a thin film of viscous oil covers a planar solid surface. We propose to remove the oil by using a high speed jet of water, directed against the surface as suggested in Fig. 1.1.4. The contaminated solid surface is a disk of finite radius r_W.

If this cleaning method works, the amount of oil on the surface will decrease with time. Our goals are to predict the average oil film thickness as a function of time and to determine how this relationship depends on jet characteristics and oil properties. Do you think the nature of the solid surface plays a role? Would it be easier to remove an oil from a Teflon surface than from an equally smooth metal surface? These are good questions to think about before proceeding further.

Some experimental data on the average oil film thickness remaining on a surface as a function of flushing time are shown in Fig. 1.1.5, where $\bar{h}^*$ is a dimensionless film thickness defined as $\bar{h}^* = h(t)/h_o$, where h_o is the initial film thickness. A dimensionless time t^* is defined as

$$t^* = \frac{0.089\rho\nu^{0.2}U_j^{1.8}h_o}{\mu_f r_w^{1.2}} t \qquad (1.1.2)$$

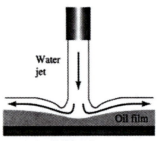

Water
jet

Oil film

Solid surface

Figure 1.1.4 A high speed jet of water is used to flush a film of oil from a solid surface.

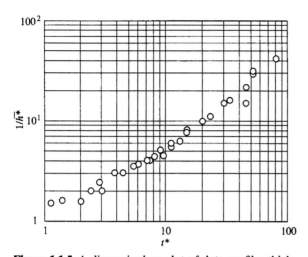

Figure 1.1.5 A dimensionless plot of data on film thickness as a function of time. Yeckel and Middleman, *Chem. Eng. Commun.*, **50**, 155 (1987).

where (note the consistent use of cgs units here)

U_j = jet velocity (cm/s)
ρ = jet fluid density (g/cm^3)
ν = jet fluid kinematic viscosity (cm^2/s) [3]
μ_f = oil film viscosity (poise = g/cm · s)
r_w = radius of disk (cm)

Does Eq. 1.1.2 appear to be a consequence of some *simple* fluid dynamic analysis? It isn't! However, it is possible to develop the model that leads to Eq. 1.1.2, based on the background provided in this textbook. (See Yeckel and Middleman, cited in the legend to Fig. 1.1.5.)

We won't proceed further with this specific example. Instead we turn to a flow problem that arises in the production of semiconductor devices.

1.1.3 Flow Field in a CVD Reactor

In semiconductor device manufacturing, thin solid films such as metals and their oxides are grown by chemical vapor deposition (CVD) from gas phase reactants on silicon wafers that sit on a heated surface, as in Fig. 1.1.6.

You might expect the flow field to influence the rate of reaction at the surface, since the flow field determines to a large degree the distribution of the reacting species in the neighborhood of the reactive surface. In fact, uniformity of film growth is essential to the success of this process, and as a consequence, control of the flow field is very important. We would like to be able to predict the response of the flow field to changes in reactor geometry, gas flowrate, the positions of the gas inlets and outlets, and operating conditions in the reactor. This turns out to be a very complex problem, but there are computational (numerical) methods for predicting the streamlines. Figure 1.1.7 shows

---

[3] The so-called kinematic viscosity is the ratio of the viscosity to the density of the fluid: $\nu = \mu/\rho$.

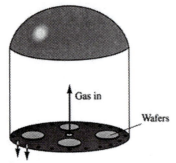

Figure 1.1.6 A reacting gas flows through a reactor in which silicon wafers sit on a heated surface. The gas flow distribution can affect the rate and uniformity of film growth.

an example of the changes that occur in the streamlines (the paths followed by imaginary inertialess tracer particles) as the flowrate increases. You will not learn how to produce these particular streamlines in your study of this book. The details are beyond the normal content of an introductory course in fluid dynamics. However, you will learn some of the principles that are used to develop the methods for producing Fig. 1.1.7.

An important issue in reactor design is that of "scale-up," the design of a larger capacity reactor based on the observed performance of a small-scale system. Imagine, for example, that we learn (perhaps through trial and error) how to design and operate a small-scale reactor of the shape illustrated in Fig. 1.1.6. When we can grow solid film of acceptable quality, we move on to design a larger reactor that will produce more product (e.g., more wafers), but the quality produced in the smaller system must be maintained. Based on the comments offered above, we anticipate that we can achieve this if the streamlines have the same shape in both the small and large systems.[4]

What might the streamline shapes depend on? Certainly the geometry of the reactor. Let's suppose that in the scale-up we maintain the same geometrical *shape* of the reactor: one reactor is simply larger than the other. We expect the magnitude of the flowrate (call it Q) of gas into the reactor to affect the streamlines. Of course the size of the reactor (some length scale L) will be important, since a low flowrate into a large reactor and a high flowrate into a small reactor should produce very different flow patterns. When we think about our day-to-day experiences with fluid dynamics, we anticipate a role for the *viscosity* of the fluid. Fluids very high in viscosity exhibit flow patterns much different from those very low in viscosity.

At a preliminary stage, then, we expect that the streamline patterns depend on the following:

Gas flowrate Q with dimensions L^3/t
Reactor length scale L with dimensions L
Fluid viscosity μ with dimensions m/Lt

---

[4] In this oversimplified discussion, we acknowledge the great importance of the chemistry of the film growth reactions by implying that we operate at the same temperature and pressure with the same feed gas composition.

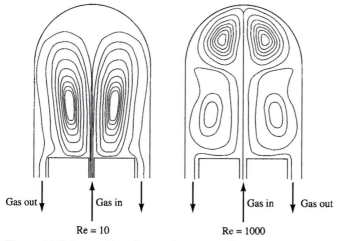

Gas out   Gas in   Gas in   Gas out

Re = 10   Re = 1000

Figure 1.1.7 Streamlines for gas flow through an isothermal bell jar reactor.

A streamline pattern is not "measured" by an amount of something; it is a curve characterized by a shape. A shape is dimensionless. Then the shapes of the streamlines must depend on some dimensionless group or groups that characterize the design and operation of the reactor. Can we define a dimensionless group based on Q, L, and μ? By a dimensionless group (let's call it G_1) we mean a combination of these parameters in the form

$$G_1 = Q^a L^b \mu^c \tag{1.1.3}$$

Upon introducing the dimensions of these parameters, we see that if G_1 is to be dimensionless, we require that the exponents a, b, and c be such that the dimensions L, m, and t disappear. That is,

$$G_1 = \left(\frac{L^3}{t}\right)^a L^b \left(\frac{m}{Lt}\right)^c = L^0 m^0 t^0 \tag{1.1.4}$$

There is clearly no way to satisfy this requirement except for the trivial case $a = b = c = 0$. Moreover, we have another clue to why G_1 cannot be dimensionless, as defined by Eq. 1.1.3: since μ contains the dimension of mass, if the mass dimension is to disappear from G_1, there must be another parameter with the dimension of mass in G_1. Hence one "solution" to this dilemma is to find another parameter with mass in it that is relevant to the physics of the flow of interest. One reasonable speculation is that the mass density ρ of the fluid would also be of importance in determining the nature of the flow through the reactor. With this hypothesis we may rewrite Eq. 1.1.3 slightly, to a form (let's call it G_2 now) that includes ρ:

$$G_2 = Q^a L^b \mu^c \rho^d \tag{1.1.5}$$

Now Eq. 1.1.4 becomes

$$G_2 = \left(\frac{L^3}{t}\right)^a L^b \left(\frac{m}{Lt}\right)^c \left(\frac{m}{L^3}\right)^d = L^0 m^0 t^0 \tag{1.1.6}$$

which leads to the following set of algebraic equations:

$$\text{for } L: \quad 3a + b - c - 3d = 0 \tag{1.1.7}$$

$$\text{for } t: \quad -a - c = 0 \tag{1.1.8}$$

$$\text{for } m: \quad c + d = 0 \tag{1.1.9}$$

Note that we have four unknowns (a, b, c, d) and only three equations. We can solve for b, c, and d in terms of a and find

$$c = -a \tag{1.1.10}$$

$$d = a \tag{1.1.11}$$

$$b = -a \tag{1.1.12}$$

Then we may write G_2 as

$$G_2 = \left(\frac{Q\rho}{L\mu}\right)^a \tag{1.1.13}$$

It is not difficult to verify that G_2 is dimensionless, regardless of the value of the exponent a. Hence we may simply take $a = 1$ arbitrarily, and define G_2 as

$$G_2 = \frac{Q\rho}{L\mu} \tag{1.1.14}$$

Later we will see that G_2 is of a form that we identify as a Reynolds number (See Problem P1.5).

At this stage we are essentially *predicting* that the streamlines depend only on the Reynolds number. To test this prediction we would carry out a series of experiments in reactors of the same shape, but of differing sizes and at various flowrates. We could use various fluids. We would somehow (we don't address this issue here) find the shapes of the streamlines, and we would test the hypothesis that the streamlines are identical as long as the Reynolds number is fixed, regardless of the individual values of Q, L, μ, and ρ.[5]

Let's look next at a problem that arises in the field of human physiology. It does not appear to be an engineering problem at all. In fact, though, it is related to some important engineering phenomena.

1.1.4   Dry-Eye Syndrome

Figure 1.1.8 shows the cornea of an eye covered by a tear film. This thin liquid film lubricates the surface of the eye with respect to the motion of the eyelid. In "dry-eye syndrome," the thickness of the tear film decreases so rapidly between blinks that dry patches form on the cornea. This can lead to discomfort, and to damage to the corneal surface.

What factors control the coherence or stability of a thin liquid film? What physical properties of the lubricating liquid do you think are important? Can pathological conditions be attributed to changes in easily measured fluid properties? This book will show you how to define and measure some of the most important physical properties that control the behavior of a liquid.

Similar questions regarding the stability of a liquid film arise in the consideration of bubble coalescence in an aerated fermentation vessel, as suggested in Fig. 1.1.9. For two small bubbles to coalesce into one larger bubble the thin liquid film that separates

---

[5] We would find that the hypothesis holds as long as there are no strong buoyancy effects (due, e.g., to density differences that in turn result from large temperature or composition changes in the system). In most real commercial reactors, these buoyancy effects are quite important.

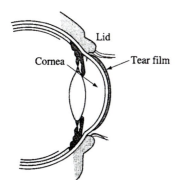

Figure 1.1.8 The cornea of the normal eye is protected by an aqueous liquid tear film.

them must rupture. Bubble coalescence plays a very important role in the behavior of gas–liquid contactors such as fermentors.

Later we will develop the tools necessary to write mathematical models for flow in thin liquid films. The tear film *stability* problem, however, is more difficult than those that we will learn to solve in this textbook. Hence we now turn to a discussion of another complex problem that we *will* learn how to model.

1.1.5   Absorption of an Ink Drop by Paper

In ink-jet printing on paper, a very small droplet of ink hits the surface of the paper and then is absorbed into the paper. It is important to be able to control the size of the "spot" left on the surface, as well as the rate at which the drop is "sucked into" the paper. Figure 1.1.10 is drawn from video images of a small drop being absorbed into a sheet of paper (this is not an ink-jet system).

We will learn how to predict the rate of "wicking," or the disappearance of an isolated drop into the surface, using a very simple mathematical model of flow into a porous medium, in which surface tension is the primary driving force for the flow and viscosity retards the flow. The model we develop will be tested against experimental data.

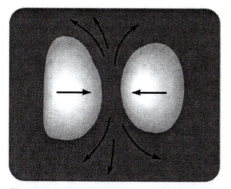

Figure 1.1.9 The fluid dynamics in the thin liquid film separating two approaching bubbles determines whether they will coalesce into a larger bubble.

Figure 1.1.10 A sequence of images of a liquid drop as it is absorbed by a sheet of paper.

Whenever feasible, mathematical models are tested against observations. These tests are carried out most conveniently when we recast the results into forms that involve dimensionless groupings of the relevant parameters. Hence we turn now to a brief discussion of dimensional analysis. The topic is taken up again in Chapter 5.

1.2 DIMENSIONAL ANALYSIS

It has been implied that dimensional analysis is an aid to understanding the physics of a given problem. In many simple problems, the number of parameters is expected to be relatively small and we can form dimensionless groups quite easily by inspection and trivial algebraic manipulation. It is appropriate at this point to examine an *organized* procedure for selecting dimensionless groups. Its foundation, the so-called Buckingham pi theorem, is a statement that in a list of P parameters that together include D fundamental dimensions (length, time, mass, temperature, electric charge, etc.), the number of independent dimensionless groups is simply the difference between P and D.

Let's consider the problem of a drop of liquid formed at the lower end of a vertical capillary in still air. There is a slow flow of the liquid through the capillary. The drop grows larger, slowly, until its weight exceeds the balancing force due to surface tension. Our interest is in the size of the drop when it breaks away from the capillary tip. This is a problem to be considered shortly (Section 2.1.1). The development of a set of independent dimensionless groups that characterize this problem can be achieved in the following steps.

Step 1 *Make a list of the relevant parameters.* This step is essential, and it requires a good sense of the physics of the process under consideration. If we leave out an important parameter, we will fail to obtain all the relevant dimensionless groups. For the case at hand, a reasonable list would begin with the dependent variable—the drop size. We will select the drop volume V_{drop}, although we could as well select the drop diameter (i.e., the diameter of a spherical drop of volume V_{drop}). Now let's name the parameters V_{drop} is likely to depend on. My list would be as follows: V_{drop} depends on D_c, ρ, σ, g, μ, and U (capillary diameter, liquid density, liquid/air surface tension, acceleration of gravity, liquid viscosity, and the average linear velocity of efflux of the liquid as it crosses the end of the capillary and forms the drop).

Step 2 *List the fundamental dimensions of each parameter.*

$$V_{\text{drop}}\,[=]\,L^3 \quad D_c\,[=]\,L \quad \rho\,[=]\,m/L^3 \quad \sigma\,[=]\,m/t^2$$
$$g\,[=]\,L/t^2 \quad \mu\,[=]\,m/Lt \quad U\,[=]\,L/t$$

We denote the dimensions of length, time, and mass by L, t, and m. The notation $[=]$ means "has the dimensions of."

At this point we might wonder about temperature, since the physics could be expected to be quite different at different temperatures. While this is true, the effect of temperature arises only through its effect on the physical properties of the system, particularly ρ, σ, and μ.

Step 3 *Determine, from the Buckingham pi theorem, the number of dimensionless groups that characterize this problem.* In this case, $P = 7$ (remember to count the dependent variable V_{drop} itself among the P parameters) and $D = 3$, so $P - D = 4$ dimensionless groups.

Step 4 *From the list of the independent parameters (i.e., all but V_{drop}) select a number equal to D (= 3, in this case) that will be used as "recurring parameters" in all subsequent attempts to form dimensionless groups.* It is usually wise to pick a set of parameters that include all the dimensions. For example, select D_c, ρ, and σ, which together include L, m, and t. The rest of the parameters, called "nonrecurring parameters," are V_{drop}, g, U, and μ in this example. Either the recurring parameters must, among them, have dimensions in common with those in each nonrecurring parameter or a dimension not in the nonrecurring parameter must appear in two of the recurring parameters. Only in this way can the dimensions cancel. This point will be clearer as we work through the details of the procedure.

Step 5 *Form, in turn, dimensionless groups that are proportional to each of the remaining nonrecurring parameters.* We do this by first writing the appropriate equations (for this example, Eqs. 1.2.1–1.2.4) and then solving for the exponents that render each group dimensionless.

For example, a dimensionless drop volume will be formed from

$$
\overset{\text{recurring parameters}}{\underset{\text{nonrecurring parameter}}{V_{\text{drop}}^* = V_{\text{drop}} \left[D_c^a \sigma^b \rho^c \right]}}
\tag{1.2.1}
$$

and the task is to solve for the exponents a, b, and c. From the remaining nonrecurring parameters we form

$$
g^* = g D_c^a \sigma^b \rho^c
\tag{1.2.2}
$$

$$
U^* = U D_c^a \sigma^b \rho^c
\tag{1.2.3}
$$

$$
\mu^* = \mu D_c^a \sigma^b \rho^c
\tag{1.2.4}
$$

The asterisk * denotes the dimensionless group proportional to the nonrecurring parameter chosen for each group. The specific values of a, b, and c are different for each of these four equations.

Step 6: *For each of the four equations above, solve for the set of exponents a, b, and c.* We will get a different solution for each equation involving V_{drop}^*, g^*, U^*, and μ^*; we carry the step out by inserting the dimensions into each parameter and requiring that the equation be dimensionally homogeneous. For example, from

$$
V_{\text{drop}}^* = V_{\text{drop}} D_c^a \sigma^b \rho^c
\tag{1.2.5}
$$

we must have

$$
m^0 L^0 t^0 = L^3 L^a (m/t^2)^b (m/L^3)^c
\tag{1.2.6}
$$

from which it follows by equating the sums of the powers on m, L, and t that

$$
m: \quad 0 = b + c
\tag{1.2.7}
$$

$$
L: \quad 0 = 3 + a - 3c
\tag{1.2.8}
$$

$$
t: \quad 0 = -2b
\tag{1.2.9}
$$

Thus we find

$$b = 0, c = 0, a = -3 \tag{1.2.10}$$

and

$$V_{\text{drop}}^* = \frac{V_{\text{drop}}}{D_c^3} \tag{1.2.11}$$

From the equation for g^* we write

$$m^0 L^0 t^0 = (L/t^2)\, L^a\, (m/t^2)^b (m/L^3)^c \tag{1.2.12}$$

which leads to

$$0 = b + c \tag{1.2.13}$$
$$0 = 1 + a - 3c \tag{1.2.14}$$
$$0 = -2 - 2b \tag{1.2.15}$$

Thus we find

$$b = -1 \quad c = 1 \quad a = 2 \tag{1.2.16}$$

and

$$g^* = \frac{g\rho D_c^2}{\sigma} \tag{1.2.17}$$

Continuing in this way, we find from the equation for U^*

$$m^0 L^0 t^0 = (L/t)\, L^a\, (m/t^2)^b (m/L^3)^c \tag{1.2.18}$$

which leads to

$$0 = b + c \tag{1.2.19}$$
$$0 = 1 + a - 3c \tag{1.2.20}$$
$$0 = -1 - 2b \tag{1.2.21}$$

This gives us

$$b = -\tfrac{1}{2} \quad c = \tfrac{1}{2} \quad a = \tfrac{1}{2} \tag{1.2.22}$$

and

$$U^* = U \left(\frac{D_c \rho}{\sigma} \right)^{1/2} \tag{1.2.23}$$

Finally, from the equation for μ^* we find

$$m^0 L^0 t^0 = (m/Lt)\, L^a\, (m/t^2)^b (m/L^3)^c \tag{1.2.24}$$

which leads to

$$0 = 1 + b + c \tag{1.2.25}$$
$$0 = -1 + a - 3c \tag{1.2.26}$$
$$0 = -1 - 2b \tag{1.2.27}$$

and the result is

$$b = -\tfrac{1}{2} \quad c = -\tfrac{1}{2} \quad a = -\tfrac{1}{2} \tag{1.2.28}$$

and

$$\mu^* = \mu (D_c \rho \sigma)^{-1/2} \tag{1.2.29}$$

This completes the derivation of the four independent dimensionless groups. To review, we conclude that the dimensionless drop size would depend on U^*, g^* and μ^*, and thus must satisfy a functional relationship of the general form

$$V^*_{\text{drop}} \equiv \frac{V_{\text{drop}}}{D_c^3} = f\left[U \sqrt{\left(\frac{\rho D_c}{\sigma}\right)^{1/2}}, \frac{\rho g D_c^2}{\sigma}, \frac{\mu}{(\rho \sigma D_c)^{1/2}} \right] \qquad (1.2.30)$$

Had we selected some other set of three parameters to be the recurring variables, we would have obtained a different set of groups, but they would not have been independent of the ones just presented. This is easily confirmed, but we will not present the results here. (See Problem 1.17.)

EXAMPLE 1.2.1 *Coherent Length of a Wasp Spray*

You are developing a product consisting of a reservoir (a can) of wasp killer under pressure. A nozzle–pressure system has been designed that permits the user to produce a coherent jet of the liquid and aim it at the target pest(s). The system is most effective if the target is struck by a *coherent* jet of liquid, rather than by a spray of dispersed droplets. In particular, the more coherent the jet, the more the distance from the target can be increased, with resulting safety for the user. Simply stated, a spray of droplets does not travel as far as a coherent jet.

You are responsible for an experimental program aimed at providing data, and a predictive correlation derived from the data, for the dependence of the coherent length L_c as a function of liquid properties, nozzle design, and pressure. What variables are important, what experiments would you perform, and how would you plot data to develop a correlation among the variables?

Assume that your employer makes related spray products and that you will use an existing nozzle design that is manufactured in three sizes, as defined by the nozzle diameter D_N (see Fig. 1.2.1). Three D_N values are available (0.5, 1.0, and 2.0 mm), and for each D_N value, the nozzle is available with $B/D_N = 2$ and 4. (Six nozzles all together.) We will follow the steps outlined earlier.

Steps 1 and 2 The dependent variable is the coherent length L_c, that is, the distance to the first point of appearance of drops. (Always begin by writing down the dependent variable.) There are two characteristic length scales for the nozzle design, B and D_N.

Assume that the reservoir pressure P_R that creates the flow is constant in time, is imposed immediately upon depressing a button, and can be built into the design of

Nozzle assembly

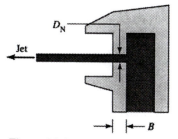

Figure 1.2.1 A nozzle for producing a coherent liquid jet.

the spray can from a range of values by controlling design parameters that won't concern us here. We will simply take P_R as a design parameter that can be varied in our experimental program, although the final form of the product will incorporate a single P_R.

What liquid properties are important? At this stage, which is really the most critical stage in experimental design, we don't know enough to make more than an intelligent guess. Let's assume that a more experienced colleague makes the following claim: the only liquid properties of importance are the surface tension and density. In particular, she counsels that the liquid viscosity will not play a role. We will accept this advice (it may be wrong!) and follow the consequences. In several homework problems you can explore alternatives.

At this stage, then, our list of parameters is

$$L_c \, [=] \, L \quad B \, [=] \, L \quad D_N \, [=] \, L \quad P_R \, [=] \, m/Lt^2 \quad \sigma \, [=] \, m/t^2 \quad \rho \, [=] \, m/L^3$$

Step 3 We have $P = 6$ parameters (counting the dependent variable L_c) and $D = 3$ fundamental dimensions (L, m, and t), so the expected number of dimensionless groups, $P - D$, is 3. This means that we expect a dimensionless dependent variable, proportional to L_c, to be a function of two other dimensionless groups.

Step 4 *For the recurring parameters, select D_N, σ, and ρ.* This is an arbitrary choice, except that we must and do include parameters that involve all three dimensions. Again, we can explore alternatives in the homework problems. Hence the *nonrecurring* parameters (which must include the dependent variable) are L_c, B, and P_R.

Step 5 *Form the following groups proportional to the nonrecurring parameters:*

$$L_c^* = L_c D_N^a \sigma^b \rho^c \tag{1.2.31}$$

$$B^* = B D_N^a \sigma^b \rho^c \tag{1.2.32}$$

$$P_R^* = P_R D_N^a \sigma^b \rho^c \tag{1.2.33}$$

Step 6 It follows by equating the sums of the powers on L, m, and t that, from Eq. 1.2.31,

$$L: \quad 0 = 1 + a - 3c \tag{1.2.34}$$

$$m: \quad 0 = b + c \tag{1.2.35}$$

$$t: \quad 0 = -2b \tag{1.2.36}$$

Thus we find

$$b = 0 \quad c = 0 \quad a = -1 \tag{1.2.37}$$

and

$$L_c^* = \frac{L_c}{D_N} \tag{1.2.38}$$

From Eq. 1.2.32 we find

$$0 = 1 + a - 3c \tag{1.2.39}$$

$$0 = b + c \tag{1.2.40}$$

$$0 = -2b \tag{1.2.41}$$

and therefore

$$b = 0 \quad c = 0 \quad a = -1 \tag{1.2.42}$$

and

$$B^* = \frac{B}{D_N} \tag{1.2.43}$$

Incidentally, these would have been natural choices, not requiring the algebraic presentation. It is quite natural to nondimensionalize the length L_c with a characteristic length of the nozzle D_N, and B/D_N is an obvious nondimensional parameter characteristic of this system.

For P_R^* we find, beginning with Eq. 1.2.33,

$$0 = -1 + a - 3c \tag{1.2.44}$$
$$0 = 1 + b + c \tag{1.2.45}$$
$$0 = -2 - 2b \tag{1.2.46}$$

and therefore

$$b = -1 \quad c = 0 \quad a = 1 \tag{1.2.47}$$

and

$$P_R^* = \frac{P_R D_N}{\sigma} \tag{1.2.48}$$

As a result of this exercise, we conclude that if we are right about which variables/ parameters affect the physics, the coherent length should satisfy a functional relationship of the form

$$L_c^* \equiv \frac{L_c}{D_N} = f\left(\frac{B}{D_N}, \frac{P_R D_N}{\sigma}\right) \tag{1.2.49}$$

An efficient experimental program based on this result would involve fixing the geometry, that is, selecting a value of the parameter B/D_N and then measuring L_c as a function of P_R over a range of σ values, for various D_N values, while holding B/D_N constant. The data should then be plotted as L_c^* versus P_R^* for that value of B/D_N. Then a new geometry (the second value of B/D_N) would be selected, and the experiments repeated. A graph like that shown in Fig. 1.2.2 should result. If a wide range of parameters

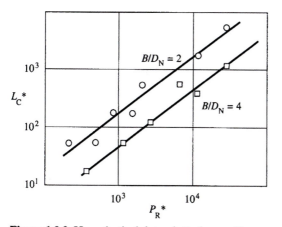

Figure 1.2.2 Hypothetical data plotted according to Example 1.2.1.

is studied, it is advantageous to plot the data on double-logarithmic coordinates, rather than arithmetically.

Let's go back to step 4 for a moment. Another possible choice of recurring parameters is the following: P_R, σ, and D_N. Among these, all three dimensions (L, m, and t) are included. Furthermore, L appears in two (D_N and P_R), m in two (σ and P_R), and t in two (σ and P_R). We will see shortly, however, that there is a problem with this choice.

According to the outlined procedure, form the dimensionless parameters

$$L_c^* = L_c P_R^a \sigma^b D_N^c \tag{1.2.50}$$

$$B^* = B P_R^a \sigma^b D_N^c \tag{1.2.51}$$

$$\rho^* = \rho P_R^a \sigma^b D_N^c \tag{1.2.52}$$

Now, from the first of these:

$$L: \quad 0 = 1 - a + c \tag{1.2.53}$$

$$m: \quad 0 = 1 + b \tag{1.2.54}$$

$$t: \quad 0 = -2a - 2b \tag{1.2.55}$$

The solution is $a = 1$, $b = -1$, $c = 0$, and

$$L_c^* = \frac{L_c P_R}{\sigma} \tag{1.2.56}$$

From the B^* equation:

$$L: \quad 0 = 1 - a + c \tag{1.2.57}$$

$$m: \quad 0 = a + b \tag{1.2.58}$$

$$t: \quad 0 = -2a - 2b \tag{1.2.59}$$

Equations 1.2.58 and 1.2.59 are not independent. We have lost an equation, and we cannot solve for the three unknowns with only two independent equations. Before we resolve this, let's look at the ρ^* equation:

$$L: \quad 0 = -3 - a + c \tag{1.2.60}$$

$$m: \quad 0 = 1 + a + b \tag{1.2.61}$$

$$t: \quad 0 = -2a - 2b \tag{1.2.62}$$

Equations 1.2.61 and 1.2.62 are inconsistent! Equation 1.2.61 says that $a + b = -1$, while Eq. 1.2.62 says $a + b = 0$. Why is this happening? How do we avoid it?

There is one more rule to add to step 4. In selecting the recurring parameters, we must not choose dimensions such that those of one recurring parameter can be formed from the dimensions of combinations of the others. In other words, the dimensions must be independent. The choice of P_R, σ, and D_N violates that rule. The dimensions of P_R can be formed from the ratio of the dimensions of σ and D_N. That is, we cannot include among the recurring parameters a set that can, without even considering the nonrecurring parameter, be formed into a dimensionless group. In the case at hand, $P_R D_N / \sigma$ is itself dimensionless. Thus the selection P_R, σ, and D_N turns out to be unacceptable, and the result is the dilemma illustrated here.

Dispersion (Breakup) of an Oil Stream in an Aqueous Pipe Flow

We will examine the system sketched in Fig. 1.2.3. Water is flowing turbulently through a long pipe of diameter D_p (m) at a volumetric flowrate Q (m³/s). An *immiscible* oil is

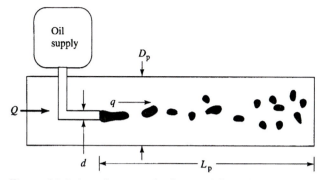

Figure 1.2.3 An oil stream is dispersed into droplets by a surrounding aqueous flow.

introduced into the flow at a rate $q \ll Q$, through a small tube of diameter d aligned with the pipe axis, as shown. Far downstream we observe that the flow consists of oil drops carried along with the water phase. Our interest is in predicting the mean droplet diameter $\overline{D}$ as a function of the parameters that characterize this flow. Of course, to do so, we must understand something about the physics by which droplets are produced as the oil phase enters the surrounding aqueous fluid.

One of the first questions we might raise about the physics of this flow is: Does the oil phase enter the aqueous phase as droplets that break off from the end of the small tube or does the oil "jet" into the water and then get broken into droplets by the interaction of the flowing water with the oil jet? Actually, either might occur, depending on the flowrate q of the oil phase. We will make the following assumption or speculation about the breakup process:

> *The drop size observed downstream is a consequence of the ability of the energy associated with the turbulence in the water flow to disrupt the oil jet (or the first oil drops) that comes off the exit tube. Opposing this energy is the cohesive energy of the jet or drops, which arises form the interfacial tension at the oil/water interface. This interfacial tension gives rise to a force that tends to keep the oil phase together; that is, it resists the disruption of the oil phase into smaller drops.*

On the assumption that this speculation has some relationship to the real physics, we can now use the methodology illustrated earlier in this section.

Step 1 *Make a list of parameters.* The dependent variable is the mean drop size $\overline{D}$. Now, we list the parameters that we think $\overline{D}$ depends on. My list would be: pipe diameter D_p, pipe length L_p, oil inlet tube diameter d, the continuous (water) and dispersed (oil) phase density and viscosity ρ, μ, and ρ', μ', the interfacial tension σ, and the two flowrates q and Q.

Step 2 *List the fundamental dimensions.*

$$\overline{D} \, [=] \, L \quad D_p \, [=] \, L \quad L_p \, [=] \, L \quad d \, [=] \, L \quad \rho \, [=] \, m/L^3 \quad \sigma \, [=] \, m/t^2$$

$$\mu \, [=] \, m/Lt \quad Q \, [=] \, L^3/t \quad q \, [=] \, L^3/t \quad \rho' \, [=] \, m/L^3 \quad \mu' \, [=] \, m/Lt$$

Step 3 *Use the Buckingham pi theorem.* In this case, $P = 11$ (remember to count the dependent variable $\overline{D}$ itself among the P parameters) and the number of fundamental dimensions is $D = 3$, so $P - D = 8$ dimensionless groups.

Step 4 *Select the recurring parameters.* We will select D_p, σ, and ρ, which together include all three dimensions: L, m, and t. The rest of the parameters are nonrecurring.

Hence the nonrecurring parameters are the mean drop size $\overline{D}$ and L_p, μ, μ', ρ', d, q, and Q.

Step 5 *Form, in turn, dimensionless groups that are proportional to each of the remaining nonrecurring parameters, starting with the dependent variable:*

$$\overline{D}* = \overline{D}D_p^a\sigma^b\rho^c \tag{1.2.63}$$

$$L_p^* = L_p D_p^a\sigma^b\rho^c \tag{1.2.64}$$

$$\mu* = \mu D_p^a\sigma^b\rho^c \tag{1.2.65}$$

$$\mu'* = \mu' D_p^a\sigma^b\rho^c \tag{1.2.66}$$

$$\rho'* = \rho' D_p^a\sigma^b\rho^c \tag{1.2.67}$$

$$d* = d D_p^a\sigma^b\rho^c \tag{1.2.68}$$

$$q* = q D_p^a\sigma^b\rho^c \tag{1.2.69}$$

$$Q* = Q D_p^a\sigma^b\rho^c \tag{1.2.70}$$

Step 6 Solve for the coefficients a, b, and c for each of these seven equations. For example, for the first equation,

$$\overline{D}* = \overline{D}D_p^a\sigma^b\rho^c \tag{1.2.71}$$

we must have

$$m^0L^0t^0 = LL^a(m/t^2)^b(m/L^3)^c \tag{1.2.72}$$

from which it follows by equating the sums of the powers on m, L, and t that

$$m: \quad 0 = b + c \tag{1.2.73}$$

$$L: \quad 0 = 1 + a - 3c \tag{1.2.74}$$

$$t: \quad 0 = -2b \tag{1.2.75}$$

Thus we find

$$b = 0 \quad c = 0 \quad a = -1 \tag{1.2.76}$$

and

$$\overline{D}* = \frac{\overline{D}}{D_p} \tag{1.2.77}$$

From the equation for $\mu*$ we find

$$m^0L^0t^0 = (m/Lt)\, L^a\, (m/t^2)^b(m/L^3)^c \tag{1.2.78}$$

which leads to

$$0 = 1 + b + c \tag{1.2.79}$$

$$0 = -1 + a - 3c \tag{1.2.80}$$

$$0 = -1 - 2b \tag{1.2.81}$$

and the result is

$$b = -\tfrac{1}{2} \quad c = -\tfrac{1}{2} \quad a = -\tfrac{1}{2} \tag{1.2.82}$$

and

$$\mu* = \mu\, (D_p\rho\sigma)^{-1/2} \tag{1.2.83}$$

It is not hard to see that similar results will follow for $d*$, L_p^*, and $\mu'*$:

$$d* = \frac{d}{D_p} \qquad L_p^* = \frac{L_p}{D_p} \tag{1.2.84}$$

and

$$\mu'* = \mu'(D_p\rho\sigma)^{-1/2} \tag{1.2.85}$$

For ρ' we begin with

$$m^0L^0t^0 = (m/L^3)\, L^a\, (m/t^2)^b\, (m/L^3)^c \tag{1.2.86}$$

which leads to

$$0 = 1 + b + c \tag{1.2.87}$$
$$0 = -3 + a - 3c \tag{1.2.88}$$
$$0 = -2b \tag{1.2.89}$$

and the result is

$$b = 0 \quad c = -1 \quad a = 0 \tag{1.2.90}$$

and

$$\rho'* = \frac{\rho'}{\rho} \tag{1.2.91}$$

For $Q*$ we begin with

$$m^0L^0t^0 = (L^3/t)\, L^a\, (m/t^2)^b(m/L^3)^c \tag{1.2.92}$$

which leads to

$$0 = b + c \tag{1.2.93}$$
$$0 = 3 + a - 3c \tag{1.2.94}$$
$$0 = -1 - 2b \tag{1.2.95}$$

and the result is

$$b = -\tfrac{1}{2} \quad c = \tfrac{1}{2} \quad a = -\tfrac{3}{2} \tag{1.2.96}$$

and

$$Q* = Q\left(\frac{\rho}{\sigma D_p^3}\right)^{1/2} \tag{1.2.97}$$

In the same manner we will find

$$q* = q\left(\frac{\rho}{\sigma D_p^3}\right)^{1/2} \tag{1.2.98}$$

To summarize these results, we expect to find that

$$\overline{D}* = \frac{\overline{D}}{D_p} = F\left[\mu(D_p\rho\sigma)^{-1/2}, \mu'(D_p\rho\sigma)^{-1/2}, \frac{\rho'}{\rho}, \frac{d}{D_p}, \frac{L_p}{D_p}, Q\left(\frac{\rho}{\sigma D_p^3}\right)^{1/2}, q\left(\frac{\rho}{\sigma D_p^3}\right)^{1/2}\right] \tag{1.2.99}$$

While this result is formally correct, it is helpful to rewrite some of these groups in simpler terms. For example, instead of both $Q(\rho/\sigma D_p^3)^{1/2}$ and $q(\rho/\sigma D_p^3)^{1/2}$, we may use $Q(\rho/\sigma D_p^3)^{1/2}$ and q/Q, since the latter is just the ratio of the two former groups.

In the same way, it is simpler to use $\mu'(D_p \rho \sigma)^{-1/2}$ and μ'/μ. Hence an alternative formulation of Eq. 1.2.99 is

$$\overline{D}* = \frac{\overline{D}}{D_p} = F\left[\mu'(D_p\rho\sigma)^{-1/2}, \frac{\mu'}{\mu}, \frac{\rho'}{\rho}, \frac{d}{D_p}, \frac{L_p}{D_p}, Q\left(\frac{\rho}{\sigma D_p^3}\right)^{1/2}, \frac{q}{Q}\right] \qquad (1.2.100)$$

Other alternative formulations are possible as well. See Problem 1.26.

It is important to keep in mind that the formulation of Example 1.2.2 depends on the validity of our initial speculation about the physics of drop breakup, since this speculation leads to the listing of relevant parameters. We may have omitted one or more parameters. For example, is gravity irrelevant? Alternatively, we may have included parameters that have such a minor effect on the physics that they could be ignored. For example, experimental studies suggest that if the viscosity in the oil phase is comparable to that of water, the viscosity of the dispersed (oil) phase is not a significant factor in the drop breakup. Furthermore, it is observed that as long as the inlet tube diameter d is large compared to the ultimate mean drop size, the value of d is of no significance. Since most liquid densities lie in a narrow range, we do not observe an effect due to the density ratio ρ/ρ'. If q/Q is small, as is the case in many systems of practical interest, this ratio does not affect the drop breakup. This is because the volume fraction of drops (which is in fact equal to q/Q) is so small that the drops break up independently of one another. Finally, we usually observe that as long as $L_p/D_p \gg 1$, the mean drop size reaches equilibrium some distance downstream from the oil phase entrance and no longer changes further downstream from this point. Under the circumstances described, a number of the parameters become irrelevant, and we would predict the following functional relationship:

$$\overline{D}* = \frac{\overline{D}}{D_p} = F\left[Q\left(\frac{\rho}{\sigma D_p^3}\right)^{1/2}\right] \qquad (1.2.101)$$

In Problem 1.27 we can compare some experimental data to this result.

1.3   CLASSIFICATION OF PROBLEMS IN FLUID DYNAMICS

All the problems sketched in Section 1.1 share some common features, but they also are distinct.

Problem 1 is a *turbulent two-phase flow* problem. It is very complex, primarily because turbulence is a random, chaotic phenomenon. The velocity components and the pressure fluctuate randomly in time, at every point within the fluid, and they cannot be described by simple mathematical models. Turbulence is usually associated with high speed flows of low viscosity fluids. We will have to discuss turbulence more carefully later. For now, we can think of turbulent flow as probabilistic, rather than deterministic.

Problem 2 involves *laminar flow* in a thin film (the oil layer) driven by the shear forces of the impinging jet. The jet itself could be turbulent, but the flow in the oil film could be "smooth," a loose term to describe a laminar flow. Since the oil film thickness changes with time, this is an unsteady-state flow. Nevertheless, laminar flows with such simple geometrical boundaries can often be described by simple mathematical models that mimic our physical observations.

Problem 3 involves a single-phase (gas) flow of a multicomponent *reactive* gas. Flow reactors often require a strong heat source to promote the reaction (as in the case of CVD), and as a consequence the flow is strongly nonisothermal. In such a complex case, a mathematical model that leads to an analytical solution for the variables of interest may not be writable, and numerical methods will have to be applied.

Problem 4 is a *stability* problem in a thin viscous film. Stability problems are usually quite complicated and are not typically studied in an introductory course in fluid dynamics. We will, however, treat a stability problem later and develop a mathematical model that leads to an analytical solution that mimics some observations for the system selected.

Problem 5 involves flow through a porous medium (paper) with solid boundaries that are so complex geometrically that they cannot be described analytically. Chapter 10 describes successful approaches to treating flows in porous media.

Analyses of various flows are essentially applications of fundamental physical principles of *mechanics*. The primary principle is that of *conservation of momentum,* but the principle takes a different mathematical form from that in which it appears in rigid body mechanics. This is because:

> *A fluid is a continuous medium—its mass is distributed throughout its volume, and forces that act on a body of fluid cannot be replaced by a single force acting on the center of mass of the body.*

> *A fluid is deformable—obviously because it flows—and we will have to derive the form taken by the principle of conservation of momentum within a deformable continuous medium (i.e., within a fluid).*

We derive the mathematical forms of the basic principles of conservation of momentum (and mass) in Chapter 4. Before we attack such complex ideas, we want to review some simple mechanical principles and apply them to relatively simple but important problems in fluid statics and dynamics. We will do this in Chapter 2.

SUMMARY

Engineers are interested in fluid dynamics problems that cover a wide range of physical phenomena. The development of a mathematical model of a particular flow involves a number of stages. In this chapter we examine, by example, two of the early steps toward model building. The first involves thinking about the physics of the flow of interest. We ask questions like: What's happening? and What would happen if I made certain changes to the system, or to physical properties? The latter question leads to the next stage: we list the physical properties, geometrical factors, and operating conditions that we believe affect the physics. These are parameters that must appear in the mathematical model we ultimately create, which we hope will mimic the observed behavior of the system. Once the parameters have been established, an algebraic procedure based on the Buckingham pi theorem indicates how many independent dimensionless groups determine the behavior of the system and provides a means to "discover" these groups. Later we learn that this knowledge provides guidance to the design of experiments, and to the correlation of data for the particular flow system of interest. This guidance is especially important in cases of fluid dynamics so complex that we have little hope of producing a purely *theoretical* mathematical model for the functional relationships among the parameters. In such situations, which are very common, we can often achieve our goals through the guidance provided by dimensional analysis.

PROBLEMS

1.1 Demonstrate that Eq. 1.1.2 is dimensionless.

1.2 Demonstrate that the Weber number, as defined in Fig. 1.1.3, is dimensionless.

1.3 Surface tension between a gas and a liquid, or between two immiscible liquids, is often described in either of two ways: as a force per unit of contact length between the two phases, or as the energy per unit of interfacial area separating the two phases. Show that in either case these descriptions are consistent with the stated units of σ, namely, $\sigma\,[=]\,m/t^2$.

1.4 In Section 1.1 we found that by playing around with the parameters introduced in that discussion, we can convince ourselves that the grouping of parameters $(N^2D_1^3\rho_1/\sigma_{12})$ is dimensionless. We can "derive" this group in the following manner. We assume that a grouping of these parameters, and only these parameters, of the form

$$G = N^a D_1^b \rho_1^c \sigma_{12}^d$$

is dimensionless. Then the task is to find values of the constants a, b, c, and d consistent with this assumption.

Now we introduce the dimensions of each individual parameter:

$$N\,[=]\,t^{-1}\quad D_1\,[=]\,L\quad \rho_1\,[=]\,m/L^3\quad \sigma_{12}\,[=]\,m/t^2$$

where $[=]$ is read as "has the dimensions of." Then it must be the case that

$$G = t^{-a}L^b\,(m/L^3)^c(m/t^2)^d$$

is dimensionless. This means, for example, that the dimension t must have a zero exponent, or

$$-a - 2d = 0$$

and similarly with the exponents on L and m. Thus we obtain *three* linear equations, but we have *four* unknowns a, b, c, and d. How do we then end up with $N^2D_1^3\rho_1/\sigma_{12}$ when we are short one equation?

1.5 Figure 1.1.7 shows streamlines for two cases that correspond to low and high flowrates of the gas into the reactor. The two cases are labeled Re = 10 and Re = 10^3, but Re is not defined. In fact, Re is a dimensionless group called the Reynolds number. If the streamlines are known to depend on the average velocity of the gas as it enters the reactor, the density and viscosity of the gas, and the diameter of the gas inlet pipe to the reactor, and only on these parameters for reactors of various sizes but all of the same shape (geometrically similar reactors), a dimensionless group involving these parameters can be defined. Give that definition. Your (correct) result is uniquely related to the Reynolds number used in Fig. 1.1.7. Can you make this definition compatible with that given in Eq. 1.1.14?

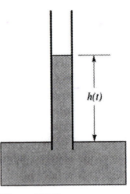

Figure P1.6

1.6 The lower end of a vertical glass capillary tube of small inside diameter D is dipped and held slightly below the free surface of a pan of liquid of viscosity μ, surface tension σ, and density ρ. The liquid is observed to "wick" into the capillary and reach an equilibrium height (Fig. P1.6) after a time t_e, called the wicking time. Assume that the only other parameter of importance is the gravitational constant g.

Define a *dimensionless* wicking time by forming a dimensionless group proportional to t_e and involving some (not necessarily all) of the other parameters. Define some other dimensionless groups from these parameters, but do not use t_e in any of the other definitions. Show that only two of these groups are independent of the others you define. This is an example of the Buckingham pi theorem, which states that the number of *independent* dimensionless groups that can be formed from P parameters that involve D fundamental units or dimensions (e.g., mass, length, time, temperature) is given by the difference $P - D$. Hence, in this case, we expect three groups (the one with wicking time t_e and two others), since there are six parameters (D, μ, σ, ρ, g, t_e) with three fundamental dimensions (mass, length, time).

1.7 Figure P1.7 shows a schematic of an *airlift reactor*, a vertical vessel containing liquid into the bottom of which air is pumped continuously. Rising air bubbles induce liquid flow and provide circulation and mixing, thus improving the performance of the reactor. In the schematic, we illustrate the case of gas rising through a concentric inner tube (called the "riser") and the recirculating flow moving down the outer annular region (called the "downcomer").

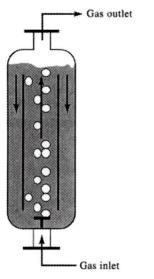

Figure P1.7 An airlift reactor.

Define the "liquid circulation velocity" V_L as the volume flowrate of the induced liquid flow in the riser, divided by the cross-sectional area of the riser. Make a list of the parameters you think V_L depends on. Give the dimensions of each parameter.

An important characteristic of the system is the volume fraction of gas in the riser at any time, ε_G. Make a list of the parameters you think ε_G depends on. Give the dimensions of each parameter.

1.8 An "airlift pump" is similar to an airlift reactor (see Problem 1.7) except that the "plumbing" is designed to ensure a continuous overflow of liquid Q_L (volume/time) from the top end of the system. Gas is introduced into the bottom of a pipe of diameter D at a distance S (the "submergence") below the external liquid surface, as shown in Fig. P1.8. The height L (the "lift") can be fixed by cutting the pipe at some distance above the surface. Make a list of the parameters you think Q_L depends upon. Give the dimensions of each parameter.

1.9 A sphere of diameter D is driven toward a second identical sphere by a force **F** that acts along the line joining the centers of the two spheres. The spheres are completely surrounded by a large body of a viscous liquid. We are interested in the time $t_{1/2}$ required for the separation of the sphere centers to be reduced to half of its

initial value. Make a list of the parameters you think this time depends on. Give the dimensions of each parameter.

Define a *dimensionless* half-time by forming a dimensionless group proportional to $t_{1/2}$ and involving some (not necessarily all) of the other parameters. Define some other dimensionless groups from these parameters, but do not use $t_{1/2}$ in any of the other definitions. How many independent dimensionless groups can you form from these parameters? Note the comments regarding the Buckingham pi theorem, presented in Problem 1.6.

Assume that this half-time depends upon the parameters p_i of your list in an explicit algebraic expression of the form

$$t_{1/2} = p_1^a p_2^b p_3^c \cdots \qquad \textbf{(P1.9)}$$

Give the values of the coefficients a, b, c, $\ldots$ for your list of parameters p_i.

1.10 Return to the example of drop breakup in a stirred tank (Section 1.1.1). Define a dimensionless stirring speed using N, D_I, σ_{12}, and μ_1.

1.11 An investigator is studying the breakup of drops of a fluorocarbon liquid in a stirred tank of water. The tank of interest is geometrically similar to those used in the study that led to Fig. 1.1.3. The impeller diameter D_I is 0.06 m. The liquid has a density of 1300 kg/m³ and an interfacial tension with water of $\sigma_{12} = 0.016$ N/m. The

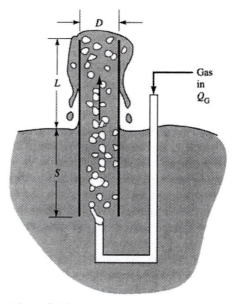

Figure P1.8 An airlift pump.

investigator is skeptical that Fig. 1.1.3 is applicable and wants to obtain some data to indicate whether the correlation of Fig. 1.1.3 holds for the system at hand. Over what range of rotational speeds should the investigator generate data?

1.12 Calabrese and his coworkers [*AIChE J.*, **32**, 657 (1986)] have studied the problem described in Section 1.1.1, but with *high* viscosity liquids as the oil phase. In that case, the dimensionless drop size is not exclusively a function of the Weber number. Speculate on the physics of drop breakup of viscous drops, and suggest an additional dimensionless group that would be important. What kind of correlation might you anticipate in such a system? Specifically, how would you plot data if water were the continuous phase in each case, but you studied oil phases of three widely different viscosities?

1.13 Think about the problem of the wicking of a small drop of liquid into a sheet of paper. List the parameters that might be important. (*Hint:* Paper is porous. Include a length scale characteristic of the pores of the paper.) If you had data for wicking time as a function of drop size, how might you plot these variables against each other in a dimensionless format?

1.14 A liquid initially fills the inside of a vertical tube of length L_o and inside diameter D. The tube is capped at both ends. Then the caps are suddenly removed, and the liquid flows out of the bottom of the tube as a continuous stream until the tube is nearly empty. List the parameters you think affect the rate of flow of the liquid from the tube. Give the dimensions of each parameter.

Define a dimensionless time $t^*_{1/2}$ for half the liquid to drain from the tube. On what dimensionless groups of parameters would this dimensionless time depend?

1.15 A liquid initially fills the inside of a vertical tube of length L and small inside diameter D. The tube is capped at both ends. Then the caps are suddenly removed, and the liquid *drips* out of the bottom of the tube until the tube is nearly empty. List the parameters you think affect the rate of flow of the liquid from the tube. Give the dimensions of each parameter.

Define a dimensionless time $t^*_{1/2}$ for half the liquid to drain from the tube. On what dimensionless groups of parameters would this dimensionless time depend?

1.16 A viscous drop of syrup sits on a planar horizontal Teflon surface. The drop does not spread across the surface. It has an equilibrium shape. List the parameters you think determine the shape of the drop. Give the dimensions of each parameter. What dimensionless groups determine the shape of the drop?

Define a dimensionless height of the drop, and list the dimensionless groups that determine this height.

1.17 Repeat the analysis given in Section 1.2, but select as the recurring parameters ρ, σ, and μ. Show that the resulting dimensionless groups can be formed from combinations of those given in Section 1.2.

1.18 In Section 1.2 we assumed that the liquid viscosity was important in forming the parameters for nondimensionalization. It is not obvious that there is any physical basis for this assumption. If we had left μ out of the list of parameters, we would have obtained one less dimensionless group. Give a set of dimensionless groups corresponding to this case.

1.19 Dimensionless groups often used to characterize a problem such as the one presented in Section 1.2 are the Weber number, We = $\rho U^2 D_c/\sigma$, the Bond number, Bo = $\rho g D_c^2/\sigma$, and the Ohnesorge number $\mu/(\rho \sigma D_c)^{1/2}$. Show that these groups can be formed from U^*, g^*, and μ^* given in Section 1.2.

1.20 Bubbles of various sizes rise through a viscous liquid. When each bubble "pops" at the surface, small droplets are flung upward from the surface. We are concerned with the maximum height above the surface that can be achieved by such droplets.

Suppose experiments with gas bubbles are carried out over a range of sizes and that liquid properties can be varied. How would you plot data in order to find a correlation of maximum drop height with fluid properties and bubble size?

1.21 In Example 1.2.1, account for the possibility that the liquid viscosity is an important parameter.

1.22 Return to Example 1.2.1. Keep the assumed variables given, but select as the recurring variables D_N, σ, and B. Why is this an unacceptable choice?

1.23 A small liquid drop, falling freely under gravitational acceleration, impacts on a smooth surface. The drop "pancakes" (i.e., spreads radially), usually attaining a maximum radius and then retracting somewhat to a final static equilibrium shape. The viscosity and surface tension of the liquid are expected to play an important role in determining the maximum spreading radius. Define a dimensionless maximum spreading radius. On what dimensionless groups would you expect the dimensionless spreading radius to depend? Suggest a method of plotting the data obtained for such a system.

1.24 Suppose, in considering the physics of Problem 1.23, you were convinced that for large values of viscosity, surface tension would exert minimal influence on the maximum spreading radius. Now, on what dimensionless groups would you expect the dimensionless spreading radius to depend?

1.25 Data for the type of physics described in Problem 1.23 have been obtained by Scheller and Bousfield [*AIChE J.*, **41**, 1357 (1995)]. Test data sets A and B against the ideas that follow from solutions to Problems 1.23 and 1.24. What conclusions do you draw? The "dimensionless maximum spreading radius" $R*$ is defined as the measured maximum spreading radius divided by the diameter D of the drop prior to its impact, at speed U.

1.26 Show that Eq. 1.2.99 may be written

$$\overline{D}* = \frac{\overline{D}}{D_p} = F\left(We, Re, \frac{\mu'}{\mu}, \frac{\rho'}{\rho}, \frac{d}{D_p}, \frac{q}{Q}, \frac{L_p}{D_p}\right) \quad \textbf{(P1.26)}$$

and give the resulting definitions of the Reynolds and Weber numbers.

1.27 Figure P1.27 shows data for the mean drop size resulting from the breakup of a stream of

Data Set A

ρ(kg/m³)	μ(mPa·s)	σ(N/m)	U(m/s)	D(m)	$R*$
1236	296	0.065	1.44	0.00319	1.47
			2.52		1.73
			3.25		1.88
			3.70		1.91
1154	32.0	0.0405	1.42	0.00285	2.28
			2.39		2.52
			3.20		2.67
			3.93		2.80
			4.96		2.97
1121	12.77	0.0415	1.37	0.00287	2.45
			2.36		2.99
			3.17		3.26
			3.93		3.38
			4.96		3.78

Data Set B

ρ(kg/m³)	μ(mPa·s)	σ(N/m)	U(m/s)	D(m)	$R*$
999	1.05	0.0731	1.34	0.00365	3.11
			2.44		4.28
			3.18		4.44
			3.85		4.99
			4.74		5.39
1029	2.35	0.0474	1.95	0.00308	3.72
			2.87		4.16
			3.58		4.47
			4.27		4.78
			4.88		4.90

organic liquid dispensed into a continuous flow of water through pipes of two different diameters. Physical properties of the organic phase are tabulated below. To what extent are these results consistent with the expectations we derive from Example 1.2.2?

Properties of the Dispersed Phase Liquids at 25°C

	ρ' (kg/m³)	μ' (mPa · s)	σ (N/m)
Anisole	990	1.0	0.026
Benzene	870	0.6	0.040
Benzyl alcohol	1000	5.0	0.005
Cyclohexane	760	0.8	0.046
Oleic acid	900	26.0	0.016
Toluene	870	0.6	0.032

1.28 After collecting and correlating the data shown in Fig. P1.27, a designer specifies a system for producing a dispersion of cyclohexane in water. Water at 25°C flows at a rate of 9 L/min through a pipe of inside diameter 5 cm and axial length 1 m. The volume fraction of cyclohexane in water is 0.005. The inlet tube diameter is 10 mm. What mean drop size do you predict?

a. Suppose the volumetric flowrate is doubled. By what factor does the mean drop size change?

b. Suppose the diameter of the pipe is increased by a factor of 2, without any change in flowrate

(i.e., still at 9 L/min). By what factor does the mean drop size change?

1.29 A chemical process is being designed that involves a large mixing tank, agitated by an impeller of diameter D_I that rotates at an angular velocity N rev/s. The process requires that a batch of solid particles of diameter D_p and density ρ_p be well suspended throughout the vessel. It is observed that if N is too small, the particles remain at the bottom of the vessel. There is a critical rotational speed N_{crit} above which adequate suspension is maintained. On what design and operating conditions do you expect N_{crit} to depend? Define a *dimensionless* N_{crit}, and list the *dimensionless* groups that N_{crit}^* should depend on.

1.30 The following equation is recommended, based on experimental data [Zwietering, *Chem.*

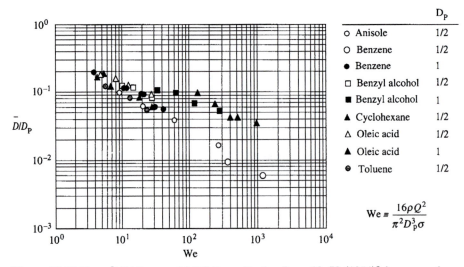

Figure P1.27 Data [Middleman, *I&EC Proc. Design Dev.*, **13**, 78 (1974)] for mean drop size in turbulent pipe flow. Note that the group labeled We is a Weber number and is proportional to the square of the group Q^* defined in Eq. 1.2.97.

Eng. Sci., **8**, 244 (1958)], for calculating the critical rotational speed N_{crit} required to completely suspend heavy particles in an agitated tank:

$$N_{crit} D_I^{0.85} = S \nu^{0.1} D_p^{0.2} \left(\frac{g \Delta \rho}{\rho} \right)^{0.45} B^{0.13}$$

(P1.30.1)

where

N_{crit} = critical rotational speed (rev/s)
D_I = agitator diameter (m)
ν = kinematic viscosity of the liquid (m²/s)
D_p = particle diameter (m)
g = gravitational constant (9.8 m/s²)
$\Delta \rho$ = particle-to-liquid density difference (kg/m³)
ρ = liquid density (kg/m³)
B = 100 × weight solids/weight liquid
S = a dimensionless constant dependent on the specific geometry of the agitator and the shape of the vessel

a. Confirm that Eq. P1.30.1 is dimensionally correct.
b. Rearrange the equation into the form of

$$G_{N*} = S G_1^a G_2^b G_3^c B^d \qquad \text{(P1.30.2)}$$

where G_{N*} is a dimensionless critical stirring speed, and the other groups G_n do not include the stirring speed. Give your choices of the G's. Are there more than four G's? How do you know? Show that your form of Eq. P1.30.2 is consistent with Eq. P1.30.1, and give the values of a, b, and c.

c. Find the stirring speed (rpm) required to suspend particles of diameter 0.01 cm and density 3200 kg/m³ in water at 25°C. The weight ratio (solids to liquid) is 0.25. The agitator diameter is 2 feet. Experiments in a small-scale model of this system indicate that the constant S has the value 8.

1.31 Assuming the applicability of Eq. P1.30.1, find the critical stirring speed N_{crit} to suspend particles of density 1260 kg/m³ and diameter 1000 μm in water. The agitator diameter is 34 in. Take an S value of 8, and assume that B is 15. How does N_{crit} change if we double the mass of particles?

1.32 Solid suspensions often must be conveyed through horizontal pipelines. We expect to find a critical flowrate Q_{crit} at which the solids are suspended in the flow and conveyed with the liquid phase. Below this flowrate, the solids form a layer along the lower surface of the pipe, or else they move slowly and erratically in contact with the lower surface.

Make a list of the parameters you think Q_{crit} depends on. Give the dimensions of each parameter. Define a dimensionless group G^* proportional to Q_{crit} and list the independent groups, none of which contain Q_{crit}, on which this group depends.

Chapter 2

Statics, Dynamics, and Surface Tension

We begin this chapter with a discussion of hydrostatics. Then we consider the role played by surface tension in fluids. Classical fluid dynamics texts often ignore this topic because surface tension plays a minor or negligible role in so many of their illustrative cases. In systems that are useful for investigating the statics or dynamics of drops and bubbles, especially those with diameters of the order of millimeters or less, however, surface tension can dominate the interaction of the drops or bubbles with one another. More generally, flow involving gas/liquid or liquid/liquid interfaces often includes an important contribution from surface tension. Such two-phase flows are of great importance in a wide range of problems of interest to chemical, environmental, and biomedical engineers. In this chapter we lay the foundation for more detailed considerations in later chapters.

Before we have much to say about the *dynamics* of fluids it is necessary to talk about the *statics* of fluids. At first there might not seem to be much to interest us about a static fluid. We will see, however, that any fluid in a gravitational field will have a pressure distribution that arises from the presence of gravity. Furthermore, at any boundary between two immiscible fluids, as in the case of a liquid and a gas, the presence of an interfacial tension (what we normally call surface tension) has the potential to give rise to a pressure change across the interface. While this pressure is normally quite small in magnitude (compared, e.g., to atmospheric pressure), it can have a profound effect on the physics and geometry of the interface, especially in the case of small drops or bubbles and thin films.

2.1 HYDROSTATICS

We begin with a "thought experiment." Imagine a very viscous oil in a funnel that ends in a very small bore capillary tube, as shown in Fig. 2.1.1. We raise several questions about this system.

What happens if we plug the capillary at its lower end? Obviously, there can be no flow out of the tube. Does this mean that the pressure is everywhere uniform? If not, what is the pressure distribution in the fluid, in this static case?

Figure 2.1.1 Oil in a closed funnel.

Imagine that we can isolate an element of volume of the oil, as shown in Fig. 2.1.2. Think of it as a cylinder between the horizontal planes z_1 and z_2, but the boundaries of the cylinder are imaginary (i.e., they are not *material* boundaries). Let $h = z_1 - z_2$. Gravity acts downward on the fluid, as shown. If there is no motion within the fluid in the z direction, then the z components of all forces acting on the ends of the fluid column must be in *balance*.

At the top surface of this column of fluid (in the plane z_1 of Fig. 2.1.3) there is a z component of force due to pressure. We will call the pressure at the plane z_1 by the name p_1. We do not know the vaue of that pressure, but we can name it. If the imaginary volume has a vertical cylindrical surface, as we have assumed, then the pressures acting along that surface, which are perpendicular to the surface (because pressures act normal to surfaces), have no components in the z direction (the direction of gravity). Hence those pressures do not contribute to the z-component force balance.

The weight of the fluid is

$$mg = \rho Ahg \tag{2.1.1}$$

where ρ is the fluid density (assumed constant) and A is the face (end) area normal to the z axis.

Hence the lower surface is acted upon by the sum of two forces—one arising from the pressure p_1 pushing down on the whole column of fluid and the other from the weight of the fluid. What opposes these forces? From inspection of Fig. 2.1.4, we see that there must be an equal upward-directed force, arising from a pressure below the area at z_2, so that

$$F_{\text{up}} = p_2 A = p_1 A + \rho Ahg = F_{\text{down}} \tag{2.1.2}$$

or

$$p_2 - p_1 = \rho gh \tag{2.1.3}$$

Figure 2.1.2 An imaginary volume element within the oil.

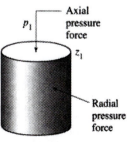

Figure 2.1.3 Forces on the upper end of the volume element.

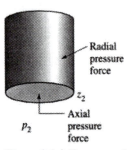

Radial
pressure
force

z_2

Axial
pressure
force

p_2

Figure 2.1.4 Forces on the lower end of the volume element.

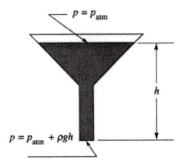

$p = p_{atm}$

h

$p = p_{atm} + \rho g h$

Figure 2.1.5 Closed funnel, no flow.

If we think of the plane z_1 as fixed, and the pressure p_1 held constant, we can see that as h increases, the pressure p_2 increases in proportion to h. *This is the fundamental law of hydrostatics for an incompressible fluid.* It is important to keep in mind that it is a *static* law. We conclude that under static conditions in a constant density fluid, the pressure will *increase* linearly with position in the direction of gravity. Hence, if we look again at the case of the funnel filled with liquid, but with no flow out of the bottom (see Fig. 2.1.5), the pressure distribution will appear as the straight line in Fig. 2.1.6.

Now we ask what happens if we unplug the bottom of the funnel so that fluid, which we take to be a liquid, can flow out. Of course this system is no longer static. We consider first the general case in which there is some finite rate of flow out of the exit tube. (Is it possible that there would be no flow if the bottom were open?) The upper surface of liquid in the funnel is open to the atmosphere, so we expect that the pressure is p_{atm} in that plane. The question is: What is the pressure in the liquid at the plane that is level with the exit end of the funnel?

With reference to the insert in Fig. 2.1.7, we argue that the fluid (the air) *outside* the tube, at the lower plane, is at atmospheric pressure. We argue now that in the liquid leaving the tube, the pressure must be continuous across the interface between the liquid and the surrounding air. If not, the forces in the *radial* direction would be unbalanced and the interface would move radially. (We will see shortly that surface tension may play a role here, but we will also shortly be able to show that for this part of the present problem, surface tension may be ignored.) Hence we conclude, at least

$z = 0$ p_{atm} p

Static
pressure
profile

z

$p_{atm} + \rho g z$

Pressure
loss due
to flow

$z = h$

Figure 2.1.6 Pressure distribution with and without flow (static pressure profile).

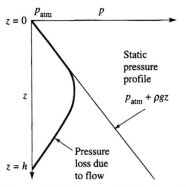

$p = p_{atm}$

$p = p_{atm}$

$p = ?$ Detail

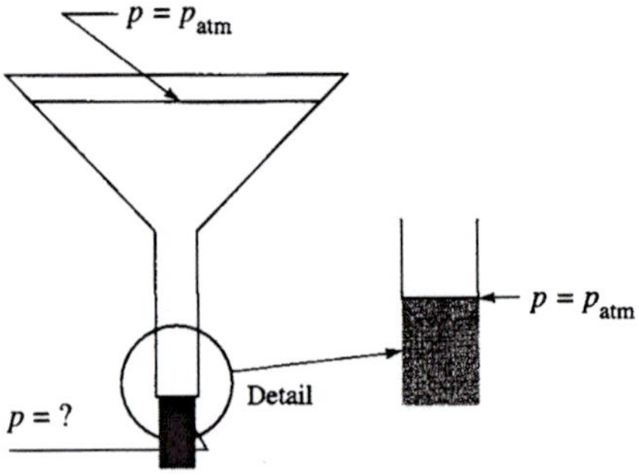

Figure 2.1.7 Open funnel with flow from the bottom.

tentatively, that the pressure in the liquid at the lower surface is atmospheric, and thus it equals the pressure at the upper surface. But in one region in the higher levels of the funnel, the large cross-sectional area results in a very low velocity of the liquid moving toward the funnel outlet. That is, there is no significant velocity in the funnel except very near the entrance to the lower tube of the funnel, and then of course along the axis of that tube. (The fact that we are making this statement does not imply that it is true. It is a *hypothesis* that ultimately will lead us to a mathematical model of the rate of flow from the funnel. We can test that model, then, for consistency with the hypothesis. Of course, this argument assumes that the details of the pressure distribution affect the flow from the funnel and that a false argument regarding pressure would lead to a measurably false conclusion regarding the exit flow velocity.)

On the basis of this picture of flow through the funnel, we conclude that the pressure distribution must look like that shown in Fig. 2.1.6. We imply that in the upper region of the funnel, where the velocity is very low, the pressure obeys (nearly) the hydrostatic law. As fluid accelerates toward the exit tube of the funnel, flow effects of some kind will cause a loss of pressure. Finally, the pressure will fall along the axis of the exit tube until it reaches atmospheric pressure at the end exposed to the air.

Now suppose that the exit flow itself is very slow, either because the liquid is very viscous or because the exit tube has a very small diameter, and that a drop begins to form slowly on the end of the capillary. We will argue now that it is possible that the flow will stop, if the surface tension of the fluid is sufficiently high, and the capillary tip is small enough. In other words, we claim that even if the funnel is open at the bottom, the liquid may remain in the funnel. To make this argument, we must talk about surface tension in some detail. We begin with two everyday observations: that small drops can cling to a surface even though gravity acts on them and that a sponge will hold a considerable weight of water—it will not *drip* completely dry. Apparently, there is some kind of adhering force acting along the boundary of a liquid with the surface with which it is in contact.

2.1.1 Surface Tension

Let's return to the drop problem. With reference to Fig. 2.1.8, we assume that the liquid is flowing through the exit tube very slowly and that a drop forms slowly on the tip of the tube. Based on common experience, we know that as the drop grows, its weight will overcome the force of surface tension that holds the drop to the tip of the capillary,

Figure 2.1.8 Drop hanging on the tip of a capillary.

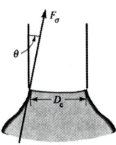

Figure 2.1.9 Forces along the contact line in the neighborhood of A in Fig. 2.1.8.

and the drop will fall off. For a simple discussion we assume that the capillary is vertical, pointed down, and that the drop forms symmetrically on the tip.

Figure 2.1.9 shows the detail of the contact region of the liquid with the capillary. The *contact line* of the liquid with the capillary is a circle of diameter D_c, and thus the contact line has a length (a perimeter) πD_c. Along this three-phase (liquid, solid, gas) contact line, surface tension may be defined as the force per unit length acting tangentially to the liquid–gas surface. We denote the surface, or interfacial, tension by the symbol σ. The surface in the near neighborhood of the contact line is not necessarily vertical. We denote the angle between the tangent to the contact surface and the vertical by θ. The magnitude of the force itself is denoted by F_σ, and it acts at an angle θ to the vertical, as shown in Fig. 2.1.9. Note that F_σ is drawn acting upward. This is the surface force exerted *by* the contact line—the perimeter of the capillary—*on* the hanging drop.

If our definition of σ makes sense, then there is a *vertical component* of force given by

$$F_\sigma \cos \theta = \sigma \pi D_c \cos \theta \tag{2.1.4}$$

If the drop is static, then this restraining force must just balance the weight of the drop, so

$$\sigma \pi D_c \cos \theta = \rho g V \tag{2.1.5}$$

where V is the drop volume. In this expression, the angle θ is determined by the forces acting on the drop. It is not a property of the liquid. In particular, θ is not the equilibrium contact angle of the liquid on the material that comprises the tip of the capillary.

We now want to determine the maximum size of a drop that will "hang" on a capillary. Physically, the critical condition arises when the weight of the drop exceeds the force of surface tension. If Eq. 2.1.5 holds, what happens if we increase the drop volume slightly? What can change on the left-hand side of the equation? The only possibility is that the angle θ changes. *The simplest speculation*[1] is that as the drop gets heavier, the angle approaches $\theta = 0$. At that point the left-hand side of Eq. 2.1.5 no longer changes, it has reached its maximum value, and so the equation could not be satisfied for any larger V values. If this idea is correct (see Problem 2.25), then the drop will fall when its volume exceeds

$$V_{max} = \frac{\sigma \pi D_c}{\rho g} \tag{2.1.6}$$

---

[1] We develop our first mathematical model here. Simple models are based on speculations or hypotheses about the physics.

*If the drop is nearly a sphere[2] of diameter D_d (note that we do not assume that the drop diameter is the same as the capillary diameter) then

$$V = \frac{\pi D_d^3}{6} \tag{2.1.7}$$

This analysis yields a simple model of the diameter of the drop that falls from a slowly dripping capillary:

$$D_d = \left(\frac{6\sigma D_c}{\rho g}\right)^{1/3} \tag{2.1.8}$$

It is useful to nondimensionalize Eq. 2.1.8 to the form

$$\frac{D_d}{D_c} = \left(\frac{6\sigma}{D_c^2 \rho g}\right)^{1/3} \tag{2.1.9}$$

Note that we do this simply by dividing both sides of the equation by a characteristic length scale of the system—the capillary diameter. Since the left-hand side of this equation has no units (i.e., it is dimensionless), it follows that the right-hand side must also be dimensionless. Certain dimensionless groupings of parameters appear so often in certain classes of problems that we give the groupings specific names, usually in honor of individuals who made classic contributions to the study of the problems in which these groups arise. An important dimensionless group that determines the nature of the interaction of gravitational and surface forces is the *Bond number,* defined as

$$Bo = \frac{D_c^2 \rho g}{\sigma} \tag{2.1.10}$$

In terms of the Bond number, our model for drop size takes the form

$$\frac{D_d}{D_c} = \left(\frac{6}{Bo}\right)^{1/3} = 1.82\, Bo^{-1/3} \tag{2.1.11}$$

Notice two things: (1) by nondimensionalizing equations, we can write mathematical models in very simple-looking formats and (2) this is a *testable* model. Let's do it!

2.1.2 Test of the Model of Critical Drop Size

Experiments have been conducted with a variety of liquids, over as wide a range of surface tensions as practical, to test the model of critical drop size. A constant and measured flow of liquid was initiated into the capillary, and the falling drops were counted, collected, and weighed. From the measured density of the liquids it was then possible to calculate the volume, hence the drop diameter D_d. The data were presented in terms of the ratio D_d/D_c, and the model was tested by calculating the function

$$N_d = \left(\frac{D_d}{D_c}\right) Bo^{1/3} \tag{2.1.12}$$

If the model is correct we should find that N_d is independent of the liquid flowrate, denoted by Q, and that from Eq. 2.1.11, N_d has a value of 1.82 regardless of the liquid tested.

---

[2] This is another speculation or hypothesis, and so this is a part of the process of developing the mathematical model.

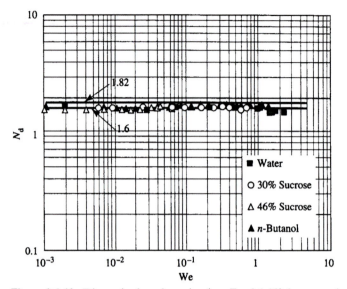

Figure 2.1.10 Dimensionless drop size (see Eq. 2.1.12) for several liquids, as a function of the Weber number.

Figure 2.1.10 shows the data obtained in our study. The results are in agreement with this model, except that the coefficient has a value of 1.6 instead of 1.82. This lower value arises because when the drop falls from the capillary tip it leaves some residual liquid attached to the tip (see Fig. 2.1.11). Hence the drop volume is lower than predicted by this model.

The data in Fig. 2.1.10 are plotted as a function of the dripping rate Q (the flowrate of the liquid). The flowrate is made dimensionless through the introduction of the *Weber number* (recall our comment in the discussion of Fig. 1.1.3), which is defined in this problem as

$$\text{We} = \frac{\rho(4Q/\pi D_{ci}^2)^2 D_{co}}{\sigma} \qquad (2.1.13)$$

Figure 2.1.11 A drop falling from a capillary.

where ρ is the liquid density and the subscripts *ci* and *co* refer to the inner and outer diameters of the capillary, respectively. This distinction reflects the observation that the drops in the experiments that lead to Fig. 2.1.10 fall from the tube while in contact with the *outer* diameter, while the flow through the tube is obviously *within* the inner diameter. The term $4Q/\pi D_{ci}^2$ arises because it is the average linear velocity of flow through the tube. (The Weber number is always of the form $\rho U^2 L/\sigma$, where L is some appropriate length scale for the flow and U is a characteristic velocity in the system.)

Figure 2.1.10 demonstrates that the drop size is nearly independent of the drip rate until the Weber number reaches order unity. Beyond that point, the drop has enough inertia to be able to separate from the capillary at a lower weight, hence at a lower drop size. When the Weber number exceeds a value of about 3, we observe that dripping ceases and the liquid issues as an initially continuous jet. This jet will then break down into droplets, often of two very disparate diameters. Drop breakup of a liquid jet, another very interesting problem, is discussed in Chapter 7.

We conclude that the factors that affect whether an open tube will drip at all, and (if it does) the diameter of drops that fall under the force of gravity, can be understood in terms of a very simple model based on a balance of *static* forces at the tip of the capillary. The theory that results is not exact, but it does capture the essential physics of the process, failing only to yield the exact value of the coefficient in Eq. 2.1.11. Figure 2.1.10 may be used with confidence for the prediction of drop size from a knowledge of the liquid properties and the capillary diameters. Equation 2.1.11 is limited to small values of D_c and flowrate Q—actually to small values of the Bond number as well as to small Weber numbers.

An important practical consequence of this discussion is that we may determine the surface tension of an unknown liquid by performing a drop size experiment. If we invert Eq. 2.1.11 (and use the empirical coefficient 1.6 instead of the theoretical value of 1.82), we find

$$(\mathrm{Bo})^{-1} = \left(\frac{D_d}{1.6\,D_c} \right)^3 \tag{2.1.14}$$

or

$$\sigma = 0.244 \, \frac{\rho g (D_d)^3}{D_c} \tag{2.1.15}$$

It follows that careful measurements of the drop diameter D_d will yield a value for σ.

EXAMPLE 2.1.1   *Surface Tension by the "Drop Weight" Method*

Water at 20°C drips slowly from a capillary, and 20 drops are collected and weighed. The total weight (mass) is found to be 0.448 g. (We will use cgs units consistently in this example.) Close observation indicates that the drops wet the face of the capillary and are in contact with the outside diameter of the capillary, which is found to be 0.133 cm. Hence, in Eq. 2.1.15 we use $D_c = 0.133$ cm. To find the drop diameter, we use the following relationship:

$$\frac{\pi \rho D_d^3}{6} = \text{mass/drop} = \frac{0.448 \text{ g}}{20 \text{ drops}} = 0.0224 \text{ g/drop} \tag{2.1.16}$$

Taking the density of water at the temperature of the experiment (20°C) to be 1 g/cm³, we find a drop diameter from Eq. 2.1.16 of $D_d = 0.35$ cm. Substituting this value into Eq. 2.1.15, we find

$$\sigma = 0.244 \frac{(1.0)(980)(0.35)^3}{0.133} = 77 \text{ dyn/cm} \tag{2.1.17}$$

This value is higher than expected. Very clean water, studied in an apparatus that itself is very clean, would have a surface tension of 72 dyn/cm at 20°C. We would expect a lower value (probably in the range of 65–70) for water handled under less than pristine conditions. Keeping in mind that the coefficient (1.6) appearing in Eq. 2.1.14 is an average for the data of Fig. 2.1.10 and is probably not precise, we can expect an error of the order of 10% through the use of this method, and Eq. 2.1.15. Hence the measured value may be considered to be a reliable approximation, but certainly not a precise value. In Problem 2.27 we describe an empirical correction method that, when care is taken in the experimental procedure, leads to values of surface tension more precise than what we would calculate through the use of Eq. 2.1.15.

EXAMPLE 2.1.2 *Design of a Drop Dispenser*

In a packaging process we require a system that will dispense drops of a liquid subject to the following constraints:

Each drop is to have a volume $V_{\text{drop}} = 0.014 \text{ cm}^3$.
Drops are to be produced in air at a frequency of 1 drop every 2.5 s.
The liquid has a viscosity of twice that of water.
The liquid has a surface tension, with respect to air, of 45 dyn/cm.

In this example, a "design" begins with the specification of the dimensions of the delivery tube, and the flowrate of the liquid that will permit these constraints to be met.
  As our design equation we will use Eq. 2.1.15, in a form explicit for capillary diameter:

$$D_c = 0.244 \frac{\rho g (D_d)^3}{\sigma} \tag{2.1.18}$$

For the drop diameter we have

$$D_d = \left(\frac{6 V_{\text{drop}}}{\pi}\right)^{1/3} = \left(\frac{6 \times 0.014}{\pi}\right)^{1/3} = 0.3 \text{ cm} \tag{2.1.19}$$

From Eq. 2.1.18 this yields a value for the capillary diameter of

$$D_c = 0.244 \frac{1.0 \times 980 (0.3)^3}{45} = 0.144 \text{ cm} \tag{2.1.20}$$

Note that in the absence of any information, we assume that the density of the liquid is 1.0 g/cm³. Any deviation from that value will require a corresponding deviation from the value of D_c calculated here.
  We will assume that the liquid in this case does not wet the outside of the capillary, and so we regard D_c as the *inside* diameter.
  The required flowrate follows from the drop volume and the frequency of production:

$$Q = \frac{0.014 \text{ cm}^3}{2.5 \text{ s}} = 5.6 \times 10^{-3} \text{ cm}^3/\text{s} \tag{2.1.21}$$

The use of Eq. 2.1.15 is based on the experimental data of Fig. 2.1.10. That is, the design at hand is assumed to operate in the same range of parameters used to generate the data of that figure. Hence we must calculate a value for the Weber number, defined in this case as

$$\mathrm{We} = \frac{16\rho Q^2}{\pi^2 \sigma D_c^3} \qquad\qquad (2.1.22)$$

with the result

$$\mathrm{We} = \frac{16 \times 1 \times (5.6 \times 10^{-3})^2}{\pi^2 \times 45 \times (0.144)^3} = 3.8 \times 10^{-4} \qquad\qquad (2.1.23)$$

Since this value lies to the left of the experimental range, the flowrates are even lower than those used to generate the experimental data. This result implies that the liquid will drip slowly, with no inertial effects that could cause jetting. Figure 2.1.10 should be a reliable design guide. If the Weber number is large, however, a coherent jet will issue from the capillary and drops will be formed *far downstream* by a completely different mechanism. The conditions under which this occurs are discussed in Chapter 7.

Now that the capillary diameter has been specified, there remain two important design and operating parameters to deal with. One is the length of the capillary, and the other—related to the length—is the pressure required to maintain the flow through this capillary, at the specified flowrate. These issues will be addressed after Chapter 3 has developed some necessary background. First, we need to learn some more about the effect of surface tension on the fluid inside a drop.

2.2 CAPILLARY HYDROSTATICS: THE YOUNG–LAPLACE EQUATION

In this section we derive a very basic and important equation related to the physics of fluid interfaces. We want to show that interfacial tension can give rise to a pressure difference across an interface that separates a pair of immiscible fluids. In addition, we want to show how that pressure difference can be calculated from the shape and size of the interface. First, however, a few comments about how the shape of a surface is characterized in space. In particular, we need to talk about the "curvature" of a surface.

We begin by examining two simple surfaces: the surface of a sphere and the surface of a cylinder (Fig. 2.2.1). Lines A and B on the surface of the *sphere* are any pair of lines *in the surface* that meet at right angles. We call them orthogonal curves. Each of these curves is the arc of a circle, and that circle has a radius identical to that of the sphere. For a sphere, the radii of curvature of the surface in any pair of orthogonal directions are identical, and they are the same as the radius of the sphere.

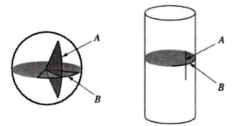

Figure 2.2.1 Two simple curved surfaces.

Lines A and B on the surface of the *cylinder* are a specific pair of lines which are orthogonal. Line B is the arc of a circle whose radius is that of the cylinder. Line A is a straight line in the surface, parallel to the axis of the cylinder. It is not a curve, except in a kind of degenerate sense. We may regard a straight line as the arc of a circle with an *infinite* radius. Hence we may say that for a cylindrical surface, the radii of curvature of the surface in the chosen pair of orthogonal directions are different. One radius is that of the cylinder, and the other is infinite.

When we examine surfaces of arbitrary shape (rather than the simple surfaces of Fig. 2.2.1), the situation is a little more complex. We omit the details here, making only a loose statement about surfaces.

> *For any surface there is a pair of orthogonal curves at any point on the surface, and if the curves are of small (differential) dimensions, each curve may be approximated as the arc of a circle. The radii of the two circles are called the* principal radii *of curvature. The curvature of a surface, at a point, is described in terms of the radii R_1 and R_2 of the curves corresponding to that point. The radii may vary from point to point on a complex surface.*

When we talk of an interface or boundary between two fluids, we speak mathematically. We imply that it is possible to identify one fluid or the other on either side of a surface. If the fluids are *miscible,* the concept loses its utility because the fluids will interpenetrate, by diffusion and convection, and no clearly defined boundary separates them. If the fluids are *immiscible,* the surface exhibits behavior consistent with the idea that the surface is in tension. If we could somehow make in the surface a small slit of length ds, the surface would open up (spread apart, like the skin of a balloon) unless we pulled the edges back together with a force F. The magnitude of the force is characteristic of the fluids on either side. We define a property of the interface, called surface tension, or interfacial tension, as the force per unit length acting along any line in the surface, in a direction tangential to the surface and normal to the line. We can show that this interfacial tension can give rise to a pressure *difference* across an interface, if that interface has finite curvature.

We begin by relating the pressure difference due to interfacial tension to the geometry of the surface. Figure 2.2.2 shows a differential element of area within the boundary surface separating a pair of immiscible fluids. The area element is bounded by two pairs

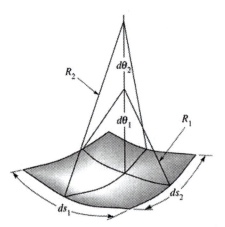

Figure 2.2.2 Differential area in the surface separating two immiscible fluids.

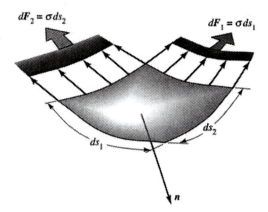

Figure 2.2.3 Surface tension forces acting along the edges of the differential area. If you sliced the surface along the line ds_2, the edges would separate. The force $\sigma \, ds_2$ is required to keep the edges together. The same is true for the line ds_1.

of parallel arcs (of lengths ds_1 and ds_2), having radii R_1 and R_2. *If ds_1 is small enough,* it can always be represented as the arc of some circle having a radius R_1 and an included angle $d\theta_1$.

Now, how does interfacial tension manifest itself? By definition, the interfacial tension σ *can be thought of as a force per unit length* acting across a line element in a surface, the line of action of the force being normal to the line and tangential to the surface.

If we isolate the differential area for the purpose of writing a force balance, we may consider it in terms of Fig. 2.2.3. Along each of the four arcs that define the area, a distributed force acts uniformly, and it has magnitude $\sigma \, ds$. Let us now sum the components of these four forces in the direction of the normal **n** to the surface. The **n** component of those forces acting along the pair of parallel lines of length ds_2 is most easily calculated by using Fig. 2.2.4, which is a view normal to ds_1 and parallel to ds_2. Along each of the lines ds_2 is a force $\sigma \, ds_2$ tangent to the surface. The **n** component of each force is simply $(\sigma \, ds_2) \sin (d\theta_1/2)$.

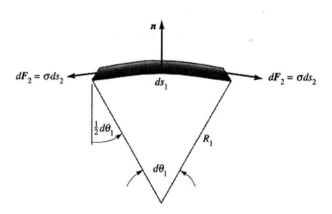

Figure 2.2.4 Sketch for the force balance on the area separating two immiscible fluids.

The angle $d\theta_1$ is related to the arc ds_1 through the radius of curvature of the arc R_1:

$$ds_1 = R_1 d\theta_1 \tag{2.2.1}$$

For *differential* (i.e., *small*) *angles,* a good approximation to the sine function is

$$\sin\left(\tfrac{1}{2}\,d\theta_1\right) = \tfrac{1}{2}\,d\theta_1 \tag{2.2.2}$$

Using these results, and summing the *two* forces along the *pair* of ds_2 lines, we find

$$dF_2 = \frac{\sigma}{R_1}\,ds_1 ds_2 \tag{2.2.3}$$

By identical argument, the **n** component of forces acting along the pair of ds_1 lines is

$$dF_1 = \frac{\sigma}{R_2}\,ds_1 ds_2 \tag{2.2.4}$$

Thus the net **n** component of force due to interfacial tension is the sum

$$dF_\sigma = \sigma\,ds_1 ds_2 \left(\frac{1}{R_1} + \frac{1}{R_2}\right) \tag{2.2.5}$$

Note that this force is in the $-\mathbf{n}$ direction. If the fluids are *static,* then the only stresses on either side of the surface separating the two fluids are static pressures, say, p_o and p_i, where p_i is the pressure on the concave side ("inside") of the interface (Fig. 2.2.5). These pressures also give rise to forces in the **n** direction of magnitudes $p_o\,ds_1 ds_2$ and $-p_i\,ds_1 ds_2$. Hence, at equilibrium, the force balance on the differential surface becomes (taking forces as positive down)

$$\left[p_o - p_i + \sigma\left(\frac{1}{R_1} + \frac{1}{R_2}\right)\right]ds_1 ds_2 = 0 \tag{2.2.6}$$

But this result is valid for *any* area $ds_1 ds_2$ as long as it is a differential area. Hence it follows that the bracketed term vanishes on any small region of the surface. Therefore Eq. 2.2.6 gives the condition of equilibrium of normal stresses across a static interface separating a pair of immiscible fluids:

$$p_i - p_o = \sigma\left(\frac{1}{R_1} + \frac{1}{R_2}\right) \tag{2.2.7}$$

This relation is known as the Young–Laplace equation.

Physically, the equation tells us that interfacial tension causes an increased pressure on the "inside" of a surface, the magnitude depending on the radii of curvature of the surface. The geometrical boundary separating two immiscible fluids defines a region

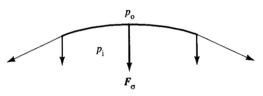

Figure 2.2.5 Pressures and forces acting on a curved interface.

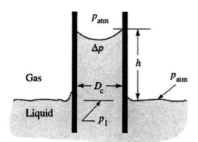

Figure 2.2.6 Rise of a liquid in a capillary.

across which there is a discontinuity in fluid physical properties. Equation 2.2.7 shows that there is a corresponding discontinuity in normal stress[3] across the boundary.

EXAMPLE 2.2.1   *Pressure Inside a Gas Bubble*

For a *spherical* gas bubble inside a liquid, each of the two principal radii of curvature is equal to the bubble radius:

$$R_1 = R_2 = R \qquad\qquad (2.2.8)$$

Hence, there is a pressure difference across the bubble/liquid interface given by (see Eq. 2.2.7)

$$\Delta p = \frac{2\sigma}{R} \qquad\qquad (2.2.9)$$

A typical surface tension for a liquid is $\sigma = 0.05$ newton/m (N/m). For a bubble of radius $R = 10^{-3}$ m, we find

$$\Delta p = \frac{2(0.05)}{0.001} = 100\ \text{N/m}^2 \qquad\qquad (2.2.10)$$

Clearly, since 1 atm $\approx 10^5$ N/m^2, this is a very small pressure. Nevertheless, we will find that a pressure of only 100 N/m^2 can have a significant effect on the fluid dynamics near the interface. Indeed, this pressure can create flow, as indicated in Example 2.2.2.

EXAMPLE 2.2.2   *Liquid Rise in a Capillary*

A narrow capillary tube is dipped into a liquid that wets the tube. We observe that the liquid rises in the tube, above the level of the free surface. We want to derive a mathematical model that relates the height of capillary rise h to the surface tension σ and the inner capillary diameter D_c. This model (Fig. 2.2.6) will suggest another method for measuring the surface tension.

---

[3] The normal stress acts perpendicular (normal) to the surface. By contrast, any stress that acts in the plane of the surface (i.e., *tangential* to the surface) is called a *shear* stress.

If we approximate the meniscus as a hemisphere of diameter D_c, then the two radii of curvature of this hemispherical surface are equal to each other, and are given by

$$R_1 = R_2 = \tfrac{1}{2}D_c \qquad\qquad\qquad (2.2.11)$$

From the Young–Laplace equation (Eq. 2.2.7), we find the pressure difference across the meniscus as

$$\Delta p = -\frac{2\sigma}{R} = -\frac{4\sigma}{D_c} \qquad\qquad\qquad (2.2.12)$$

(Note the negative sign. The pressure on the liquid side of the meniscus is *below* atmospheric. The pressure is always higher on the concave side of the surface, which in this case is in the atmosphere.)

Since the system is in *static equilibrium,* the pressure p_1 must equal p_{atm}, for these two pressures occur at points that lie in the same horizontal plane in the same fluid. By the hydrostatic law (Eq. 2.1.3) then, we can write

$$p_{atm} + \Delta p + \rho g h = p_1 = p_{atm} \qquad\qquad\qquad (2.2.13)$$

Hence we find

$$\rho g h = -\Delta p = \frac{4\sigma}{D_c} \qquad\qquad\qquad (2.2.14)$$

or

$$h = \frac{4\sigma}{\rho g D_c} \qquad\qquad\qquad (2.2.15)$$

We may make this nondimensional (note Eq. 2.1.10) and write

$$\frac{h}{D_c} = \frac{4\sigma}{\rho g D_c^2} = \frac{4}{\text{Bo}} \qquad\qquad\qquad (2.2.16)$$

where Bo is the Bond number defined in Eq. 2.1.10. For the water/air interface, the surface tension is known to be $\sigma = 0.072$ N/m. The density that appears here is that of the liquid, water, so we use $\rho = 10^3$ kg/m³. In SI units we use g = 9.8 N/kg = 9.8 m/s². For the example, we choose a small diameter capillary, say $D_c = 10^{-3}$ m, and we find that the capillary rise is

$$h = 0.029 \text{ m} = 2.9 \text{ cm} \qquad\qquad\qquad (2.2.17)$$

This example suggests a technique by which we may measure the surface tension of a gas/liquid interface. Indeed, the capillary rise method is often used for the determination of surface tension values.

We now have discussed two methods by which we claim to be able to measure an important property of a gas/liquid interface—the surface or interfacial tension. We need to pause for a moment and recall that we have "built" these models based on a concept of surface tension we have taken as a "given." The concept is that an interface between two immiscible fluids (gas/liquid or liquid/liquid) behaves as if there is a property of the interface—surface tension—such that a kind of cohesive force exists and acts tangentially to the surface. The force manifests itself if the surface has curvature, and the magnitude of the force per unit length along a line in the surface is the property we call surface tension. Do we have to accept this concept? This is a very general kind of philosophical question that we must raise when we "do physics." A simple answer is that if the concept does not produce any violations of established physical principles, and if it leads to mathematical models that reflect observations, it is a useful concept that we may hold onto until confronted with evidence to the contrary.

What would be an example of evidence against our concept of surface tension? If our concept makes sense, we should be able to measure surface tension by our two methods: the drop size method (Eq. 2.1.15) and the capillary rise method (Eq. 2.2.16 in a form explicit in σ), obtaining the same value for a specific pair of fluids—at least within the precision of the experiment. Failure to obtain equal values in such a trial would imply either that the concept of a physical property called surface tension, as we have defined it, is not consistent with good physics, or that the mathematical models for drop size and/or capillary rise may be in error. The latter is certainly a strong possibility, and this would have to be carefully examined before we abandoned the physical concept of surface tension.

Mathematical models may be in error if we have introduced assumptions that prove to be poor representations of the details of the physics. In the analysis of the capillary rise, for example, we assumed that the meniscus is a hemisphere, with a radius equal to that of the capillary. This is a good assumption for liquids that wet the inside of the capillary with a contact angle near zero, in very small capillaries, say of the order of 1 mm or less in diameter, but it is easy to see that the meniscus in a large tube (we probably would no longer call it a "capillary") is nearly flat. If we were to do a series of experiments on capillary rise in tubes of diameters in the range of 0.5 to 10 mm, we would measure an "apparent" surface tension, by applying Eq. 2.2.16 to the data on the capillary rise h, and the resulting value of σ would turn out to depend on the size of the tube. We would conclude from that observation that the property we are measuring is not strictly a property of the interface, and we would be obliged to reexamine the analysis that had led us to that point.

Surface tension, as we have introduced it, is a well-defined material property, and Table 2.2.1 lists values for several pairs of immiscible fluids. Note that the SI units of surface tension are newtons per meter, and the cgs units (commonly cited in older tables of data) are dynes per centimeter.

Table 2.2.1 Surface Tensions of Common Pairs of Fluids at 25°C

Pair	σ (N/m)[a]
Air/water	0.072
Air/benzene	0.029
Air/methyl alcohol	0.023
Air/cyclohexane	0.025
Air/dibutyl phthalate	0.032
Air/silicone oil (2 St)	0.019
Air/glycerol	0.063
Air/mineral oil	0.030
Water/benzene	0.040
Water/toluene	0.032
Water/oleic acid	0.016
Water/isoamyl alcohol	0.0048

[a]To get cgs units (dyn/cm), multiply by 1000.

EXAMPLE 2.2.3 *Pressure Inside a Growing Drop*

In Example 2.1.2 we considered the design of a system for producing drops from a capillary. The required flowrate was found from the specified drop volume and drip frequency. We did not address the issue of flow production. One common approach makes use of a syringe pump, which is essentially a cylindrical reservoir, and a close-fitting piston that moves at constant speed, thereby displacing the fluid and producing a constant flowrate. The plumbing between the reservoir and the end of the capillary resists the flow, and as a consequence a back pressure is exerted on the piston. The mechanism that moves the piston must exert a driving force that corresponds to this pressure.

In view of the comments made earlier in this section, we realize that an additional back pressure arises as the drop grows. With reference to Fig. 2.2.7 we see that initially the drop has a large radius of curvature. As the drop grows larger, its radius of curvature *decreases*. If the drop is always a sector of a sphere, it is possible to calculate the radius of curvature as a function of the external (but attached) drop volume. The minimum radius of curvature occurs when the drop is a hemisphere. Further growth on the tip leads to further increase in the radius of curvature, until the drop falls off the tip.

In view of the Young–Laplace equation, we can see that the maximum back pressure corresponds to the minimum radius of curvature:

$$\Delta p_{\text{max}} = \frac{4\sigma}{D_c} \tag{2.2.18}$$

Under most circumstances, the magnitude of this pressure is quite small. In the case of Example 2.1.2 we have (in cgs units)

$$\Delta p_{\text{max}} = \frac{4 \times 45}{0.144} = 1250 \text{ dyn/cm}^2 \tag{2.2.19}$$

Since atmospheric pressure is 10^6 dyn/cm^2, the pressure found in Eq. 2.2.19 is a small one. This effect may therefore lack practical consequence in the type of flow system (constant flowrate) described here. If, however, the flow is produced from a reservoir at *constant pressure,* the pressure difference available to overcome the flow resistance

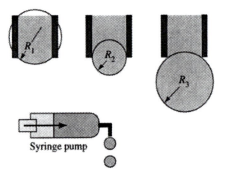

Figure 2.2.7 Flow from a syringe pump. As liquid issues from the capillary, the radius of curvature decreases to a minimum (R_2) and then increases thereafter.

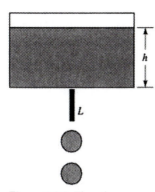

Figure 2.2.8 Flow from a constant head reservoir.

in the "plumbing" of the system will change as the drop grows on the tip. One way to achieve such a constant pressure is to use a "gravity-fed" flow, as suggested in Fig. 2.2.8. By maintaining the "head" of liquid in the reservoir (h in Fig. 2.2.8) constant, we can keep the pressure at the entrance to the capillary constant. If there were no back pressure due to surface tension and the curvature of the drop surface, the pressure drop across the capillary length L would be constant, and the resulting flowrate would be constant. However, if the back pressure (as calculated from the Young–Laplace equation) is a significant fraction of the hydrostatic pressure, the flow will slow down as the maximum back pressure is approached, increasing again as the drop grows beyond the hemispherical shape. The flow would be unsteady (time dependent). This is a more complex flow than we can consider at this point, but eventually we will be able to develop a model of dripping from such a system.

2.3 PRESSURE IN RESPONSE TO EXTERNAL FORCES

In one interpretation of the hydrostatic law given as Eq. 2.1.3, we say that if a mass of fluid is subject to a force with the *potential* to produce accelerated motion, such as gravity, a pressure distribution is set up in the line of action of the force. This interpretation is consistent with Newton's laws of motion, and it is consistent with observation.

Consider, as another example, a liquid in a cylindrical container, as shown in Fig. 2.3.1. The cylinder has been rotating for a long time about its axis at a steady rotational speed N (rev/s), and the axis is vertical. This is a *static* system *at equilibrium*—there is no acceleration of the fluid.[4]

We will impose a simple force balance on a volume element of liquid in the container. We assume that the system is in static equilibrium, and in particular that the liquid is rotating within its container as a rigid body. Specifically, any "particle" of liquid has the same rotational speed N as any other particle. Simple Newtonian mechanics for a rigid body asserts that

$$\Sigma \mathbf{F} = m\mathbf{a} \tag{2.3.1}$$

---

[4] This is a loose statement. The forces are in balance, and so the speed of a fluid "particle" does not vary in time. However, the angular position varies with time (rotation) and since the velocity is a *vector*, there really is acceleration.

Figure 2.3.1 A container of liquid rotates about its axis.

If we impose this relationship in the vertical direction over a volume element of height h, we recover the hydrostatic law

$$p_2 - p_1 = \rho g h \tag{2.3.2}$$

We rewrite this law for a *differential* vertical distance dz, and we call the pressure difference by the name dp to reflect the differential nature. By "differential" we simply mean that the difference (in distance or pressure) is very small. Hence we have

$$dp_z = -\rho g \, dz \tag{2.3.3}$$

The minus sign reflects our convention that the z axis is positive upward. The subscript z on dp reminds us that we are calculating the difference dp in the z direction.

To impose Eq. 2.3.1 in the radial or horizontal direction, we must first state the nature of the force that results from rotation. Of course we are talking about the concept of centripetal acceleration due to steady rotation, and the *magnitude* of the acceleration experienced by a particle of liquid at radial position r is simply $r\omega^2$, where ω is the rotational speed in radians per second rather than in revolutions per second. (One revolution is 2π radians.) Note, however, that the fluid is in static equilibrium—it is not accelerating. The rotation has the *potential* to produce acceleration, but the boundaries of the fluid prevent the motion.

Then, by using arguments identical to those that led to Eq. 2.1.3 (i.e., a force balance on a cylinder of liquid), but this time on a cylinder with its axis oriented along a radius of the vessel, we would find

$$dp_r = \rho r \omega^2 \, dr \tag{2.3.4}$$

The subscript r reminds us that we are calculating the difference dp in the r- direction. From differential calculus it should be clear that the pressure distribution must be a function $p(r, z)$ such that

$$dp = \frac{\partial p}{\partial z} dz + \frac{\partial p}{\partial r} dr = dp_z + dp_r \tag{2.3.5}$$

From Eqs. 2.3.3 and 2.3.4 it is easy to see that a function that satisfies these equations is

$$p = -\rho g z + \frac{\rho r^2 \omega^2}{2} \tag{2.3.6}$$

Now that we have a model of the pressure field within a rotating body of liquid, what do we do with it? The answer to this kind of question is always the same: **we examine testable consequences of the model.**

Surface of a Rotating Liquid

Consider the upper surface (the free surface) of the liquid in the vessel. That surface is exposed to atmospheric pressure. Hence the upper boundary of the liquid is a surface of constant pressure. Let us take atmospheric pressure to have the arbitrary value of zero. Then if we set $p = 0$ in Eq. 2.3.6, we find the curve in the rz plane that defines the free surface. That is, we find the *shape* of the free surface, which we will call $Z(r)$. From Eq. 2.3.6, for $p = 0$, we solve for $z(r)$, change its name to $Z(r)$, and find

$$Z(r) = \frac{r^2\omega^2}{2g} \qquad\qquad (2.3.7)$$

This is an easily tested prediction, and it is found to be consistent with experience. The shape of a free surface in a rotating vessel of liquid, under conditions ensuring that the liquid rotates as a rigid body, is a parabolic function of r with a shape related to rotational speed ω in accordance with Eq. 2.3.7.

Design of a Parabolic Mirror

Mathematical models have implications! A good engineer examines these models for their potential to trigger ideas for new products or processes. Let's look at Eq. 2.3.7 with that challenge in mind. The equation tells us that we can create a parabolic surface by rotating a liquid in a gravitational field and that the shape of the surface can be controlled by the rotational speed. Thus we should be able to make a "dynamic parabolic mirror" by rotating mercury in a cylindrical vessel. From knowledge of analytic geometry, we may define the position of the focal point f of a parabola by writing the surface in the form

$$y^2 = 4fx \qquad\qquad (2.3.8)$$

or, in the notation of the rotating liquid example,

$$r^2 = \frac{2g}{\omega^2} Z_v \qquad\qquad (2.3.9)$$

where Z_v is the height of the free surface relative to the vertex at $r = 0$. Hence the focal point is at a distance $g/2\omega^2$ from the vertex, and we can control the focal length of the mirror by controlling the rotational speed ω. (See Problem 2.19.)

Pressure in an Accelerating Liquid; Shape of the Free Surface

A related problem arises when we consider a fluid that is subjected to an unbalanced force in a way that does in fact cause it to accelerate, though as a rigid body. Suppose we have a liquid in a container, as shown in Fig. 2.3.2. The container moves in the $+x$ direction at a velocity that increases linearly with time—hence the acceleration $\mathbf{a}_o$ of the container is constant. Let's look at the application of Newton's second law to this case. A force balance of the form $\Sigma\mathbf{F} = m\mathbf{a}$ can be written in the direction of motion on the volume element shown, and the individual terms are of magnitudes

$$p_x dA - p_{x+dx} dA = \rho(dA)(dx)a_o \qquad\qquad (2.3.10)$$

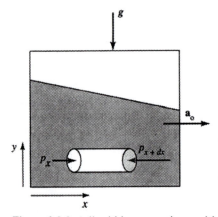

Figure 2.3.2 A liquid in a container, with a free surface open to the atmosphere, accelerates in the x direction.

Note that we take the acceleration $\mathbf{a_o}$ to be positive: that is, to be acting in the $+x$ direction. It follows then that

$$\frac{\partial p}{\partial x} = -\rho a_o \tag{2.3.11}$$

Thus we predict that the pressure decreases through the fluid as we move toward the right: that is, in the $+x$ direction. One of the consequences of this is easily derived. If we integrate Eq. 2.3.11 with respect to x we find

$$p(x) = -\rho a_o x + A(y) \tag{2.3.12}$$

and the integration "constant" $A(y)$ is unknown at this point.

Equation 2.3.11 is the result of writing Newton's second law in the x direction. We may (and must) also write this force balance in the y direction, but we did this earlier (Eq. 2.3.3), and the result is the hydrostatic law:

$$\frac{\partial p}{\partial y} = -\rho g \tag{2.3.13}$$

Since the acceleration is in a direction normal to the y axis (i.e., there is no component of acceleration in the direction of gravity), the hydrostatic law is unaffected by the x-directed acceleration. Integration of Eq. 2.3.13 with respect to y yields

$$p(y) = -\rho g y + B(x) \tag{2.3.14}$$

where $B(x)$ is unknown. Equations 2.3.11 and 2.3.13 are both satisfied by a solution of the form

$$p(x, y) = -\rho a_o x - \rho g y + D \tag{2.3.15}$$

where D is a constant.

We would like to know the shape of the free surface. The surface is in contact with the atmosphere, and we can take atmospheric pressure to be constant, say p_o. By writing the surface shape as $Y(x)$ and setting $p = p_o$ at the surface we find:

$$p_o = -\rho a_o x - \rho g Y(x) + D \qquad (2.3.16)$$

or

$$Y(x) = -\frac{(p_o - D) + \rho a_o x}{\rho g} \qquad (2.3.17)$$

We see that the surface is planar (since the profile is linear in x) and has a negative slope equal to $-a_o/g$. The constant D is determined by the initial height of the liquid in the system, before acceleration begins (see Problem 2.33).

2.4 THE SHAPE OF INTERFACES

If we want to describe the shape of a static interface in detail, for situations more complicated than those we have looked at so far, we must use the mathematical tools acquired in a course in differential equations. In this section we will engage in some mathematical analysis that is at a slightly higher level than we will ordinarily need through most of this text.

Let's return to the static interface beween two immiscible fluids. To be specific, we will look at the meniscus at the top of a column of liquid in a capillary. Figure 2.4.1 shows the geometry. We ask two questions:

1. Why is there a curved meniscus?
2. Can we calculate its shape?

We take the pressure on the gas side to be p_o, and we take the gravity vector as positive in the $+z$ direction, which is shown as positive *down* in Fig. 2.4.1. By virtue of the hydrostatic law (Eq. 2.1.3), we may write the pressure $p_i(r)$ *just inside the liquid* at any position r as

$$p_i(r) = p_{1o} + \rho g[H(r) - H(r = 0)] \qquad (2.4.1)$$

The pressure p_{1o} is the pressure on the *liquid* side of the meniscus, at the point $r = 0$, $z = H_{(r=0)} = H_o$. The maximum depth of the meniscus below the contact line, $H_{(r=0)} = H_o$, is not known at this point; ρ is the *liquid* density. We might be tempted to set p_{1o} equal to the outside (gas side) pressure, p_o, but we will avoid this because we note that the interface is curved. Hence surface tension must play a role here.

From the Young–Laplace equation (Eq. 2.2.7) we may write

$$p_o = p_i(r) + \sigma\left(\frac{1}{R_1} + \frac{1}{R_2}\right) \qquad (2.4.2)$$

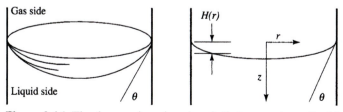

Figure 2.4.1 The free surface (meniscus) that separates a liquid from a gas when the liquid is confined to a capillary tube.

at any position r on the interface. [The gas side pressure p_o will be assumed to be independent of position, but the liquid side pressure $p_i(r)$ will depend on r through Eq. 2.4.1.]

Our first task is to find out how to write the radii of curvature for this surface. Because of the axial symmetry of the capillary, the free surface is a surface of revolution. It can be shown that for a surface of revolution *with a small slope,* a good approximation to the curvature is

$$\left(\frac{1}{R_1} + \frac{1}{R_2}\right) = \nabla^2 H(r) = \frac{d^2 H}{dr^2} + \frac{1}{r}\frac{dH}{dr} \tag{2.4.3}$$

(The small slope restriction is a serious one. It means that this model would not be accurate for small contact angles, which correspond to a *large* slope near the wall. Many liquids of interest have small contact angles. Water on clean glass in air, for example, has a contact angle of about 10°.)

Combining these Eqs. 2.4.1–2.4.3, we find

$$p_o = p_{1o} + \rho g[H(r) - H_o] + \sigma\left[\frac{d^2 H}{dr^2} + \frac{1}{r}\frac{dH}{dr}\right] \tag{2.4.4}$$

If we define

$$h(r) = H(r) - H_o \tag{2.4.5}$$

we may write the following equation for the surface shape function $h(r)$:

$$\left[\frac{d^2 h}{dr^2} + \frac{1}{r}\frac{dh}{dr}\right] + \frac{\rho g}{\sigma} h = \frac{p_o - p_{1o}}{\sigma} \tag{2.4.6}$$

The right-hand side of this equation is some constant. (It is an *unknown* constant at this point.) Thus a "particular solution" of the differential equation, obtained by solving Eq. 2.4.6 with the derivative terms set to zero, is simply the constant

$$h_p = \frac{p_o - p_{1o}}{\rho g} \tag{2.4.7}$$

The solution to the homogeneous part of the differential equation, which is obtained by setting the right-hand-side to zero and is called a "zero-order Bessel equation," is then added to the particular solution to yield

$$h(r) = AJ_0\left[\left(\frac{\rho g}{\sigma}\right)^{1/2} r\right] + BY_0\left[\left(\frac{\rho g}{\sigma}\right)^{1/2} r\right] + h_p \tag{2.4.8}$$

The functions J_0 and Y_0 are called Bessel functions. Before proceeding further, let's deal with a few of the mysteries of Bessel functions.

Certain differential equations occur so frequently in physical problems that their solutions are called special functions, named after a mathematician on occasion, and tabulated and plotted. An example is the "harmonic equation," which has the form

$$\frac{d^2 h}{dr^2} + \omega^2 h = 0 \tag{2.4.9}$$

where ω is a constant. This equation has two independent solutions, sometimes called the "C function" and the "S function," and each may be expressed as an infinite series.

The respective series representations can be shown to have the forms

$$C = \sum_{n=0}^{\infty} \frac{(-1)^n}{(2n)!} (\omega r)^{2n} = 1 - \frac{(\omega r)^2}{2!} + \frac{(\omega r)^4}{4!} - \cdots \qquad (2.4.10)$$

and

$$S = \sum_{n=0}^{\infty} \frac{(-1)^n}{(2n+1)!} (\omega r)^{2n+1} = \omega r - \frac{(\omega r)^3}{3!} + \frac{(\omega r)^5}{5!} - \cdots \qquad (2.4.11)$$

We would write the general solution to the (homogeneous) equation (Eq. 2.4.9) as

$$h(\omega r) = AC(\omega r) + BS(\omega r) \qquad (2.4.12)$$

where A and B are constants of integration. With a graph of the functions C and S, we could then plot the function $h(\omega r)$, once the boundary conditions had been established by the physics of the problem.

The functions C and S occur so often that we give them a name, and their values (calculated from the series expressions given above) are tabulated and plotted in many reference books. We may call them the harmonic functions of the first and second kind, though most people call C the cosine function and S the sine function, because that is exactly what they turn out to be!

The situation with the Bessel functions is exactly analogous. These functions are solutions to a specific differential equation; they have a series representation; they are tabulated and graphed; and they are given a special name. Let's return now to Eq. 2.4.8, where the constants of integration, A and B are to be found from some boundary conditions on the meniscus shape. One condition is that, by definition,

$$h = 0 \qquad \text{at} \quad r = 0 \qquad (2.4.13)$$

Before we can understand the implication of this boundary condition, we have to look at Fig. 2.4.2, which is a graph of the Bessel functions. We see that Y_0 becomes unbounded (it approaches $-\infty$) as r approaches zero. But the function $h(r)$ in Eq. 2.4.8 must be bounded if it is to be physically meaningful. Hence we must have

$$B = 0 \qquad (2.4.14)$$

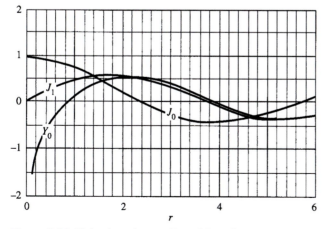

Figure 2.4.2 Behavior of some Bessel functions.

and it follows from the fact (see Fig. 2.4.2) that $J_0 = 1$ at $r = 0$ that

$$A = -h_p \qquad \text{(2.4.15)}$$

Thus, at this stage of the analysis, we find

$$h(r) = h_p \left\{ 1 - J_0 \left[\left(\frac{\rho g}{\sigma} \right)^{1/2} r \right] \right\} \qquad \text{(2.4.16)}$$

What is our second boundary condition? It comes from the fact (observation) that at a static contact line, the liquid meets the solid surface at an angle that depends on the nature of the two fluids and on the nature of the solid surface. This so-called static contact angle, then, is a characteristic property of the liquids and the solid. This is the angle θ shown in Fig. 2.4.1. Knowing this, we see that our second boundary condition entails the slope of the contact line: specifically, it must satisfy

$$\cot \theta = -\frac{dH}{dr} \qquad \text{at} \quad r = R \qquad \text{(2.4.17)}$$

Now we need to know another property of the Bessel function J_0. It is that

$$\frac{d}{dr} J_0(ar) = -a J_1(ar) \qquad \text{(2.4.18)}$$

where a is any constant and J_1 is just another kind of Bessel function, which is also plotted in Fig. 2.4.2. (Since we know the series forms of the Bessel functions, although we don't display them here, it is easy to find J_1 by differentiating J_0.) Applying Eq. 2.4.17 and using Eq. 2.4.18, we find

$$\frac{dh}{dr} = h_p a J_1(aR) = -\cot \theta \qquad \text{(2.4.19)}$$

where we now define the parameter a as

$$a = \left(\frac{\rho g}{\sigma} \right)^{1/2} \qquad \text{(2.4.20)}$$

This yields h_p in the form

$$h_p = -\frac{\cot \theta}{a J_1(aR)} \qquad \text{(2.4.21)}$$

Finally (well—almost) we may write the surface shape as

$$h = -\frac{\cot \theta}{a J_1(aR)} [1 - J_0(ar)] \qquad \text{(2.4.22)}$$

so, going back to $H(r)$, we have

$$H(r) = H_o - \frac{\cot \theta}{a J_1(aR)} [1 - J_0(ar)] \qquad \text{(2.4.23)}$$

Do we know H_o? No! But we can now calculate it, since the z axis was set up so that the plane $z = 0$ was the plane of the contact line. Hence, if we arbitrarily set

$$H(R) = 0 \qquad \text{at} \quad r = R \qquad \text{(2.4.24)}$$

we find

$$H_o = \frac{\cot \theta}{a J_1(aR)} [1 - J_0(aR)] \qquad \text{(2.4.25)}$$

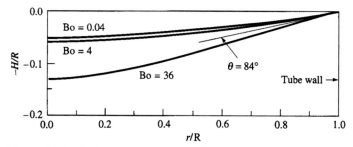

Figure 2.4.3 Meniscus shapes at several Bond numbers, for $\theta = 84°$. (The contact angle does not appear to be as large as 84° because the vertical scale is expanded relative to the horizontal scale. These menisci are really very shallow compared to the tube radius.)

and after a little algebra we end with

$$H(r) = \frac{\cot \theta}{aJ_1(aR)} [J_0(ar) - J_0(aR)] \qquad (2.4.26)$$

Now we would like to calculate the shapes of a few menisci as a function of relevant parameters. We can present and interpret the results in a very efficient manner if we first nondimensionalize Eq. 2.4.26. We begin by noting that the argument of the Bessel function (aR) must be dimensionless. If we take the definition of a (Eq. 2.4.20) and recall an earlier definition (Eq. 2.2.16) of the Bond number, we find that

$$\alpha \equiv aR = \left(\frac{Bo}{2}\right)^{1/2} \qquad (2.4.27)$$

where the Bond number is now defined as

$$Bo = \frac{4\rho g R^2}{\sigma} \qquad (2.4.28)$$

This lets us write the shape of the meniscus in the form

$$\frac{H(r)}{R} = \frac{\cot \theta}{\alpha J_1(\alpha)} \left[J_0\left(\alpha \frac{r}{R}\right) - J_0(\alpha) \right] \qquad (2.4.29)$$

Figure 2.4.3 shows several meniscus shapes. We observe that the meniscus is less shallow for large Bond numbers than for small Bond numbers. All menisci must satisfy the contact angle constraint (84°) at the capillary wall. It is important to recall here that the model, which is based on Eq. 2.4.3 as an approximation to the curvature, will be inaccurate for small contact angles because the curvature in the neighborhood of the solid surface will be large when the contact angle is small. (The approximation should be quite good for an angle as large as 84°, as in Fig. 2.4.3.) We will not look at a more exact method of calculating meniscus shapes here.[5] Our main point is that the

---

[5] For an axisymmetric surface, the right-hand side of Eq. 2.4.3 is replaced by the expression

$$\left(\frac{1}{R_1} + \frac{1}{R_2}\right) = \frac{d^2H/dr^2}{\left[1 + \left(\frac{dH}{dr}\right)^2\right]^{3/2}} + \frac{1}{r}\frac{dH/dr}{\left[1 + \left(\frac{dH}{dr}\right)^2\right]^{1/2}} \qquad (2.4.3')$$

shape depends on the Bond number, and the shape is determined by a static balance between hydrostatic pressure and the pressure due to the curvature of the surface.

We conclude this chapter with an inference about the meniscus shapes as a function of Bond number. In the limit of vanishing Bond number (no gravity, or very small capillary radius), there is no hydrostatic term. Then the pressure inside the liquid must be everywhere uniform (this is a consequence of the hydrostatic law in the limit of vanishing gravity), and this condition can be compatible with the Young–Laplace equation only if the *curvature is everywhere uniform*. The only axisymmetric surface with this property is the spherical surface. (The planar surface has uniform curvature, also, but it could not satisfy an arbitrary contact angle other than 90°.) Thus in the absence of gravity, or at very small Bond number, the meniscus is a section of a sphere of such a diameter that the intersection of the sphere and the cylindrical tube meets the contact angle constraint.

SUMMARY

Our main concern in this chapter is with the forces that act on fluids when they are static, or nearly so. In many situations these same forces can dominate the *dynamics* of the flow, even when the flow is far beyond the static state. In fluid flows characterized by two distinct phases (gas/liquid, or liquid/liquid), the *interfacial tension* plays a central role. With a definition of interfacial tension as the force (per unit of length) acting tangentially to the interface that separates two fluids, we are able to derive a simple mathematical model for the size of drops that fall from a tube of small diameter. Recalling the discussion in Chapter 1, we cast the model into dimensionless groups, and when we test the model against experimental data, we find that it works very well.

We then proceed with a derivation of the Young–Laplace equation, from which we can find the pressure difference that exists across the interface between two fluid phases when the interface is curved. This provides a tool with which we can produce additional models of static fluids that yield, for example, predictions of the rise of a liquid in a capillary partially immersed in that liquid, or the shape of a liquid drop sitting on a surface in a gravitational field. With such models we can develop methods for the *measurement* of interfacial tension.

PROBLEMS

2.1 A process requires the delivery of drops of volume 3.2×10^{-8} m³. The liquid has a density of 900 kg/m³ and a surface tension σ of 0.03 N/m. What size capillary would you recommend for forming these drops, and what would be your estimate of an upper limit on the production rate from a single capillary, in units of kilograms per hour of drops? The drops are formed in air.

2.2 Lu and Huang [*AIChE J.*, **35**, 1573 (1989)] found data represented by Fig. P2.2. In their experiment, a liquid flows through a vertical capillary at a flowrate that causes it to *drip* from the end of the capillary; that is, it does not leave the capillary as a steady stream or jet. They measured the volume flowrate Q, (cm³/min) and the dripping frequency f, (drops/min).

Instead of plotting Q versus f for each combination of capillary and fluid, Lu and Huang found that all their data correlated as shown in Fig. P2.2 if they defined a new parameter X as

$$X \equiv \frac{\sigma \pi d_o}{\rho g} f \qquad \textbf{(P2.1)}$$

where σ is surface tension in millinewtons per meter (mN/m, which is dyn/cm), d_o is the outer diameter of the capillary (cm), ρ is the density of the liquid (g/cm³), and g is 980 cm/s². They state that X has units of cubic centimeters per minute.

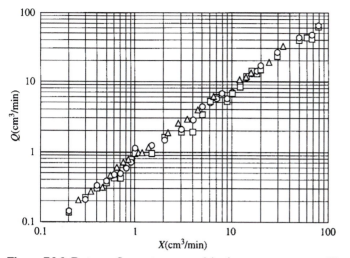

Figure P2.2 Data on flow rate versus dripping rate parameter X.

a. Based on the material in this chapter, develop a mathematical model of Q as a function of X, and plot your model on the graph of the data in Fig. P2.2.

b. What would be an appropriate *nondimensional* form of plotting of these data?

2.3 Lu and Huang (see Problem 2.2) present the data shown on Fig. P2.3 for flow rate versus drop frequency f in units of drops per minute for distilled water dripping from two different tubes. Water properties are $\sigma = 0.072$ N/m, $\rho = 998$ kg/m^3, and $\mu = 9 \times 10^{-4}$ Pa·s. Values of

the outside and inside diameters of the two tubes are shown in Fig. P2.3. Test these data against Eq. 2.1.11, and Fig. 2.1.10.

2.4 When the cells that generate insulin in the body cease to function properly, diabetes results. A potential therapy involves encapsulation of healthy cells within microscopic protective gelatin capsules that can be injected into the body. A method for manufacturing these capsules from liquid gelatin containing the cells has been described (*NASA Tech Briefs,* November 1989, p. 88). The liquid gelatin is formed into drops that

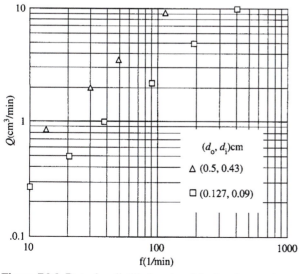

Figure P2.3 Data for distilled water dripping from tubes.

fall through a "hardening solution," which cross-links the gelatin into a hard capsule.

Can the dripping method described in Section 2.1 be used to make these capsules? What are the limitations on production rate (capsules/s)? The capsule diameter must be in the range of 250 to 350 μm. The liquid gelatin/cell slurry has a viscosity of 0.5 Pa·s, a density of 1100 kg/m^3, and a surface tension of 0.07 N/m.

a. What size capillary would you use? (Distinguish between the inside and outside diameters.) What problems do you foresee in using a capillary of this size in a commercial operation?

b. If the Weber number is to be below unity (see Eq. 2.1.13) what is the maximum number of capsules per second that can be produced from a single capillary?

c. In the reference cited, it is stated that the droplets have a diameter of 300 μm, and are produced at a rate of several thousand per second, with the liquid ejected from the capillary at a velocity of a few hundred centimeters per second. What is the order of magnitude of the diameter of the capillary being used under these conditions? What is the order of magnitude of the Weber number? Can the mechanism of drop formation described in Section 2.1 explain these results?

2.5 Water is said to travel from the roots of a tree to the uppermost branches through capillary action. Assume that the smallest capillaries observed in plant material have diameters of the order of 20 μm and that the surface tension of the water is 0.065 N/m. How high can water rise in such a capillary? What does your answer imply about the mechanism in question?

2.6 A long capillary tube of inside diameter 0.001 m is immersed vertically in water at 25°C, and the water level is 0.1 m above the upper tip of the tube. An air bubble is formed at the top of the tube. The bubble has a diameter of 0.005 m.

a. Can the bubble remain on the tip, or will buoyancy cause it to float up?

b. What is the air pressure inside the largest bubble that will remain on the capillary tip?

c. The bubble appears to be a sector of a sphere; that is, every point on the surface has the same radius of curvature. Is the bubble shape really spherical, or only approximately so? Explain your answer, and argue why the bubble appears to be spherical.

2.7 In the capillary rise method of measuring surface tension, we normally measure the rise height to the bottom of the meniscus, rather than to the contact line. How different are these two heights, relative to the total measured rise height?

a. Give a theoretical answer by using Eq. 2.4.26. What are the practical limitations of this answer? For a liquid of surface tension 0.05 N/m in a tube of inside diameter 0.001 m, what is the percentage error in the height rise measurement if the contact line position is measured instead of the bottom of the meniscus? Assume a contact angle of 85°. Calculate the Bond number (Eq. 2.4.28) for this case.

b. Give a simpler, but approximate, answer by assuming that the meniscus is a hemisphere within the tube. For a liquid of surface tension 0.05 N/m in a tube of inside diameter 0.001 m, what is the percentage error in the height rise measurement if the contact line position is measured instead of the bottom of the meniscus?

2.8 Air is slowly forced through a capillary tube, the tip of which is immersed in a liquid of height H above the tip. A gas bubble forms, grows on the tip, and ultimately becomes so large that buoyancy lifts it away from the tip. Derive a simple model from which to estimate the interfacial tension from a measurement of the released bubble size. Make a qualitative plot of the pressure inside the bubble as a function of time, during the growth of the bubble. For what bubble size is the pressure a maximum? Derive a simple model from which to estimate the interfacial tension from a measurement of the maximum bubble pressure. How sensitive a pressure measuring device would be necessary to build a practical experimental system?

2.9 A capillary tube is brought into contact with a liquid, as suggested in Fig. P2.9.1. Match the curves of height versus time (see Fig. P2.9.2) with the corresponding cases A, B, and C shown in Fig. P2.9.1 and described as follows:

Case A: The water is in a large reservoir. What is the value of the *equilibrium* height for this case?

Case B: The capillary is placed in contact with a drop of initial diameter 0.3 cm and remains in contact as the liquid is transferred into the capillary.

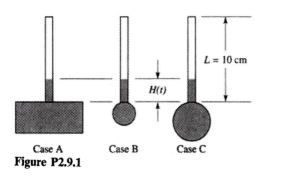

Case A   Case B   Case C

Figure P2.9.1

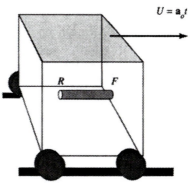

$U = a_o t$

R   F

Figure P2.10

Case C: The capillary is placed in contact with a drop of initial diameter 0.4 cm and remains in contact as the liquid is transferred into the capillary.

Explain your conclusion briefly for each case. In *all* cases the liquid is water, and the capillary has an inner diameter of 0.001 m. Time $t = 0$ is defined as the time at which the lower end of the capillary is first brought in contact with the water. Gravity acts down in all cases.

2.10 A rectangular cart on wheels is completely filled with a viscous liquid, as shown in Fig. P2.10. A solid lid covers the cart and is in contact with the liquid within. The cart moves along a smooth horizontal track at a speed that increases linearly with time: $U = a_o t$. A tube, immersed in the liquid and open at both ends, is rigidly attached to the body of the cart. The tube axis is horizontal, and the tube is filled with air before the cart is set in motion. If the cart did not move, the air would stay in the tube. As the cart moves, bubbles are forced out of the tube. Other data

are as follows:

Tube diameter $D = 1$ cm   length $L = 10$ cm
$$\mu = 1 \text{ Pa·s} \quad \rho = 1000 \text{ kg/m}^3$$
$$\sigma = 0.050 \text{ N/m}$$

a. Which end (F or R) do bubbles come out of?
b. What is the minimum acceleration (a_o) required to force bubbles out of the tube?

2.11 Air is bubbled at a very slow rate through an upward-pointing capillary submerged in water, as shown in Fig. P2.11. A pressure measuring device gives a periodic signal as shown. Each period of the pressure signal corresponds to the growth and liftoff of a single bubble.

a. What is the magnitude of the pressure P_{min} (N/m²)?
b. What is the magnitude of the pressure P_{max} (N/m²)?
c. What is the radius of the bubbles formed (mm)?

2.12 Two glass microscope slides are separated by a spacer in such a way that the slides are parallel to each other, with a separation of 500 μm. This glass "sandwich" is placed in a vertical orientation, with one open end slightly below the surface of a reservoir of silicone oil at 25°C. What is the expected height of rise of the oil into the open space between the glass slides? Take $\sigma = 0.019$ N/m and $\rho = 900$ kg/m³. Assume that silicone oil has a zero contact angle on glass.

2.13 A pair of concentric glass cylinders, with diameters of 1 and 1.05 cm, respectively, are held together in such a way that both annular ends of the pair of cylinders are open to the flow of

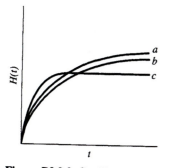

Figure P2.9.2 Capillary rise as a function of time.

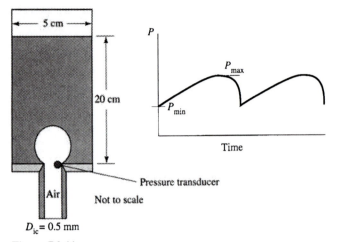

5 cm

20 cm

P

P_{max}

P_{min}

Time

Pressure transducer

Air

Not to scale

$D_{ic} = 0.5$ mm

Figure P2.11

fluids. The pair of cylinders, with its axis vertical, is so placed that one open end is slightly below the surface of a reservoir of silicone oil at 25°C. What is the expected height of rise of the oil into the annular space between the cylinders? Use the data of Problem 2.12.

2.14 Write a brief denial to the following patent proposal:

> *Figure P2.14 shows a means of supplying a continuous flow of liquid to an external surface; there are no moving parts, and no expenditure of energy is required. Liquid is drawn into the capillary tube by the action of surface tension. The tube is of an axial length less than the capillary rise height defined by*

$$h = \frac{4\sigma}{\rho g D_c} \qquad \textbf{(P2.14)}$$

> *Hence the liquid will spontaneously rise to the end of the capillary, from which it can be delivered to an external surface.*

2.15 An aquatic insect called a water strider (Fig. P2.15) walks across the surface of a pond supported by surface tension. Draw a sketch that shows the forces acting on each "leg" of the insect. If each of the joints of the four large (posterior) legs of the insect that are in contact with the water surface can be considered to be cylinders of radius 500 μm, and length 5 mm, what is the most the creature can weigh and still deserve its name?

2.16 Review Problem 2.15, and then design a pair of "water strider boots" that will support the weight of a 70 kg person. Do not use buoyancy as a means of supporting weight.

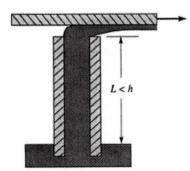

$L < h$

Figure P2.14

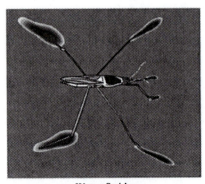

Water Strider

Figure P2.15 A water strider.

2.17 A straight metal pin can be floated on the surface of water, even though the pin is much denser than water. This is accomplished by taking advantage of surface tension: for example, by laying the pin horizontally on a small piece of tissue paper, and then placing the paper-and-pin combination on the water surface. The paper quickly absorbs the water and falls beneath the surface, leaving the pin "floating." If the density of the pin is 7000 kg/m^3, what is the largest diameter of a pin that can be floated in this way?

2.18 Read "Surface Tension," by James Blish (*Science Fiction Hall of Fame,* Vol. 1, R. Silverberg, Ed., Doubleday, New York, 1970). Make up a homework problem based on the story.

2.19 Under NASA sponsorship, a rotating mercury telescope was built for use in galactic surveys. It consists of 300 pounds of liquid mercury in a dish of diameter 0.7 m. The dish spins at 10 rpm. What is the mercury thickness when the dish is not spinning? What is the mercury thickness at the center when the dish is spinning? What is the focal distance of this "mirror"?

2.20 Liquid initially fills a very long vertical capillary tube of inside diameter D, then drains slowly into the surrounding air under the force of gravity. Find the length L of liquid that remains in the tube when dripping ceases. Why is this not the same as the equilibrium rise height given by Eq. 2.2.15?

2.21 Air initially fills a very long vertical capillary tube of inside diameter D. The tube is suddenly immersed in a large body of water, still in the vertical position. Water wets the tube surface and, as soon as the ends of the tube are submerged, enters the tube. When equilibrium is reached, what is the length, if any, of the air column that remains in the tube?

2.22 For the geometry shown in Fig. P2.22, find the height H as a function of the relevant geometrical and physical parameters. **A** is a solid cylinder of radius R_A. **B** is a circular tube of inside radius R_B. The system is in static equilibrium.

2.23 Find the force required to pull a horizontal solid cylinder free from a liquid/air interface, in terms of the relevant geometrical and physical parameters.

2.24 Find the force required to pull a wire loop or ring of cylindrical cross section free from a liquid/air interface, in terms of the relevant geometrical and physical parameters. See Fig. P2.24. Assume $R_o - R_i \ll R_i$.

2.25 Boucher and Evans [*Proc. R. Soc. London,* **A346,** 349 (1975)] have presented numerical solutions for the shape of pendent (hanging) drops, as well as an analysis of the maximum stable drop size that can hang from a capillary. Their results are plotted in Fig. P2.25. Boucher and Evans define a dimensionless capillary radius as $X = R_c/a$ and a dimensionless maximum drop volume as $V = V_{max}/a^3$. The length scale a is defined as

$$a \equiv \left(\frac{2\sigma}{\rho g}\right)^{1/2} \qquad \textbf{(P2.25)}$$

Compare these numerical results to our simplistic model (Eq. 2.1.6), which implies that the maximum volume corresponds to $\theta = 0$.

2.26 Garandet et al. [*J. Colloid Interface Sci.,* **165,** 351 (1994)] have obtained data for the volume of molten metal drops that fall from a cylindrical support of contact diameter D_c, as in Fig. P2.26, when their weight exceeds the surface tension force. Compare the data below to the prediction of Eq. 2.1.6.

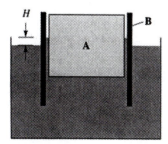

Figure P2.22

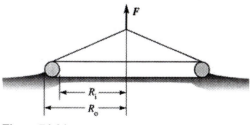

Figure P2.24

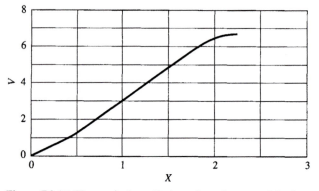

Figure P2.25 Theoretical prediction of maximum stable drop size for a pendent drop on a capillary.

RHENIUM: $\rho = 18,000$ kg/m^3 $\sigma = 2.52$ N/m

D_c (mm)	V (m^3 × 10^{-8})
0.31	1.29
0.5	1.96
1	3.69
1.77	6.13
4	12.1

ZIRCONIUM: $\rho = 6050$ kg/m^3 $\sigma = 1.435$ N/m

D_c (mm)	V (m^3 × 10^{-8})
0.76	5.16
1	6.54

2.27 Equation 2.1.6 gives an oversimplified model of the size of the drop that falls from a capillary when the drop weight exceeds the restraining force of surface tension. With this model as a basis, we may write an improved expression for the falling drop volume as

$$V_{drop} = \frac{\pi D_c \sigma}{\rho g} \psi \qquad \text{(P2.27.1)}$$

The factor ψ is called the Harkins–Brown correction factor. An empirical correlation of data leads to the following expression:

$$\psi = 0.6 + 0.4 (1 - 0.488 \, Bo^{1/3})^{2.2} \quad \text{(P2.27.2)}$$

with the Bond number defined as in Eq. 2.1.10. Use this expression to predict the diameter of the drop that falls from a capillary of inside diameter $D_c = 1$ mm when $\sigma = 0.06$ N/m and $\rho = 1000$ kg/m^3.

2.28 The Harkins–Brown correction factor defined in Problem 2.27 is a correction to the drop *volume*. With Eq. P2.27.2 as a basis, give an analytical expression for the correction factor for the drop diameter, based on Eq. 2.1.11

$$\frac{D_d}{D_c} = \psi_d \left(\frac{6}{Bo} \right)^{1/3} \qquad \text{(P2.28)}$$

For the case of $D_c = 1$ mm, $\sigma = 0.06$ N/m, and $\rho = 1000$ kg/m^3, what is the error if Eq. 2.1.11 is used for the prediction of the drop diameter?

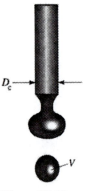

Figure P2.26

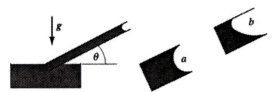

Figure P2.29

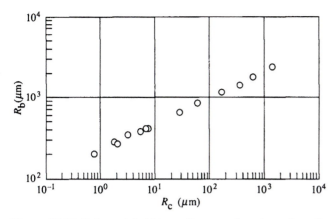

Figure P2.30 Released bubble radius as a function of capillary radius.

2.29 A capillary tube is inclined at an angle θ to the horizontal, and a liquid has risen under capillary forces to a height h above the reservoir of the liquid. What determines the shape of the meniscus? Define a dimensionless group whose magnitude determines the meniscus shape between the extremes (*a* or *b*) suggested in Fig. P2.29.

2.30 Blanchard and Syzdek [*Chem. Eng. Sci.*, **32**, 1109 (1977)] present the data shown in Fig. P2.30. Bubbles of air were formed very slowly in very clean water at 21°C. The capillaries were very clean glass tubes. The original 20-year-old reference is worth reading because it exemplifies the care that is required to produce very precise results in the field of interfacial fluid dynamics. Test the ability of Eq. 2.1.11 to describe these data.

2.31 Find the integration constant D in Eq. 2.3.17 in terms of the initial height of liquid H.

2.32 A container of liquid such as shown in Fig. 2.3.2 has a length in the x direction of $L = 1$ m and a vertical height H of 0.5 m. It is half-filled with liquid. What is the maximum acceleration a_o that can be imposed on a cart carrying this container without any liquid spilling over the rear end?

2.33 Why does no constant of integration appear in Eq. 2.3.6? What boundary condition is implied by writing the solution in the form shown?

2.34 Return to Example 2.1.2. What is the value of the Weber number needed to produce drops at a higher rate of 1 drop/s? What is the required capillary diameter? At what drip rate (drops/s) is We = 1?

2.35 Filters that are used for purification in the pharmaceutical industry must be tested for loss of integrity after sterilization. Consider a filter whose structure can be represented as in Fig. P2.35. Integrity is tested by "soaking" the filter

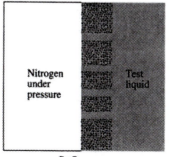

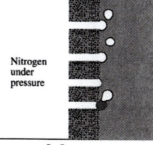

$P < P_{\text{bubble}}$

$P > P_{\text{bubble}}$

Figure P2.35

in a liquid that wets the interior pores of the filter. It is assumed that all the pores of the filter are filled with liquid after soaking. Then a pressure differential is applied across the filter, using purified nitrogen as the pressure source. No nitrogen leaks through an intact filter until a critical pressure differential is reached. At that point, gas is observed to bubble continuously from the wetted side of the filter. This is called the "bubble point" of the filter. Estimate the bubble point for the following conditions:

a. Liquid/nitrogen surface tension = 28 dyn/cm.
b. Contact angle of the liquid/nitrogen interface on the filter material = 75°.
c. All pores are cylindrical, with a diameter of 0.2 μm.

2.36 Some experimental observations indicate that the highest flowrate at which we can observe dripping of a liquid from the exit of a vertical capillary corresponds to a Weber number of unity:

$$\text{We} = \frac{16\rho Q^2}{\pi^2 D_c^3 \sigma} = 1$$

a. Using this constraint, derive an explicit expression for the maximum dripping rate f (drops/s) in terms of the relevant parameters of the system. *Put your result in a dimensionless format.*
b. Give the expected value of the maximum drip rate for water from a capillary of inside diameter $D_c = 1$ mm. Assume that the capillary is made of a material that is not wetted by water.

Chapter 3

Forces on, and Within, a Flowing Medium

In this chapter we begin a careful discussion of the internal stresses that arise when a viscous fluid is deformed. The analysis is simplified by the choice of very simple geometries that bound the flow. As a consequence, the resulting flows considered are one-dimensional. We show that application of Newton's second law leads to a differential equation that the stress must satisfy. Addition of an assumed relationship between the stress and the velocity gradient (Newton's law of viscosity) leads to a differential equation for the velocity profile. Several examples show how we can solve these equations and use the solutions to answer questions of engineering significance.

The viscosity of a Newtonian fluid is defined, and several methods are presented that permit us to estimate the viscosity of fluids of interest.

Now we are ready to investigate the dynamics of a deforming fluid. The most difficult issue entails description of the forces acting *within* a continuous deformable medium. Both the mathematics and the physics can be confusing at first. While there are some obvious relationships to the treatment of forces and momentum in rigid body mechanics, the transition to *continuum* mechanics introduces many significant differences.

3.1 CONCEPTS OF SHEAR STRESS AND MOMENTUM FLUX

When we attempt to slide a solid along another solid surface, a resistive force, usually referred to as "sliding friction," retards the relative motion of the two surfaces. In a fluid a similar phenomenon occurs. Relative motion of two "elements" of fluid is retarded by intermolecular interactions that are due to *viscosity*, the property of a fluid that determines the ease with which elements of the fluid may be moved relative to one another through the action of some external force. We will give a quantitative definition of viscosity later. For the present, we note that the term "viscous friction" can be applied to interaction of the second type.

With reference to Fig. 3.1.1, we suppose that a liquid film is flowing down a long, wide, inclined plane under the action of gravity. We would observe a nonuniform

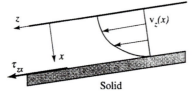

Figure 3.1.1 A liquid film flows down an inclined plane.

velocity profile $v_z(x)$ under these conditions. "Nonuniform" implies that the velocity vector varies with position in the liquid film, as suggested in Fig. 3.1.1. We observe that a liquid does not normally "slip" at a solid boundary. It maintains continuous contact with the solid. Hence the velocity would vanish at the stationary solid boundary and increase as the liquid/gas interface is approached. Viscous friction is exerted by the faster moving fluid on the adjacent slower moving fluid. Conversely, we could say that the slower moving fluid, and ultimately the stationary surface, retards the flow down the plane. Somewhere downstream of the region where the liquid is continuously deposited onto the plane, these frictional viscous forces just balance the force of gravity acting on the liquid, and the velocity profile becomes independent of position z downstream, and depends only on the x coordinate. When the velocity profile becomes independent of the axis in the direction of flow, we refer to the flow as "fully developed." Because the velocity varies in the x direction, the flow is still nonuniform. We say that the fluid is being "deformed." The viscous liquid exerts a force on the solid stationary surface. This force acts in the plane of the liquid/solid interface, in the direction of motion. *We define the force per unit of area upon which it acts as the shear stress.*

The shear stress depicted in Fig. 3.1.1 and labeled τ_{zx} illustrates a standard notational convention: the second subscript refers to the direction *normal* to the surface on which the shear force acts; the first subscript refers to the direction in which the shear force acts. Thus the double subscripting carries and conveys two pieces of information—the orientation of the surface and the direction of the force. (We note that some texts use the opposite convention on subscripting—the second subscript gives the force direction and the first refers to the normal to the surface.)

Look for a moment at the *dimensions* of shear stress:

$$\tau_{zx} = \frac{\text{force}}{\text{area}} [=] \frac{mL/t^2}{L^2} = \frac{m}{Lt^2} \tag{3.1.1}$$

In writing this equation we used the conventions introduced in Chapter 1: [=] means "has the dimensions of" and m, L, and t refer to mass, length, and time, respectively.

We will soon be led to the relationship of stresses to momentum. Thinking of momentum (which is a vector) as $m\mathbf{v}$ for the moment, we see that the dimensions of momentum are

$$m\mathbf{v} [=] mL/t \tag{3.1.2}$$

The time rate of change of momentum has dimensions

$$\frac{\partial}{\partial t} mv [=] mL/t^2 \tag{3.1.3}$$

We define a momentum *flux* as the rate at which momentum crosses (is transferred to) a boundary, per unit of area. Hence it has dimensions

$$\text{momentum flux} [=] \frac{mL/t^2}{L^2} [=] m/Lt^2 \tag{3.1.4}$$

which is the same as the dimensions of stress. In addition to this equivalence of dimensions, the shear stress is *mechanically* equivalent to a momentum flux, as we will see later.

If the shear stress is really a reflection of the viscous friction, as we have asserted, it should depend on some measure of how rapidly the adjacent liquid layers are sliding relative to each other. The *velocity gradient* (often called the "shear rate") is such a measure: for the flow shown in Fig. 3.1.1 the velocity gradient is simply

$$\dot\gamma = \frac{\partial v_z}{\partial x} \tag{3.1.5}$$

($\dot\gamma$, read "gamma dot," is a commonly used notation for shear rate).

The simplest mathematical model of friction within a viscous liquid takes the shear stress to be proportional to the shear rate:

$$\tau_{zx} = -\mu\dot\gamma = -\mu\frac{\partial v_z}{\partial x} \tag{3.1.6}$$

We claim that the proportionality constant μ is the fluid property called "viscosity." As with the surface tension σ, introduced earlier, we will have to support the claim that μ is really a material property. We may regard it as an assumption at this point. Alternatively, we could regard this equation as a *definition* of viscosity, but then the issue is whether our definition is consistent with the one used by the rest of the scientific community.

The minus sign used in Eq. 3.1.6 comes from the arbitrary convention for assigning a positive or negative sign to a stress. Most fluid dynamics texts write Eq. 3.1.6 without the minus sign. We will explain our reasons for doing the opposite in a later section. Equation 3.1.6 is often called Newton's law of viscosity, and any fluid that obeys this *linear* relationship is said to be a Newtonian fluid. In fact, however, certain fluids that obey this linear relationship between shear stress and shear rate exhibit phenomena not associated with Newtonian fluids. We consider only Newtonian fluids in this text.

3.2 PROBLEM SOLVING/MODEL BUILDING

One major goal of this text is to teach the methodology for creating mathematical models of flows of interest and importance to engineers. There is a rational procedure for building mathematical models. The steps are as follows:

1. Identify and describe the phenomenon of interest.
2. Give a clear statement of the goal(s) of the model. (What is expected from the model?)
3. State clearly the assumptions *you choose* to make regarding the physics and the geometry. *Your* choices define *your* model.
4. Apply the appropriate physical (mechanical, thermodynamic) principles.
5. Solve the resultant equation(s).
6. Compare the predictions of your model to reality. (Do an experiment, or find a set of appropriate and reliable experimental data.)
7. If necessary, modify the assumptions in the hope of improving the degree to which your model mimics reality.

We will illustrate these concepts and guidelines by developing a mathematical model for the flow of a viscous liquid through a long straight tube of circular cross section.

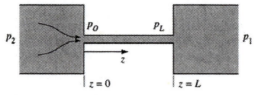

Figure 3.2.1 Liquid flows at a constant rate from one reservoir to the other.

3.2.1 A Model of Laminar Flow Through a Tube

The Phenomenon

A tube of circular cross section connects two reservoirs, each of which contains the same viscous liquid (Fig. 3.2.1). The liquid is forced to flow at a constant rate through the tube, from one reservoir to the other.

Goal

Develop a model for the relationship of the pressure difference $P_o - P_L$ imposed across the ends of a fluid-filled cylindrical tube to the flowrate that results.

Assumptions

Geometry

- The tube is of uniform circular cross section along its axis.

Physics

- The fluid is Newtonian (Eq. 3.1.6).
- The flow is isothermal.
- The fluid is incompressible (density ρ is not a function of pressure).
- The flow is at *steady state,* (i.e., it is unchanging in time).
- the flow is *laminar and unidirectional,* specifically, the velocity *vector* has only the single component[1]:

$$\mathbf{v} = (0, 0, v_z) \tag{3.2.1}$$

- The flow field does not vary along the tube axis; that is,

$$v_z \neq v_z(z) \text{ but } v_z = v_z(r) \tag{3.2.2}$$

This is called *fully developed* flow.
- The axis of the tube is collinear (parallel) with the gravity vector.

---

[1] "Laminar flow" also implies a flow that is not turbulent. In a turbulent flow the components of $\mathbf{v}$ fluctuate randomly about mean values. The flow is chaotic at some level. Observations demonstrate that a criterion for laminar flow in a long tube of circular cross section is based on the value of the Reynolds number, defined as $\text{Re} = 4\rho Q/\pi D\mu$. The usual criterion is that the flow will be laminar for $\text{Re} < 2000$.

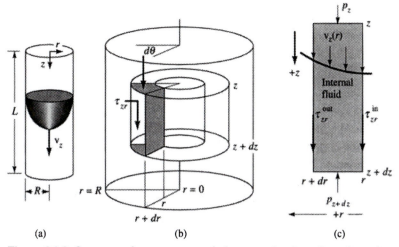

Figure 3.2.2 Geometry for momentum balance on laminar flow through a tube: (a) the expected velocity profile within the tube, (b) a volume element in a cylindrical coordinate system, and (c) a cross section in the rz plane. In (b), the volume lies between the cylinders of radii r and $r + dr$, the planes z and $z + dz$, and the planes θ and $\theta + d\theta$. The pressure force acts in the z direction on the faces normal to the z axis. The force associated with the shear stress acts in the z direction on the faces normal to the r axis.

Physical Principles

The fundamental physical principles that we will impose are the statements of conservation of mass and conservation of momentum.

The principle of conservation of momentum will be applied to a *fixed* volume element through which fluid is *flowing* and on which forces are acting. The forces must be in balance, consistent with Newton's second law: any changes in momentum of fluid in the volume element must be offset by some net force(s) acting on the element. If there is no change in momentum, the sum of all the forces must vanish. Figure 3.2.2 shows the geometry for the analysis.

Our first task is to distinguish between forces exerted by the fluid that is *outside* the volume element acting *on* the fluid *within* the volume element from the (equal and opposite) forces exerted by the fluid within the volume element acting on the fluid outside the volume element.[2] *We will consider forces acting on the fluid inside the volume element.*

Read the preceding paragraph again. It expresses an important idea.

Next we must keep in mind that force is a *vector* quantity. We must write the force balance for a specific *direction*. We will begin with force components in the axial (z) direction.

There are two kinds of force exerted by the external fluid on the fluid inside the volume element: *shear* forces and *pressure* forces. The force associated with the shear stress τ_{zr} acts in the z direction, on the curved surface normal to the r direction. The pressure p acts in the z direction only on the surfaces normal to the z direction. (Note that pressure, by definition, acts equally in all directions. For this reason we refer to pressure as an *isotropic* stress.) But the pressure acting on (normal to) the cylindrical

---

[2] If you read this sentence aloud you can hear that it is a real earful! We're going to have some sentences like this periodically, because we have to be very precise.

surfaces (surfaces of constant r) does *not* exert a force in the z direction. It has a component only in the r direction. That is why we consider only the pressures p_z and p_{z+dz} labeled in Fig. 3.2.2: it is only the pressure on *these* surfaces that gives rise to forces in the z direction.

The shear force component (acting in the z direction) exerted by the fluid external to the volume element on the surface of the element normal to the r direction at radius r is

$$dF_z|_r = +2\pi r \tau_{zr}|_r \, dz \tag{3.2.3}$$

Note that this force component is the product of a stress times an area. Since the area is a differential area ($2\pi r \, dz$), the force is also a differential force component (dF_z). We are considering the cylindrical surface at the radial position r and so we use $|_r$ to denote where the stress is evaluated. Because the flow field is assumed to be axisymmetric (independent of the angular position θ), we apply the same stress to the whole cylindrical surface, and we let $d\theta$ in Fig. 3.2.2 be the full circle, or $d\theta = 2\pi$.

We do not know, at this point, whether τ_{zr} is positive or negative. According to Eq. 3.1.6, the shear stress has the opposite sign of the velocity gradient. This will determine the sign of this stress. Whatever the sign of the stress itself, however, we write a $+$ sign in front of the stress in Eq. 3.2.3 to accommodate the following sign convention:

> *When passing across the boundary of the volume element, from the external fluid to the internal fluid, if we move in the $+$ direction of the coordinate axis, we use a $+$ sign on the force acting on that boundary.*

This convention will help us to keep the signs consistent as we write the various contributions to the force balance.

Next we consider the surface at the position $r = r + dr$, where a shear stress $\tau_{zr}|_{r+dr}$ acts. On this surface

$$dF_z|_{r+dr} = -2\pi(r + dr)\,\tau_{zr}|_{r+dr}\, dz \tag{3.2.4}$$

Notice the minus sign. (When we pass from the external fluid to the internal fluid, across the cylindrical surface at $r = r + dr$, we move in the *minus r* direction.) Note also that the area is larger, since the radius of the cylindrical surface is $r + dr$.

Now consider the pressures $p|_z$ and $p|_{z+dz}$ acting on the two annular-shaped surfaces at z and $z + dz$. (These surfaces are annular because we again let $d\theta$ go to the full circle, $d\theta = 2\pi$.) *The pressure $p|_z$ exerted by the fluid external to the volume element gives rise to a z-directed force component

$$+p|_z\, 2\pi r \, dr \tag{3.2.5}$$

while the pressure $p|_{z+dz}$ gives

$$-p|_{z+dz}\, 2\pi r \, dr \tag{3.2.6}$$

(Note again that the plus and minus signs have to do with our direction of motion when we move from external fluid to internal fluid across the boundary of the volume element.)

A body force due to gravity acts uniformly on all the fluid within the volume element. It is not transmitted across the surfaces. (That is why it is called a *body* force.) The body force is

$$\rho g 2\pi r \, dr \, dz \tag{3.2.7}$$

where ρ is the mass density of the fluid. Note that we write the volume as $2\pi r \, dr \, dz$, instead of as $\pi[(r + dr)^2 - r^2]\, dz$. We are using an approximation based on the idea that dr is small compared to r itself.

Now we consider the rate of flow of momentum across the boundaries of the volume element. Since we have *assumed* that $v_r = 0$ (the laminar unidirectional flow assumption imposed earlier), the only surfaces across which there is any flow are the surfaces normal to the z axis.

The (differential) volume flowrate across the area $2\pi r \, dr$ is just

$$dQ = v_z 2\pi r \, dr \qquad (3.2.8)$$

We assume that dr is so small that we can use $v_z(r)$ for the velocity *anywhere* on this segment of surface. Otherwise we would have to write

$$dQ = 2\pi \int_r^{r+dr} v_z r \, dr \qquad (3.2.9)$$

How do we get a rate of flow of momentum from this? In point mass mechanics, momentum is just $m\mathbf{v}$. In *continuum fluid* mechanics we must talk about momentum per unit of fluid volume. This is just the product of the mass *density* ρ times the velocity, or $\rho \mathbf{v}$. Hence the z component of momentum per unit volume ρv_z, times the volume flowrate dQ, gives the *rate* at which momentum enters the face at z:

Rate of flow of momentum in: $\quad \rho v_z (v_z \, 2\pi r \, dr)|_z \qquad (3.2.10)$

Then it follows that

Rate of flow of momentum out: $\quad -\rho v_z (v_z 2\pi r \, dr)|_{z+dz} \qquad (3.2.11)$

(Note the sign.)

Is $v_z|_z = v_z|_{z+dz}$? The rate of flow of mass into the volume is the mass density ρ times dQ, or

$$\rho v_z 2\pi r \, dr|_z \qquad (3.2.12)$$

The rate of flow of mass out is

$$\rho v_z 2\pi r \, dr|_{z+dz} \qquad (3.2.13)$$

The *net* rate of flow of mass across these two boundaries (and they are the *only* boundaries across which there is flow) is

$$2\pi r \, dr(\rho v_z|_z - \rho v_z|_{z+dz}) \implies zero \qquad (3.2.14)$$

There is no way that the amount of mass within the volume element can vary (assuming it is always filled by fluid). It must follow that the expression in Eq. 3.2.14 is zero. But we have assumed that the fluid is isothermal and incompressible, so $\rho|_z = \rho|_{z+dz}$. We conclude that

$$v_z = constant \; at \; a \; fixed \; value \; of \; r \qquad (3.2.15)$$

We can conclude that Eq. 3.2.2, which we included as an assumption about the flow field, is in fact a consequence of the other assumptions already applied. In particular, Eq. 3.2.2 really follows from the assumption given in Eq. 3.2.1.

Now if we write conservation of momentum as

$$\begin{matrix} \text{Sum of forces} \\ \text{in the } z\text{-direction} \end{matrix} = \begin{matrix} \text{Rate of change of} \\ \text{momentum in the} \\ z\text{-direction} \end{matrix} \qquad (3.2.16)$$

we find

$$dF_z|_r + dF_z|_{r+dr} + p_z \, 2\pi r \, dr + (-p|_{z+dz} \, 2\pi r \, dr) + \rho g 2\pi r \, dr \, dz = 0 \qquad (3.2.17)$$

Note that from Eq. 3.2.15 the two momentum flux terms (Eqs. 3.2.10 and 3.2.11) are identical, and cancel out of (hence don't appear in) this momentum balance.

Using Eq. 3.2.3 we find

$$2\pi \, dz \, [r\tau_{zr}|_r - (r + dr) \, \tau_{zr}|_{r+dr}] + 2\pi r \, dr \, (p|_z - p|_{z+dz} + \rho g \, dz) = 0 \quad (3.2.18)$$

We can write this as

$$(r\tau_{zr})_r - (r\tau_{zr})_{r+dr} + r \, dr \left(\frac{(p)_z - (p)_{z+dz}}{dz} + \rho g \right) = 0 \quad (3.2.19)$$

(We have used the definition $r|_{r+dr} = r + dr$ and the notation $p|_z = (p)_z$.)

Now we divide by $r \, dr$ and take the limit as the volume element shrinks to zero volume:

$$\lim_{dr \to 0} \left\{ \frac{(r\tau_{zr})_r - (r\tau_{zr})_{r+dr}}{r \, dr} \right\} = \lim_{dz \to 0} \left\{ -\left(\frac{(p)_z - (p)_{z+dz}}{dz} + \rho g \right) \right\} \quad (3.2.20)$$

By using the definition of a derivative, we find that this may be written as

$$-\frac{1}{r}\frac{\partial}{\partial r}(r\tau_{zr}) = \frac{\partial p}{\partial z} - \rho g \quad (3.2.21)$$

We have derived a partial differential equation that expresses our application of the principle of conservation of momentum to this particularly simple flow through a tube. But we have only a single equation, and it has two unknowns: τ_{rz} and p. Fortunately we have additional information, in the form of the assumption listed earlier that the flow field does not vary down the z axis (Eq. 3.2.2). If this is so, then the velocity gradient $\partial v_z / \partial r$ would not vary in the z direction. (That is, since v_z is not a function of z, its derivative with respect to r is not a function of z either.) On the assumption that the shear stress depends only on the velocity gradient, we conclude that τ_{zr} is not a function of z. We have also assumed that the flow field satisfies Eq. 3.2.1. In particular, we assume that the flow is laminar and unidirectional which means that there is no radial component of velocity. Neither is there any net force or acceleration acting in the radial direction. Hence there can be no pressure variation in the radial direction. Thus we conclude that $\partial p / \partial z$ cannot depend on r. Now let us find the consequence of these statements.

If we apply these ideas to Eq. 3.2.21, we conclude that the left-hand side of the equation is a function only of r, not of z, while the right-hand side is not a function of r. This can be so only if both sides are equal to a constant, C. Thus we conclude that

$$-\frac{1}{r}\frac{d}{dr}(r\tau_{zr}) = C = \frac{dp}{dz} - \rho g \quad (3.2.22)$$

[Notice that we now use ordinary rather than partial derivatives, since τ_{zr} and p are functions of single (but different) variables.] Equation 3.2.22 really corresponds to *two* differential equations.

We can integrate the shear stress equation once and find

$$-(\tau_{zr}) = \frac{Cr}{2} + \frac{E}{r} \quad (3.2.23)$$

where E is a constant of integration. If E is nonzero, we have a problem along the axis $r = 0$, where Eq. 3.2.23 would predict an infinite shear stress. This does not make sense physically—there is no reason for a stress to become infinite in this flow. Hence we conclude that $E = 0$. *Notice how we have used a physical statement to find a constant of integration.* Thus we find that the shear stress has the form

$$\tau_{zr} = -\frac{Cr}{2}$$ (3.2.24)

Equation 3.2.24 gives a shear stress that goes to zero at $r = 0$. Does *that* make sense? Yes, in view of our expectation that the shear stress depends on the velocity gradient. Since the velocity profile is *symmetrical around the axis,* we see that $dv_z/dr = 0$ at $r = 0$. Hence we expect that τ_{rz} should vanish at $r = 0$.

Now we must solve the other half of Eq. 3.2.22 for the pressure distribution along the tube axis. We integrate the equation with respect to z and find

$$p = \rho g z + C z + G$$ (3.2.25)

where G is another constant of integration. We suppose that the tube has a total axial length L, and that the pressures at the ends of the tube are p_0 at $z = 0$, and p_L at $z = L$. Then we impose these boundary conditions on p in Eq. 3.2.25 and it follows that $G = p_0$, and

$$-C = \frac{p_0 - p_L}{L} + \rho g$$ (3.2.26)

At this point we have an expression for the shear stress in the form

$$\tau_{zr} = \frac{p_0 - p_L + \rho g L}{2L} r$$ (3.2.27)

If we introduce into Eq. 3.2.27 the assumption that the fluid is Newtonian,

$$\tau_{zr} = -\mu \frac{dv_z}{dr}$$ (3.2.28)

we have a new differential equation to solve:

$$\frac{dv_z}{dr} = -\frac{p_0 - p_L + \rho g L}{2\mu L} r = \frac{C}{2\mu} r$$ (3.2.29)

The solution is simply obtained by integration to give

$$v_z = \frac{C}{4\mu} r^2 + F$$ (3.2.30)

where F is another constant of integration. To find F we impose another physical statement. We assert that the liquid in contact with the stationary wall of the tube is itself stationary. This is usually called the "no-slip" assumption. Hence we write this statement in the mathematical form

$$v_z = 0 \quad \text{at} \quad r = R$$ (3.2.31)

When this is imposed as a condition on Eq. 3.2.30, we find

$$F = -\frac{C}{4\mu} R^2$$ (3.2.32)

Finally, we may write the velocity profile in the form

$$v_z = \frac{-CR^2}{4\mu}\left[1 - \left(\frac{r}{R}\right)^2\right]$$ (3.2.33)

with $-C$ given by Eq. 3.2.26.

This *nearly* completes our mathematical model of the laminar flow of a viscous liquid through a long straight tube, subject to the assumptions made at the beginning of the analysis, as well as periodically along the way leading to this result. The specific goal stated at the beginning—deriving a relationship between pressure drop and flowrate—has not yet been achieved, although we are poised to finish the job. The model in its present form is testable in several respects, however, and we do indeed find that Eq. 3.2.33 describes observations, as long as all the restrictions of the model are met. *(We have made an important connection: the assumptions introduced in developing a mathematical model are restrictions on its validity.)*

To achieve the modeling goal stated, we need only integrate the velocity profile across the cross-sectional area of the tube (note Eq. 3.2.8) to find the volumetric flowrate Q:

$$Q = \int_0^R 2\pi v_z r \, dr = \frac{\pi R^4}{8\mu} \frac{\Delta \mathscr{P}}{L} \qquad (3.2.34)$$

where

$$\Delta \mathscr{P} = p_o - p_L + \rho g L \qquad (3.2.35)$$

We have defined $\Delta \mathscr{P}$ as a pressure difference that includes the hydrostatic term. This simplifies the format of Eq. 3.2.34.

Equation 3.2.34 is the relationship sought between the pressure drop and the flowrate. Again we may test this result in the laboratory, and we will find that for a given fluid, the volumetric flowrate is indeed proportional to the *fourth power* of the tube radius and depends on the pressures and tube length as predicted here. Equation 3.2.34 is known as *Poiseuille's law,* or sometimes as the *Hagen–Poiseuille law.*

If we perform tube flow experiments on a fluid in a way that meets the restrictions of the model, we may find the parameter μ, which we earlier identified (without proof) as the *viscosity* of the fluid. One test of the internal consistency of the modeling procedure carried out here is to confirm that the property we call viscosity, which we claim to be a material property, is the same viscosity that is measured by any of several other techniques (some of which are described later). We will assume this to be the case as we proceed.

EXAMPLE 3.2.1   *Determination of Viscosity from a Poiseuille Flow Experiment*

Data for flow of a 33% sugar/water solution through a horizontal, smooth, straight capillary were presented by Back [*J. Biomech.*, **27**, 169 (1994)] as part of a study of flow through catheterized coronary arteries. He used a capillary tube with an inside diameter of 1.93 mm. Two small pressure taps in the wall of the capillary were separated by a length L of 6.526 cm. Pressure was imposed on the reservoir containing the solution, and the resulting mass flowrate was determined as a function of the pressure drop ΔP across the length L. Back's data are plotted in Fig. 3.2.3. A solution density of 1150 kg/m³ was used to convert the measured mass flowrates to the plotted volumetric flowrates. We wish to find the viscosity of the solution from these data.

A least-squares, straight-line fit of these data yields a slope S of 0.0149 (mmHg/cm)/(mL/min). The line has been forced through the origin, to conform to Poiseuille's law. There is no data point at the origin. From Poiseuille's law (Eq. 3.2.34), the slope corresponds to

$$S = \frac{128\mu}{\pi D^4} \qquad (3.2.36)$$

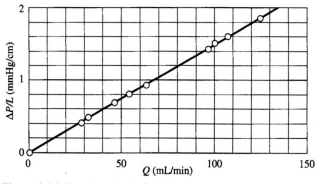

Figure 3.2.3 Results of a Poiseuille flow experiment.

We will convert the slope to SI units:

$$S = \left(0.0149 \frac{mmHg/cm}{cm^3/min}\right)(133.3 \, Pa/mmHg)(60 \, s/min) \times (10^8 \, cm^4/m^4)$$

$$= 1.2 \times 10^{10} \, Pa \cdot s/m^4$$

(3.2.37)

From Eq. 3.2.36, using $D = 1.93 \times 10^{-3}$ m, we find a viscosity of

$$\mu = 4.1 \times 10^{-3} \, Pa \cdot s$$

(3.2.38)

This value is within a few percent of the viscosity reported by Back for this solution.

Let's turn to an example of the application of the Hagen–Poiseuille law to a technological problem that arises in assessing the safety of a designed system.

EXAMPLE 3.2.2   *Passage of a Viscous Drop Through a Long Tube*

A chemical reaction is being carried out in a viscous liquid that is confined to a stirred vessel, as shown in Fig. 3.2.4. The vessel has a capacity of one liter of liquid. The reaction creates a gaseous by-product, and the vessel is vented to prevent the buildup of excessive pressure. The vent is a capillary tube of small inner diameter, to minimize the release

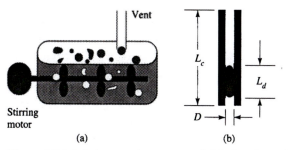

Figure 3.2.4 (a) A vented reactor and (b) the clogged vent tube.

of the gaseous reaction product. Bubbles of the product gas form within the body of the reacting liquid, and these bubbles burst as they rise to the surface of the liquid. If a drop of liquid ejected from a bursting bubble were to cover the entrance to the vent, the only means of pressure relief would be through the flow of that liquid drop through the vent line. If the flow were too slow, pressure would build up before the vent line cleared of liquid, and the vessel might rupture. Let's carry out a model of this phenomenon, using our newfound knowledge of viscous flow through a tube.

This is an example of the application of analysis to the evaluation of a design concept. In this case we are evaluating some features of the design of the vent tube.

When we attempt to model a physical process, we first *think about the physics*. What are the essential features of the physics that we have to account for? Once this question has been answered, we can begin an attempt to *simplify* the physics by making assumptions that are reasonable and lead to a solvable problem.

We want to write a model that produces an estimate of the time required for a trapped drop of viscous liquid to travel the length of the vent tube. We assume that drops occlude the vent infrequently and that we can consider the behavior of a single drop. We need to have some estimate of the drop size, and we need physical property data for the liquid. We assume that the geometry of the vent tube is specified. In our case it is a long capillary.

When a drop occludes the tube entrance, pressure will build in the reactor. Thus there will be a pressure drop across the ends of the drop, and the drop will begin to flow into and through the tube. We want to derive a model from which we can estimate the time required for a drop of a given volume and physical properties to pass through the tube. Let's get started before questions of "What if?" and "How about?" cause us to lose our momentum.

Suppose we have sought physical property data and have developed the following relevant information:

Liquid viscosity $\mu = 0.3$ Pa·s at reactor temperature (this is a "syrupy" liquid)
Liquid density $\rho = 900$ kg/m^3 at reactor temperature
Surface tension $\sigma = 0.04$ N/m at reactor temperature

For the vent tube (see Fig. 3.2.4), we find $L_c = 0.1$ m and $D = 0.003$ m.

What else do we need to know? We need some information regarding the size of the drop that must be pushed through the tube. From Fig. 3.2.4a, we expect the pressure drop–flowrate relationship to depend on the length of the wetted section of the tube, the drop length L_d. This would follow from knowledge of the drop volume. When bubbles burst at the surface of low viscosity liquids such as water, beer, and carbonated soft drinks, the liquid droplets form an aerosol of very small diameter, probably of the order of several hundred micrometers, but smaller than a millimeter. In a *viscous* liquid, could larger drops form if the evolution of gas through the chemical reaction were very violent? At this stage we do not understand some of the essential physics of this system. In one *possible* scenario, however, there would be infrequent drops having diameters of a few millimeters, or even more likely, small drops would accumulate at the face of the vent tube, ultimately to coalesce, occluding the opening. It seems reasonable, then, to consider the fate of a drop of a diameter equal to a small multiple (say a factor of 2) of the diameter of the vent tube. Thus we consider a drop of volume

$$V_d = \frac{\pi D_d^3}{6} = \frac{\pi (2 \times 0.003)^3}{6} \approx 11 \times 10^{-8} \, \text{m}^3 \tag{3.2.39}$$

In the tube this volume corresponds to an axial length (assuming the drop is a "perfect" cylinder):

$$L_d = \frac{4V_d}{\pi D^2} = \frac{4 \times 11 \times 10^{-8}}{\pi (0.003)^2} \approx 0.015 \text{ m} \qquad \textbf{(3.2.40)}$$

Of course the drop will take some time to "squeeze" into the tube before it reaches its full length. We will calculate how long the drop will take to move through the tube, once it is entirely inside. For this calculation we need an estimate of the pressure that builds up in the system, an essential feature of its physics. This is not available. However, we should be able to find the manufacturer's statement of the highest pressure at which the vessel can be safely operated, the so-called pressure limitation. We suppose, then, that this value is known, and we take 10^6 N/m^2 (about 10 atm) as the pressure limitation. We now can examine the following problem:

> At one-tenth *of the upper limit on pressure, how long will it take to move this droplet through the vent tube?*

If this time is short, in comparison to the rate of pressure buildup (another piece of essential, and at this point unknown, information), and if the accumulated assumptions of the analysis are physically reasonable, we may expect that droplets clogging the vent will not create a safety hazard.

> *When safety is an issue, however, we must be especially careful to evaluate these assumptions, and we must be prepared to spend the time and money required to check the validity of our conclusions.*

Note how much important information is unspecified in our problem statement. This is characteristic of an evaluation of an engineering design. Note, as well, that we are moving ahead with an analysis, in spite of the uncertainties. This is characteristic of what engineers do. Once we have arrived at crude estimates of the behavior of the system, we can decide how important these uncertainties are.

We now formulate and solve the following problem:

> *How long is required for a drop (having the characteristics assumed above) to pass through the tube under a pressure difference of 10^5 N/m^2? (Recall that we are picking a pressure limitation of one-tenth of the maximum permissible pressure.)*

To solve this problem, we go directly to Poiseuille's law (Eq. 3.2.34) in the form

$$Q = \frac{\pi D^4}{128\mu} \frac{\Delta P}{L_d} = \frac{\pi D^2 U}{4} = \frac{\pi D^2 L_c}{4t_c} \qquad \textbf{(3.2.41)}$$

We have ignored the effect of gravity, since the pressure is so much greater than the hydrostatic "head" associated with a few millimeters of liquid. The volume flowrate is replaced by the average velocity U (times the cross-sectional area of the tube). This velocity is in turn related to the time t_c required for the drop to move a distance L_c. Everything is known in Eq. 3.2.41, except the time to clear the drop through the tube, t_c. Hence we solve for that time. The result is

$$t_c = \frac{32\mu L_d L_c}{D^2 \Delta P} \qquad \textbf{(3.2.42)}$$

and we find that the time is 1.7×10^{-2} s! According to this result, if the pressure should rise to *one-tenth* of the limiting pressure, and a drop of the size assumed were trapped at the vent entrance, that drop would be squeezed out in a very short time. Thus we could conclude that drops of this size would not occlude the vent; there is no safety problem.[3]

However, we must always assess the assumptions used in developing a model as simple as the one displayed here, searching especially for any gross errors in our physical intuition that might have created a misleading, highly inaccurate model.[4] One of the least tenable assumptions of the model is the one with regard to drop size; the assumed infrequency of occlusion is open to question, as well. We simply do not know enough about the dynamics of bubbling and splattering from a viscous liquid to support the assertions made. In such a situation, it is sometimes useful to examine the worst-case scenario. What would that be?

Suppose we inverted the reactor, with the vent now at the bottom, and completely covered with the viscous liquid. (In practice this worst-case scenario could be triggered if someone installed the reactor upside-down! This happens.) How long would it take to empty the reactor contents completely, at a pressure of one-tenth the maximum permitted pressure? From Eq. 3.2.41 we find that the volumetric flowrate would be

$$Q = \frac{\pi D^4}{128\mu} \frac{\Delta P}{L_c} = \frac{\pi (0.003)^4}{128 \times 0.3} \frac{10^5}{0.1} = 6.6 \times 10^{-6} \text{ m}^3/\text{s} \qquad (3.2.43)$$

or 6.6 cm^3/s. Since the vessel has a capacity of one liter, it would take about 150 s to empty the entire contents of the vessel, at a pressure of one-tenth of the limiting pressure.

This approach yields a *conservative* answer, in the following sense. Some liquid would be vented before the pressure reached 10^5 N/m^2, although at a flowrate lower than that given by Eq. 3.2.43. As the pressure exceeded 10^5 N/m^2 and rose toward the critical level, the flowrate would increase in proportion to the pressure. When the pressure was at 7.5×10^5 N/m^2, for example the flowrate would have reached a level of nearly 50 cm^3/s. Based on these crude arguments, we might conclude that we could vent all the liquid before the pressure reached the critical level. This conclusion, however, neglects the *rate of rise* of pressure. Hence we must seek additional information regarding the kinetics of the reaction, from which we could estimate the rate of production of gaseous product, and the rate of rise of pressure. In particular, we must know whether the reaction could be considered explosive, corresponding to an extremely rapid increase in pressure. This is probably the most essential piece of information from a safety stand-point.

Another concern is related to the viscosity of the liquid. We have taken a value of 0.3 Pa · s and assumed that it holds within the vent tube. If, for example, the reactor is at a highly elevated temperature, and the vent tube is exposed to a cold ambient medium, the liquid might cool down in the vent tube, with a consequent significant increase in viscosity. It is important to know, then, something about the thermal conditions surrounding this reactor, as well as the temperature dependence of the viscosity.

We leave this example at this point, having illustrated how we manipulate a simple flow model, and how we go about "thinking about" the physics, making assumptions,

---

[3] In fact, when a viscous liquid moves through a tube as a drop, it trails a thin film behind it, and the drop decreases in volume. If the tube is long enough, the drop "disappears" as a thin film coating the inside of the tube. See Problem 3.23.

[4] For example, is the flow laminar, as assumed? See Problem 3.34.

and worrying about their validity. We have used a mathematical model to *guide* our quantitative thinking about this problem. We have not "solved" the problem.

The postanalysis evaluation is especially important, and it often leads to the conclusion that additional information, or additional laboratory studies, are required. In this example, our crude analysis leads us to define the specific areas of information that are essential to obtaining a better assessment of the dynamics of the system. Such additional effort is time-consuming and costly, and of course decisions must be put off until the information is available. Economic considerations will often dictate how many additional studies will be supported before a design is finalized.

When safety is involved, economics must not take precedence.

EXAMPLE 3.2.3   *A Model of Laminar Flow Through a Lubricated Tube*

One lesson of the model just presented is that significant pressure drops are required to force viscous liquids through tubes of small diameter. In this example we look at a proposal to reduce the required pressure by creating a thin lubricating layer of liquid at the inner surface of the tube. A mathematical model like the one presented here permits the evaluation of proposals for process improvement. Figure 3.2.5 indicates the model geometry.

We will (as usual) introduce a number of assumptions into the analysis of this system. Then we will develop the model. A review and assessment of some of the more tenuous assumptions concludes this example.

To begin, we adopt the assumptions used in our analysis of laminar flow through a tube (see Section 3.2.1), except that we will ignore gravity. These assumptions will hold for *both* fluids, but we add one assumption about fluid II: the lubricant layer thickness h is taken to be independent of axial position and independent of time.

Assumptions are an integral part of the model-building process. First, they reflect not what we know about a system, but what we think or hope may be true about the physics of the system. Assumptions can be based on experience. We often call such tentative conclusions "intuition." Sometimes assumptions have no obvious physical basis, and they simply reflect our awareness that *if* the assumed point were true, then the mathematical model would be a lot simpler to handle mathematically. In other words, sometimes we introduce assumptions because they lead to simple, solvable equations. Of course, this is all nonsense if the assumptions are so far from reality that they lead us to formulate a mathematical model that does not really reflect the essential physics of the system.

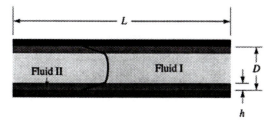

Figure 3.2.5 Laminar flow through a lubricated tube: fluid II is the lubricant, with a film thickness h.

Then it should be clear that an important part of the art of building mathematical models of physical systems is developing the judgment to distinguish between reasonable assumptions and those that violate good physical sense.

Such judgment is not easy to come by, and one way to learn the art of model building is to study many examples of mathematical models and assess their degree of correspondence to reality.

Let's return, then, to the example at hand. If we review Section 3.2.1 we find that the analysis that leads to Eq. 3.2.23 holds in both fluids of the present example. Equation 3.2.30 is still the general solution for the velocity profile in fluid I because the boundary condition that the shear stress must be finite along $r = 0$ still holds in fluid I. Hence we have the following solution for the velocity profile in fluid I:

$$v_z^{\mathrm{I}} = \frac{C^{\mathrm{I}}}{4\mu^{\mathrm{I}}}r^2 + F^{\mathrm{I}} \qquad \text{over} \quad 0 \le r \le \frac{D}{2} - h \tag{3.2.44}$$

In fluid II, however, the constant E that appears in Eq. 3.2.23 does not vanish, since Eq. 3.2.23 does not include $r = 0$ for fluid II. Thus we must integrate Eq. 3.2.23, keeping E as an unknown constant. We still introduce Eq. 3.2.28, so we must solve for the velocity profile in the lubricant film from

$$\mu^{\mathrm{II}}\frac{dv_z^{\mathrm{II}}}{dr} = \frac{C^{\mathrm{II}}r}{2} + \frac{E^{\mathrm{II}}}{r} \qquad \text{on the region} \quad \frac{D}{2} - h \le r \le \frac{D}{2} \tag{3.2.45}$$

The solution is

$$v_z^{\mathrm{II}} = \frac{C^{\mathrm{II}}r^2}{4\mu^{\mathrm{II}}} + \frac{E^{\mathrm{II}}}{\mu^{\mathrm{II}}}\ln r + F^{\mathrm{II}} \qquad \text{on the region} \quad \frac{D}{2} - h \le r \le \frac{D}{2} \tag{3.2.46}$$

Before we get mired in the algebra, let's look at Eqs. 3.2.44 and 3.2.46, which have five unknown constants between them. These constants will have to come from boundary conditions—further statements about the physics of this assumed flow field. What physical statements can we make at this point?

First of all, the statements that led us to Eq. 3.2.22, and ultimately to Eq. 3.2.26, are still valid in this "two-fluid" problem. Hence we still find

$$-C^{\mathrm{I}} = \frac{\Delta P^{\mathrm{I}}}{L} \qquad \text{and} \qquad -C^{\mathrm{II}} = \frac{\Delta P^{\mathrm{II}}}{L} \tag{3.2.47}$$

Furthermore, a single pressure drop ΔP is imposed across the ends of the tube, and there is no mechanism that would allow the pressure to be different in the two fluids, at a fixed axial position. [We ignore the very small difference in pressure that would arise from surface tension and curvature (see Eq. 2.2.7) if the two fluids were immiscible.] Hence it follows that

$$-C^{\mathrm{I}} = -C^{\mathrm{II}} = \frac{\Delta P}{L} \tag{3.2.48}$$

The remaining three equations we need arise from boundary conditions on velocity and shear stress. First, we still satisfy a no-slip condition (see Eq. 3.2.31) in the fluid adjacent to the wall of the tube (which is only fluid II), so (letting $v_z^{\mathrm{II}} = 0$ at $r = R = D/2$), we write

$$0 = \frac{C^{\mathrm{II}}R^2}{4\mu^{\mathrm{II}}} + \frac{E^{\mathrm{II}}}{\mu^{\mathrm{II}}}\ln R + F^{\mathrm{II}} \tag{3.2.49}$$

We also satisfy a no-slip condition at the interface between the two fluids. In this case, the velocities in the two fluids are taken to be equal at their boundary $r = R - h$, so we write

$$v_z^{II}\bigg|_{R-h} = \frac{C^{II}(R-h)^2}{4\mu^{II}} + \frac{E^{II}}{\mu^{II}}\ln(R-h) + F^{II} = v_z^{I}\bigg|_{R-h} = \frac{C^{I}}{4\mu^{I}}(R-h)^2 + F^{I}$$

(3.2.50)

We need one more boundary condition that reflects something we know (or imply to be true) about the flow.

Equation 3.2.50 is really a statement that the velocity field is continuous through the fluids. In many physical problems we argue that physical variables (velocity, temperature, pressure) vary *continuously* through a fluid, rather than undergoing sudden jumps from one value to another. The no-slip condition is such a statement. We also expect that the *shear stress* will vary continuously as we pass from one fluid into the other, or

$$\mu^{II}\frac{dv_z^{II}}{dr} = \mu^{I}\frac{dv_z^{I}}{dr} \quad \text{at} \quad r = R - h$$

(3.2.51)

After performing the differentiation of Eq. 3.2.44 and using Eq. 3.2.45, we find the following requirements

$$\frac{C^{I}(R-h)}{2} = \frac{C^{II}(R-h)}{2} + \frac{E^{II}}{(R-h)}$$

(3.2.52)

We now have the required five equations for the five constants of integration in the solutions for the two velocity fields. The rest is tedious algebra. After completing the algebra we can write the two velocity expressions in the following formats:

$$\frac{v_z^{I}}{\Phi} = [(1-\eta)^2 + M\eta(2-\eta)] - s^2$$

(3.2.53)

$$\frac{v_z^{II}}{\Phi} = M(1-s^2)$$

(3.2.54)

where, to simplify the format, the following notations were introduced:

$$\Phi = \frac{-CR^2}{4\mu^{I}} = \frac{\Delta PR^2}{4\mu^{I}L} \qquad s = \frac{r}{R} \qquad \eta = \frac{h}{R} \qquad M = \frac{\mu^{I}}{\mu^{II}}$$

(3.2.55)

Note that s is a dimensionless radial variable, while η and M are dimensionless parameters. We will commonly write our mathematical models in terms of dimensionless variables and parameters. As noted, this is sometimes done to simplify the format—the appearance—of the resulting model. There are other reasons as well, which we will discuss later. The grouping of parameters called Φ is *not* dimensionless. It has the dimensions of velocity.

Now we must pause to recall the goal of this analysis. It is to investigate the possibility of increasing the flowrate through a tube by providing a "lubricating" layer of a second fluid at the wall of the tube. How can we examine this feature of the model we have here? We start by calculating the volume flowrate of the inner fluid, which is the fluid we want to convey through the tube. This follows by integrating the velocity field across the region occupied by fluid I, so we calculate

$$Q^{I} = \int_0^{R-h} 2\pi r v_z^{I}\, dr$$

(3.2.56)

If there were no lubricating layer ($h = 0$), the volume flowrate would be given by Eq. 3.2.34 (with ΔP instead of $\Delta\mathscr{P}$, since we are assuming the axis of the tube is horizontal). Performing the integration indicated in Eq. 3.2.56, we find

$$\frac{Q^{\mathrm{I}}}{2\pi R^2 \Phi} = \frac{(1 - \eta)^4}{4} + \frac{M\eta(2 - \eta)(1 - \eta)^2}{2} \tag{3.2.57}$$

The flowrate without the lubricating layer is obtained from this equation simply by setting $\eta = 0$, and the desired ratio of flowrates with, and without, the lubricating layer is then found to be

$$\Psi \equiv \frac{Q^{\mathrm{I}}}{Q^{\mathrm{I}}(\eta = 0)} = (1 - \eta)^4 + 2M\eta(2 - \eta)(1 - \eta)^2 \tag{3.2.58}$$

We show the dependence of Ψ on M and η in Fig. 3.2.6. The results are consistent with our expectations. Only for $M > 1$, when the lubricating layer is less viscous than the mainstream fluid, is there an enhancement of the flowrate ($\Psi > 1$). Furthermore, we see that when η approaches unity (even for large M) the flow is reduced, since as the lubricating layer grows thicker, the region available for flow of fluid I is significantly decreased. For a given value of $M > 1$ there is an optimum film thickness η, at which the flow enhancement is a maximum.

One final point needs mentioning here. A major assumption is that the lubricating layer thickness h remains constant along the tube axis and is independent of time. We might hope to achieve these conditions, by providing a means of supplying a thin film of fluid II along the circumference of the tube, at the tube entrance. However, there is another issue here, and it involves the *stability* of a thin liquid film in this type of flow. Without further analysis, or experimental confirmation, we have no guarantee that a thin film of low viscosity liquid would remain as an annulus of constant thickness in the flow field assumed in our model.

We will not deal with the stability problem here except to point out that in the absence of experimental confirmation, we cannot yet know whether this model is at all useful to us. This awareness should encourage us to design an experimental program of study of a lubricated tube flow. The model we have would serve as an aid in the design of the experiments.

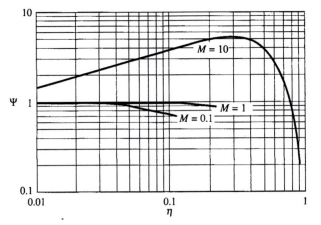

Figure 3.2.6 Effect of lubricating layer on flow through a tube (Eq. 3.2.58).

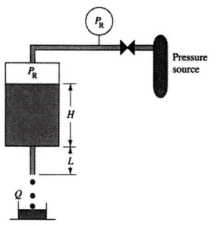

Figure 3.3.1 Conceptual design of a capillary viscometer.

3.3   ENGINEERING DESIGN: THE ROLE OF ANALYSIS

We can use a mathematical model (such as the Hagen–Poiseuille law) as an aid in engineering design. In a design problem, we typically have a number of unspecified parameters. The designer is free to specify some of these parameters in accordance with considerations of economics, availability of materials, ease of manufacture, and so on. Design problems are open-ended in a sense, and there is no *single* solution, nor is there a *correct* solution. There is, however, a set of solutions, some of them "good" and some of them "poor." Let's look at an example of a design problem. Notice how we break down the procedure of "solving" a design problem.

TASK (What is our goal?) Design a capillary viscometer that will be useful for fluids with viscosities of the order of 1000 poise.[5]

INTERPRETATION (What do we mean by "design?") Specify values for the radius R and length L of the capillary. Provide estimates of the required pressure to operate the viscometer.

CONCEPTUAL DESIGN (We start with an approach, or concept, that we think is worth exploring.)

Fluid will be held in a pressure vessel to which the capillary (see Fig. 3.3.1) is attached. Measurements will consist of P_R and Q. The pressure source could be a tank of compressed air. By opening a control valve on the tank, a pressure P_R can be imposed on the liquid in the reservoir. Liquid flows through the capillary and is collected. A volume flowrate is determined.

DESIGN EQUATION (With what mathematical model do we relate the measured to the desired quantities?) We invert Poiseuille's law, solve for viscosity, and write

$$\mu = \frac{\pi R^4}{8Q} \frac{\Delta \mathscr{P}}{L} \tag{3.3.1}$$

---

[5] The cgs unit of viscosity, the poise, is the unit commonly found in many currently available compilations of viscosity data. The SI unit of viscosity is the pascal-second (Pa · s). We multiply the viscosity in pascal-seconds by a factor of 10 to get the viscosity in units of poise. For comparative purposes, it is useful to know that the viscosity of water at room temperature is about 1 centipoise (i.e., 0.01 poise or 10^{-3} Pa · s). Hence a 1000-poise liquid is more viscous than water by a factor of 10^5.

We take

$$\Delta \mathscr{P} = P_R + \rho g H + \rho g L - p_L \qquad (3.3.2)$$

where p_L is the pressure at the outlet of the capillary.

CONSTRAINTS (Under what conditions is this design equation valid?) To use Eq. 3.3.1, we must have a flow that is laminar, fully developed, and isothermal. The fluid must be Newtonian.

Stainless steel tubing is readily available with radii in the range of 0.1–1 cm. We require a long L/R to eliminate the "end effects" that arise primarily from the transition of the flow from the reservoir into the small capillary. It is within this transition region that the fluid attains the parabolic velocity profile given in Eq. 3.2.33. There is an entrance length, usually several tube radii in length, over which this rearrangement takes place. A reasonable rule of thumb cites $L/R = 100$, yielding an entrance length that is a small fraction of the total length of the tube. Then Poiseuille's law should be valid over most of the length of the tube.

Let's suppose that the entrance length L_e is known from prior studies of laminar flows to be given approximately by

$$\frac{L_e}{R} = 1.2 + 0.16 \frac{Q\rho}{\pi R \mu} = 1.2 + 0.08 \, \text{Re} \qquad (3.3.3)$$

A dimensionless group called the *Reynolds number* appears on the right-hand side of this equation. For flow through a tube, it is defined as

$$\text{Re} = \frac{2Q\rho}{\pi R \mu} \qquad (3.3.4)$$

ROUGH DESIGN PROCEDURE (Let's get started. We can modify the design later.) Make some "ballpark" or "order of magnitude" estimates of some of the parameters that enter into the model.

For example, we should measure Q by collecting a "reasonable" volume of liquid over a "reasonable" length of time. A ballpark estimate is that we should measure about 100 cm³ over about 100 s, since these will be easy to measure accurately. Hence we choose $Q = 1$ cm³/s as a basis for design calculations that follow. For these preliminary purposes, we say that "all" liquids have densities of the *order* of 1 g/cm³ (one obvious exception is mercury), and we pick a capillary radius R of 0.1 cm.

Our task statement suggests that μ will be of the order of 1000 poise. Hence we expect that typical operating conditions of this design correspond to Reynolds numbers of about

$$\text{Re} = \frac{2(1)(1)}{\pi(0.1)(1000)} = 0.01 \qquad (3.3.5)$$

(Note that $2/\pi = 1$ for the purpose of a rough design calculation.) From Eq. 3.3.3 we conclude that we do not have to worry much about fully developed flow. The flow will be fully developed almost from the tube entrance.

We want the entrance length to be very small in comparison to the capillary length, say

$$\frac{Le}{L} = 0.01 \qquad (3.3.6)$$

Then we choose $L = 100 \, L_e \approx 100 \, R = 10$ cm. (This explains the rule of thumb noted above, that we should use $L/R = 100$.)

We also note here that the magnitude of the Reynolds number determines whether the flow is laminar or turbulent. As another rule of thumb, laminar flow can be maintained in Poiseuille flow as long as the Reynolds number is less than about 2000. The Reynolds number for this flow is so small that we can be assured that the flow is not turbulent.

We have now specified the following design parameters:

$$R = 0.1 \text{ cm} \qquad \text{and} \qquad L = 10 \text{ cm}$$

What pressure P_R is needed to drive the liquid through this viscometer? From Eq. 3.3.1 we find

$$\Delta \mathscr{P} = \frac{8\mu Q L}{\pi R^4} = \frac{8(1000)(1)(10)}{\pi(0.1)^4} \tag{3.3.7}$$

Hence

$$\Delta \mathscr{P} = \frac{8}{\pi} \times 10^8 \text{ dyn/cm}^2 = \frac{800}{\pi} \text{ atm} \tag{3.3.8}$$

$$= 4000 \text{ psi} \quad \text{(all numbers rough, because we want quick estimates)}$$

EVALUATION (Based on these rough numbers, do we need to modify any of the choices we made?)

This design requires a very high pressure, hence would require an expensive pressure vessel. To keep costs down, we need to relax some of the specifications. Let's keep $L/R = 100$ and rewrite Eq. 3.3.1 as

$$\Delta \mathscr{P} = \frac{8\mu Q}{\pi} \frac{L}{R} \frac{1}{R^3} \tag{3.3.9}$$

We can significantly reduce $\Delta \mathscr{P}$ by manipulating R, since $\Delta \mathscr{P}$ depends so strongly (third power) on R, at fixed L/R. For example, if we increase R by a factor of 2, and keep $L/R = 100$, so that L is also increased by two times, then $\Delta \mathscr{P}$ is reduced by $2^3 = 8$ to

$$\Delta \mathscr{P} = \frac{4000}{8} = 500 \text{ psi} \tag{3.3.10}$$

This is a much more manageable pressure drop.

Note that in Eq. 3.3.2 we have two hydrostatic terms: $\rho g H$ and $\rho g L$. For $\rho = 1 \text{ g/cm}^3$ and $g = 10^3 \text{ cm/s}^2$ (since we are doing rough design calculations, we do not bother to use the more exact value of $g = 980$), and with $L = 20$ cm and H of the same magnitude, we can write

$$\rho g H \approx \rho g L = (1)(10^3)(20) = 2 \times 10^4 \text{ dyn/cm}^2 \approx 0.3 \text{ psi} \tag{3.3.11}$$

In other words, hydrostatic effects are negligible when P_R is large.

We will assume that the capillary exit is at atmospheric pressure. We can just take $p_L = 0$ and regard all pressures as *gage pressures*—that is, as relative to atmospheric pressure.

At this stage we have some rough design criteria: $R = 0.2$ cm and $L = 20$ cm. The hydrostatic pressure associated with the head of liquid in the reservoir is not significant if we restrict the use of the viscometer to viscous liquids. The required pressure will be several hundred psi. If we reread the original task statement, and our interpretation of it, we see that we have completed the rough design of the viscometer. Our mathematical model of Poiseuille flow was an aid in the design process.

EXAMPLE 3.3.1   *Treatment of Data from a Capillary Viscometer*

The following data were obtained by means of a viscometer designed according to the discussion above. Data were obtained at 25°C. Find the viscosity of the liquid. P_R is the gage pressure (*i.e.*, the pressure above atmospheric), in units of pounds per square inch.

P_R(psi)	60	120	340	1000	2000
Q(cm³/s)	0.052	0.11	0.32	0.85	1.6

The data are plotted in Fig. 3.3.2, and a least-squares linear fit to the data yields

$$Q = 8 \times 10^{-4} \, P_R$$

From Eq. 3.3.1 we may solve for the viscosity in the form

$$\mu = \frac{\pi R^4}{8L} \frac{P_R}{Q} = \frac{\pi (0.2)^4}{8(20)} \frac{6.9 \times 10^4}{8 \times 10^{-4}} = 2710 \, \text{poise}$$

Note that the gravitational (hydrostatic) pressure contribution is ignored for these data. (The factor of 6.9×10^4 in the numerator of this expression converts the pressure in psi to cgs units, dyns/cm².)

3.3.1   Classification of Simple Laminar Flows

It is convenient to classify flows in certain ways that emphasize the distinctive features of the flows. We can illustrate this with one simple geometry, shown in Fig. 3.3.3. We consider a pair of *concentric* cylinders, of radii R and κR ($\kappa < 1$). We can generate several distinctive flow fields in this geometry. In one case, the cylinders are stationary with respect to each other, and an axial pressure difference is imposed across the ends. This is the Poiseuille flow, except in this case the tube's cross section is *annular* rather than circular. A sketch of the expected flow field is given in Fig. 3.3.4. (The maximum velocity does not necessarily fall at the midpoint between the surfaces at R and κR. See Problem 3.9.) We call this flow *axial annular pressure flow*.

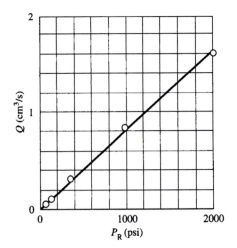

Figure 3.3.2 Capillary viscosity data.

Flow is down the axis and
between the two cylinders

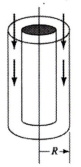

Figure 3.3.3 Geometry for annular flow.

Another possibility (that is, a different type of flow) occurs if the inner cylinder translates in the axial direction. The expected flow field is shown in Fig. 3.3.5. Notice that we have *assumed* the absence of an *imposed* pressure gradient, which allows us to say that there is no pressure difference along the axis. Hence this is not a pressure flow. Fluid is entrained or dragged by the moving surface. We call this *axial annular drag flow*.

A variant on this nonpressure flow is suggested in the sketch in Fig. 3.3.6, where the fluid is *confined* to a finite annular volume. We assume that there is no leakage flow. In this sketch we suggest the streamlines and the velocity profile. We will come back to confined flows shortly.

Another variation, similar to but distinct from the preceding case, occurs when a cylinder of *finite* length falls along the axis of a coaxial cylinder, as sketched in Fig. 3.3.7. This is a more complex problem, since the faces of the inner cylinder might alter the flow field near its ends.

Thus we see that within the confines of a specific geometry (e.g., coaxial cylinders) there are several different flow fields that can be generated, depending on what forces or motions are imposed on the boundaries of the system.

3.3.2 Axial Annular Flow in a Closed Container

Now we turn to analysis of the flow depicted in Fig. 3.3.6. We will ignore the flow at the ends because it is so complex. Over most of the length L, the flow is strictly axial laminar flow; that is, $\mathbf{v} = (v_z, 0)$. We expect to get away with this idealization if L is large compared to R. We begin with Eq. 3.2.21, which is valid for laminar axisymmetric axial flow. (We will neglect gravity in this model, or assume that the cylindrical axis is horizontal.)

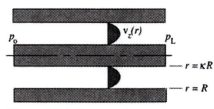

Figure 3.3.4 Cross section through pressure-driven flow in an annulus.

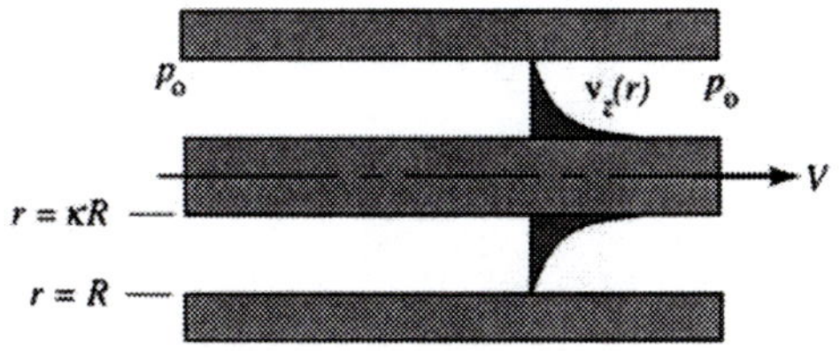

Figure 3.3.5 Cross section through drag flow in an annulus.

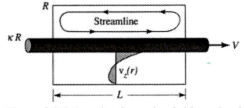

Figure 3.3.6 Annular flow of a fluid confined between two cylinders.

$$-\frac{1}{r}\frac{d}{dr}(r\tau_{zr}) = \frac{\Delta P}{L} \tag{3.3.12}$$

In this problem we do not *impose* ΔP by external means, as we do with Poiseuille flow. But since there *might* be a finite ΔP across the ends of the annular region, we leave ΔP in the force (momentum) balance. If there is a ΔP, it will come out of the analysis. If ΔP vanishes, that will follow from the analysis.

We assume Newtonian behavior, so we write the shear stress as

$$\tau_{zr} = -\mu\frac{dv_z}{dr} \tag{3.3.13}$$

and we take no-slip boundary conditions on v_z at the surfaces κR and R:

$$v_z = V \quad \text{at} \quad r = \kappa R \tag{3.3.14}$$
$$v_z = 0 \quad \text{at} \quad r = R \tag{3.3.15}$$

Substitution of Eq. 3.3.13 into Eq. 3.3.12 yields a second-order ordinary differential equation. After some calculus and a lot of algebra we solve Eq. 3.3.12 for $v_z(r)$ to find:

$$v_z(r) = \frac{-\Delta PR^2}{4\mu L}\left(1 - \frac{r^2}{R^2}\right) + \frac{\ln(R/r)}{\ln(1/\kappa)}\left[V + \frac{\Delta PR^2}{4\mu L}(1 - \kappa^2)\right] \tag{3.3.16}$$

We will save some aggravation if we introduce the following nondimensional variables:

$$\frac{v_z(r)}{V} = \varphi(s) \qquad \frac{r}{R} = s \qquad \frac{\Delta PR^2}{4\mu LV} = \Phi \tag{3.3.17}$$

Then Eq. 3.3.16 takes the form:

$$\varphi(s) = -\Phi(1 - s^2) + \frac{\ln s}{\ln \kappa}[1 + \Phi(1 - \kappa^2)] \tag{3.3.18}$$

Figure 3.3.7 Annular axial flow induced by the motion of a cylinder of finite length.

Note that we have now buried our ignorance of ΔP in the parameter Φ. Where does Φ (i.e., ΔP) come from? Is there anything about the physics of this system that remains to be worked into our analysis? We note that the fluid motion induced by the motion of the cylinder is "closed." Whatever fluid is dragged in the direction of motion of the moving cylinder, the same amount of fluid must return. Otherwise we would lose fluid. But we have ignored loss through leakage. Hence we are led to conclude that across any surface normal to the z axis there is no *net* flow:

$$Q = 2\pi \int_{\kappa R}^{R} v_z(r) r \, dr = 0 \tag{3.3.19}$$

We impose this constraint, in the dimensionless form

$$\int_{\kappa}^{1} \varphi(s) s \, ds = 0 \tag{3.3.20}$$

and we find

$$0 = -\Phi \int_{\kappa}^{1} (1 - s^2) s \, ds + [1 + \Phi(1 - \kappa^2)] \frac{1}{\ln \kappa} \int_{\kappa}^{1} (\ln s) s \, ds \tag{3.3.21}$$

Algebra aside, this equation is of the functional form

$$0 = F(\Phi, \kappa) \quad \text{or} \quad \Phi = G(\kappa) \tag{3.3.22}$$

After doing the indicated integrations and the algebra, we find the functional form of $G(\kappa)$, and write Φ, which is a function of κ, as

$$\Phi(\kappa) = -\frac{1 - \kappa^2 + 2\kappa^2 \ln \kappa}{(1 - \kappa^4) \ln \kappa + (1 - \kappa^2)^2} \tag{3.3.23}$$

Hence, given the geometry (κ) we can find $\Phi(\kappa)$. This permits us to calculate ΔP as a function of operating parameters (V) and design parameters (R and L). Of course ΔP depends on μ as well.

Figure 3.3.8 shows the function $\Phi(\kappa)$. Note that we have defined ΔP to be the

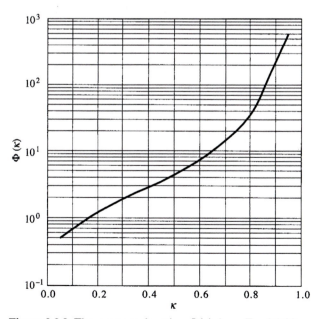

Figure 3.3.8 The pressure function $\Phi(\kappa)$ from Eq. 3.3.23.

difference between the pressure where the moving inner cylinder *exits* the outer cylinder and the pressure at the entrance. We find that Φ (and so ΔP) exceeds 0, which means that *the pressure is higher at the exit end.*

An important characteristic of the function Φ is that when κ is close to unity, Φ changes by an order of magnitude (a factor of 10) when κ changes by only about 10%. For this reason we plot Φ on a logarithmic scale. This makes it possible to read values of Φ with uniform precision over the whole range of Φ.

<table>
<tr><td>EXAMPLE 3.3.2</td><td>*Pressure in an Enclosure*</td></tr>
</table>

To get an idea of the magnitude of the pressure difference developed by the flow just analyzed, let's do a calculation for the following conditions:

$$\mu = 0.1\,\text{Pa}\cdot\text{s} \qquad L = 0.1\,\text{m} \qquad R = 3\,\text{mm} \qquad \kappa = 0.8 \qquad V = 1\,\text{m/s}$$

From Fig. 3.3.8 (or Eq. 3.3.23) we find, for $\kappa = 0.8$, $\Phi = 38$. From the definition of Φ we then find

$$\Delta P = \frac{4L\mu V}{R^2}\Phi = \frac{4(0.1)(0.1)(1)}{(0.003)^2} \times 38 = 1.69 \times 10^5\,\text{N/m}^2 \approx 1.67\,\text{atm} = 24.6\,\text{psi}$$

$$(3.3.24)$$

This is a relatively small pressure difference. If the upstream (entrance) end were somehow vented to the atmosphere, the downstream end would be about 25 psi above atmospheric pressure. It is easy to see, however, that either a small reduction in R, or a small increment in κ would lead toward the development of a much higher pressure.

3.3.3 Axial Annular Flow in an Open Tube

Let us go back to *axial annular pressure flow* (Fig. 3.3.4). An analysis *could* begin with Eq. 3.3.12, which is generally valid for any axisymmetric axial flow confined within concentric cylinders (of course, including the assumptions of laminar Newtonian isothermal flow, as well). If we have pure pressure flow (no axial motion of either boundary), then the boundary condition of Eq. 3.3.14 is changed to

$$v_z = 0 \qquad \text{at} \quad r = \kappa R \qquad\qquad (3.3.25)$$

Equation 3.3.15 still holds as the second boundary condition.

We can solve Eq. 3.3.12, using Eq. 3.3.13 to relate the shear stress to the velocity gradient, subject to conditions given by Eqs. 3.3.25 and 3.3.15. But there is an easier way to get a solution. It begins by realizing that Eq. 3.3.16 should be valid for *any* value of V, including $V = 0$. Hence we should be able to write the solution by inspection and modification of Eq. 3.3.16. Note that since the presence of *closed ends* does not enter the analysis until the imposition of Eq. 3.3.19, the result that follows is still valid for an *open* outer cylinder. When we set $V = 0$ in Eq. 3.3.16, and now define $\Delta P = p(0) - p(L)$ the result is found to be

$$v_z(r) = \frac{\Delta P R^2}{4\mu L}\left[(1 - s^2) - (1 - \kappa^2)\frac{\ln s}{\ln \kappa}\right] \qquad\qquad (3.3.26)$$

Note that we introduce $s = r/R$ in this expression. Since there is no V in this problem (the inner cylinder is taken as stationary), we cannot nondimensionalize $v_z(r)$ as we did in Eq. 3.3.17. Note also that there is no function Φ as defined in Eq. 3.3.17 either, since $V = 0$. We can, however, define a nondimensional velocity function as

$$\varphi_p(s) = \frac{4\mu L v_z(r)}{\Delta P R^2} \tag{3.3.27}$$

and write Eq. 3.3.26 as

$$\varphi_p(s) = (1 - s^2) - (1 - \kappa^2)\frac{\ln s}{\ln \kappa} \tag{3.3.28}$$

(We use a subscript p on φ to remind us that this is the solution for pure *pressure* flow. There is no drag flow component. Note also that we are using a lowercase φ here, whereas before we had the capital letter Φ.)

In some applications we are interested primarily in the volume flowrate Q, which is calculated as

$$Q = 2\pi \int_{\kappa R}^{R} v_z(r)\, r\, dr \tag{3.3.29}$$

Using Eq. 3.3.26 we find (after some algebra)

$$Q_p = \frac{\pi \Delta P R^4}{8\mu L}\left[1 - \kappa^4 + \frac{(1 - \kappa^2)^2}{\ln \kappa}\right] \tag{3.3.30}$$

While we are looking at Q, let's also get Q for the purely *drag flow* case (Fig. 3.3.5), which we can obtain from Eq. 3.3.16 simply by setting $\Delta P = 0$. If the fluid is dragged by the inner cylinder, then we find (setting $\Delta P = 0$ in Eq. 3.3.16)

$$v_z(r) = V\frac{\ln s}{\ln \kappa} \tag{3.3.31}$$

and with Eq. 3.3.29 it follows that

$$Q_d = \pi R^2(1 - \kappa^2) V\left[\frac{2\kappa^2 \ln \kappa + 1 - \kappa^2}{-2(1 - \kappa^2)\ln \kappa}\right] \tag{3.3.32}$$

We now have two simple models of axial flow in an annulus: Q_p gives the flowrate induced strictly by pressure (*no drag*) and Q_d gives the flowrate induced strictly by drag (*no pressure*). Since in both cases it is implied that the axial length of the annulus is long in comparison to the gap ($R - \kappa R$), we can ignore any "end effects" associated with the ends of the inner cylinders. We can now examine some practical applications of these models to an engineering process.

3.3.4 The Design of a Wire Coating Die and Analysis of Its Performance

It is common in industry to apply electrical insulation continuously to electrical wire on a long reel. Typically, molten polymer (e.g., nylon) is fed to a circular die, along whose axis the wire is continuously drawn. The geometry is shown in Fig. 3.3.9. The goal of this analysis is to find the relationship between the downstream coating thickness δ and the other parameters that characterize the performance of the system.

As the wire in the coating system moves steadily along its axis, it entrains the polymer and subsequently carries the fluid through the die and out of it. This is an axial annular drag flow. We will assume steady state isothermal flow of a Newtonian fluid, although often the fluids that are used for coating are non-Newtonian. Downstream of the die the polymer solidifies and is transported as a rigid coating. Hence at that point and beyond, the coating has a velocity identical to that of the wire, V. The mass flowrate

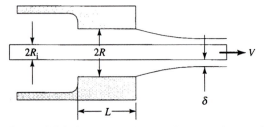

Figure 3.3.9 Geometry of a continuous wire coating system (not to scale). Usually $R - R_i \ll R$ and $R - R_i \ll L$.

of coating may be written in the form (using the notation $R_i = \kappa R$)

$$\dot{m} = \rho\pi[(\kappa R + \delta)^2 - \kappa^2 R^2]V \tag{3.3.33}$$

where ρ is the density of the solid polymer. Conservation of mass requires that $\dot{m}$ be identical to the mass flowrate due to the drag flow through the die:

$$\dot{m} = \rho' Q \tag{3.3.34}$$

where ρ' is the density of the fluid within the die.

We are assuming isothermal flow. Why do we have two different densities? Because the isothermal assumption is needed for the flow within the die boundaries and can be attained with good temperature control on the die block. But once outside the die, the coating on the wire is exposed to ambient temperature, which may be 100°C or more below that of the molten polymer. Clearly, there could be a significant density difference between molten and solid polymer over so broad a temperature range.

Assuming that there is no imposed pressure component to this flow, and that $R - R_i \ll L$, we take Q as given in Eq. 3.3.32. We can eliminate $\dot{m}$ between Eqs. 3.3.33 and 3.3.34 and obtain a quadratic equation for δ in the form

$$\delta'^2 + 2\delta' - \frac{\rho'}{\rho}\frac{1 - \kappa^2}{\kappa^2} H(\kappa) = 0 \tag{3.3.35}$$

where $\delta' = \delta/R_i$ and $H(\kappa)$ is the term in brackets in Eq. 3.3.32. Solving for δ', we can find

$$\delta' = \frac{\delta}{R_i} = \left[1 + \frac{\rho'}{\rho}\frac{1 - \kappa^2}{\kappa^2} H(\kappa)\right]^{1/2} - 1 \tag{3.3.36}$$

Figure 3.3.10 shows δ' as a function of the parameter κ, for $\rho/\rho' = 1$. (Again, notice the use of a *logarithmic* scale for δ', since it changes by several orders of magnitude.)

We have been advocating the importance of nondimensionalization of our mathematical models. One advantage of this practice is that the resulting equations have a much simpler format because they contain fewer variables and parameters. Another benefit is more compact graphical representations of the models. Figure 3.3.10, for example, is a simple plot of δ' versus κ, valid for any choices of R_i, R, and V. Sometimes, however, nondimensionalization hides the physical meaning of the results or clouds simple interpretations that might be drawn from the resultant equations and graphs. For example, suppose we ask the question: What is the relationship of coating thickness to the gap between the wire and the die? By "gap" we mean the difference $R - R_i$. One way to answer this is to look at the ratio

$$\frac{\delta}{R - R_i} = \frac{\kappa\delta'}{1 - \kappa} \tag{3.3.37}$$

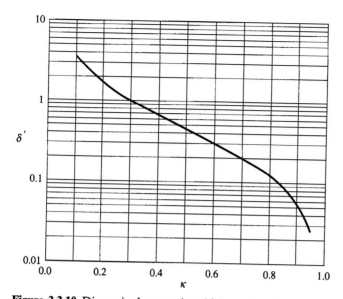

Figure 3.3.10 Dimensionless coating thickness in wire coating (no pressure) for $\rho/\rho' = 1$.

If we examine the limiting behavior of Eq. 3.3.36 as κ approaches 1, we can show that

$$\delta' \to \frac{1 - \kappa}{2\kappa} \tag{3.3.38}$$

and if we superimpose this curve onto Fig. 3.3.10, we will find that for all κ, δ' is always below this limiting value. Still, Eq. 3.3.38 serves as a very good approximation for values of κ greater than 0.5, and it is considerably simpler, algebraically, than Eq. 3.3.36.

Now it is advantageous to go back to the real variables, and this will show us that Eq. 3.3.38 is equivalent to

$$\delta = \frac{R - R_i}{2} \quad \text{in the limit of } \kappa \to 1 \tag{3.3.39}$$

Hence we conclude that the maximum coating thickness possible when we rely on drag flow is half the gap dimension. This is a very simple result, easy to remember, and it provides a rule of thumb for die design.

Another important implication of this model is the independence of δ from the wire velocity. For a given fluid, δ could be varied only through changes in the geometry of the die. But if we designed a die, based on a model, and then found that the coating thickness was not *exactly* at the desired level, we would have wasted quite a bit of time and money. This design does not provide the flexibility for control of δ without a complete die change. Inflexibility is a poor design feature.

There is a means of introducing the desired operating flexibility while using the same die. It requires us to impose a positive pressure on the fluid in the reservoir upstream of the die. The flow will then appear as in Fig. 3.3.11. An excess pressure ΔP is imposed on the fluid, and the total flow is the sum of a drag flow and the pressure flow. For a Newtonian fluid, we can prove that these two contributions are additive. Then we begin by adding the Q_p for the pressure flow (Eq. 3.3.30) to Q_d for drag flow (Eq. 3.3.32).

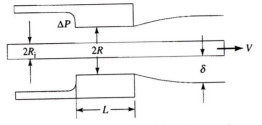

Figure 3.3.11 Wire coating with an imposed pressure.

The result is

$$Q = \frac{\pi R^2 V}{2}\left(\frac{1 - \kappa^2}{\ln(1/\kappa)} - 2\kappa^2\right) + \frac{\pi \Delta P R^4}{8\mu L}\left[1 - \kappa^4 - \frac{(1 - \kappa^2)^2}{\ln(1/\kappa)}\right] \qquad (3.3.40)$$

If Eq. 3.3.33 for the rate of the mass flow is equated to $\rho'Q$, as in the pure drag flow case earlier, solving for the dimensionless coating thickness will give

$$\delta' = \frac{\delta}{R_i} = (1 + f_d + f_p)^{1/2} - 1 \qquad (3.3.41)$$

where

$$f_d = \frac{\rho'}{\rho}\left(\frac{1 - \kappa^2}{2\kappa^2 \ln(1/\kappa)} - 1\right) \quad \text{and} \quad f_p = \frac{\rho'}{\rho}\Phi\frac{1}{2\kappa^2}\left[1 - \kappa^4 - \frac{(1 - \kappa^2)^2}{\ln(1/\kappa)}\right] \qquad (3.3.42)$$

and Φ is given in Eq. 3.3.17. For a specific die geometry (κ) and a given operating pressure drop (Φ) one may find δ' from Eq. 3.3.41. On the other hand, if the coating thickness is specified, as is more likely, Eqs. 3.3.41 and 42 can be rearranged to solve for the required pressure term in the form

$$\Phi = \frac{2\kappa^2}{1 - \kappa^4 - \frac{(1 - \kappa^2)^2}{\ln(1/\kappa)}}\frac{\rho}{\rho'}(\delta'^2 + 2\delta' - f_d) \qquad (3.3.43)$$

Figure 3.3.12, which shows this function, displays an important design feature that relates to the sensitivity of the coating thickness to the pressure and velocity fluctuations that always exist in real commercial systems. The nearly vertical character of the $\Phi(\delta')$ curves implies a small slope for $d\delta'/d\Phi$. This means that coating thickness is relatively insensitive to the parameter Φ, which (from its definition above) is seen to involve ΔP, μ, and V, three parameters that may vary (perhaps unintentionally) in a process. Thus the system is fairly insensitive to inadvertent process fluctuations; we can take advantage of this property. On the other hand, to change δ' for a specific die geometry, it is necessary to make a large (order of magnitude) change in Φ. We find, then, that imposing a pressure onto the drag flow gives a more flexible operating system.

Our model reveals still another feature of the performance of this system. Look at the curve in Fig. 3.3.12 for the parameter $\kappa = 0.5$. As we move along that curve in the direction of decreasing Φ we see, as expected, that the coating thickness decreases. But as Φ continues to decrease, this curve becomes asymptotically vertical, and so do the others for other values of κ. This means that as Φ goes toward very small values, a lower limit to the coating thickness is reached. In other words, it is impossible to achieve a coating thickness below a limiting value, even in the absence of a positive pressure. For the case $\kappa = 0.5$ we find that the lower limit to δ' is $\delta' = 0.471$; for $\kappa = 0.7$ the lower limit is $\delta' = 0.208$.

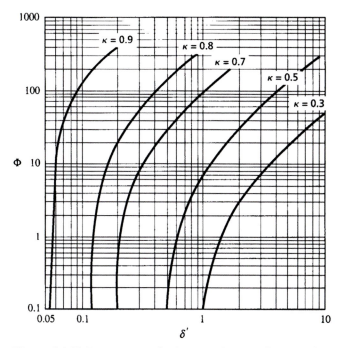

Figure 3.3.12 Pressure required to produce a given coating thickness.

But what happens if we impose a *negative* pressure on this system? We can show that if f_p is negative, Eq. 3.3.41 permits solutions for δ' values below the limits shown in Fig. 3.3.12. In theory, there is some value for a negative pressure that would prohibit the deposition of any coating onto the wire. In practice, if the liquid wets the wire surface, we would expect some residue of coating to appear on the wire, though it might be quite thin.

EXAMPLE 3.3.3 *Conditions to Achieve a Specific Coating Thickness*

Suppose the dimensions of a coating die are $R = 0.1$ cm, $R_i = 0.07$ cm, and $L = 1$ cm. The polymeric fluid is Newtonian and has a viscosity, measured at die conditions, of 10 poise. The required coating thickness is $\delta = 0.021$ cm. The wire speed is 100 ft/s. (Note the mixed units here. Such combinations are common in engineering problems.)

We find $\delta' = \delta/R_i = 0.021/0.07 = 0.30$, and $\kappa = R_i/R = 0.7$. Then from Fig. 3.3.12 we find

$$\Phi = 7.5 = \frac{\Delta P R^2}{4\mu V L} \tag{3.3.44}$$

Finally, we solve for ΔP (remember to convert V to centimeters per second) and find $\Delta P = 90 \times 10^6$ dyn/cm$^2 \approx 1300$ psi.

EXAMPLE 3.3.4 *Maximum Shear Stress in a Wire Coating Die*

In polymeric coating systems it sometimes happens that a critical shear stress is exceeded within the die, preventing the polymer from coating smoothly onto the wire. Instead, a very rough surface is observed. This condition is to be avoided if possible. Hence it is important to be able to calculate the maximum shear stress to which the fluid is subjected in the die flow. A model for the velocity profile in this type of flow will enable us to make such calculations.

For the case of isothermal Newtonian flow, the velocity profile can be shown (see Problem 3.50) to be the sum of the pressure and drag flow solutions obtained (Eqs. 3.3.26 and 3.3.31). Hence we begin with

$$v_z(r) = \frac{\Delta P R^2}{4\mu L}\left[(1 - s^2) - (1 - \kappa^2)\frac{\ln s}{\ln \kappa}\right] + V\frac{\ln s}{\ln \kappa} \tag{3.3.45}$$

We define the shear rate as

$$\dot{\gamma} = \frac{\partial v_z(r)}{\partial r} = \frac{\Delta P R}{4\mu L}\left[-2s - \frac{1 - \kappa^2}{s\ln \kappa}\right] + \frac{1}{R}\frac{V}{s\ln \kappa} \tag{3.3.46}$$

and a dimensionless shear rate is

$$\dot{\gamma}^* = \frac{R}{V}\dot{\gamma} = -\Phi\left[2s + \frac{1 - \kappa^2}{s\ln \kappa}\right] + \frac{1}{s\ln \kappa} \tag{3.3.47}$$

Figure 3.3.13 shows the profile of the dimensionless shear rate across the gap between the wire surface and the die surface. For small values of Φ, the flow is close to pure drag flow, and the shear rate is nearly uniform across the gap. For large values of Φ, the velocity profile is more nearly parabolic; there is a maximum velocity within the gap, and the shear rate vanishes at the point of maximum velocity. In particular, the shear rate changes sign across the gap.

For the data given in Ex. 3.3.3 we can calculate the shear rate profile. We find that the shear rate achieves its maximum *magnitude* on the die surface $s = 1$, and the value is

$$\dot{\gamma}^* = \frac{R}{V}\dot{\gamma} = -7.5\left[2 + \frac{1 - 0.7^2}{\ln 0.7}\right] + \frac{1}{\ln 0.7} = -7.08 \tag{3.3.48}$$

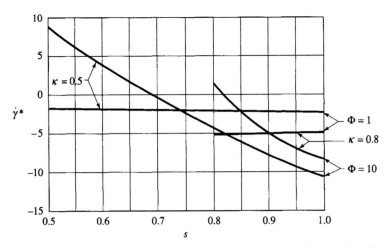

Figure 3.3.13 Dimensionless shear rate profiles across the wire coating die.

It follows that the absolute magnitude of the shear rate at the die surface is

$$\dot{\gamma} = \frac{V}{R}\dot{\gamma}^* = 7.08\frac{3048}{0.1} = 2.2 \times 10^5\,\mathrm{s}^{-1} \qquad (3.3.49)$$

and the magnitude of the shear stress is

$$\tau = \mu\dot{\gamma} = 10 \times 2.2 \times 10^5 = 2.2 \times 10^6\,\mathrm{dyn/cm}^2 \qquad (3.3.50)$$

3.4 THE VISCOSITY OF FLUIDS

We have been using the concept of viscosity throughout this chapter. Qualitatively, we imply that viscosity is a material property, related to the resistance a fluid exhibits to being deformed by the imposition of stresses. Alternatively we may think of viscosity as a measure of the stresses exerted by a fluid on the surrounding media when the fluid is undergoing deformation. We took a quantitative definition of viscosity in Eqs. 3.1.6 and 3.3.13, as well. In Chapter 4 we will show that these two definitions are compatible, that they look different because they refer to flows in different geometries, and that there is a more general definition of viscosity that does not depend on the geometry of the flow field.

Here we simply point out that the viscosity of a Newtonian fluid is a well-defined material property and that viscosity data for various fluids are available. For liquids there is an enormous range of values of viscosity. Among commonplace industrially important liquids, water is one of the least viscous liquids: at room temperature, it has a viscosity of approximately 1 centipoise, or, in SI units, 10^{-3} Pa · s. At the same temperature sulfuric acid has a viscosity of about 0.02 Pa · s, or 20 times that of water, while glycerol has a viscosity of about 1 Pa · s, or 1000 times that of water.

3.4.1 The Viscosity of Liquids

The viscosity of liquids is typically a strong function of temperature, with the viscosity decreasing as the temperature increases. Data for a few common liquids are presented in Fig. 3.4.1; note the form of plotting these data. It is observed that for many liquids, the temperature dependence of viscosity is of the form

$$\mu = Ae^{B/T} \qquad (3.4.1)$$

where A and B are characteristic constants for the liquid and T is the *absolute temperature in kelvins*. Hence the form of plotting in Fig. 3.4.1 yields nearly straight lines, which is an aid to accurate interpolation and to extrapolation over a limited range beyond the data.

It is difficult to predict the viscosity of a pure liquid from a small number of well-defined parameters such as the molecular weight of the liquid. There are empirical correlations available, many of them based on parameters that characterize the molecular structure. A detailed discussion of various methods of predicting viscosities is found in *The Properties of Gases and Liquids* by Reid, Prausnitz, and Poling (McGraw-Hill, 1987). A method given in that reference, which offers a compromise between accuracy and simplicity, is that of Morris, who suggested

$$\log\frac{\mu}{\mu^+} = J\left(\frac{1}{T_r} - 1\right) \qquad (3.4.2)$$

This model involves two parameters, a so-called pseudocritical viscosity μ^+, and a structure function J. The T_r is the "reduced temperature," which is the ratio T/T_c of

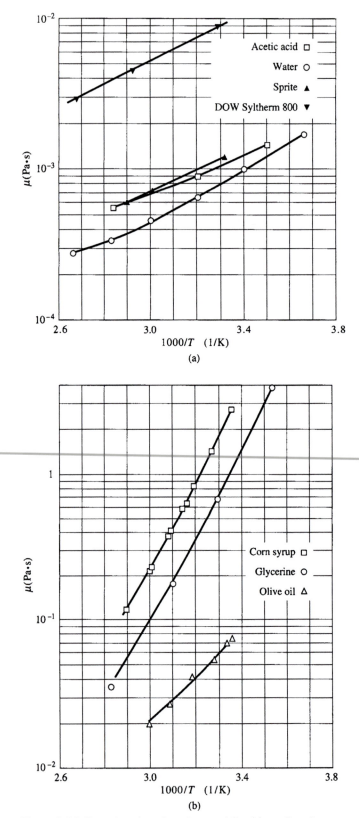

Figure 3.4.1 Data for viscosity of several liquids at $P = 1$ atm as a function of temperature: (a) low viscosity liquids and (b) high viscosity liquids.

Table 3.4.1 Pseudocritical Viscosity μ^+ for Some Common Liquids[a]

Type of liquid	μ^+ (Pa·s × 10^{-5})
Hydrocarbons	8.75
Halogenated hydrocarbons	14.8
Benzene derivatives	8.95
Halogenated benzene derivatives	12.3
Alcohols	8.19
Organic acids	11.7
Ethers, ketones, aldehydes, and acetates	9.6
Phenols	1.26

[a]Each table entry must be multiplied by 10^{-5}.

Source: Reid et al., 1987, Table 9.8.

the temperature to the critical temperature of the liquid (both are absolute temperatures, in kelvins). Thus it is necessary to have data for T_c for various liquids. Table 3.4.1 gives some selected values for μ^+.

The structure function J is calculated from

$$J = \left[0.0577 + \sum_i b_i n_i \right]^{1/2} \tag{3.4.3}$$

where the b_i are group contributions given in Table 3.4.2 and n_i is the number of times that group appears in the molecule.

With an accompanying table of critical temperature data (see Appendix A of Reid *et al.* Some T_c values are given later in Table 3.4.3), one may estimate liquid viscosities by this method. The average error in using this method for about 40 common organic liquids is about 15%, but errors as large as 50–70% are observed for some liquids.

Table 3.4.2 Group Contributions to J

Group	b_i	Group	b_i
CH_3, CH_2, CH	0.0825	Additional H in ring	0.1446
Halogen-substituted CH_3	0	CH_3, CH_2, CH adjoining ring	0.0520
Halogen-substituted CH_2	0.0893	NH_2 adjoining ring	0.7645
Halogen-substituted CH	0.0667	F, Cl adjoining ring	0
Halogen-substituted C	0	OH for alcohols	2.0446
Br	0.2058	COOH for acids	0.8896
Cl	0.147	C=O for ketones	0.3217
F	0.1344	O=C—O for acetates	0.4369
I	0.1908	OH for phenols	3.4420
Double bond	−0.0742	—O— for ethers	0.1090
C_6H_4 benzene ring	0.3558		

Source: Reid et al., 1987, Table 9.10.

EXAMPLE 3.4.1   *Viscosity of Liquid Propane*

We will use the method of Morris to calculate the viscosity of liquid propane at $-150°C$. Propane is a three-carbon alkane: C_3H_8. Its critical temperature is 369.8 K. At $-150°C$ the reduced temperature is

$$\frac{T}{T_c} = \frac{273 - 150}{369.8} = 0.333 \tag{3.4.4}$$

Since propane is a hydrocarbon we have, from Table 3.4.1, $\mu^+ = 8.75 \times 10^{-5}$ Pa·s. Propane has two CH_3 groups and one CH_2 group. From Eq. 3.4.3 and Table 3.4.2 we find

$$J = [0.0577 + 3 \times 0.0825]^{1/2} = 0.552 \tag{3.4.5}$$

Then Eq. 3.4.2 gives us

$$\log \frac{\mu}{8.75 \times 10^{-5}} = 0.552\,(3 - 1) = 1.104 \tag{3.4.6}$$

and we find

$$\mu = 1.11 \times 10^{-3}\,\text{Pa·s} \tag{3.4.7}$$

The measured value [Swift et al., *AIChE J.*, **5**, 98 (1959)] is 1.34×10^{-3} Pa·s, so our prediction is only 20% low.

3.4.2   The Viscosity of Gases

For pure gases, at least at low pressures, or not near the critical pressure, the kinetic theory of gases provides a very useful and accurate method of predicting viscosities. It is based on the so-called Chapman–Enskog theory, and it takes the form

$$\mu = 2.67 \times 10^{-6} \frac{(MT)^{1/2}}{\sigma^2 \Omega_v}\,\text{Pa·s} \tag{3.4.8}$$

where M and T are molecular weight and absolute temperature (K). The parameters σ and Ω_v have molecular interpretations, but they are actually determined by fitting Eq. 3.4.8 to viscosity data. The "collision diameter," σ is sometimes referred to as the molecular diameter.[6] We use angstrom units for σ in Eq. 3.4.8 (1 Å $= 10^{-10}$ m). To estimate Ω_v, which is called the "collision integral," we need one more empirical parameter, the "characteristic energy" ε. Tables of σ and ε are available (see, e.g., Reid et al.). Table 3.4.3 presents an abbreviated set of these data. With a value for ε, one calculates a dimensionless temperature T^* as

$$T^* = \frac{kT}{\varepsilon} \tag{3.4.9}$$

where k is Boltzmann's constant ($= 1.38 \times 10^{-23}$ N·m/K). Figure 3.4.2 plots Ω_v versus T^*.

---

[6] Recall that we used the symbol σ for surface tension in Chapter 2. If you pay attention to the context of the discussion you should have no trouble figuring out which of the two meanings applies.

Table 3.4.3 Parameters for the Chapman–Enskog Equation
(Eq. 3.4.8) and T_c Values

Compound		σ (Å)	ε/k (K)	M	T_c(K)
Air		3.71	78.6	29	132
Carbon tetrachloride	CCl_4	5.95	323	154	556
Chloroform	$CHCl_3$	5.39	340	119	536
Methanol	CH_3OH	3.63	482	32	512.6
Methane	CH_4	3.76	149	16	191
Carbon monoxide	CO	3.69	91.7	28	133
Carbon dioxide	CO_2	3.94	195	44	304
Ethane	C_2H_6	4.44	216	30	305
Propane	C_3H_8	5.12	237	44	370
n-Butane	C_4H_{10}	4.69	531	58	408
Acetone	CH_3COCH_3	4.6	560	58	508
Benzene	C_6H_6	5.35	412	78	562
Hydrogen	H_2	2.83	59.7	2	33.2
Water	H_2O	2.64	809	18	647
Ammonia	NH_3	2.9	558	17	405.6
Nitrogen	N_2	3.8	71.4	28	126.2
Oxygen	O_2	3.47	107	32	154.6
Silane	SiH_4	4.08	208	18	269.5
Helium	He	2.58	10.2	4	5.2

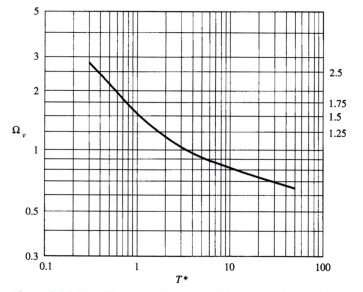

Figure 3.4.2 The Chapman–Enskog collision integral as a function
of T^*.

EXAMPLE 3.4.2   *Viscosity of Gaseous Ammonia*

We will use the Chapman–Enskog theory to calculate the viscosity of ammonia at 220°C and atmospheric pressure, at which conditions it is a gas. We need the following parameters for NH_3:

$$\sigma = 2.9 \text{ Å} \qquad \varepsilon/k = 558 \text{ K} \qquad M = 17$$

Then $T^* = (220 + 273)/558 = 0.884$. From Fig. 3.4.2 we find $\Omega_v = 1.7$. From Eq. 3.4.8, then,

$$\mu = 2.67 \times 10^{-6} \frac{(17 \times 493)^{1/2}}{2.9^2 \times 1.7} = 1.71 \times 10^{-5} \text{ Pa} \cdot \text{s} \tag{3.4.10}$$

The reported value (Reid *et al.*) is 1.69×10^{-5} Pa·s, in very good agreement with this estimate.

From inspection of Eq. 3.4.8 we can expect the temperature dependence of the viscosity of gases to be quite different from that of liquids. First, we see that the viscosity *increases* with increasing temperature—just opposite to the behavior of liquids. Second, the dependence is approximately a "power law," rather than an exponential, function. From Fig. 3.4.2 we see that except for small values of T^*, the Ω_v function is a weak function of temperature—approximately $T^{-1/2}$. Hence the viscosity of gases should increase approximately linearly with absolute temperature. Furthermore, as Eq. 3.4.10 suggests, the viscosity is independent of pressure. This latter point holds as long as the gas is well below its critical pressure. Some typical data are plotted in Fig. 3.4.3, showing the extent to which this behavior is observed. (See Problem 3.67.)

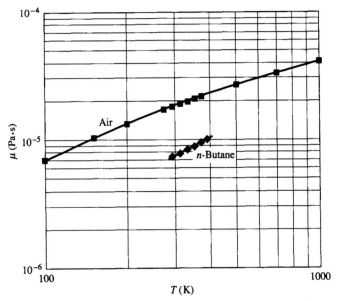

Figure 3.4.3 Gas viscosity as a function of temperature, at $P = 1$ atm.

3.4.3 The Design of a System for Measurement of the Viscosity of Gases

We described a capillary viscometer in Section 3.3. Could we use such a system for the measurement of the viscosity of gases? The primary issue arises from an obvious property of gases: the viscosity is so small that we might expect to need a very long capillary to yield a pressure drop we could measure with any precision. We shall see that there is a second issue—namely, gases are *compressible*; but our model for the capillary viscometer (Poiseuille's law) was derived using the assumption that the fluid is incompressible.

Many processes of concern to us involve gas flows. If the linear velocities are small, we do not ordinarily worry about compressibility phenomena such as shock waves that we associate with Mach numbers[7] much larger than unity. Nevertheless, the variation of the fluid density with pressure must be accounted for, as we will see.

We begin with an analysis of *viscous* flow of a compressible fluid, to reveal the effect of compressibility on the pressure drop–flowrate relationship. We select as the geometry a long tube of circular cross section. For an *incompressible* fluid of constant viscosity in such a geometry, we have the classical solution for laminar viscous flow, the Hagen–Poiseuille law (Eq. 3.2.34), which we write in the form

$$q = -\frac{\pi D^4}{128\mu}\frac{dp}{dz} \tag{3.4.11}$$

In this equation we are asserting that the *local* volumetric flowrate q at some point z along the axis of the tube is proportional to the *local* pressure gradient dp/dz, in the form of Poiseuille's law. We use what we call "local" values because we know that for a compressible fluid, the volumetric flowrate will change as the absolute pressure changes along the axis of the tube. In a *compressible* fluid, however, we know that since the *mass* flowrate is constant, it is the product

$$\dot{m} = \rho q \tag{3.4.12}$$

that is constant along the tube axis. Since the fluid density is a function of pressure, through an equation of state such as the ideal gas law, then, *at constant temperature,*

$$\rho = \frac{\rho_0}{p_0}p \tag{3.4.13}$$

where the subscripts refer to some specific point along the tube where the pressure and some density are known. Clearly then, if p varies along the tube axis, ρ, and so q, must also vary along the axis.

We may derive an approximate model of *compressible* laminar viscous flow by assuming that the Hagen–Poiseuille law is valid *locally* (Eq. 3.4.11), and upon introducing the constant mass flowrate we find

$$\frac{\dot{m}}{\rho(z)} = q(z) = -\frac{\pi D^4}{128\mu}\frac{dp}{dz} = \frac{\dot{m}p_0}{\rho_0 p(z)} \tag{3.4.14}$$

at any axial position z, even if the fluid is compressible.

This gives us a differential equation for $p(z)$, the solution of which is

$$p^2 = p_0^2 - \frac{2\dot{m}p_0}{\rho_0 K}z \tag{3.4.15}$$

---

[7] The Mach number is the ratio of the velocity to the speed of sound in the gas at the temperature of the flow. The speed of sound, in turn, is essentially independent of gas pressure for an ideal gas. For air at 273 K, the speed of sound is 3.3×10^4 cm/s.

where

$$K = \frac{\pi D^4}{128\mu} \qquad \text{(3.4.16)}$$

We have used as a boundary condition the definition that at the tube entrance

$$p = p_0 \qquad \text{at} \quad z = 0 \qquad \text{(3.4.17)}$$

At the downstream end of the tube, we define the pressure as

$$p = p_L \qquad \text{at} \quad z = L \qquad \text{(3.4.18)}$$

From Eq. 3.4.15, it follows then that the *volumetric* flowrate at the tube entrance is

$$q_0 = \frac{\dot{m}}{\rho_0} = \frac{K}{2L}\frac{p_0^2 - p_L^2}{p_0} \qquad \text{(3.4.19)}$$

and the *mass* flowrate (which is constant) is[8]

$$\dot{m} = \frac{\rho_0 K}{L}\left(\frac{p_0 + p_L}{2}\right)\left(\frac{p_0 - p_L}{p_0}\right) \qquad \text{(3.4.20)}$$

Note that q_0 and $\dot{m}$ are both flowrates; $\dot{m}$ is a *mass* flowrate, and it does not vary along the tube axis. By contrast, q_0 is the *volumetric* flowrate at the tube entrance ($z = 0$).

We may now use Eq. 3.4.20 as a design equation for a capillary viscometer for the study of gases.

Design Specifications for a Capillary Viscometer for Gases

In this example we will apply the information we have developed to this point to get an idea of how analysis can be integrated with design. Our goal is to design a simple viscometer for the measurement of the viscosity of gases. We have several "working equations" that constrain the design. First, we will use Eq. 3.4.20 as the basic flow model. Hence we are assuming that the flow will be maintained isothermal and that the Reynolds number will be kept in the laminar range (Re < 2000).

The design concept is sketched in Fig. 3.4.4. The gas of interest is maintained at constant pressure in a reservoir. A long section of capillary tubing will serve as the "working section" of the viscometer. A means is provided for measuring the pressure p_0 at the entrance to the tubing section. The tubing is immersed in a constant temperature water bath. Flowrate will be measured volumetrically by permitting the flow from the tube exit to bubble through the water bath into an inverted graduated cylinder. The time required to displace a specific volume of water from the cylinder can be converted to the volumetric flowrate. Since volume is a function of pressure (and temperature) in a gas, we need to know the pressure at which the gas is collected. If the head of displaced water H (see Fig. 3.4.4) is kept small, say less than 10 cm, then the pressure in the gas collection cylinder is within a few percent of atmospheric pressure, since

$$\frac{\rho_{\text{water}} g H}{P_{\text{atm}}} \leq \frac{1000 \text{ kg/m}^3 \times 9.8 \text{ m/s}^2 \times 0.1 \text{ m}}{10^5 \text{ N/m}^2} \leq 0.01 \qquad \text{(3.4.21)}$$

---

[8] Using the ideal gas law, we may write this in the form $\dot{m} = (K/L)\,\bar{\rho}\,\Delta p$, which is identical to Poiseuille's law (Eq. 3.2.24) except that the density is evaluated at the mean pressure.

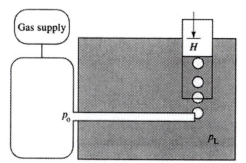

Figure 3.4.4 Design concept for a capillary viscometer for gases.

Hence from Eq. 3.4.19, the expected flowrate into the collection cylinder is

$$q_L = \frac{K}{2L} \frac{p_0^2 - p_L^2}{p_0} \frac{p_0}{p_L} = \frac{K}{2L} \frac{p_0^2 - p_L^2}{p_L} \qquad (3.4.22)$$

Note that we have used the ideal gas law to "correct" the entrance flow q_0 to the exit flow q_L. Note also that when we use the gas law, we must use *absolute,* not gage, pressures—we cannot take atmospheric pressure as a zero pressure in this kind of calculation.

In this design problem a number of operating and design parameters are unspecified. We are free to make choices for some of these. We have to take care not to select parameters that violate any constraints (such as laminar flow) or lead to poor operating conditions (e.g., a gas flow of a million cubic meters per second!). One way to proceed is to make some arbitrary but reasonable choices, and then determine whether the operating conditions are indeed reasonable. Unless we are very good at guessing the most convenient sets of parameters, this approach can involve a lot of trial and error. An alternative, which we illustrate here, is to look at the range of possible designs in a "parameter space" of key variables. One variable we can control, and may expect to exert a strong effect on the performance of the system, is the diameter of the capillary tubing, D. We call D an "independent variable," in this case. The key "dependent variable" is the measured volumetric flowrate q_L. We can cast some of the equations that define the system in terms of these two (and other) variables.

Before we write down equations, though, let's think about the pressures that drive the flow. The downstream (exit) pressure will be taken as $p_L = 1$ atm $= 10^5$ N/m^2. We will choose an upstream (entrance) pressure of twice this value: $p_0 = 2$ atm $= 2 \times 10^5$ N/m^2. This yields a pressure difference that is easily measured. Another factor in this choice comes from thermodynamic considerations. If the gas expands too much in passing from the pressure p_0 to the exit, the rate of heat transfer required to maintain the assumed isothermal flow may not be achievable without special precautions. (This point is dealt with in detail much later, in Chapter 11.) By keeping the pressure ratio small, we can avoid this problem. Hence a preliminary design specification names an upstream pressure of 2 atm as our operating pressure.

With the foregoing points in mind, and making note of the definition of K (Eq. 3.4.16), we may write one of the design equations in the form

$$q_L = \frac{\pi D^4}{256 \mu L} \frac{p_0^2 - p_L^2}{p_L} \qquad (3.4.23)$$

To design this system, we need an estimate for the magnitude of the viscosities that will be measured. We will use the viscosity of air at room temperature ($\mu_{air} \approx 2 \times 10^{-5}$ Pa·s; see Fig. 3.4.3) for this purpose. Then Eq. 3.4.23 can be written, incorporating μ_{air} and our choices for the pressures, as follows

$$q_L = 1.84 \times 10^8 \frac{D^4}{L} \tag{3.4.24}$$

(All units are SI.) Let's proceed to plot this equation on a q_L–D parameter space, with L as a parameter to be determined later. The result is shown in Fig. 3.4.5, where each line corresponds to a specific choice of the tubing length L.

As noted earlier, a constraint on the design is the need to keep the Reynolds number in the laminar regime. Since

$$Re = \frac{4q\rho}{\pi D \mu} \tag{3.4.25}$$

we may write another relationship between q and D in the form

$$q_L = \frac{\pi D \mu}{4 \rho L} Re = 1.3 \times 10^{-5} \, Re \, D \tag{3.4.26}$$

Lines are shown on Fig. 3.4.5 for several values of Re well into the laminar regime.

If we are going to measure the gas flowrate, we must think of the range of values that can be conveniently measured. We probably will want flowrates larger than 0.01

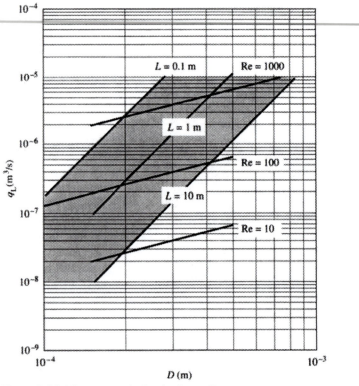

Figure 3.4.5 Viscometer design in the q–D parameter space.

mL/s (10^{-8} m³/s) but no larger than 10 mL/s (10^{-5} m³/s). Hence we will aim for a design that yields flowrates in that range. It would be difficult to do any "plumbing" with a capillary length less than 10 cm, and equally difficult to cope with a length in excess of 10 m. Hence there are some practical limitations to capillary length. Finally, it would be difficult to work with tubing much smaller than $D = 100$ μm (10^{-4} m).

All these constraints are approximate, but they suggest that in a q–D space, the best design would lie in the region shaded in Fig. 3.4.5. Let's pick a point in that space and evaluate the design.

Suppose

$$D = 150\ \mu m\ (1.5 \times 10^{-4}\ m) \qquad \text{and} \qquad L = 10\ m$$

The Reynolds number is clearly in the laminar regime (Re $\approx$ 7). The flowrate that we will have to measure is about

$$q_L = 10^{-8}\ m^3/s\ (0.01\ mL/s)$$

This seems to be a very low flowrate, but since we would be able to collect 1 mL in only 100 s, it might be acceptable given a precisely graduated collection vessel.

Let's think about another aspect of this design. Bubbles are being dispersed from the tip of the capillary. We should examine two features of the bubble formation: What is the bubble size, and does the bubble formation process give rise to a measurable back pressure on the system?

Data are available for the formation of bubbles from small capillaries in water. The results for slow bubble formation are presented in Fig. P2.30 in the Homework Problems section of Chapter 2. We see that at a capillary diameter of 150 μm ($R_c = 75$ μm) the expected bubble radius is $R_b = 940$ μm. This corresponds to a bubble volume of

$$V_b = \frac{4\pi}{3} R_b^3 = \frac{4\pi}{3} (940 \times 10^{-6})^3 = 3.5 \times 10^{-9}\ m^3 \qquad \textbf{(3.4.27)}$$

With $q_L = 10^{-8}$ m³/s, this yields about three bubbles per second forming at the capillary tip. The reference cited for Fig. P2.30 suggests that this is a moderately large bubble formation rate. Nevertheless, we will use this value as a guide to the expected performance of the system.

The second issue, back pressure, is assessed by recalling that surface tension can increase the pressure inside a small bubble over that of the surrounding fluid according to

$$\Delta p = \frac{2\sigma}{R} \qquad \textbf{(3.4.28)}$$

What radius of curvature do we use? The maximum bubble pressure occurs when the bubble has the minimum radius. This corresponds to the case of a bubble that has just reached a hemispherical shape, so that $R = R_c$. Then

$$\Delta p = \frac{2\sigma}{R_c} = \frac{2 \times 0.072}{75 \times 10^{-6}} = 1900\ N/m^2 \qquad \textbf{(3.4.29)}$$

This is only 2% of the exit pressure, so we will not expect any problem from the periodic back pressure that will accompany the bubble formation process.

With Fig. 3.4.5 as a guide, we could examine other designs. We see rather quickly, however, that if we reduce the capillary length L without reducing the capillary diameter (which is already marginal), we will have to accept very high bubbling rates, which may make the flowrate measurement uncertain. A characteristic of engineering design is the continual iterative approach to finding a set of design parameters that yield performance in an acceptable range. In this case, we might very well decide that the design concept or a gas viscometer investigated in this example is not worth pursuing.

3.5   HYDROSTATICS AND BODY FORCES: ANOTHER LOOK

Newcomers to fluid dynamics often have some conceptual difficulties with the distinction between the pressure associated with a hydrostatic head and the effect of a body force (usually gravity) on the dynamics. In hydro*statics* the relationship is one-to-one. The gravitational field (the body force) gives rise to the hydrostatic pressure field. In hydro*dynamics,* by distinction, the situation is not always so clear. This section presents two examples designed to clarify the issue.

EXAMPLE 3.5.1   *Motion of a Planar Sheet Through a Submerged Restriction*

Consider the flow sketched in Fig. 3.5.1. A pair of parallel wide planes of length L in the x direction are completely immersed in a large body of viscous liquid. Another wide, very long planar sheet moves along the midplane of this system, at constant speed V parallel to the two surrounding planes. We want to answer two questions:

1. What is the velocity profile in the region $a < y < b$ on either side of the moving sheet?
2. Does gravity have any effect on this flow?

We must always be careful in defining the system we wish to model. In the case at hand, we will assume that the surrounding liquid is motionless initially. Then the inner plane is set in motion at the constant speed V. After some time a steady state flow will be set up in the region $a < y < b, 0 < x < L$. Outside the confining planes the liquid in the reservoir will circulate, if the reservoir is not infinite, since fluid dragged out of the lower end of the system must return to enter the top $(x = 0)$. If the surrounding reservoir is quite large compared to the region $a < y < b, 0 < x < L$, we can expect the liquid to be nearly still throughout most of the reservoir. Obviously, then, there will be some transition of the flow as it crosses the planes $x = 0$ and $x = L$, from (or to) a nearly stagnant region to (or from) a dynamic region.

We will begin the development of a model by neglecting the flow near the entrance and exit of the system. A control volume in Cartesian coordinates may be set up within

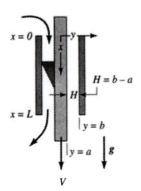

Figure 3.5.1 Flow between parallel planes.

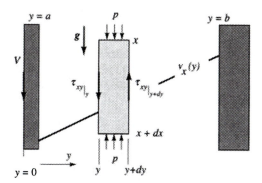

Figure 3.5.2 Control volume for momentum balance.

the flow, as sketched in Fig. 3.5.2. We assume that as long as our control volume is not near the entrance and exit, the velocity field corresponds to unidirectional laminar (fully developed)[9] flow. This implies that the velocity may be written in the form

$$\mathbf{v}(y) = (v_x(y), 0, 0) \tag{3.5.1}$$

With this assumption, the shear stress τ_{xy} is a function of the y coordinate only, since Newton's law of viscosity, in Cartesian coordinates, would be

$$\tau_{xy}(y) = -\mu \frac{dv_x(y)}{dy} \tag{3.5.2}$$

Now let's write a momentum balance in the direction of motion, x. The volume element has dimensions dx, dy, as shown in Fig. 3.5.2, and dz in the direction normal to the page. We assume that there is no flow, and no forces act, in the z direction. That is, we are assuming a *two*-dimensional flow in the sense that the velocity is only in the x direction and varies only in the y direction.

The net pressure force in the x direction is

$$F_{px} = p_x dy\, dz - p_{x+dx} dy\, dz \tag{3.5.3}$$

The shear stresses on the two vertical sides of the volume element give rise to forces that sum to

$$F_{\tau x} = \tau_{xy_x} dx\, dz - \tau_{xy_{x+dx}} dx\, dz \tag{3.5.4}$$

The signs of these two terms are consistent with the sign convention introduced following the presentation of Eqs. 3.2.3 and 3.2.4.

Since v_x is not a function of the x coordinate, the flow of momentum into the volume element exactly balances the flow of momentum out. Thus we do not write a rate of change of momentum term in this force balance.

Finally, there is a body force, given by

$$F_g = \rho g\, dx\, dy\, dz \tag{3.5.5}$$

---

[9] We learned in Section 3.2.1 that unidirectional laminar flow implies fully developed flow. Hence we put "fully developed" in parentheses to make clear that this modifier is really not a separate assumption about the flow.

Putting these terms together in the statement that the sum of all these forces vanishes [since there is no net flow of momentum into (or out of) the volume element], we find

$$p_x \, dy \, dz - p_{x+dx} \, dy \, dz + \tau_{xy_x} dx \, dz - \tau_{xy_{x+dx}} dx \, dz + \rho g \, dx \, dy \, dz = 0 \qquad (3.5.6)$$

Now we divide this equation by the volume $dx \, dy \, dz$ and take the limit as the volume shrinks to a point within the fluid. The result is

$$-\frac{\partial p}{\partial x} + \rho g = \frac{\partial \tau_{xy}}{\partial y} \qquad (3.5.7)$$

Consistent with the assumptions already described, we may separate this equation into two equations, since the left-hand side is a function of x only, and the right-hand side is a function of y only:

$$-\frac{dp}{dx} + \rho g = C \qquad (3.5.8a)$$

$$\frac{d\tau_{xy}}{dy} = C \qquad (3.5.8b)$$

The constant C may be obtained from boundary conditions on the pressure distribution.

What physical statements can we make about pressure at the boundaries? Figure 3.5.3 may be of some help as we argue for an approximation regarding pressure in this system. We want to know what we can say about the pressures at the entrance and exit planes $x = 0$ and $x = L$. In the liquid *far outside* the neighborhood of the system, the flow may be assumed to be (nearly) static if the reservoir is large relative to the volume of liquid in the region $0 < x < L$, $a < y < b$. Under such static conditions, the pressures in the planes $x = 0$ and $x = L$ would differ by a hydrostatic term, $\rho g L$. Because of entrance effects that occur as the liquid is dragged into the narrow restriction at $x = 0$, the pressure p_0 will not be equal to the static pressure at the same level, P_{stat}. Unfortunately, we have no basis for calculating this entrance pressure effect exactly. In the absence of a viable alternative, let's see what happens when we assume that entrance effects are unimportant. We write the boundary conditions on pressure as

$$p_0 = P_{\text{stat}} \qquad \text{at} \quad x = 0 \qquad (3.5.9)$$

and

$$p_L = P_{\text{stat}} + \rho g L \qquad \text{at} \quad x = L \qquad (3.5.10)$$

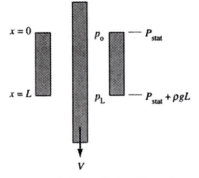

Figure 3.5.3 Detail for discussion of boundary conditions on pressure.

The solution for $p(x)$ from Eq. 3.5.8a is

$$p(x) = \rho g x - Cx + D \qquad (3.5.11)$$

and Eqs. 3.5.9 and 3.5.10 are satisfied only if

$$C = 0 \qquad (3.5.12)$$

Now we can return to the equation for the stress field, Eq. 3.5.8b, and find, since $C = 0$,

$$\tau_{xy} = \text{constant} = d \qquad (3.5.13)$$

Hence the shear stress is constant across the liquid. With the aid of Newton's law of viscosity (Eq. 3.5.2), we can solve for the velocity profile that satisfies no-slip conditions on the solid planar boundaries. After a little algebra, the result is easily found to be

$$\frac{v_x(y)}{V} = \frac{b - y}{b - a} \qquad (3.5.14)$$

What is most interesting about this result is the absence of gravity from the model. We would find exactly the same flow field if the motion were horizontal. Hence we are tempted to conclude that gravity plays no role in altering the nature of the flow field—rather, we might suggest, it simply affects the pressure distribution. This conclusion is misleading, as the next example demonstrates.

EXAMPLE 3.5.2   *Motion of a Wetted Planar Sheet Through a Restriction*

We look now at a slightly different flow problem, depicted in Fig. 3.5.4. The moving sheet is coated with a thin film of viscous liquid, and its thickness is such that the region $a < y < b, 0 < x < L$ is always filled with the liquid. Otherwise, the system shown is surrounded by air at some uniform ambient pressure. We could again make a momentum balance on a control volume within the restricted region, but there is nothing different between the two systems—in Figs. 3.5.4 and Fig. 3.5.1 that would change the bookkeeping on forces. Hence Eq. 3.5.7 would be obtained for this system as well, and Eqs. 3.5.8 and 3.5.11 would still follow. The difference, if there is one, will lie in the boundary

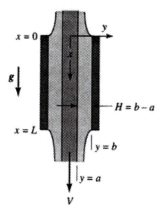

Figure 3.5.4 A long wetted planar sheet moves through a parallel restriction.

conditions on pressure, since clearly there is no difference between these two systems with respect to boundary conditions on velocity.

Again, one could anticipate entrance and exit effects at the planes $x = 0$ and $x = L$. Additionally, in this case, and depending on the length scales involved, surface tension might affect the pressures in these regions through the curvatures of the entrance and exit interfaces. But as in the preceding case, we have no firm basis for estimating the magnitude of these effects. Hence we ignore them, to produce a model that could be subjected to experimental verification. Therefore, the boundary conditions on pressure become

$$p = P_{atm} \quad \text{at} \quad x = 0 \quad \text{and} \quad x = L \tag{3.5.15}$$

because *both* ends of the system are exposed to the ambient air, and there is no significant hydrostatic pressure difference over a length L in air. Equation 3.5.11 can satisfy these boundary conditions only if

$$C = \rho g \tag{3.5.16}$$

Now Eq. 3.5.8b leads to

$$\tau_{xy} = Cy + E = \rho g y + E \tag{3.5.17}$$

Introduction of the Newtonian viscosity model, followed by integration and application of the no-slip boundary conditions, leads to the velocity profile in the form

$$\frac{v_x(y)}{V} = \frac{b - y}{b - a} + \frac{\rho g}{2\mu V}[b^2 - y^2 - (a + b)(b - y)] \tag{3.5.18}$$

We see that in this case gravity does affect the velocity profile. The first term on the right-hand side is the solution given earlier, as Eq. 3.5.14. It is instructive to rewrite Eq. 3.5.18 in the form

$$\frac{v_x(y)}{V} = \frac{b - y}{b - a}\left[1 + \frac{\rho g(a - b)^2}{2\mu V}\left(\frac{y - a}{b - a}\right)\right] \tag{3.5.19}$$

and look at the term

$$\beta = \frac{\rho g(b - a)^2}{2\mu V} = \frac{\rho g(b - a)WL}{[2\mu V/(b - a)]\, WL} \tag{3.5.20}$$

The numerator of β is (half) the weight of the liquid within the restricted region. The denominator is approximately equal to the shear force acting on each side of the moving plane. Hence we conclude that if the weight of the confined liquid is very small compared to the shear force (i.e., $\beta \ll 1$), the velocity profile is very close to that of the fully submerged case, and gravity does not alter the flow profile.

We conclude from Examples 3.5.1 and 3.5.2 that the effect of gravity in a vertical flow depends on the pressure boundary conditions, which themselves reflect the physics in the *surrounding* medium. For a fully submerged system (the first case), the body force is effectively canceled by the (adverse) hydrostatic head. In the second case there is only a body force, which aids the downward directed flow, and there is no hydrostatic pressure to counter it.

3.6 MOLECULAR FLOW

When gases flow through tubes at *ultralow pressures,* viscous effects no longer occur as they do at "normal" pressures. This is because at very low pressure the mean free path of the molecules in a gas may be comparable to the dimensions of a tube or orifice. For example, at one millitorr (1 mtorr: approximately 1 atm $\times$ 10^{-6}), the mean free path in O_2, N_2, or A is about 5 cm at 300 K, and about 10 and 15 cm, respectively, in H_2 or He. To a good approximation, the mean free path is proportional to absolute temperature and inversely proportional to pressure.

An important characteristic parameter of such a low pressure flow is the ratio of the mean free path of the molecules to the separation between confining surfaces (such as a tube diameter). When this ratio (called the Knudsen number) is greater than unity, momentum transfer is dominated by wall collisions rather than by collisions among the molecules in the gas. This regime of flow is usually referred to as *molecular flow*, or *Knudsen flow*.

In the molecular flow regime, the relationship of flow rate to pressure difference is quite distinct from that of the viscous flow regime. For example, the mass flow through a long circular tube is given theoretically by Knudsen as

$$\dot{m} = \frac{\pi}{12} \frac{D^3}{L} \frac{p_0 - p_L}{p_0} \left(\frac{8R_G T}{\pi M_w}\right)^{1/2} \rho_0 \tag{3.6.1}$$

where ρ_0 is the mass density measured at the upstream pressure p_0, and D and L are the tube diameter and length, respectively. This result is quite different from the corresponding *viscous* flow equation (Eq. 3.4.20). Note especially the changed dependence of flowrate on tube diameter.

For the gas constant R_G we use 8.314 J/K · mol, and it is understood that we use the gram-mole unit here. Hence for nitrogen, for example, $M_w = 0.028$ kg/mol. Alternatively, we can use $R_G = 8314$ J/K · (kg)mol, in which case for nitrogen, for example, $M_w = 28$ kg/(kg)mol. This latter choice is less likely to cause confusion.

The kinetic theory of gases gives the following expression for calculating the mean free path:

$$\lambda = \frac{\pi}{4} \frac{\mu \overline{v}_m}{\overline{p}} \tag{3.6.2}$$

and the mean molecular speed is given, again through the kinetic theory, as

$$\overline{v}_m = \left(\frac{8R_G T}{\pi M_w}\right)^{1/2} \tag{3.6.3}$$

EXAMPLE 3.6.1   *Characteristics of a Gas at Low Pressure*

For nitrogen at 26°C, find the mean molecular speed and the mean free path at a pressure of 1 mtorr. What is the Knudsen number, defined as the ratio of the mean free path to the tube diameter, for a tube of diameter 1 mm under these conditions?

First we will convert pressure to SI units. There are 760 torr at one atmosphere pressure. Hence[10]

$$1 \text{ mtorr} = \frac{0.001}{760} \times 10^5 \text{ Pa} = 0.132 \text{ Pa} \tag{3.6.4}$$

---

[10] Generally throughout this text we take one atmosphere to be 10^5 Pascals, rather than the exact value of 1.013×10^5 Pa.

Next, we find the mean molecular speed as

$$\overline{v}_m = \left(\frac{8 \times 8314 \times 299}{28\pi}\right)^{1/2} = 476 \text{ m/s} \tag{3.6.5}$$

Then λ follows as

$$\lambda = \frac{\pi}{4}\frac{\mu\overline{v}_m}{\overline{p}} = \frac{\pi}{4}\frac{(1.8 \times 10^{-5}) \times 476}{0.132} = 0.051 \text{ m} \tag{3.6.6}$$

Finally, the Knudsen number in a 1 mm diameter tube is

$$Kn = \frac{\lambda}{D} = \frac{0.05}{0.001} = 50 \tag{3.6.7}$$

We shall find that this is a sufficiently large Knudsen number to ensure that molecular (Knudsen) flow will prevail.

When the Knudsen number is neither small nor large, the flow rate is affected by both wall collisions and intermolecular collisions and neither Poiseuille's law for a compressible fluid (Eq. 3.4.20) nor Eq. 3.6.1 holds. Based on an examination of experimental data, an empirical relationship has been developed of the form

$$\frac{\dot{m}}{\dot{m}_p} = F(K') \tag{3.6.8}$$

where $\dot{m}_p$ is the flow given by Eq. 3.4.20, and

$$K' = \frac{\pi}{8}\frac{D}{\lambda} = \frac{D\overline{p}}{2\mu\overline{v}_m} \tag{3.6.9}$$

Thus, except for the coefficient $\pi/8$, we have an *inverse* Knudsen number. Figure 3.6.1

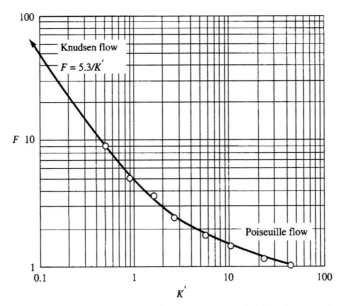

Figure 3.6.1 The function $F(K')$ recommended by Scott and Dullien [*AIChE J.*, **8**, 3 (1962)] compared to experimental data [Q] of Brown *et al.* [*J. Appl. Phys.*, **17**, 802 (1946)].

shows the recommended form of $F(K')$, with a comparison to some experimental data. $F(K')$ is calculated from

$$F = \left(\frac{1.144}{K'} - 1\right) \exp\left(\frac{-8K'}{\pi}\right) + \frac{4.189}{K'} + 1 \tag{3.6.10}$$

EXAMPLE 3.6.2   *Treatment of Some Data for a Gas Flow at Low Pressure*

Brown *et al.* [*J. Appl. Phys.*, **17**, 802 (1946)], who measured gas flow at very low pressures, reported the following. For air at 299 K flowing through a pipe of inside radius 1.3 cm and length 3.32 m, with upstream and downstream pressures of 18 and 8 μmHg, respectively, "the speed U in units of micron cu. ft. per sec. per micron" is 0.029. Is this measurement consistent with the equations presented above?

Before we can address the question, we must decipher the units (micron cu. ft. per sec. per micron) on U. What does this mean? These strange units arise from the literature of vacuum technology, which is greatly concerned with the amount of gas being pumped through a system, or in some cases the amount leaking from (or into) a system. The word "amount" is ambiguous. Volume is not precise unless we specify a pressure. Hence *volume* flowrate is not a useful measure of flow in compressible flow systems. "Mass flow" would make more sense. Historically, the vacuum flow literature adopted the convention of reporting flows in units proportional to *molecular* flow, rather than *mass* flow.

Let's begin by going back to the Poiseuille flow equation for a compressible fluid, Eq. 3.4.20, which we now write in the form

$$\dot{m}_p = \frac{\pi R^4}{8\mu L} \frac{\Delta p \, \bar{p}}{p_0} \rho_0 \tag{3.6.11}$$

We divide the mass flowrate by the density ρ_0, obtaining the volume flowrate at the pressure p_0:

$$q_0 = \frac{\dot{m}_p}{\rho_0} = \frac{\pi R^4}{8\mu L} \frac{\Delta p \, \bar{p}}{p_0} \tag{3.6.12}$$

If we multiply both sides of this equation by the pressure p_0, we find

$$p_0 q_0 = \frac{\pi R^4 \Delta p \, \bar{p}}{8\mu L} \tag{3.6.13}$$

Next we recognize that for an ideal gas

$$p_0 q_0 = \dot{n} R_G T \tag{3.6.14}$$

where $\dot{n}$ is the *molar* flowrate, which is independent of pressure but does depend on the absolute temperature of the flow. Hence $p_0 q_0$ is a measure of the flowrate that is independent of pressure—it is a measure of how many molecules (actually, moles) are flowing through the pipe or tube. The custom in the vacuum literature is to define a "specific" flowrate as this molecular flow per unit of pressure difference that is driving the flow, that is,

$$\frac{p_0 q_0}{\Delta p} = \frac{\pi R^4 \bar{p}}{8\mu L} \tag{3.6.15}$$

This quantity has the units of pressure times volume flowrate per pressure difference. While the pressure units cancel, the pressures are different: one is the upstream pressure and the other is the pressure difference. Hence the pressure units are reported, and the older vacuum technology literature, and often today's publications, as well, use microns [micrometers] of mercury as a pressure unit. This practice reflects the use in vacuum technology, 50 years ago, of pressure gauges equipped with mercury manometers in which head differences were measured in microns. Thus, in Brown et al., U is the quantity on the left-hand side of Eq. 3.6.15, with pressures in units of microns of mercury and volume flowrate reported in cubic feet per second.

Now we return to the data of the example. The mean pressure is 13 μmHg. In SI units this is (using the density of mercury)

$$\bar{p} = \rho_{Hg}gh = 13{,}600 \text{ kg/m}^3 \times 9.8 \text{ m/s}^2 \times 13 \times 10^{-6} \text{ m} = 1.73 \text{ Pa} \qquad \textbf{(3.6.16)}$$

We calculate the parameter K' (using the result for mean speed given in Eq. 3.6.5) as

$$K' = \frac{D\bar{p}}{2\mu\bar{v}_m} = \frac{0.026 \times 1.7}{2 \times (1.8 \times 10^{-5}) \times 476} = 2.6 \qquad \textbf{(3.6.17)}$$

From Eq. 3.6.10 we find $F = 2.6$. From Eq. 3.6.15

$$U \equiv \frac{p_0 q_0}{\Delta p} = \frac{\pi R^4 \bar{p}}{8\mu L} = \frac{\pi (0.013)^4 \, 1.73}{8 \times (1.8 \times 10^{-5}) \times 3.32} = 3.25 \times 10^{-4} \frac{\text{Pa} \cdot \text{m}^3}{\text{s} \cdot \text{Pa}} \qquad \textbf{(3.6.18)}$$

This is the "speed" predicted from Poiseuille's law. With the factor F, the expected flow is

$$FU = 2.6 \times (3.25 \times 10^{-4}) \frac{\text{Pa} \cdot \text{m}^3}{\text{s} \cdot \text{Pa}} = 8.44 \times 10^{-4} \frac{\text{Pa} \cdot \text{m}^3}{\text{s} \cdot \text{Pa}} \qquad \textbf{(3.6.19)}$$

The measured flow was reported as "0.029 micron cu. ft. per sec. per micron." We convert cubic feet to cubic meters and recognize that the pressure units cancel, hence can be replaced by pascal pressure units. The result is

$$FU = 0.0283 \text{ m}^3/\text{ft}^3 \times 0.029 \frac{\mu\text{m} \cdot \text{ft}^3}{\text{s} \cdot \mu\text{m}} = 8.21 \times 10^{-4} \frac{\text{Pa} \cdot \text{m}^3}{\text{s} \cdot \text{Pa}} \qquad \textbf{(3.6.20)}$$

which is in very good agreement with the predicted value, Eq. 3.6.19.

One of the assumptions implicit in this discussion is that the flow is subsonic. This state of flow is determined by the value of the Mach number, defined as the ratio of the linear velocity of the gas to the speed of sound in the gas. The speed of sound in air at 298 K (close enough to the value of 299 K in this example) is 350 m/s. We can find the linear velocity at the inlet pressure p_0 from the volume flowrate:

$$\frac{\Delta p U}{p_0} = q_0 = \frac{0.029 \times 10}{18} = 0.016 \text{ ft}^3/\text{s} \qquad \textbf{(3.6.21)}$$

At the downstream end, where the pressure has fallen to 8 μmHg, the flowrate is increased to

$$q_L = 0.016 \frac{18}{8} = 0.036 \text{ ft}^3/\text{s} = 1030 \text{ cm}^3/\text{s} \qquad \textbf{(3.6.22)}$$

With a tube radius R of 1.3 cm, the linear velocity at the tube outlet is

$$v_L = \frac{1030}{\pi(1.3)^2} = 194 \text{ cm/s} \qquad \textbf{(3.6.23)}$$

This yields a Mach number of

$$M = \frac{194}{3.5 \times 10^4} = 5.5 \times 10^{-3} \ll 1 \qquad \textbf{(3.6.24)}$$

Hence the flow is clearly subsonic. This is an important issue because, as we will see in Chapter 11, when flows approach sonic conditions the analysis becomes much more complex.

A Vapor Delivery System

In semiconductor fabrication it is often necessary to convey a reactive gas at a controlled flowrate to a low pressure "furnace" in which the gas reacts to form a solid film on a set of silicon wafers. A liquid source is often used, heated to a temperature that yields a vapor pressure large enough to provide the pressure difference required to sustain the flow.

An important liquid source material for growing silicon oxide films is TEOS [tetra-ethoxysilane: $Si(OC_2H_5)_4$]. Viscosity and vapor pressure data are shown in Fig. 3.6.2. Suppose that a liquid source vapor delivery system has been designed to operate in the following manner. Liquid TEOS is in a reservoir maintained at 50°C. The "headspace" above the reservoir, which fills with pure TEOS vapor at the vapor pressure of the liquid, is connected to a reactor through a series of plumbing elements. The dominant resistance to flow of the vapor through the plumbing is a bundle of 100 capillaries in parallel, each of length 4 cm and inside diameter 0.1 mm. Assume that the plumbing lines are isothermal, at 50°C. What mass flowrate is delivered to the reactor if the reactor is at a pressure of 150 mtorr? What is the volume flowrate, in standard liters per minute (SLM)?

First we should determine whether the flow of the vapor is in the viscous or the molecular regime. To do so, we calculate

$$\overline{v_m} = \left(\frac{8R_G T}{\pi M_w}\right)^{1/2} = \left[\frac{8(8314)323}{\pi(208)}\right]^{1/2} = 181 \text{ m/s} \qquad (3.6.25)$$

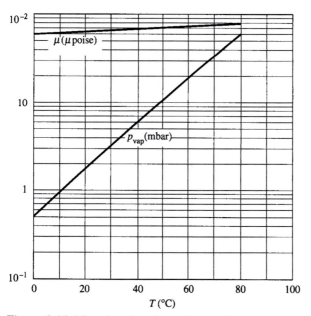

Figure 3.6.2 Viscosity, in micropoise, and vapor pressure, in millibars, of TEOS vapor.

At the upstream (source) end, the vapor pressure is found from Fig. 3.6.2 to be 10 mbar or 10^3 Pa. At the downstream (low pressure) end the pressure is 150 mtorr, or

$$p_L = \frac{0.150 \times 10^5}{760} = 19.7 \, \text{Pa} \qquad (3.6.26)$$

Hence the mean pressure along the capillary is 510 Pa.

The viscosity of the TEOS vapor is found from Fig. 3.6.2 as 60 μpoise = 60×10^{-7} Pa. Now we can calculate

$$K' = \frac{D\bar{p}}{2\mu\bar{v}_m} = \frac{10^{-4} \times 510}{2\,(60 \times 10^{-7}) \times 181} = 23.5 \qquad (3.6.27)$$

Looking at Fig. 3.6.1, we conclude that at this average pressure the flow is essentially viscous, in the sense that the correction factor to the viscous model is close to unity. (From Eq. 3.6.10 we find $F = 1.2$.)

With Eq. 3.6.11 (in terms of D instead of R) we find

$$\dot{m}_p = \frac{\pi D^4 \Delta p}{128\mu L} \frac{\bar{p}\rho_o}{p_o} = \frac{\pi D^4 \Delta p}{128\mu L} \frac{\bar{p}M_w}{R_G T}$$

$$= \frac{\pi (10^{-4})^4 (10^3 - 19.7)}{128\,(60 \times 10^{-7}) \times 0.04} \frac{510 \times 208}{8314 \times 323} = 3.9 \times 10^{-10} \, \text{kg/s} \qquad (3.6.28)$$

With the factor $F = 1.2$ we predict a mass flowrate of

$$\dot{m} = F\dot{m}_p = 1.2 \times (3.9 \times 10^{-10}) = 4.7 \times 10^{-10} \, \text{kg/s} \qquad (3.6.29)$$

This is the mass flow through each of the 100 capillaries in parallel, so the total flow is

$$\dot{m}_{total} = 4.7 \times 10^{-8} \, \text{kg/s} \qquad (3.6.30)$$

To find the volume flowrate, we need the density of the gas, but since we want the volume flowrate in *standard* liters per minute we must use the density at standard temperature and pressure: one atmosphere and 273 K. Assuming that TEOS is an ideal gas at these conditions, we calculate

$$\rho_{STP} = \frac{M_w p_{atm}}{R_G T} = \frac{208 \times 10^5}{8314 \times 273} = 9.2 \, \text{kg/m}^3 \qquad (3.6.31)$$

and the volume flowrate at standard conditions is

$$q_{STP} = \frac{4.7 \times 10^{-8}}{9.2} = 5.1 \times 10^{-9} \, m^3/s = 5.1 \times 10^{-6} \, \text{L/s} = 3.1 \times 10^{-5} \, \text{L/min} \qquad (3.6.32)$$

This value usually is reported as 3.1×10^{-5} SLM for "standard liters per minute."

In performing these calculations we have assumed that the flow is laminar. A quick check on the Reynolds number confirms this, since

$$\text{Re} = \frac{4\dot{m}}{\pi D \mu} = \frac{4(4.7 \times 10^{-10})}{\pi \times 10^{-4} \times (60 \times 10^{-7})} = 1 \qquad (3.6.33)$$

Note that the Reynolds number is calculated for the flow in a *single* capillary and is not based on the total flowrate. We have also assumed that the flow is subsonic. The speed of sound in TEOS can be estimated from

$$c = \left(\frac{R_G T \gamma}{M_w}\right)^{1/2} = \left(\frac{8314 \times 323 \times 1.4}{208}\right)^{1/2} = 134 \, \text{m/s} \qquad (3.6.34)$$

where γ is the ratio of heat capacities for TEOS vapor. We don't know this number, but for the sake of getting a quick answer, we will assume that it is close to the value (1.4) for air. The speed of the vapor at the capillary outlet is

$$v_L = \frac{4q_L}{\pi D^2} = \frac{4 \dot{m} R_G T}{\pi D^2 M_w p_L} = \frac{4(4.7 \times 10^{-10}) \times 8314 \times 323}{\pi (10^{-4})^2 \times 208 \times 19.7} = 39 \text{ m/s} \qquad \textbf{(3.6.35)}$$

Thus we find a Mach number of

$$M = \frac{39}{134} = 0.29 \qquad \textbf{(3.6.36)}$$

This is subsonic flow.

SUMMARY

When a fluid is in motion, its viscosity gives rise to internal stresses and pressures. The relationship of the *dynamics* (the forces arising from these stresses) to the *kinematics* (the motion itself) comes out of an application of the principle of conservation of momentum. In continuum fluid mechanics this relationship takes the form of a differential equation valid everywhere within the fluid. To solve this differential equation, we must impose boundary conditions that reflect what we know about the motion or forces imposed on the boundaries of the flow. In addition, we must make an assumption about the nature of the fluid, and in this text we consider only Newtonian fluids, for which the stresses are *linearly* related to the velocity gradients. Through a series of examples we gain some practice in the derivation and solution of the equations that describe a variety of relatively simple flows. (The simplifications lie largely in the restrictions to flows confined by very uniform boundaries, such that the flows are one-dimensional.)

Since viscosity is the key physical property that determines the nature of the flow, we must be able to calculate or estimate the viscosity of gases and liquids. For gases we use the Chapman–Enskog equation (Eq. 3.4.8); for liquids, the method of Morris (Eq. 3.4.2) is suggested.

When we consider *gas* flow, two factors need to be examined: the compressibility of the gas, and the possibility that at sufficiently low pressures the continuum approximation is not valid and the flow is dominated by molecular rather than viscous interactions. Thus Poiseuille's law (Eq. 3.2.34), central to the prediction of the relationship of the flow rate through a long tube to the pressure causing the flow, must be modified to account for compressibility (Eq. 3.4.20) or molecular interactions (Knudsen flow, described by Eq. 3.6.1).

PROBLEMS

3.1 Return to Problem 2.10 in Chapter 2. Estimate the bubble size, and the frequency of bubble formation, when the tube is half-filled with the liquid. State clearly your assumptions regarding the physics of release of a bubble from the capillary tip. Take $a_0 = 2a_0^{min}$.

3.2 Predict the volumetric flowrates for the two systems shown in the sketch in Fig. P3.2. The liquid is glycerin at 25°C, which has a density of 1200 kg/m³. In both cases the capillary tube has a radius of 0.002 m and a length of 0.1 m. The

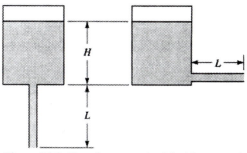

Figure P3.2 Flow from a tank with either a vertical or a horizontal outlet tube.

liquid "head" in the reservoir, H, remains constant, at $H = 1$ m. The reservoir is a cylinder of radius 0.5 m. Use Fig. 3.4.1b for viscosity data.

3.3 Referring to Problem 3.2, suppose the head H is not constant, but falls as a result of the efflux of liquid from the capillary. Find the time for the reservoir to empty, for each of the systems shown. Take the initial height $H = H_0 = 1$ m. *Hint:* Use a "quasi-steady" analysis, which assumes that Poiseuille's law (Eq. 3.2.34) holds, with $Q = Q(t)$, and that the pressure is time-dependent through the varying hydrostatic head $H(t)$.

3.4 In Example 3.2.3 we find the factor by which the flowrate of the inner fluid is increased as a consequence of the presence of an annular layer of a second fluid. The results are given in Eq. 3.2.58 and Fig. 3.2.5. The comparison is for the case of *constant applied pressure gradient,* $\Delta P/L$. Give the result for the case of *constant flowrate of the inner fluid, Q^1.* That is, give the factor by which the required pressure gradient is reduced, at a *fixed* value of Q^1, as a function of η and M.

3.5 A Newtonian oil of viscosity 3 Pa·s must be pumped through a pipe of diameter 0.1 m and length 10 m at a flowrate of 10 m³/h.

a. What is the required pressure drop across the ends of the pipe?

b. What thickness of lubricating liquid, of viscosity 0.3 Pa·s, will reduce the required pressure to the maximum possible extent?

3.6 Using Eq. 3.2.58, find and plot the value of η that maximizes Ψ, as a function of M.

3.7 The viscometer designed in Section 3.3 is going to be used for measuring the viscosity of a waterlike liquid. If it is necessary to keep the Reynolds number below 2000 to ensure laminar flow, and if it is necessary to keep the entrance length L_e less than 10% of the total tube length, what is the upper limit on flow rate for this design? What is the expected pressure drop at the upper limit? Is the viscometer well designed for this task?

3.8 With reference to the problem solved in Section 3.3.2, calculate the pressure developed when the following conditions hold:

$$R = 0.01 \text{ m} \qquad L = 0.1 \text{ m} \qquad \kappa = 0.5$$

$$V = 3 \text{ m/s} \qquad \mu = 10 \text{ Pa·s}$$

3.9 Using Eq. 3.3.26, find the radial position at which the axial velocity achieves its maximum value, for axial annular pressure flow. Is the maximum achieved midway between the two surfaces?

3.10 Figure 3.3.8 is not accurate when $\kappa \approx 1$ because of the very high slope of the curve. Show that Eq. 3.3.23 behaves as

$$\Phi \rightarrow 1.5 \, (1 - \kappa)^{-2} \qquad \textbf{(P3.10)}$$

when κ is very close to unity. (Write $\kappa = 1 - \varepsilon$ and expand $\ln \kappa$ and powers of κ in power series in ε. You will need to retain higher than linear powers of ε to get the desired result.)

3.11 A wire is being drawn through a coating die similar to that shown in Fig. 3.3.9. No pressure is imposed to aid the flow. Find the force required to draw the wire through the die under the following conditions:

$$D_i = 1 \text{ mm} \qquad \kappa = 0.5 \qquad L = 0.01 \text{ m}$$

$$\mu = 50 \text{ Pa·s} \qquad V = 10 \text{ m/s}$$

The ratio of the force to the cross-sectional area of the wire is the tension in the wire. Compare your result to the yield stress of steel ($\sim 10^5$ psi).

3.12 Molten nylon is to be coated onto a wire of diameter 1 mm at a line speed of 25 m/s. The die has $D_0 = 2$ mm and $L = 0.01$ m. Under the coating conditions, nylon may be regarded as a Newtonian fluid with a viscosity of 10 Pa·s. The required coating thickness is 0.5 mm. What pressure is required to provide this coating thickness?

3.13 A wide sheet of smooth planar film, of thickness $2B$, is pulled through the region between a pair of wide parallel planes separated by a distance $2A$. Figure P3.13 shows the geometry. The entire space surrounding the film is filled with a Newtonian liquid of viscosity μ. Following the method illustrated in Section 3.2.1, derive a model for the velocity profile $v_x(y)$. Assume that there is no pressure variation in the x direction, and that gravity acts parallel to the y axis.

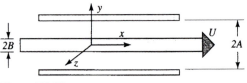

Figure P3.13

3.14 Estimate the viscosity of hydrogen and of silane at both 700 and 1000 K, and atmospheric pressure. (Both are gases at these conditions.)

3.15 Calculate the viscosities of silane and hydrogen over the temperature range from 300 to 1000 K. Show that an empirical model of the form $\mu = AT^b$ gives a reasonably good fit to the behavior, and give the values of b for the two gases. Note that T is *absolute* temperature.

3.16 Predict the viscosity of the liquids carbon tetrachloride and trichloroethane over the temperature range of 0 to 30°C. Make a copy of Fig. 3.4.1a and plot your results on the resulting grid. Rescale the y axis if necessary.

3.17 Predict the viscosities of air and n-butane in the gaseous state, and compare your predictions with the data in Fig. 3.4.3.

3.18 Find an expression for the force required to withdraw a wire from a container of viscous liquid as shown in Fig. P3.18. (Neglect the fact that some of the liquid is removed from the reservoir because it adheres to the wire as a coating.) Give the magnitude of the force for the case $U = 10$ cm/s, $L = 10$ cm, $\mu = 10$ poise, $R_i = 0.1$ cm, $R_0 = 0.125$ cm.

3.19 A viscous liquid is dripping from a reservoir, open at the top to the atmosphere as shown in Fig. P3.19. The drip rate is observed to be 3 drops/min. Drops break away from the tip of the capillary while in contact with the inside of the tip, but they do not wet the outside of the capillary. The liquid has a surface tension of 0.06 N/m and a density of 900 kg/m³. What is the liquid viscosity? Make use of the Harkins–Brown correction factor described in Chapter 2, (Eq. P2.27.2 in Problem 2.27).

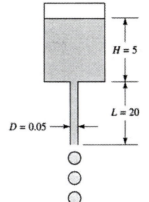

Figure P3.19 Dimensions (cm) for dripping reservoir.

3.20 In Problem 3.19 we would normally neglect surface tension effects at the capillary exit, hence neglecting as well the effect any back pressure, due to surface tension, might have on the flowrate. Is this shortcut based on a reasonable assumption? Suppose water is confined in the system shown in Fig. P3.19, with the geometrical dimensions given, except that the capillary diameter D may be varied. Take surface tension into account in answering the following questions.
a. Plot the drip rate (cm³/s) against capillary diameter for 0.1 mm $< D <$ 4 mm. At what capillary diameter is the effect of surface tension first noticeable?
b. What is the lower bound on diameter that will effectively close off the flow?
c. For what D does the flow make a transition between dripping and continuous flow?
d. What portion (if any) of the plot in part a is reliable?

3.21 Based on the concept sketched in Fig. P3.21, design a viscometer suitable for measuring liquid viscosities in the range of 10 to 1000 poise. In the drawing, D is a large reservoir containing a viscous fluid A and E is a cylindrical tube, open at both ends, surrounding and coaxial with B, which is a cylindrical "bob" that is dragged up through the fluid by a frictionless pulley system (C) utilizing a weight W. A "design" requires specification of all geometrical dimensions of the system. In addition, present a design equation that relates viscosity to the geometry of the system and describe the measurements you would

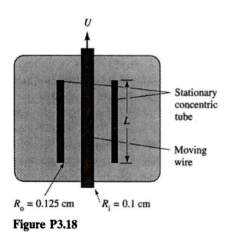

Figure P3.18

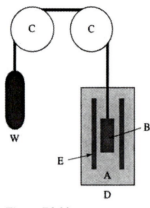

Figure P3.21

make of the behavior of the system. What range of W values (in grams mass) do you require?

3.22 Data for the viscosity of molten lithium as a function of temperature are given below. Plot these data in a form suitable for determination of the coefficients in Eq. 3.4.1, and give values of the coefficients.

T (°C)	μ (Pa·s)
204	0.000542
316	0.000446
427	0.000393
538	0.000347

3.23 In Example 3.2.2 we considered a viscous drop being squeezed through a capillary. When such a drop moves through a long capillary, a fraction of the liquid is left behind as a coating on the inside of the tube. Experimental studies have shown that to a good approximation, the residual film thickness h is given by

$$\frac{2h}{D_i} = 1.4 \, Ca^{2/3} \quad \text{for} \quad Ca < 0.1 \quad \textbf{(P3.23.1)}$$

$$\frac{2h}{D_i} = 0.33 \quad \text{for} \quad Ca > 0.1 \quad \textbf{(P3.23.2)}$$

where Ca is the capillary number, defined as

$$Ca = \frac{\mu U}{\sigma} \quad \textbf{(P3.23.3)}$$

and U is the average velocity of flow through the tube.

Consider a drop of volume $V_0 = 10^{-8}$ m³ and viscosity $\mu = 0.01$ Pa·s, surface tension $\sigma = 0.05$

N/m, in a tube of inside diameter $D_i = 0.25$ mm and length $L = 0.1$ m, being "pushed" by a pressure difference $\Delta P = 1000$ N/m².

a. Is the drop able to leave the downstream end of the tube?

b. What fraction of the initial drop volume leaves the tube exit? Be approximate: More than 50%? More than 67%?

c. Does surface tension impede the rate of flow of the drop through the tube? How? To a significant extent?

d. Answer questions a–c with the tube length increased to 0.5 m.

3.24 Equation 3.3.32 has a peculiar format, since we could cancel the $1 - \kappa^2$ factor that appears in both the numerator and denominator. It must have been written in this form for a reason. The explanation lies partly in the fact that the coefficient of the bracketed term is just the product of the velocity V and the cross-sectional area open to flow. The bracketed term is a measure of the extent to which viscous effects prevent us from dragging all the liquid through the annular region as a rigid body at the velocity V.

a. Find the limiting behavior of the bracketed term as $\kappa \to 1$. You will need to apply L'Hôpital's rule twice. (You may need to buy back your calculus book.)

b. Derive a model for a problem similar to pure annular drag flow, specifically, pure *planar* drag flow. In this case the flow is generated by the movement of a very wide planar surface between two parallel planes: see Fig. P3-13. Go back to Fig. 3.3.5, but let the solid boundaries be infinitely wide planes instead of surfaces of cylinders. Find an expression for Q_d, put it in the form of Eq. 3.3.32 (i.e., as a product of the velocity V and the cross-sectional area open to flow). Compare the bracketed term for the planar solution to the result you find in part a. Does this make sense?

c. Based on these results, offer a criterion based on the value of κ that will allow you to approximate annular drag flow by the algebraically simpler *planar* drag flow model. State clearly your definition of a "good" approximation.

3.25 Rework Problem 3.21 for the geometry shown in Fig. P3.13. Specifically, replace the cylindrical bob B by a planar rigid sheet of length L in the x direction, width W in the z direction, and thickness t, and replace the tube E by a pair of planes parallel to and surrounding the sheet B. The distance between this pair of planes is $2A$ in the y direction, and they are of width W in the z direction. This pair of planes is of length L_0 in the x direction. The sheet is controlled in such a way that it remains centered between the planes E. In your design, make $L \ll L_0$ and $W \gg A$.

3.26 A commercial viscometer has been used to study the shear stress–shear rate behavior of Heinz tomato ketchup at 20°C. The data are shown in Fig. P3.26. Is ketchup Newtonian? What is the viscosity of ketchup at a low shear rate of 0.1 s^{-1}? What is the viscosity of ketchup at a high shear rate of 100 s^{-1}?

3.27 A capillary viscometer has been used to measure the viscosity of whole blood at 25.5°C. From the data given in Fig. P3.27 for flow rate Q as a function of pressure gradient $\Delta P/L$, plot the viscosity as a function of the shear stress at the wall of the tube, $r = R$. Is blood Newtonian?

3.28 Some shear stress–shear rate data are given in Fig. P3.28 for whole blood. Measurements were made in a viscometer that uses the annular flow between concentric rotating cylinders. Are these data compatible with the data given in Problem 3.27? Is "What is the viscosity of blood?" a well-defined question?

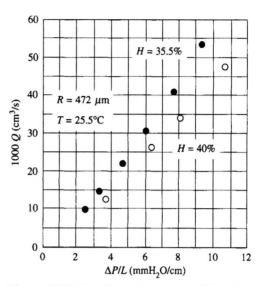

Figure P3.27 Capillary viscometry of blood: the hematocrit H is the volume percentage of red cells. [Thurston, *Biorheology*, **31**, 179 (1994)].

3.29 Give the derivation of Eq. 3.3.16, from Eqs. 3.3.12–3.3.15.

3.30 Give the derivation of Eq. 3.3.23.

3.31 Return to Example 3.4.3. Evaluate the characteristics of a viscometer design that uses $D = 100$ μm and $L = 1$ m. Specifically, plot volume flowrate, at atmospheric pressure, as a function of the upstream pressure p_0. Use the properties of air at 25°C for your calculations.

3.32 A gas is being studied in a capillary viscometer designed according to the discussion in Section 3.4.3. The design parameters are $D = 100$ μm and $L = 1$ m. The upstream pressure is $p_L = 3$ atm and the exit of the capillary is essentially at atmospheric pressure. The system is isothermal. The observed flowrate is 0.004 mL/s. What is the viscosity of the gas?

3.33 Calculate values of the viscosity of air at a pressure of 7.6 torr, and at temperatures of 100 and 1000 K.

3.34 For the data given in Example 3.2.2, calculate the Reynolds number for flow through the tube. Is the flow laminar? Suppose $D = 1.5$ mm, and $D_d = 3$ mm. Is the flow laminar?

3.35 We wish to design an experiment on the passage of viscous drops through a long capillary, under the action of a constant pressure difference ΔP across the ends of the capillary. Two

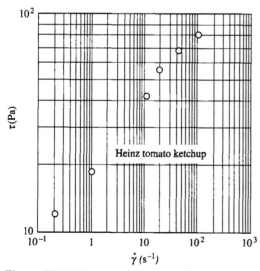

Figure P3.26 Shear stress versus shear rate for ketchup. Macklet et al., *Chem. Eng. Sci.*, **49**, 2551 (1994).

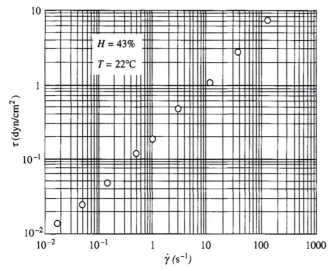

Figure P3.28 Rotational viscometry of blood: the hematocrit H is the volume percentage of red cells. [Thurston, *Biorheology*, **31,** 179 (1994)].

conditions are to be met in the experimental design:

1. Less than 20% of the liquid is to be deposited inside the capillary as a residual film (see Problem 3.23).

2. The effect of surface tension on the available pressure drop ΔP is to be less than 10% of that pressure.

Show that these conditions require that the capillary have a length such that $L/D = 10^4$.

3.36 Wehbeh et al. [*Phys. Fluids,* **A5,** 25 (1993)] present, without derivation, an expression for the dimensionless drag force on a long cylinder moving at constant speed through a closed annular reservoir:

$$\frac{\text{force}}{2\pi\mu VL} = \frac{\dfrac{1 - 3\kappa^2}{1 + \kappa^2} - \dfrac{4\kappa^4 \ln \kappa}{1 - \kappa^4}}{\ln \dfrac{1}{\kappa} - \dfrac{1 - \kappa^2}{1 + \kappa^2}} \quad \textbf{(P3.36)}$$

Derive this result, beginning with the model presented in Section 3.3.2. The algebra is awful (except in Southern California, where it is totally awesome).

3.37 A long thin cylindrical rod is to be placed along the axis of a long tube of circular cross section, through which a Newtonian liquid is in steady laminar pressure flow. If we want a reduction in throughput (at constant $\Delta P/L$) of less than 10%, due to the presence of the rod (relative to a tube with no rod), what must the value of the ratio of the inner to outer diameters be?

3.38 In the molecular flow regime, the relationship of mass flowrate to pressure difference across a long circular tube is given theoretically by Knudsen as

$$\dot{m} = \frac{\pi}{12} \frac{D^3}{L} \frac{p_0 - p_L}{p_0} \left(\frac{8R_G T}{\pi M_w}\right)^{1/2} p_0 \quad \textbf{(P3.38.1)}$$

where p_0 is the mass density measured at the upstream pressure p_0, and D and L are the tube diameter and length, respectively.

a. Derive an expression for the ratio

$$R \equiv \frac{\dot{m}}{\dot{m}_p} \quad \textbf{(P3.38.2)}$$

where the denominator is the mass flowrate given by Eq. 3.4.20 (Poiseuille flow for a compressible fluid).

b. Consider the flow of nitrogen under the following conditions:

$$D = 1 \text{ mm} \qquad L = 0.1 \text{ m} \qquad T = 300 \text{ K}$$

$$p_0 - p_L = 10 \text{ mtorr} \qquad p_0 = 20 \text{ mtorr}$$

Calculate the mass flowrate according to Eq. P3.38.1, and compare your result to that which would be predicted through the (mis)use of Eq. 3.4.20.

3.39 Brown et al. [*J. Appl. Phys.*, **17**, 802 (1946)] give the following data for the flow of hydrogen through a copper pipe with a radius $R = 1.30$ cm and length $L = 332$ cm, at 299 K. U is the "speed" defined in Eq. 3.6.18. Produce a plot similar to Fig. 3.6.1 for these data. Use $\mu = 8.8 \times 10^{-6}$ Pa·s for the viscosity of hydrogen. The units for the pressures are micrometers of mercury, and U is in the complex units discussed earlier: (micrometers of mercury) − (cubic foot) per second − (micrometer of mercury).

p_0	p_L	U	p_0	p_L	U
4	3	0.093	40	36	0.143
8	6	0.079	61	55	0.162
9	7	0.091	125	116	0.271
16	12	0.091	327	314	0.620
17	15	0.093	681	665	1.19
25	22	0.103	1017	1000	1.8

3.40 We have assumed that the viscosity of liquids is independent of pressure. This is a good approximation at moderate absolute pressures, but liquids subjected to high pressure service, such as lubricants, show a pressure dependence of viscosity $\mu(P)$ that can be expressed in the form

$$\mu = \mu_0 e^{bP} \qquad \text{(P3.40.1)}$$

where μ_0 and b are material parameters, and P is absolute pressure. Rework the derivation of Poiseuille's law (see Section 3.2, ending with Eq. 3.2.34) and show that

$$Q = \left[\frac{1 - e^{-b\Delta P}}{b\Delta P}\right]\frac{\pi \Delta P R^4}{8L\mu_0} \qquad \text{(P3.40.2)}$$

a. Make a plot of the viscosity correction factor (the bracketed term in Eq. P3.40.2) as a function of $b\Delta P$ over the range $0 < b\Delta P < 5$.
b. A liquid has a viscosity of the order of 10^4 poise at 120°C and atmospheric pressure. The parameter b has a value of 0.005 bar^{-1}. Assume that b is independent of temperature. We wish to measure the viscosity of this liquid accurately at 120°C, using a capillary viscometer. What is the order of magnitude of the pressure required to extrude this liquid through a capillary of inside diameter $D = 0.03$ mm and length 2 mm for operation at a maximum shear rate (defined here as $32Q/\pi D^3$) of 100 s^{-1}? What is the magnitude

of the pressure correction factor under these conditions?
3.41 It is often difficult to measure the inside radius of very small capillaries. One method proposed is to measure the pressure drop–flowrate relationship, and use Poiseuille's law to calculate the radius. Find the capillary radius R for the following experimental observation:

$$L = 50 \text{ cm } (\pm 1 \text{ mm})$$
$$\mu = 0.05 \text{ Pa·s } (\pm 0.005 \text{ Pa·s})$$
$$Q = 0.3 \text{ cm}^3/\text{s } (\pm 0.01 \text{ cm}^3/\text{s})$$
$$\Delta P = 50 \text{ Pa } (\pm 1 \text{ Pa})$$

In view of the uncertainties of the measurements, as stated above for each parameter, what is the expected percent error in the "measurement" of R?
3.42 Equation 3.2.34 gives the Hagen–Poiseuille law for laminar flow through a long capillary of uniform circular cross section:

$$Q_P = \frac{\pi R^4}{8\mu}\frac{\Delta P}{L} \qquad \text{(P3.42.1)}$$

Equation 3.3.30 gives the corresponding result for flow through a long tube of annular cross section:

$$Q_P = \frac{\pi \Delta P R^4}{8\mu L}\left[1 - \kappa^4 + \frac{(1 - \kappa^2)^2}{\ln \kappa}\right] \qquad \text{(P3.42.2)}$$

Derive the form of Poiseuille's law for flow through a long rectangular duct whose planar surfaces are separated by a distance $2B$ in one direction and W in the other, for $W \gg B$. Your result should be

$$Q_P = \frac{2B^3W}{3\mu}\frac{\Delta P}{L} \qquad \text{(P3.42.3)}$$

3.43 With reference to Problem 3.42, prove that Eq. P3.42.2 yields Eq. P3.42.3 in the limit as $\kappa \to 1$.
3.44 Go back to the design problem in Chapter 2 (Example 2.1.2). Suppose that the capillary specified has an axial length of 2 cm. What is the required pressure drop across the capillary? Will surface tension affect the steadiness of the efflux from the capillary if the system is designed to operate at this constant pressure drop? In fact, can the specified flow be produced at this pressure drop?
How would you modify the design to meet the flowrate constraint, but under conditions of

steady flow? Outline a procedure for calculating the steady pressure needed to yield the specified *average* flowrate in the event that surface tension strongly affects the flow from the capillary but does not prohibit the flow.

3.45 Prove the statement presented in footnote 8 with respect to Eq. 3.4.20.

3.46 A capillary tube, initially empty, is held in a vertical position and touched to the surface of a pool of water at 25°C. The water rises in the tube until an equilibrium height (the "capillary rise") is achieved. Develop a model for the dynamics of this capillary rise phenomenon.

a. Derive a differential equation for $h(t)$.
b. Give the solution to the equation.
c. Plot h versus t for the following parameters:

$$\rho = 1000 \text{ kg/m}^3 \qquad \mu = 0.001 \text{ Pa} \cdot \text{s}$$

$$\sigma = 0.07 \text{ N/m} \qquad R_0 = 2.42 \times 10^{-4} \text{ m}$$

3.47 Calculate and plot the viscosity of TEOS [tetraethoxysilane: $Si(OC_2H_5)_4$] vapor as a function of temperature in the 40–70°C range. Compare the value used in Example 3.6.3 to your plot. You will need the following property data:

$$T_c = 593 \text{ K} \qquad \varepsilon/k = 522 \text{ K}$$

$$M = 208 \text{ g/g} \cdot \text{mol} \qquad \sigma = 7.6 \text{ Å}$$

3.48 A wire coating die is designed as shown in Fig. 3.3.9. It is necessary to produce a coating 2 mm thick on the wire, which has a radius of 2 mm and is moving through the die, of radius 4 mm, at a speed of 250 cm/s. Assuming that all the pressure drop occurs across the "land

length" $L = 20$ mm, find the pressure ΔP required to achieve this coating thickness. What is the *magnitude* of the maximum shear stress that the coating fluid undergoes in the die? The coating liquid has a viscosity of 1000 poise.

3.49 With reference to Problem 3.48, suppose the wire radius is changed to 1.5 mm, but all other conditions are unchanged, including the requirement of a 2 mm thick coating. Find the pressure ΔP required to achieve this coating thickness.

3.50 Prove the assertion regarding the additivity of the velocity profiles made at the beginning of Example 3.3.4.

3.51 With reference to the wire coating analysis presented in Section 3.3, show that in the limit of vanishing pressure ($\Phi \to 0$) the smallest value of coating thickness corresponds to

$$\delta' = \frac{1/\kappa - 1}{2} \qquad \textbf{(P3.51)}$$

3.52 With reference to the wire coating analysis presented in Section 3.3, find an expression for the value of the *negative* pressure (in terms of the function Φ) that just prevents deposition of any coating onto the wire. Present a plot of $\Phi(\kappa)$ for this case.

3.53 A glass filament is drawn through a coating system, and the filament is centered by being constrained in a closely fitting annular guide, as shown in Fig. P3.53. The filament is lubricated to prevent damage to its surface due to slight motion causing contact with the inner surface of the guide. However, it is necessary to keep the

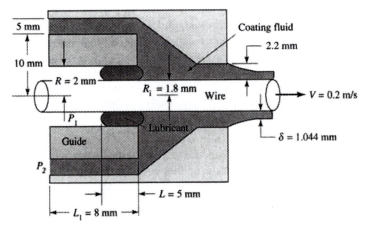

Figure P3.53 Coating die with lubricated wire (not to scale).

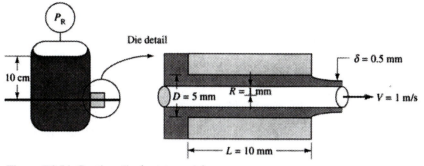

Figure P3.54 Coating die (not to scale).

lubricant from coating the filament as it passes beyond the guide. What negative pressure P_1 is required to achieve this? The lubricant viscosity is $\mu = 0.1$ Pa·s. Other parameters are as shown in Fig. P3.53. The coating fluid has a low viscosity, and only a small hydrostatic pressure P_2 is required upstream to supply the coating fluid to the die.

3.54 In a wire coating system (Fig. P3.54), the coating liquid supplied from a large reservoir has a viscosity of 1.5 Pa·s and a density of 900 kg/m^3. The head in the reservoir is essentially constant during the operation of the system. The design and operating parameters are shown in the figure. What pressure P_R is required to achieve the desired coating thickness of $\delta = 0.5$ mm?

3.55 Go back to Example 3.3.4. Suppose we are not confident that the value given for the viscosity of the liquid is reliable, other than as an order-of-magnitude estimate. Therefore, we want to *measure* the viscosity at a shear rate that is comparable to the maximum shear rate in the coating flow of interest, and we choose to use a capillary viscometer. Specify a suitable capillary diameter and length, and the required pressure drop, to achieve this goal. As part of your response, give the Reynolds number for the viscometer operating at the required shear rate and the volume flowrate exiting the capillary.

3.56 A long solid cylindrical weight of length L has a small hole of diameter D drilled from end to end along the cylinder axis. A wire of radius R is strung through the hole in the weight, and the wire is stretched vertically between two supports. The wire is wetted with a viscous liquid. When the weight is released from a high position,

it falls under the action of gravity, along the wire. The viscous liquid always fills the annular region between the weight and the wire, and as a consequence there is a viscous resistance to the motion of the weight. See Fig. P3-56. Derive an expression for the steady state speed of fall, V, of the weight.

3.57 Rework the analysis of axial annular pressure flow to account for the possibility of slip at the inner surface only. As a slip model, replace the no-slip boundary condition at the inner surface with the following expression:

$$V_s = -\alpha \tau_{zr} \quad \text{at} \quad r = \kappa R \quad \textbf{(P3.57)}$$

where the "slip coefficient" α is a characteristic of the liquid and the solid surface. This slip model is based on the assumption that the slip velocity is simply proportional to the shear stress at the liquid/solid interface.

Find an expression, in nondimensional form, for the ratio of the slip velocity to the maximum

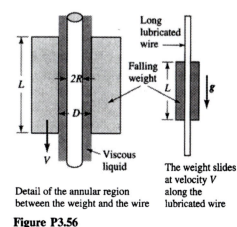

Figure P3.56

velocity in the annulus, as a function of a dimensionless slip coefficient α^*. Present a plot of that ratio as a function of α^*, for $\kappa = 0.5$ and 0.707.

Suppose slip also occurred at the *outer* boundary, $r = R$. Would you make any change in Eq. P3.57 in writing the slip boundary condition at $r = R$? Before answering this question, make a qualitative sketch of the shear stress profile across the annular region.

3.58 Consider axial annular flow of a Newtonian liquid under the following conditions: the inner cylinder moves at a velocity V and the pressure gradient pumps liquid in the same direction as the motion of the cylinder. Write a model for the drag on the inner moving cylinder. Put the model in a dimensionless format, and present a plot of the dimensionless drag on the inner moving cylinder as a function of a dimensionless inner cylinder speed, with κ as a parameter. Give an analytical expression for the speed at which the drag on the inner moving cylinder vanishes.

3.59 At some initial instant of time a viscous drop of volume V_0 just covers the opening of a capillary of length L and inside diameter D_i (Fig. P3.59). A constant pressure difference $\Delta P > 0$ is maintained across the length L. Write and solve a model that yields the time to completely transfer the entire drop from one end of the capillary to the other. State clearly any assumptions you make in developing your model, especially with regard to the early and final portions of the transfer of the drop. Give a criterion for the neglect of surface tension in the model.

3.60 Derive Eq. 3.5.18.

3.61 In Example 3.5.2 we assume that the moving sheet is coated with just enough liquid to ensure that the region between the restrictions is filled but not overflowing. What liquid film thickness δ is required to achieve this? Give your answer in the form of $\delta/(b - a)$.

3.62 It can be shown that the velocity profile for steady, laminar, incompressible, isothermal, fully developed Newtonian flow through a tube of elliptical cross section takes the form

$$v_z = \frac{\Delta\mathcal{P}a^2b^2}{2\mu L(a^2 + b^2)}\left[1 - \left(\frac{x}{a}\right)^2 - \left(\frac{y}{b}\right)^2\right]$$

$$(\text{P3.62.1})$$

where

$$\left(\frac{x}{a}\right)^2 + \left(\frac{y}{b}\right)^2 = 1 \qquad (\text{P3.62.2})$$

describes the perimeter of the ellipse, in Cartesian coordinates. From this result, derive the analog to Poiseuille's law in the form

$$Q = \frac{K(a, b)\Delta\mathcal{P}}{\mu L} \qquad (\text{P3.62.3})$$

and give the form of the coefficient K.

3.63 Using the results given in Problem 3.62, find the ratio of the volume flowrate through an elliptical tube relative to that through a tube of circular cross section, with the constraint that the cross-sectional areas be equal. Assume that we are comparing the same fluid at equal pressure gradients.

Suppose a tube of circular cross section is deformed into an elliptical tube in such a way that the tube perimeter is unchanged. Plot the reduction in flow as a function of the ratio a/b of the major to minor axes.

3.64 Under the action of gravity, a viscous liquid flows steadily as a thin laminar film down a vertical planar surface that occupies the plane $y = 0$. (See Fig. P3.64.) At some point $z = 0$, the

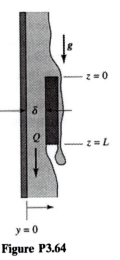

Figure P3.64

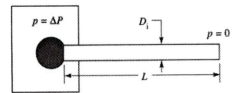

Figure P3.59

liquid encounters a planar narrow restriction in the region $0 < z < L$ that is parallel to the surface at $y = 0$. Develop a model for the volume flowrate through the narrow restriction as a function of δ, the separation between the planar surfaces. In particular, give the differential equation the velocity component u_z must satisfy. Write the boundary conditions that u_z must satisfy. Solve the equation for u_z, and give the function $u_z(y)$. Write an expression for the volume flowrate Q, per unit width W in the x direction, in terms of the other parameters of the system.

3.65 What does dimensional analysis tell us about the system described in Problem 3.64? List the parameters the volume flowrate per unit width W, call it Q_W, is likely to depend on. Give the dimensions of each parameter you list. Form a dimensionless group proportional to Q_W. Call it G_Q. What independent dimensionless groups does G_Q depend on?

3.66 The following data [Federspiel et al., *AIChE J.*, **42**, 2094 (1996)] are available for the flow of air through a bundle of parallel hollow fibers connecting two manifolds. Each fiber has an inside diameter of 207 μm, and is 11.5 cm long; there are 93 parallel fibers in the bundle. Flow is isothermal, at 25°C, and in all cases the upstream pressure is atmospheric (760 mmHg) and various vacuums are pulled at the downstream end of the system. The flowrates are given in liters per minute measured at 25°C and 760 mmHg pressure. Plot the data as a test of Poiseuille's law for an incompressible fluid, and then test the applicability of Eq. 3.4.20 for laminar compressible isothermal flow. What conclusions do you draw? At what pressure ratio (upstream to downstream) are deviations from the incompressible model observed?

$p_0 - p_L$ (mmHg)	50	100	150	200	250	300	350	400
q_0 (L/min)	0.75	1.55	2.2	2.8	3.3	3.8	4.25	4.55

3.67 Using Fig. 3.4.3, give values for the coefficient b in the expression

$$\mu = AT^b \qquad \text{(P3.67)}$$

for air and for *n*-butane at $T = 300$ K.

Chapter **4**

Conservation of Mass and Momentum in a Continuous Fluid

In this chapter we move toward modeling flows that are more complex than those treated in Chapter 3. We begin with a derivation of the partial differential equation that expresses the principle of conservation of mass in a continuous flowing material. A specific definition of the rate of deformation within a fluid comes next, followed by a derivation of the equations that correspond to the principle of conservation of momentum in a flowing material. When this material is coupled with a generalized definition of a Newtonian fluid, we obtain the Navier–Stokes equations—the central equations of fluid dynamics. The remainder of the chapter illustrates how to pass from these general conservation equations to models for specific flow problems of interest to us.

In Chapter 3 we illustrated the application of the principles of conservation of mass and momentum. We derived the equations that permitted us to obtain analytical solutions for the flow fields in several *simple* problems. We did this by writing a force balance on an element of volume within the fluid. Let's review the features of these problems that made them simple:

Dimensionality The velocity vector had only one nonzero component, and that component was a function of only one space variable.
Geometry The solid boundaries of the systems studied were always the coordinate surfaces of a simple coordinate system—cylindrical or planar in the cases examined.

These simple analyses led us to equally simple models. Despite their simplicity, these models are very useful. But they are limited. We are often concerned with flow fields bounded by surfaces of complex shape. The velocity field is often a function of at least two space variables, as well as time. The velocity vector often has more than a single component. We need to examine, now, the more general equations that are appropriate to these more complex flows. Because we will be considering general three-dimensional, unsteady-state flow fields, the algebra gets very messy. Just *writing* the generalized flow equations becomes tedious. Before we can begin, however, we have to discuss a few very basic concepts stemming from the great complexity of the mechanics of a fluid, as compared with the mechanics of rigid solids.

In rigid body mechanics we can resolve forces acting on the surface of a body into a single force vector that affects the motion of the center of mass of the body, which in turn gives rise to rotation of the body about some axis. The application of Newton's laws of motion yields ordinary differential equations for the position vector of the center of mass as a function of time. In *fluid* mechanics we must begin by recognizing that a fluid is a continuous medium, or *continuum,* and if there is deformation of the fluid, there will be a continuous distribution of velocities, pressures, and viscous stresses throughout the body of fluid. Application of the principle of conservation of momentum (Newton's second law for a continuum) will yield, in general, *partial* differential equations for the velocities and stresses as functions of position and time.

Our discussion of forces in a continuum is preceded by a consideration of the dynamics of mass flow in a continuum. Then we present a generalized momentum principle for flow of a continuous fluid, followed by a series of examples of applications of our equations to the development of mathematical models of flows, some of which are more complex than those encountered in Chapter 3.

4.1 THE DISTRIBUTION OF MASS IN A CONTINUOUS FLUID (THE CONTINUITY EQUATION)

A *continuum* may be defined as a region of material space throughout which properties such as velocity, temperature, density, and composition vary in a mathematically continuous manner. Discontinuities may occur at phase boundaries that separate, for example, the gas in a bubble from the surrounding liquid. We regard such properties as density to be definable at every point within the continuum through a mathematical limiting process. That is, if we examine some small volume ΔV within a material, at some instant of time, and measure a mass Δm in that volume, as we examine smaller and smaller volumes centered about that point, the measured ratio of mass to volume approaches a limiting value, which we define as the mass density ρ:

$$\rho \equiv \lim \frac{\Delta m}{\Delta V} \qquad \text{at any point} \quad \text{as } \Delta V \to 0 \qquad \textbf{(4.1.1)}$$

Equation 4.1.1 simply states that density is well defined; that restriction, however, does not prevent the density from varying spatially as well as with time, within a flowing medium. We examine first the effect of nonuniform flow on the local density distribution within a fluid. We are going to invoke the principle of conservation of mass, which states that if we neglect conversion of mass to energy, then mass can be moved about but it cannot be created or destroyed by flow.[1] In Fig. 4.1.1, which gives a simple way to derive the consequences of this principle, we set up a Cartesian coordinate system with a vector of axes x_i, and place a rectangular parallelepiped in the system aligned with the axes as shown. This is a "control volume" whose surfaces are not material surfaces. The surfaces are open, and fluid passes freely across them. We might call them "bookkeeping surfaces" for reasons that will be clear in a moment. The volume element is of small dimensions which we label dx_i. Hence its volume is $dV = dx_1 dx_2 dx_3$.

Fluid flows across the surfaces of dV, carrying mass in and out of the volume. We

---

[1] Chemical reactions could be occurring, but this possibility does not alter the principle. Reactions will change the distribution of molecular structures within the fluid, but not the total mass.

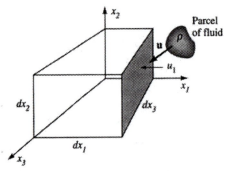

Figure 4.1.1 Volume element for derivation of the continuity equation.

denote any change of mass within the volume by

$$dm = \left[\left(\frac{\partial}{\partial t} \right) \bar{\rho} \, dV \right] dt \qquad (4.1.2)$$

(We use average density $\bar{\rho}$ in this equation because the density may vary from one part of the volume to the other.) Within dV, mass can change only through the mechanism of unbalanced flows across its boundaries. Let's examine a "parcel" of fluid (see Fig. 4.1.1) that is small enough to be characterized by a velocity vector $\mathbf{u}$ and a local density ρ. We want to calculate the rate at which that parcel carries mass across the surface shown in the figure, which is the surface in the plane $x_1 = dx_1$, and which has an area $dA = dx_2 dx_3$. The *volumetric* rate of flow across any surface is the product of the velocity *normal to the surface* times the area of the surface. Hence the *mass* rate of flow across a surface is the product of the density of the fluid at that point times the volume rate of flow. We conclude that the mass rate of flow across the surface at dx_1 is expressible in the form

$$-[\rho u_1 dx_2 dx_3]_{x_1 = dx_1}$$

The minus sign arises because if the velocity u_1 is positive (as drawn in the figure it is negative), then mass would be *leaving* the volume. (This is the same sign convention introduced in Chapter 3. At this face, when we go from the fluid external to the volume element into the volume element, we move in the $-x_1$ direction. Hence the term has a minus sign in front of it.)

By an identical argument we write the rate of flow of mass across the parallel surface, in the plane $x_1 = 0$, as

$$[\rho u_1 dx_2 dx_3]_{x_1 = 0}$$

These two rates may differ because the velocities and densities at $x_1 = 0$ and $x_1 = dx_1$ may differ. That is why we label the bracketed terms as shown. Thus the net rate of flow of mass across the pair of parallel faces normal to the x_1 axis is the difference

$$[\rho u_1 dx_2 dx_3]_{x_1 = 0} - [\rho u_1 dx_2 dx_3]_{x_1 = dx_1}$$

By similar arguments we may do the bookkeeping on the mass flows across the pairs of faces normal to the x_2 and x_3 axes. The two difference terms that arise are

$$[\rho u_2 dx_1 dx_3]_{x_2 = 0} - [\rho u_2 dx_1 dx_3]_{x_2 = dx_2}$$

and

$$[\rho u_3 dx_2 dx_1]_{x_3=0} - [\rho u_3 dx_2 dx_1]_{x_3=dx_3}$$

These differences represent the only sources of mass change within the volume element. Hence our mass balance, which is our statement of conservation of mass, takes the form

$$\left(\frac{\partial}{\partial t}\right) \bar{\rho} \, dV = [\rho u_1 dx_2 dx_3]_{x_1=0} - [\rho u_1 dx_2 dx_3]_{x_1=dx_1}$$

$$+ [\rho u_2 dx_1 dx_3]_{x_2=0} - [\rho u_2 dx_1 dx_3]_{x_2=dx_2}$$

$$+ [\rho u_3 dx_2 dx_1]_{x_3=0} - [\rho u_3 dx_2 dx_1]_{x_3=dx_3} \qquad \textbf{(4.1.3)}$$

The volume element is regarded as fixed in space, and of constant volume dV. Hence we may divide both sides of Eq. 4.1.3 by $dV = dx_1 dx_2 dx_3$, with the result

$$\frac{\partial \bar{\rho}}{\partial t} = \frac{[\rho u_1]_{x_1=0} - [\rho u_1]_{x_1=dx_1}}{dx_1} + \frac{[\rho u_2]_{x_2=0} - [\rho u_2]_{x_2=dx_2}}{dx_2} + \frac{[\rho u_3]_{x_3=0} - [\rho u_3]_{x_3=dx_3}}{dx_3} \qquad \textbf{(4.1.4)}$$

Now we let the volume dV become very small, and we ask how the terms in the equation above behave in the limit of $dV \to 0$. If the medium is continuous (i.e., if its properties such as density and velocity vary continuously throughout the volume), then each term in this equation is mathematically well behaved, and by the definition of the derivative as a limit, we conclude that Eq. 4.1.4 becomes

$$\frac{\partial \rho}{\partial t} = -\left(\frac{\partial \rho u_1}{\partial x_1} + \frac{\partial \rho u_2}{\partial x_2} + \frac{\partial \rho u_3}{\partial x_3}\right) \qquad \textbf{(4.1.5)}$$

(We no longer need to use the overbar on ρ since the density that appears on both sides of this equation is the density at the point (x_1, x_2, x_3) to which the volume dV has shrunk.)

We can conserve space if we write this equation in a vector format. Recalling the definition of the divergence operator $\nabla\cdot$ we may write Eq. 4.1.5 as

$$\frac{\partial \rho}{\partial t} = -\nabla \cdot \rho\mathbf{u} \qquad \textbf{(4.1.6)}$$

Equation 4.1.6 is called the continuity equation. It states that the density in the neighborhood of a point in a continuous medium may change only through unbalanced flows in that region.

For some fluids in certain flows, the density is independent of time and position. The most common example is an incompressible fluid—(one whose density is independent of pressure)—in an isothermal flow. (We need the latter restriction because the density of an incompressible fluid is temperature dependent.) For a constant density fluid, or more properly, for a constant density *flow*, the continuity equation takes the simpler form

$$\nabla \cdot \mathbf{u} = 0 \qquad \textbf{(4.1.7)}$$

This equation represents a constraint on possible relations among the components of velocity. In effect, it says that since mass must be conserved, a flow field must adjust itself to meet this requirement. Thus the velocity components must satisfy this equation.

Keep in mind, when comparing Eqs. 4.1.5 and 4.1.6 that they are the same equation. The vector form is simply the more general expression, since it does not assume any particular coordinate system, while Eq. 4.1.5 is the form of Eq. 4.1.6 in Cartesian coordinates. Since we will want to be able to write fluid dynamic equations in several

Table 4.1.1 Components of the Divergence Operator
(for any vector **A**)

Cartesian

$$\nabla \cdot \mathbf{A} = \frac{\partial A_x}{\partial x} + \frac{\partial A_y}{\partial y} + \frac{\partial A_z}{\partial z} \qquad \textbf{(4.1.8a)}$$

Cylindrical

$$\nabla \cdot \mathbf{A} = \frac{1}{r}\frac{\partial (rA_r)}{\partial r} + \frac{1}{r}\frac{\partial A_\theta}{\partial \theta} + \frac{\partial A_z}{\partial z} \qquad \textbf{(4.1.8b)}$$

Spherical

$$\nabla \cdot \mathbf{A} = \frac{1}{r^2}\frac{\partial (r^2 A_r)}{\partial r} + \frac{1}{r\sin\theta}\frac{\partial (A_\theta \sin\theta)}{\partial \theta} + \frac{1}{r\sin\theta}\frac{\partial A_\phi}{\partial \phi} \qquad \textbf{(4.1.8c)}$$

coordinate systems, it is useful to have the components of the divergence operator $\nabla\cdot$ available to us in the most commonly used coordinate systems: Cartesian, cylindrical, and spherical (Table 4.1.1). To write Eq. 4.1.6 or 4.1.7, we need only replace the components of the general vector **A** with $\rho\mathbf{u}$, or **u**, respectively.

Equation 4.1.7 is called a *kinematic* constraint, because it involves motion (kinematics) but not forces (dynamics). Since most flows of interest to us must satisfy both kinematic and dynamic constraints, we will have to examine the constraints imposed by the requirement of conservation of momentum on the velocity field, as well. Before we can proceed to these constraints, however, one other aspect of kinematics requires our attention.

4.2 DEFORMATION IN A FLUID

We again suppose that flow is occurring in a continuous medium. The velocity vector **u** with components u_i characterizes the kinematics. With reference to Fig. 4.2.1, we suppose that a coordinate system is centered at the point labeled **O**, at which the velocity is $\mathbf{u_o}$. In the near neighborhood of **O** (say, a differential distance dx) we expect to find an analytical relationship between $\mathbf{u_o}$ and **u**. If such a relationship exists, we should be able to express the components of the velocity vector **u** in terms of the corresponding components of $\mathbf{u_o}$ by a Taylor series expansion of the form

$$u_i = u_{oi} + \left(\frac{\partial u_i}{\partial x_1}\right)dx_1 + \left(\frac{\partial u_i}{\partial x_2}\right)dx_2 + \left(\frac{\partial u_i}{\partial x_3}\right)dx_3 + O(dx_j)^2 \qquad \textbf{(4.2.1)}$$

where it is understood that the partial derivatives are all evaluated at the point **O**. The notation $O(dx_j)^2$, which is read as "order of $(dx_j)^2$," is a reminder that Eq. 4.2.1 will be an approximation later, when we drop terms beyond the first order in dx_i.

Now we raise a question: What is an appropriate measure of fluid deformation in the neighborhood of **O**? A simple answer is that deformation is occurring if the velocities $\mathbf{u_o}$ and **u** are different. (We will have to qualify this shortly.) Thus the relative motion of particles of fluid in the neighborhood of point **O** is given by

$$u_i - u_{oi} = \left(\frac{\partial u_i}{\partial x_1}\right)dx_1 + \left(\frac{\partial u_i}{\partial x_2}\right)dx_2 + \left(\frac{\partial u_i}{\partial x_3}\right)dx_3 = \sum_{j=1}^{3}\left(\frac{\partial u_i}{\partial x_j}\right)dx_j \qquad \textbf{(4.2.2)}$$

[We have neglected small terms of order $(dx)^2$.]

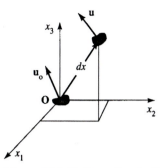

Figure 4.2.1 Deformation in the region near the point **O**.

The term $\partial u_i/\partial x_j$ is called the velocity gradient. Clearly, for there to be deformation, some terms of the velocity gradient must be nonzero. (For a three-dimensional flow there are nine such derivatives.) We can rewrite the velocity gradient in the form

$$\frac{\partial u_i}{\partial x_j} = \frac{1}{2}\left(\frac{\partial u_i}{\partial x_j} + \frac{\partial u_j}{\partial x_i}\right) + \frac{1}{2}\left(\frac{\partial u_i}{\partial x_j} - \frac{\partial u_j}{\partial x_i}\right) \tag{4.2.3}$$

If we define

$$\Delta_{ij} = \left(\frac{\partial u_i}{\partial x_j} + \frac{\partial u_j}{\partial x_i}\right) \qquad \text{and} \qquad \omega_{ij} = \left(\frac{\partial u_i}{\partial x_j} - \frac{\partial u_j}{\partial x_i}\right) \tag{4.2.4}$$

we may write

$$u_i - u_{oi} = \sum_{j=1}^{3}\left(\frac{\Delta_{ij}}{2} + \frac{\omega_{ij}}{2}\right)dx_j \tag{4.2.5}$$

This decomposes the relative velocity into two parts. Why have we chosen to do this? To find an answer, we write Eq. 4.2.5 as

$$du_i = du_{i\Delta} + du_{i\omega} = \sum_{j=1}^{3}\left(\frac{\Delta_{ij}}{2} + \frac{\omega_{ij}}{2}\right)dx_j \tag{4.2.6}$$

and examine the terms given by

$$du_{i\omega} = \tfrac{1}{2}\omega_{ij}dx_j \tag{4.2.7}$$

In this equation, and henceforth, we drop the summation sign and adopt the convention of summing over any subscript that is repeated in a product. (Thus we sum over j in Eq. 4.2.7.)

Evidently Eq. 4.2.7 describes a part of the motion in the neighborhood of the point **O**. To learn just what kind of motion this is, we first multiply both sides by dx_i and find

$$du_{i\omega}dx_i = \tfrac{1}{2}\omega_{ij}dx_jdx_i \tag{4.2.8}$$

This equation implies a double summation, since the subscript i is also repeated on the right-hand side. But the subscripts are just symbols used for counting in the summations. We could substitute k for j in Eq. 4.2.6 without changing anything but the appearance of the summation. In particular, then, we could change i to j and j to i, to find that

$$\omega_{ij}dx_jdx_i = \omega_{ji}dx_idx_j \tag{4.2.9}$$

(Note that we cannot divide both sides of this equation by the product $dx_i dx_j$, which would lead an to an incorrect conclusion: $\omega_{ij} = \omega_{ji}$. That is because both sides of the equation are summations. This point is clearer if you write out the sum first.)

From the definition given in Eq. 4.2.4 it is clear that

$$\omega_{ij} = -\omega_{ji} \tag{4.2.10}$$

It follows then that

$$\omega_{ij} dx_j dx_i = -\omega_{ij} dx_i dx_j \tag{4.2.11}$$

But we can certainly reverse the order of the product $dx_i\, dx_j$ in Eq. 4.2.11. This rearrangement leads to the conclusion that

$$\omega_{ij} dx_i dx_j = -\omega_{ij} dx_i dx_j = 0 \tag{4.2.12}$$

since a number can equal its own negative only if the number is zero. (In this case the number is a *sum* of numbers, but the conclusion holds.)

Taking note of Eqs. 4.2.8 we see that this is the same as

$$du_{i\omega} dx_i = 0 \tag{4.2.13}$$

According to our convention that repeated indices imply a summation, we may write this equation as

$$du_{i\omega} dx_i = du_{1\omega} dx_1 + du_{2\omega} dx_2 + du_{3\omega} dx_3 = 0 \tag{4.2.14}$$

But both $du_{i\omega}$ and dx_i are components of vectors. Hence we may write Eq. 4.2.14 as the scalar product

$$d\mathbf{u}_\omega \cdot d\mathbf{x} = 0 \tag{4.2.15}$$

The vanishing of a scalar product of two vectors has a simple geometrical interpretation: the two vectors are mutually perpendicular. Since $d\mathbf{x}$ is the position vector connecting the two points of interest, we conclude that the part of the motion associated with the vector $d\mathbf{u}_\omega$ is just the motion of one point relative to the other, perpendicular to the line connecting the points, *which is simply rigid rotation* of one point about the other.

What does all this mathematical manipulation mean physically? We have proven the following:

The relative motion between two points in a flowing medium is describable in terms of the components of the velocity gradient $\partial u_i / \partial x_j$.

This motion can be decomposed into two parts: a rigid rotation about the axis joining the points, described mathematically by ω_{ij}, and the rest of the motion, which we call a deformation rate *whose components are given by Δ_{ij}.*

The Δ_{ij} are referred to as the components of the *rate of deformation tensor* Δ. All the information about the internal deformation of the fluid is "encoded" in the components of Δ. While we always knew intuitively that the extent of deformation somehow depended on how the velocity field varies spatially, we now know that the proper measure of deformation rate in a fluid is given by the components of Δ. As in the case of the continuity equation, we will need to have access to the forms of Δ in several coordinate systems. These are given in Table 4.2.1.

Table 4.2.1 Components of the Rate of Deformation Tensor

Rectangular coordinates (x, y, z)

$$\Delta_{xx} = 2 \frac{\partial u_x}{\partial x} \qquad\qquad \Delta_{xy} = \Delta_{yx} = \frac{\partial u_x}{\partial y} + \frac{\partial u_y}{\partial x}$$

$$\Delta_{yy} = 2 \frac{\partial u_y}{\partial y} \qquad\qquad \Delta_{xz} = \Delta_{zx} = \frac{\partial u_x}{\partial z} + \frac{\partial u_z}{\partial x}$$

$$\Delta_{zz} = 2 \frac{\partial u_z}{\partial z} \qquad\qquad \Delta_{yz} = \Delta_{zy} = \frac{\partial u_y}{\partial z} + \frac{\partial u_z}{\partial y}$$

Cylindrical coordinates (r, θ, z)

$$\Delta_{rr} = 2 \frac{\partial u_r}{\partial r} \qquad\qquad \Delta_{r\theta} = \Delta_{\theta r} = r \frac{\partial}{\partial r} \left(\frac{u_\theta}{r} \right) + \frac{1}{r} \frac{\partial u_r}{\partial \theta}$$

$$\Delta_{\theta\theta} = 2 \left(\frac{1}{r} \frac{\partial u_\theta}{\partial \theta} + \frac{u_r}{r} \right) \qquad\qquad \Delta_{\theta z} = \Delta_{z\theta} = \frac{\partial u_\theta}{\partial z} + \frac{1}{r} \frac{\partial u_z}{\partial \theta}$$

$$\Delta_{zz} = 2 \frac{\partial u_z}{\partial z} \qquad\qquad \Delta_{zr} = \Delta_{rz} = \frac{\partial u_z}{\partial r} + \frac{\partial u_r}{\partial z}$$

Spherical coordinates (r, θ, ϕ)

$$\Delta_{rr} = 2 \frac{\partial u_r}{\partial r} \qquad\qquad \Delta_{r\theta} = \Delta_{\theta r} = r \frac{\partial}{\partial r} \left(\frac{u_\theta}{r} \right) + \frac{1}{r} \frac{\partial u_r}{\partial \theta}$$

$$\Delta_{\theta\theta} = 2 \left(\frac{1}{r} \frac{\partial u_\theta}{\partial \theta} + \frac{u_r}{r} \right) \qquad\qquad \Delta_{\theta\phi} = \Delta_{\phi\theta} = \frac{\sin\theta}{r} \frac{\partial}{\partial \theta} \left(\frac{u_\phi}{\sin\theta} \right) + \frac{1}{r\sin\theta} \frac{\partial u_\theta}{\partial \phi}$$

$$\Delta_{\phi\phi} = 2 \left(\frac{1}{r\sin\theta} \frac{\partial u_\phi}{\partial \phi} + \frac{u_r}{r} + \frac{u_\theta \cot\theta}{r} \right) \qquad\qquad \Delta_{\phi r} = \Delta_{r\phi} = \frac{1}{r\sin\theta} \frac{\partial u_r}{\partial \phi} + r \frac{\partial}{\partial r} \left(\frac{u_\phi}{r} \right)$$

EXAMPLE 4.2.1   *The Deformation Rate Tensor in Poiseuille Flow*

In Chapter 3 we derived an expression for the velocity field in Poiseuille flow (Eq. 3. 2.33). Let's examine the rate of deformation tensor Δ for this flow. We write the velocity vector as

$$\mathbf{u} = (u_z, 0, 0) \tag{4.2.16}$$

and we know that

$$u_z = \frac{-CR^2}{4\mu} \left[1 - \left(\frac{r}{R} \right)^2 \right] \tag{4.2.17}$$

Using Table 4.2.1 as our guide we find

$$\Delta_{zr} = \Delta_{rz} = \frac{\partial u_z}{\partial r} = \frac{Cr}{2\mu} \tag{4.2.18}$$

and all other components of Δ are zero. We usually write Δ as a matrix, in the form

$$\Delta = \begin{pmatrix} \Delta_{rr} & \Delta_{r\theta} & \Delta_{rz} \\ \Delta_{\theta r} & \Delta_{\theta\theta} & \Delta_{\theta z} \\ \Delta_{zr} & \Delta_{z\theta} & \Delta_{zz} \end{pmatrix} = \begin{pmatrix} 0 & 0 & \dfrac{Cr}{2\mu} \\ 0 & 0 & 0 \\ \dfrac{Cr}{2\mu} & 0 & 0 \end{pmatrix} \qquad \textbf{(4.2.19)}$$

A flow for which the Δ tensor has only *off-diagonal* components is called a "shear flow," and Δ_{zr}, called the "shear rate" (of deformation), is often denoted by the symbol $\dot{\gamma}$ ("gamma dot").

4.3   STRESSES IN A FLUID: CONSERVATION OF MOMENTUM AND THE EQUATIONS OF MOTION

Our next task is to develop the equations that relate the forces acting within a fluid to the flows that accompany those forces. Again, because we are considering a continuous medium rather than a rigid body, the form of these equations is very complex. We will begin with a definition of stress in a fluid, and then proceed to an application of Newton's second law that will lead to the desired dynamic equations for a fluid, which we often simply refer to as the equations of motion.

Let's imagine a small volume element *fixed in space*. Figure 4.3.1 defines the system. Fluid is moving across the boundaries of the volume element, and forces act on the boundaries as well. These forces may arise from viscous effects associated with the deformation of the fluid in the neighborhood of the surface of the volume element. There may be pressures acting on these surfaces, and there could also be a body force (due to gravity) acting on the mass of fluid within the volume element. Whatever the sources, imagine that we can examine a force vector $d\mathbf{F}$ acting on a small region of the surface dA, at a point P on the surface. The area is defined by its magnitude dA, but also by its *orientation* with respect to some arbitrary coordinate system. The orientation is defined by a vector $\mathbf{n}$ normal to the surface dA. Note in particular that the force vector is not necessarily aligned with the surface; that is, it neither is normal to the surface nor lies in the plane of the surface.

If we focus on the area dA, we may define a coordinate system based on this area. The axes are the normal vector $\mathbf{n}$, and any pair of orthogonal axes that lie in the plane

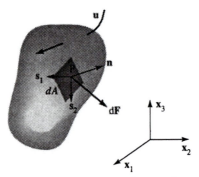

Figure 4.3.1 Forces and flows acting on some volume element in a fluid. The orientation of the surface dA is described by the orientation of the normal vector $\mathbf{n}$, with respect to the coordinate frame x_1, x_2, x_3. Axes s_1, s_2 are in the plane of dA but are not necessarily oriented in alignment with the frame x_1, x_2, x_3.

of dA, which we call $\mathbf{s}_1$ and $\mathbf{s}_2$. Then $\mathbf{n}$, $\mathbf{s}_1$, and $\mathbf{s}_2$ define a Cartesian coordinate system, and in *that* system (and specifically, not in the $\mathbf{x}_1$, $\mathbf{x}_2$, $\mathbf{x}_3$ coordinate frame) we may write the components of the force $d\mathbf{F}$ as (dF_{1n}, dF_{2n}, and dF_{nn}). We use double subscripting because the area dA is arbitrary. Hence the description of the force components relative to the area must include information about the orientation of the area. Thus one subscript (the second, n) is related to the area and is common to all three components of $d\mathbf{F}$, while the other three subscripts (in each case, the first subscript) tell which component of the force we are considering, *in the surface-based coordinate system.*

We now define the *stress vector* $\mathbf{T}$ at the point P as the limit of the force per area as the area surrounding that point shrinks to the point:

$$\mathbf{T} = \lim \frac{d\mathbf{F}}{dA} \quad \text{as} \quad dA \rightarrow 0 \tag{4.3.1}$$

Applying the same definition to the individual components of $\mathbf{T}$, we write

$$T_{in} = \lim \frac{dF_{in}}{dA} \quad \text{as} \quad dA \rightarrow 0 \tag{4.3.2}$$

and this defines the components of the stress vector on the surface defined by the normal $\mathbf{n}$.

Now let's apply these ideas to our volume element. To simplify the bookkeeping, we let the volume be a rectangular parallelepiped, as we did in deriving the continuity equation. Figure 4.3.2 shows this volume, with the stress vector components on three orthogonal faces. There are, of course, stresses on all six faces; for clarity, we show them on three orthogonal faces only. On each face, three components of the stress vector are shown. There is a so-called *normal stress,* which acts perpendicular (normal) to the face, and a pair of orthogonal *shear stresses,* which act in the plane of the face. Together, the nine stresses T_{ij} are called components of the stress *tensor.*

We have one more task before we can start the bookkeeping. We need to adopt a convention on *signs* for stress. A problem arises because by most thermodynamic conventions a pressure (a *compressive stress*) acting on a body is regarded as a *positive* stress. However, most mechanics texts take a *tensile stress* to be positive. In this text we will adopt the following convention, which is opposite that of most mechanics texts but is nevertheless self-consistent:

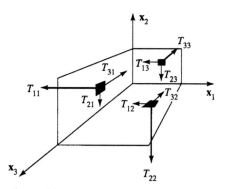

Figure 4.3.2 Resolution of the force $\mathbf{F}$ into the stress components on three orthogonal faces of the volume element dV. We are looking at the forces exerted on the surfaces by the fluid outside of the volume element.

The stress T_{ij}, due to the action of material on the positive side of the surface, act-
ing on material on the negative side, is positive if the direction of the line of action
is along negative x_i.

The stresses shown in Fig. 4.3.2 are to be understood as being exerted *by the fluid*
outside the body on the surfaces of the body. (By Newton's third law there is a set of
equal and opposite forces exerted by the fluid within dV on the fluid outside.) In all
cases shown, the direction of each stress component is in the *negative* direction of the
axis to which it is referred. In all cases, each stress is due to action by material on the
negative side of the surface (which is *outside* the volume element). Hence the stresses
shown are *negative* as drawn.

Now look at the forces arising from the nine stress vector components shown in Fig.
4.3.2. We will do this for the x_1 direction. Similar equations can be written for the other
two directions. First of all, only three of these stresses have components in the x_1
direction. In each case, the force due to each stress is the product of the stress times
the area on which it acts. Thus we may sum the forces in the x_1 direction and find

$$S_{1,o} = (T_{11}dx_2dx_3 + T_{12}dx_1dx_3 + T_{13}dx_1dx_2)_o \tag{4.3.3}$$

The subscript o on $S_{1,o}$ reminds us that this set of faces has its corner at $x = 0$. The
other three faces of the volume element are acted on by forces, and if we drew those
faces and the corresponding stresses, we would write

$$S_{1,dx} = (T_{11}dx_2dx_3 + T_{12}dx_1dx_3 + T_{13}dx_1dx_2)_{dx} \tag{4.3.4}$$

The net sum of the force components in the x_1 direction is just $(S_{1,o} - S_{1,dx})$, or

$$S_1 = (S_{1,o} - S_{1,dx_1}) = (T_{11,o} - T_{11,dx_1})dx_2dx_3 + (T_{12,o} - T_{12,dx_2})dx_1dx_3$$
$$+ (T_{13,o} - T_{13,dx_3})dx_1dx_2 \tag{4.3.5}$$

If the dx vector is small, as we take it to be, then a good approximation is

$$T_{11,o} - T_{11,dx_1} = -\frac{\partial T_{11}}{\partial x_1} dx_1 \tag{4.3.6}$$

with similar equations for T_{12} and T_{13}. This lets us write

$$S_1 = -\frac{\partial T_{11}}{\partial x_1} dx_1dx_2dx_3 - \frac{\partial T_{12}}{\partial x_2} dx_2dx_1dx_3 - \frac{\partial T_{13}}{\partial x_3} dx_3dx_1dx_2 \tag{4.3.7}$$

Thus the only contributions to forces acting in the x_1 direction that arise among the
nine stress components on each set of orthogonal faces of the volume element are those
expressed in Eq. 4.3.7.

We now look at other contributions to a force balance in the x_1 direction that arise
from flow across the boundaries. Figure 4.3.3 will be helpful. When fluid crosses the
surface of the volume element, that flow carries momentum in (or out) of the volume.
If a parcel of fluid approaches the surface with a velocity $\mathbf{u}$, that parcel has a momentum
(per unit of volume) given by $\rho\mathbf{u}$. (We call $\rho\mathbf{u}$ a *momentum density*.) Since we are
writing the x_1 component of the forces acting on dV, we also write the x_1 component
of the momentum flow across the surfaces. Thus we want the component ρu_1 of $\rho\mathbf{u}$.

What is the rate at which ρu_1 crosses the surfaces of the volume? We simply multiply
the density of momentum by the volume flowrate *normal* to the surface in question.
Hence the x_1 component of momentum flow across the surface of area dx_2dx_3 is (ρu_1)
$(u_1dx_2dx_3)$, since $(u_1dx_2dx_3)$ is the volume flowrate across the surface dx_2dx_3 normal to

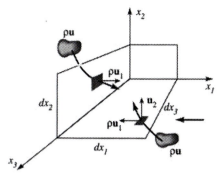

Figure 4.3.3 Momentum is carried across the surface of dV by flow.

the x_1 axis. By summing the flows across the three orthogonal faces at the corner $x = 0$, we find

$$C_{1,o} = (\rho u_1)(u_1 dx_2 dx_3) + (\rho u_1)(u_2 dx_1 dx_3) + (\rho u_1)(u_3 dx_1 dx_2) \qquad (4.3.8)$$

Note that each term (corresponding to each of the three faces shown) is expressed by the same x_1 component of momentum density (ρu_1), but since each face is at a different position in the fluid, there could be a different velocity u_1 normal to each face. In other words, u_1 is a function of position.

As before, we now account for the three terms corresponding to the other three faces. For small $d\mathbf{x}$ we make the approximation that

$$(\rho u_1 u_i)_o = (\rho u_1 u_i)_{dx_i} - \frac{\partial(\rho u_1 u_i)}{\partial x_i} dx_i \qquad \text{for} \quad i = 1, 2, \text{ or } 3 \qquad (4.3.9)$$

Then, defining $C_1 = (C_{1,o} - C_{1,dx_1})$, we find that the net rate of flow of momentum (in the x_1 direction) that crosses all six faces of the surface is

$$C_1 = -\frac{\partial(\rho u_1 u_1)}{\partial x_1} dx_1 dx_2 dx_3 - \frac{\partial(\rho u_1 u_2)}{\partial x_2} dx_2 dx_1 dx_3 - \frac{\partial(\rho u_1 u_3)}{\partial x_3} dx_3 dx_1 dx_2 \qquad (4.3.10)$$

We call C_1 the *convective flow of momentum* (in the x_1 direction). It represents the contribution of the *spatial* variation of the flow field to the rate of change of momentum within the volume element.

It is also possible for the flow field to be unsteady in time, at any point in the fluid. When that is the case, there is a time rate of change of momentum of all the fluid within the volume element, which we write as

$$A_1 = \frac{\partial}{\partial t}(\rho u_1) \, dx_1 dx_2 dx_3 \qquad (4.3.11)$$

We would call this an *acceleration* term.

Finally, we allow for the possibility that there is a body force $\mathbf{f}$ acting on the fluid. In the x_1 direction this has the component

$$B_1 = (\rho f_1) \, dx_1 dx_2 dx_3 \qquad (4.3.12)$$

We now invoke Newton's second law, in the form

time rate of change of momentum

$$= \text{sum of forces} + \text{sum of convective flow of momentum} \qquad (4.3.13)$$

or, in the 1-direction

$$A_1 = S_1 + B_1 + C_1 \tag{4.3.14}$$

If we look back at each of these terms, we see that $(dx_1 dx_2 dx_3)$ is a common factor. The result, then, is the following equation:

$$\frac{\partial}{\partial t}(\rho u_1) + \left[\frac{\partial(\rho u_1 u_1)}{\partial x_1} + \frac{\partial(\rho u_1 u_2)}{\partial x_2} + \frac{\partial(\rho u_1 u_3)}{\partial x_3} \right] = (\rho f_1) - \left[\frac{\partial T_{11}}{\partial x_1} + \frac{\partial T_{12}}{\partial x_2} + \frac{\partial T_{13}}{\partial x_3} \right] \tag{4.3.15}$$

We carried out this exercise in momentum flow and force bookkeeping for the x_1 direction. It should not be too difficult to believe that if we had selected the x_2 or x_3 direction, we would have obtained a similar equation, with a simple change in subscripts. In fact, for any direction i, we may write Eq. 4.3.15 in the form

$$\frac{\partial}{\partial t}(\rho u_i) + \left[\frac{\partial(\rho u_i u_1)}{\partial x_1} + \frac{\partial(\rho u_i u_2)}{\partial x_2} + \frac{\partial(\rho u_i u_3)}{\partial x_3} \right] = (\rho f_i) - \left[\frac{\partial T_{i1}}{\partial x_1} + \frac{\partial T_{i2}}{\partial x_2} + \frac{\partial T_{i3}}{\partial x_3} \right] \tag{4.3.16}$$

When the convective terms are manipulated by the chain rule for differentiation, we can simplify Eq. 4.3.16 by making use of the continuity equation given earlier as Eq. 4.1.6. The result may be written as

$$\rho \left[\frac{\partial}{\partial t}(u_i) + u_1 \frac{\partial u_i}{\partial x_1} + u_2 \frac{\partial u_i}{\partial x_2} + u_3 \frac{\partial u_i}{\partial x_3} \right] = (\rho f_i) - \left[\frac{\partial T_{i1}}{\partial x_1} + \frac{\partial T_{i2}}{\partial x_2} + \frac{\partial T_{i3}}{\partial x_3} \right] \tag{4.3.17}$$

We may write this in a vector format as

$$\rho \left[\frac{\partial \mathbf{u}}{\partial t} + \mathbf{u} \cdot \nabla \mathbf{u} \right]_i = \rho f_i - [\nabla \cdot \mathbf{T}]_i \tag{4.3.18}$$

The dynamic equation

Before proceeding, let's summarize our progress in developing the basic equations that constrain the flow. We have imposed two physical statements: conservation of mass and conservation of momentum. We have now derived four differential equations. One is the continuity equation, given in vector format as Eq. 4.1.6. The other *three* are represented by Eq. 4.3.18, which yields one equation for each space coordinate of any three-dimensional coordinate system selected. (The continuity equation is only a *single* equation, as the derivation makes clear. The difference arises from the physics. Mass is a scalar quantity, and Eq. 4.1.6 is a scalar equation, though it is written in a vector format. Momentum is a vector quantity, and Eq. 4.3.18 is a vector format that represents three scalar equations.)

While we are counting equations, it is instructive to count unknowns. We have many more than four unknowns, since the velocity vector u_i has three components and the stress tensor T_{ij} has nine. It is clear already that we are short on equations. To deal with this problem, we first recognize the need to introduce an assumption regarding the nature of the fluid. That is, we must relate the components of the stress in the fluid, at any point in the flow, to the components of the rate of deformation at that point. The simplest relation we can introduce is based on the following observations:

A fluid undergoing flow experiences stresses that are normal to surfaces. These stresses include a term that is independent of orientation of the surfaces. This iso-tropic component of normal stresses is what we commonly call pressure. Hence any relation between stresses and deformation must include pressure. In the ab-sence of deformation there is only pressure.

The stress components are symmetric (i.e., $T_{ij} = T_{ji}$).

There is a large class of fluids for which the components T_{ij} of the stress tensor are linearly *related to the components Δ_{ij} of the rate of deformation tensor. These are Newtonian fluids, by definition.*

A mathematical expression of the foregoing observations may be written in the following form:

$$T_{ij} = \left[p + \tfrac{2}{3}\mu (\nabla \cdot \mathbf{u}) \right] \delta_{ij} - \mu \Delta_{ij} = p'\delta_{ij} - \mu \Delta_{ij} \qquad (4.3.19)$$

General definition of the Newtonian fluid

Equation 4.3.19, which is the desired relationship between *any* component of the stress tensor T_{ij} and the corresponding component of Δ_{ij}, is called a *constitutive equation*. Note that in writing Eq. 4.3.19 we define a pressure p' that includes a contribution related to the compressibility of the fluid. For an incompressible fluid, for which $\nabla \cdot \mathbf{u} = 0$, we find $p' = p$. The form of the term with $\nabla \cdot \mathbf{u}$ is not a consequence of any additional physical assumptions. It is a mathematical consequence of the assumption of linearity of $\mathbf{T}$ on Δ.

In writing Eq. 4.3.19, we are to understand that δ_{ij} is a tensor (matrix) with components defined so that

$$
\begin{aligned}
\delta_{ij} &= 0 \quad &&\text{if} \quad i \neq j \\
&= 1 \quad &&\text{if} \quad i = j
\end{aligned}
\qquad (4.3.20)
$$

The tensor δ is simply a mathematical "device" or notation that lets us write Eq. 4.3.19 in a compact format. It has no physical significance. With the aid of δ we can represent in a single equation the several equations for the normal stresses:

$$T_{11} = p' - \mu \Delta_{11} \text{ (and similarly for 22 and 33)} \qquad (4.3.21)$$

while for the shear stresses

$$T_{12} = -\mu \Delta_{12} \text{ (and similarly for 13, 23, etc.)} \qquad (4.3.22)$$

Note that pressure does not appear in the shear stresses, since pressure is a normal stress. It does not have a component that acts *in* the plane of a surface—only *on* the surface, and normal to it.

Equation 4.3.19 is called Newton's law of viscosity. In fact, it is not a law; it is a definition of the coefficient μ, which is the viscosity of the fluid. It is a useful definition only to the extent that the flow behavior of many fluids may be modeled successfully if we assume linear behavior between stresses and rate of deformation. The choice of signs in Eq. 4.3.19 is based on our convention on the signs of stresses. In most mechanics-oriented treatments of fluid dynamics, Newton's law is written with the opposite choice of signs.

Because of the *symmetry* of Δ_{ij} with respect to its indices (i.e., $\Delta_{ij} = \Delta_{ji}$), it follows that there are only *six* stress components T_{ij}. Hence Eq. 4.3.19 really represents six equations. Returning to our count of equations versus unknowns then, we have the continuity equation (1 equation), the momentum equations (3 equations), and what we call the constitutive equations (6 equations), which constrain the following unknowns: pressure (1 unknown), the velocity components (3 unknowns), and the stress components (6 unknowns). Hence the equations and unknowns are in balance, at 10 each, and we have some hope of solving fluid dynamics problems with this set.

If we introduce the assumption of *Newtonian flow* behavior into Eq. 4.3.18 and carry out the indicated differentiations, we obtain a set of equations containing only velocity components and pressure. With the added assumption of *isothermal incompressible flow*, these equations may be written symbolically in the format

$$\rho\left[\frac{\partial \mathbf{u}}{\partial t} + \mathbf{u}\cdot\nabla\mathbf{u}\right]_i = -[\nabla p]_i + \mu[\nabla^2\mathbf{u}]_i + \rho f_i \tag{4.3.23}$$

The Navier–Stokes equations

The component forms of these equations are given in Table 4.3.1. The collective name, the Navier–Stokes equations, honors two pioneers in the development of fluid dynamics.

The equations of conservation of mass and conservation of momentum are the most general forms of the equations that describe any flow. Because they are so general, they apply to most fluid dynamics problems of interest to us. But because they are so general, their mathematical form prohibits an analytical solution in all but a few simple cases. Even when we introduce the assumption of isothermal incompressible Newtonian flow, we are left with a general set of nonlinear, coupled, partial differential equations.

Table 4.3.1 The Navier–Stokes Equations for Newtonian Isothermal Incompressible Flow

Rectangular (Cartesian) Coordinates (x, y, z)

x component

$$\rho\left(\frac{\partial u_x}{\partial t} + u_x\frac{\partial u_x}{\partial x} + u_y\frac{\partial u_x}{\partial y} + u_z\frac{\partial u_x}{\partial z}\right) = -\frac{\partial p}{\partial x} + \mu\left(\frac{\partial^2 u_x}{\partial x^2} + \frac{\partial^2 u_x}{\partial y^2} + \frac{\partial^2 u_x}{\partial z^2}\right) + \rho g_x \tag{4.3.24a}$$

y component

$$\rho\left(\frac{\partial u_y}{\partial t} + u_x\frac{\partial u_y}{\partial x} + u_y\frac{\partial u_y}{\partial y} + u_z\frac{\partial u_y}{\partial z}\right) = -\frac{\partial p}{\partial y} + \mu\left(\frac{\partial^2 u_y}{\partial x^2} + \frac{\partial^2 u_y}{\partial y^2} + \frac{\partial^2 u_y}{\partial z^2}\right) + \rho g_y \tag{4.3.24b}$$

z component

$$\rho\left(\frac{\partial u_z}{\partial t} + u_x\frac{\partial u_z}{\partial x} + u_y\frac{\partial u_z}{\partial y} + u_z\frac{\partial u_z}{\partial z}\right) = -\frac{\partial p}{\partial z} + \mu\left(\frac{\partial^2 u_z}{\partial x^2} + \frac{\partial^2 u_z}{\partial y^2} + \frac{\partial^2 u_z}{\partial z^2}\right) + \rho g_z \tag{4.3.24c}$$

Cylindrical Coordinates (r, θ, z)

r component

$$\rho\left(\frac{\partial u_r}{\partial t} + u_r\frac{\partial u_r}{\partial r} + \frac{u_\theta}{r}\frac{\partial u_r}{\partial \theta} - \frac{u_\theta^2}{r} + u_z\frac{\partial u_r}{\partial z}\right) = -\frac{\partial p}{\partial r} + \mu\left(\frac{\partial}{\partial r}\left[\frac{1}{r}\frac{\partial}{\partial r}(ru_r)\right] + \frac{1}{r^2}\frac{\partial^2 u_r}{\partial \theta^2} - \frac{2}{r^2}\frac{\partial u_\theta}{\partial \theta} + \frac{\partial^2 u_r}{\partial z^2}\right) + \rho g_r$$

$$\tag{4.3.24d}$$

θ component

$$\rho\left(\frac{\partial u_\theta}{\partial t} + u_r\frac{\partial u_\theta}{\partial r} + \frac{u_\theta}{r}\frac{\partial u_\theta}{\partial \theta} + \frac{u_r u_\theta}{r} + u_z\frac{\partial u_\theta}{\partial z}\right) = -\frac{1}{r}\frac{\partial p}{\partial \theta} + \mu\left(\frac{\partial}{\partial r}\left[\frac{1}{r}\frac{\partial}{\partial r}(ru_\theta)\right] + \frac{1}{r^2}\frac{\partial^2 u_\theta}{\partial \theta^2} + \frac{2}{r^2}\frac{\partial u_r}{\partial \theta} + \frac{\partial^2 u_\theta}{\partial z^2}\right) + \rho g_\theta$$

$$\tag{4.3.24e}$$

z component

$$\rho\left(\frac{\partial u_z}{\partial t} + u_r\frac{\partial u_z}{\partial r} + \frac{u_\theta}{r}\frac{\partial u_z}{\partial \theta} + u_z\frac{\partial u_z}{\partial z}\right) = -\frac{\partial p}{\partial z} + \mu\left(\frac{1}{r}\frac{\partial}{\partial r}\left(r\frac{\partial u_z}{\partial r}\right) + \frac{1}{r^2}\frac{\partial^2 u_z}{\partial \theta^2} + \frac{\partial^2 u_z}{\partial z^2}\right) + \rho g_z \tag{4.3.24f}$$

Table 4.3.1 (*Continued*)

Spherical Coordinates (r, θ, ϕ)

r component

$$\rho\left(\frac{\partial u_r}{\partial t} + u_r\frac{\partial u_r}{\partial r} + \frac{u_\theta}{r}\frac{\partial u_r}{\partial \theta} + \frac{u_\phi}{r\sin\theta}\frac{\partial u_r}{\partial \phi} - \frac{u_\phi^2 + u_\theta^2}{r}\right)$$

$$= -\frac{\partial p}{\partial r} + \mu\left(\left[\frac{1}{r^2}\frac{\partial^2}{\partial r^2}(r^2 u_r)\right] + \frac{1}{r^2\sin\theta}\frac{\partial}{\partial\theta}\left(\sin\theta\frac{\partial u_r}{\partial\theta}\right) + \frac{1}{r^2\sin^2\theta}\frac{\partial^2 u_r}{\partial\phi^2}\right) + \rho g_r \qquad \textbf{(4.3.24g)}$$

θ component

$$\rho\left(\frac{\partial u_\theta}{\partial t} + u_r\frac{\partial u_\theta}{\partial r} + \frac{u_\theta}{r}\frac{\partial u_\theta}{\partial \theta} + \frac{u_\phi}{r\sin\theta}\frac{\partial u_\theta}{\partial \phi} + \frac{u_r u_\theta}{r} - \frac{u_\phi^2\cot\theta}{r}\right)$$

$$= -\frac{1}{r}\frac{\partial p}{\partial\theta} + \mu\left(\frac{1}{r^2}\frac{\partial}{\partial r}\left[r^2\frac{\partial u_\theta}{\partial r}\right] + \frac{1}{r^2}\frac{\partial}{\partial\theta}\left[\frac{1}{\sin\theta}\frac{\partial}{\partial\theta}(u_\theta\sin\theta)\right] + \frac{1}{r^2\sin^2\theta}\frac{\partial^2 u_\theta}{\partial\phi^2} + \frac{2}{r^2}\frac{\partial u_r}{\partial\theta} - \frac{2\cos\theta}{r^2\sin^2\theta}\frac{\partial u_\phi}{\partial\phi}\right) + \rho g_\theta$$

$$\textbf{(4.3.24h)}$$

ϕ component

$$\rho\left(\frac{\partial u_\phi}{\partial t} + u_r\frac{\partial u_\phi}{\partial r} + \frac{u_\theta}{r}\frac{\partial u_\phi}{\partial \theta} + \frac{u_\phi}{r\sin\theta}\frac{\partial u_\phi}{\partial \phi} + \frac{u_r u_\phi}{r} + \frac{u_\theta u_\phi\cot\theta}{r}\right)$$

$$= -\frac{1}{r\sin\theta}\frac{\partial p}{\partial\phi} + \mu\left(\frac{1}{r^2}\frac{\partial}{\partial r}\left[r^2\frac{\partial u_\phi}{\partial r}\right] + \frac{1}{r^2}\frac{\partial}{\partial\theta}\left[\frac{1}{\sin\theta}\frac{\partial}{\partial\theta}(u_\phi\sin\theta)\right] + \frac{1}{r^2\sin^2\theta}\frac{\partial^2 u_\phi}{\partial\phi^2}\right.$$

$$\left. + \frac{2}{r^2\sin\theta}\frac{\partial u_r}{\partial\phi} + \frac{2\cos\theta}{r^2\sin^2\theta}\frac{\partial u_\theta}{\partial\phi}\right) + \rho g_\phi \qquad \textbf{(4.3.24i)}$$

Hence, we raise the question:

> *What is the value in having a set of equations so complex and general that we cannot solve them?*

There are two answers to this question, corresponding to the two primary uses we make of the generalized equations.

Recall how in Chapter 3 we set up the equations for the several "simple" flow problems (Eq. 3.2.22—Poiseuille flow through a circular tube, and Eq. 3.3.12—axial annular flow). In each case we *derived* a dynamic equation by making a force (momentum) balance over a differential element of volume. Because of the relative simplicity of the geometries of those one-dimensional laminar flows, it was easy to derive a differential equation for the velocity profile. With the Navier–Stokes equations in hand, we do not have to do such a derivation for future problems. All we do is *simplify* the Navier–Stokes equations by throwing out terms that vanish by virtue of the specific conditions that hold, *or that we assume hold,* for the problem at hand.

The second answer to the question above is that we may use the Navier–Stokes equations, in combination with a form of *dimensional analysis,* to guide the design and interpretation of experiments. If a specific flow field (e.g., the flow in a stirred tank) is too complex to permit us to solve the Navier–Stokes equations, we must do an experiment to learn about that flow behavior. Experiments can and should be designed based on the physics of the system. Dimensional analysis of the Navier–Stokes equations provides guidelines to experimental design. We take up this topic in earnest in Chapter 5.

4.4 SOME PROBLEMS SOLVED THROUGH SIMPLIFICATION OF THE NAVIER–STOKES EQUATIONS

We now pose several flow problems. After deleting terms from the Navier–Stokes equations on the basis of the respective problem statements, we solve the resultant dynamic equation(s), along with the continuity equation, to find the velocity and pressure profiles. We use these solutions to present a mathematical model of some aspect of the flow. Then we review the assumptions of the model and point out situations or conditions under which these assumptions might fail, causing features of the model to differ from observations. Keep in mind that the Navier–Stokes equations are strictly valid for an isothermal incompressible Newtonian flow.

4.4.1 Angular (Rotational) Drag Flow Between Long Concentric Cylinders

The geometry of angular drag flow between long concentric cylinders is given in Fig. 4.4.1. An isothermal incompressible Newtonian liquid is confined between a pair of concentric cylinders of radii κR and R, respectively. The cylinders are long compared to their radii: $L/R \gg 1$. The inner cylinder rotates steadily about its axis at Ω rad/s. This rotation induces flow in the viscous fluid that completely fills the volume between the two cylinders, and after some time a steady state (time-independent) velocity field is established. We want to find a solution for the steady state velocity vector as a function of position, and then develop a model of the relationship of the steady torque required to maintain the rotation of the inner cylinder as a function of the geometry and of the viscosity of the liquid.

We begin with the Navier–Stokes equations, since we are assuming that the fluid is an isothermal incompressible Newtonian liquid. It seems natural, in view of the geometry, to write the equations in cylindrical coordinates (Table 4.3.1: Eqs. 4.3.24d–4.3.24f). For the first few examples we will make a formal list of our assumptions, as well as the consequences implied by those assumptions.

Assumption	**Consequence**
1. Steady state has been reached.	We set all time derivatives $\partial/\partial t = 0$.
2. The cylinders are concentric and the flow is laminar.	The velocity components and pressure will be independent of angular position θ. All $\partial/\partial\theta = 0$. There is no radial velocity: $u_r = 0$.
3. The cylinders have no motion in the z direction.	There is no flow in the z direction: $u_z = 0$. The angular velocity u_θ does not vary in the z direction. Thus u_θ depends only on r.

Equation 4.3.24d takes the form

$$-\rho\frac{u_\theta^2}{r} = -\frac{\partial p}{\partial r} \tag{4.4.1}$$

Hence there is a radial pressure gradient due to centrifugal acceleration.
Equation 4.3.24e becomes

$$0 = \mu\frac{\partial}{\partial r}\left[\frac{1}{r}\frac{\partial}{\partial r}(ru_\theta)\right] \tag{4.4.2}$$

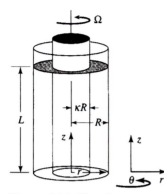

Figure 4.4.1 Flow between concentric cylinders: inner one rotating.

In Eq. 4.3.24f, only the terms

$$0 = -\frac{\partial p}{\partial z} + \rho g_z \tag{4.4.3}$$

remain. This accounts for the *hydrostatic* pressure gradient. Take note of the tremendous simplification in the Navier–Stokes equations brought about by these few assumptions.

We may solve Eq. 4.4.2 immediately.[2] Since u_θ is a function only of r, the derivatives are ordinary derivatives and we simply integrate Eq. 4.4.2 twice. The first integration gives

$$\frac{1}{r}\frac{d}{dr}(ru_\theta) = a \tag{4.4.4}$$

where a is an integration constant, and a second integration yields

$$u_\theta = \frac{ar}{2} + \frac{b}{r} \tag{4.4.5}$$

where b is a second integration constant. To evaluate these two constants, we must impose two boundary conditions based on physically plausible statements about what is going on at the boundaries. For example, it seems reasonable to assume that at the outer, stationary surface, the liquid satisfies a no-slip boundary condition. This takes the form

$$u_\theta = 0 \quad \text{at} \quad r = R \tag{4.4.6}$$

Hence

$$0 = \frac{aR}{2} + \frac{b}{R} \tag{4.4.7}$$

A no-slip condition should also hold at the inner, rotating, surface. Hence

$$u_\theta = \kappa R \Omega \quad \text{at} \quad r = \kappa R \tag{4.4.8}$$

---

[2] Once we have $u_\theta(r)$ we may integrate Eq. 4.4.1 to find $p = p_1(r) + p_2(z)$. From Eq. 4.4.3, we find $p_2(z)$ by direct integration. (See Problem P4.61.)

(Keep in mind that Ω is in units of radians per second, not revolutions per second.) We set $r = \kappa R$ in Eq. 4.4.5 and find

$$\kappa R \Omega = \frac{a \kappa R}{2} + \frac{b}{\kappa R} \tag{4.4.9}$$

We now have two linear algebraic equations with which to solve for the integration constants a and b. The results let us write the velocity profile (Eq. 4.4.5) in the form

$$u_\theta = \frac{\kappa R \Omega}{\kappa - 1/\kappa} \left(\frac{r}{R} - \frac{R}{r} \right) \tag{4.4.10}$$

It is easy to verify that this relation satisfies the two boundary conditions.

An important characteristic of this flow is the relationship of the rotational speed of the inner cylinder to the torque required to maintain that speed. If the liquid is viscous, the deformation of the liquid in the region near the moving surface must impose viscous forces on the surface. Some external torque is required to supply this force. How do we calculate this?

The only force that resists the rotation of the inner cylinder is the tangential force exerted by the fluid. This *shear* force acts in the θ direction on the surface normal to the r direction. By our notation for stresses, this force corresponds to a shear stress $T_{\theta r} = T_{r\theta}$. By our definition of the Newtonian fluid, we may find this stress (from Table 4.2.1) to be

$$T_{\theta r} = -\mu \left[r \frac{d}{dr} \left(\frac{u_\theta}{r} \right) \right] \qquad \text{at} \quad r = \kappa R \tag{4.4.11}$$

After performing the indicated differentiation, using Eq. 4.4.10, we find

$$T_{\theta r} = \frac{2\mu\Omega}{1 - \kappa^2} \qquad \text{at} \quad r = \kappa R \tag{4.4.12}$$

The torque on the inner cylinder T is obtained as the product of the tangential force ($T_{\theta r}$ times the area $2\pi\kappa R L$ on which this force acts) and the moment arm of radius κR:

$$\mathsf{T} = \frac{4\pi\mu L \kappa^2 R^2 \Omega}{1 - \kappa^2} \tag{4.4.13}$$

Because of the linearity of the stress–deformation rate law (the Newtonian fluid assumption), the torque–rotational speed relationship is linear. Thus we may use this model as a basis for the design of a viscometer, and indeed many commercial devices, called "coaxial cylinder viscometers," are based on this flow.

Several phenomena may occur and cause our observations to depart from the prediction of Eq. 4.4.13. We list some of these here, although we do not discuss them all at this point.

1. Normally the gap between the cylinders is filled with liquid and the rotation begins from a state of rest. The fluid must accelerate until the steady state velocity profile has been achieved. It would be useful to be able to estimate how long we must wait before attaining this steady state, since presumably Eq. 4.4.13 is valid only in the steady state. However, we can state here without proof that this time is of the order of

$$t_\infty = \frac{\rho R^2}{\mu} \tag{4.4.14}$$

2. The liquid might not be Newtonian. Most concentrated polymer solutions, and many particulate suspensions and emulsions, are non-Newtonian. Their stress–rate of deformation law is nonlinear, and Eq. 4.4.13 does not describe the torque–rotational speed relationship.

3. In very high viscosity liquids, viscous shear gives rise to frictional heating of the liquid. The temperature rise of the liquid will depend on the rate of rotation (among other variables); as a consequence, the data for the $T(\Omega)$ relationship may not be at a single temperature. Since viscosity is a strong function of temperature, the result would be a nonlinear $T(\Omega)$ curve, which we might mistakenly interpret as signifying non-Newtonian behavior.

4. An additional significant contribution to the torque, which arises from the shearing of the liquid between the bottom of the inner cylinder and the bottom of the outer cylinder, can be minimized by designing the bottom face of the inner cylinder to have a much smaller area than the curved cylindrical face. Hence we expect that keeping the ratio $\pi R^2/2\pi RL$ much less than 1 will serve to minimize this effect. That is why we normally design such a system with a large value of L/R. (See Problem 4.28.)

5. We have assumed laminar flow, which implies that each element of liquid moves in concentric circles about the vertical axis and that there are no other components to the velocity vector. Laminar flow might appear to be achievable as long as the cylinders are concentric. However, above a certain rotational speed (dependent on the fluid viscosity and the radii of the cylinders), the velocity field ceases to be laminar. The flow exhibits an instability, and a vortexlike structure appears in the flow. Studies of this phenomenon indicate that the assumption of laminar flow is accurate as long as the rotational speed is below a value that satisfies

$$\frac{\Omega R^2 \rho}{\mu} < \frac{40}{\kappa(1-\kappa)^{3/2}} \qquad (4.4.15)$$

4.4.2 Liquid Film on a Vertical Surface

We look next at a flow problem in which the liquid is bounded on one side by a free surface and the flow is driven strictly by a body force (gravity), not by imposition of moving boundaries. Figure 4.4.2 shows the details needed for the analysis.

Before embarking on an analysis of this flow, it is important to state clearly our reasons for doing the analysis. Our goal is to discover a relationship between the liquid

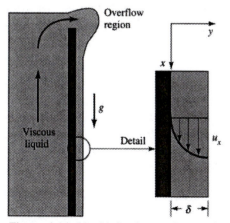

Figure 4.4.2 Liquid flowing under gravity along a vertical plate.

film thickness and the flowrate. We begin by introducing the following assumptions, which help to define the model we will develop.

Assumption	**Consequence**
1. Steady state has been reached.	We set all time derivatives $\partial/\partial t = 0$.
2. The liquid is incompressible and Newtonian, and the flow is isothermal.	We can use the Navier–Stokes equations.
3. The flow is laminar and strictly parallel to the plate. There is no flow in the y and z directions, and no derivatives $\partial/\partial x$.	The model will not hold in the overflow region. The only velocity component is u_x, downstream of the overflow region.
4. The liquid film thickness δ is not a function of position x.	u_x is a function only of y. We call this "fully developed flow."

Assumption 4 simplifies the problem enormously, but it may not be obvious that it is physically possible for this to hold. We often attempt to simplify a problem of interest by introducing assumptions for which we have no justification. Our only recourse is to follow the analysis and see whether a given assumption is so restrictive that it holds only under trivial conditions, such as zero flow. Usually an assumption that is physically overrestrictive, or even physically meaningless, will be revealed as such later, when the solution based (in part) on that assumption is critically evaluated. Hence we should not spend too much time worrying about assumptions early in the analysis.

It seems natural to use rectangular coordinates in this problem. Equation 4.3.24a becomes

$$0 = -\frac{\partial p}{\partial x} + \mu \left(\frac{\partial^2 u_x}{\partial y^2} \right) + \rho g_x \tag{4.4.16}$$

Equation 4.3.24b gives us

$$0 = -\frac{\partial p}{\partial y} \tag{4.4.17}$$

Equation 4.3.24c gives the trivial result

$$0 = 0 \tag{4.4.18}$$

since each term in the equation vanishes.

What does the continuity equation tell us, by the way? We use Eq. 4.1.6 and Table 4.1.1 (setting the vector $\mathbf{A} = \mathbf{u}$) and find

$$0 = \nabla \cdot \mathbf{u} = \frac{\partial u_x}{\partial x} \tag{4.4.19}$$

when we assume that $u_y = u_z = 0$. Hence the assumption that we can have a velocity u_x independent of x actually *follows* from the assumption that $u_y = u_z = 0$. If u_x is independent of x, it seems reasonable to suppose that δ will be independent of x, as well. In fact we can prove this by considering the volume flowrate, defined by

$$Q = W \int_0^\delta u_x dy \tag{4.4.20}$$

where W is the width of the plate in the z direction. Conservation of mass requires that Q not vary along the direction x, since no liquid crosses the boundaries at $y = 0$ and $y = \delta$. If we differentiate Q above, with respect to x, and use the Leibniz rule for differentiating an integral when the upper limit (δ, in this case) is *possibly* a function of x, we find

$$\frac{1}{W}\frac{dQ}{dx} = 0 = \int_0^\delta \frac{\partial u_x}{\partial x}\,dy + u_x(\delta)\frac{d\delta}{dx} \tag{4.4.21}$$

The integral term vanishes by virtue of the continuity equation (Eq. 4.4.19). Hence we conclude (and so we do not have to assume) that $d\delta/dx = 0$ is a *consequence* of the assumption that $u_y = u_z = 0$.

Equation 4.4.16 involves two unknowns: p and u_x. We need some additional constraint on $p(x)$. Consider the free surface ($y = \delta$). It is in contact with the atmosphere; it is planar; and in the surrounding gas the pressure just outside the free surface (say, at $y = \delta+$) is approximately independent of x, since the gas density is so small that the hydrostatic effect in the gas phase is negligible. Then the pressure along the free surface at $y = \delta+$ is independent of x, and this must be true of the pressure just inside the liquid (along $y = \delta-$). From Eq. 4.4.17 we see that the pressure is independent of the coordinate y. Hence the pressure along the solid surface ($y = 0$) must equal the pressure outside the film. It follows then that $\partial p/\partial x$ is *everywhere* zero. This eliminates p, and now we simply solve for u_x from

$$0 = \mu\left(\frac{d^2 u_x}{dy^2}\right) + \rho g_x \tag{4.4.22}$$

We integrate twice to find

$$u_x = -\frac{\rho g_x}{2\mu}y^2 + ay + b \tag{4.4.23}$$

As a boundary condition, we again assume no slip at the solid surface, or

$$u_x = 0 \quad \text{at} \quad y = 0 \tag{4.4.24}$$

The second boundary condition is less obvious. We do not know the velocity along the surface at $y = \delta$. What physical statement can we impose there that is consistent with the physical picture we have set up? Boundary conditions in fluid dynamics problems are usually in terms of the velocities on the boundaries, but sometimes we know the *stresses* on some of the boundaries. That is the case here. We know the *shear* stress on the free surface!

The shear stress exerted by the liquid along the interface $y = \delta$ must be balanced by the shear stress exerted by the gas along that surface. This is just a statement that the shear stress must be continuous across this interface. But a gas of low viscosity, as in this case, is not capable of exerting any significant shear stress unless the surface velocity is extremely large. We *assume*, then, that there is no shear stress along the surface $y = \delta$. Then our second boundary condition is

$$T_{xy} = -\mu\frac{du_x}{dy} = 0 \quad \text{at} \quad y = \delta \tag{4.4.25}$$

We may now force Eq. 4.4.23 to satisfy Eqs. 4.4.24 and 4.4.25 and solve for the integration constants a and b. This lets us write the velocity profile as

$$u_x = \frac{\rho g_x \delta^2}{2\mu}\left[\frac{2y}{\delta} - \left(\frac{y}{\delta}\right)^2\right] \tag{4.4.26}$$

It is easy to verify that this satisfies the boundary conditions given.

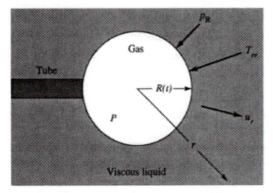

Figure 4.4.3 Gas bubble in a viscous liquid.

The volume flowrate follows from Eq. 4.4.20, and after performing the integration we find

$$Q = \frac{\rho g_x W \delta^3}{3\mu} \tag{4.4.27}$$

Equation 4.4.27 provides a model for the $Q(\delta)$ relationship that can be subjected to experimental test. Observations suggest that it is a good model for thin viscous films at low flowrates. As the flowrate increases, the free surface becomes rippled, the flow is no longer laminar, and the $Q(\delta)$ relationship is no longer as simple as that given above. As a rough criterion for the validity of Eq. 4.4.27, based on some experimental observations, we look for a flowrate that satisfies the following relationship:

$$\text{Re} = \frac{4\rho\delta\langle u_x \rangle}{\mu} = \frac{4\rho Q}{\mu W} < 10 \tag{4.4.28}$$

where $\langle u_x \rangle$ is the velocity averaged across δ. The dimensionless grouping used here is a Reynolds number,[3] but defined for this specific flow. In general a Reynolds number always has the form

$$\text{Re} = \frac{\rho L U}{\mu} \tag{4.4.29}$$

where L and U are appropriate length and velocity scales that are characteristic of the geometry and flow field of the specific problem. The factor of 4 that appears in Eq. 4.4.28 is arbitrary, but customary.

4.4.3 Growth of a Bubble in a Viscous Liquid

Our next problem has several features that are distinct from those of earlier examples. Figure 4.4.3 shows the geometry. A gas bubble is formed at the tip of a small tube that is completely immersed in a large body of liquid. The bubble remains in contact with the open end of the tube a way that allows us to control the pressure P inside the bubble. We imagine that the bubble is small, and we neglect the effect of gravity on the shape of the bubble and on the pressure distribution within the liquid in the

---

[3] Mudawar and Houpt show data for laminar appearing films at Reynolds numbers in the range 1200 to 5000 (see Problem 4.32).

neighborhood of the bubble. In this case then, buoyancy forces will not lift the bubble from the tip of the tube.

We want to develop a model for the relationship of the rate of growth of the bubble to the pressure P and to the properties of the fluid. Again we begin by listing plausible assumptions that will simplify the model.

Assumption	**Consequence**
1. The liquid is incompressible and Newtonian. The system is isothermal.	We may begin with the Navier–Stokes equations.
2. The bubble remains a sphere as it grows.	The only component of velocity in the liquid is radial, and u_r is a function only of r and t.
3. The effect of gravity is neglected.	The pressure distribution has no hydrostatic contribution. The bubble stays on the tip.

The continuity equation, in spherical coordinates, takes the form (see Table 4.1.1, Eq. 4.1.8c)

$$\frac{1}{r^2}\frac{\partial}{\partial r}(r^2 u_r) = 0 \tag{4.4.30}$$

One integration yields the immediate result that

$$u_r = \frac{A}{r^2} \tag{4.4.31}$$

Note that in Eq. 4.4.30 we continued to write the *partial* derivative, despite our assumption that the radial velocity is a function of only one coordinate, r. This is because the growth of a bubble is not a steady state dynamic problem. Clearly the radius is changing with time, and we have no reason to expect the velocity to be independent of time as long as the bubble grows. If this is the case, where is the time dependence in Eq. 4.4.31? It must be in the "constant" of integration A. In other words, $A = A(t)$, and it is an unknown function of time at this point.

At the surface of the bubble, that is, at $r = R(t)$, the velocity u_r is just the rate of growth, so

$$u_r|_{r=R} = \frac{dR}{dt} = \dot{R} = \frac{A}{R^2} \tag{4.4.32}$$

This yields a relationship for A in terms of $R(t)$, and the radial velocity then takes the form

$$u_r = \frac{R^2\dot{R}}{r^2} \tag{4.4.33}$$

Note that we have not found the function $A(t)$; we have simply replaced it by the unknown functions $R(t)$ and its time derivative.

Now we must examine the dynamic equations—the Navier–Stokes equations. Equation 4.3.24g (Table 4.3.1) is simply

$$\rho\left(\frac{\partial u_r}{\partial t} + u_r\frac{\partial u_r}{\partial r}\right) = -\frac{\partial p}{\partial r} + \mu\left[\frac{1}{r^2}\frac{\partial^2}{\partial r^2}(r^2 u_r)\right] \tag{4.4.34}$$

When we substitute u_r from Eq. 4.4.33, we end up with a single differential equation for $R(t)$, and the pressure $p(r)$. Thus we have two unknowns and only one equation, a dilemma that seems to hound us as we set up fluid dynamics problems. As before, we must come up with another relationship that involves the pressure, and we develop the following statement: the pressure inside the bubble, P, which acts normal to the surface (hence is radially directed), is equal to the radial stress component outside the bubble, at the surface of the bubble, $T_{rr}|_R$, except for a (usually) small contribution due to surface tension. Recalling the Young–Laplace equation (Eq. 2.2.7), we write this balance of radial stress as

$$P = T_{rr}|_R + \frac{2\sigma}{R} \tag{4.4.35}$$

Note that the effect of surface tension is to make the radial stress inside the bubble *greater* than that outside. Note also that the internal radial stress is strictly the internal pressure P. We imply that in the gas phase inside the bubble there are no stresses arising from the deformation of the gas as the bubble grows. This is reasonable because the gas is such a low viscosity fluid.

The radial stress outside comes from the viscous deformation in the radial direction. For an incompressible Newtonian fluid, we have

$$T_{rr} = p - \mu\Delta_{rr} \tag{4.4.36}$$

From Table 4.2.1 we find, in spherical coordinates,

$$\Delta_{rr} = 2\frac{\partial u_r}{\partial r} \tag{4.4.37}$$

Note that there is no *shear* deformation in this flow. (Refer to Example 4.2.1 for the definition of the shear flow.) In this flow, the only nonzero components of Δ_{ij} are those for which $i = j$. Such a flow is called an "extensional flow," and Δ_{rr} is called an "extension rate."

If we now substitute Eq. 4.4.33 into Eq. 4.4.37, set $r = R$ in the result, and apply Eq. 4.4.36, we find

$$T_{rr}|_R = p_R + \frac{4\mu\dot{R}}{R} = P - \frac{2\sigma}{R} \tag{4.4.38}$$

where p_R is a new notation for $p(R)$. Hence we have

$$P = p_R + \frac{2\sigma}{R} + \frac{4\mu\dot{R}}{R} \tag{4.4.39}$$

and our problem is reduced to finding $p(R) = p_R$. Now it is appropriate to substitute u_r from Eq. 4.4.33 into Eq. 4.4.34, yielding

$$\frac{\partial p}{\partial r} = \rho\left[\frac{2\dot{R}^2 R^4}{r^5} - \frac{\ddot{R}R^2}{r^2} - \frac{2R\dot{R}^2}{r^2}\right] \tag{4.4.40}$$

We integrate this equation once with respect to r and find

$$\frac{p(r)}{\rho} = -\frac{\dot{R}^2}{2}\left(\frac{R}{r}\right)^4 + \frac{\ddot{R}R^2 + 2R\dot{R}^2}{r} + B \tag{4.4.41}$$

where B is a constant of integration. We use a boundary condition according to which the pressure in the fluid, far from the bubble surface, is the ambient pressure on the body of liquid, p_∞.

We will write this boundary condition at $r = \infty$, but we will have to come back to this point, since obviously we would be interested in situations of bubbles bounded by a finite amount of liquid. We will find that the use of $r = \infty$ for this boundary condition does not cause any trouble as long as there is liquid at a distance of about $10R$ from the bubble center. Upon setting $r = \infty$ in Eq. 4.4.41, we find

$$B = \frac{p_\infty}{\rho} \tag{4.4.42}$$

Then, setting $r = R$ in Eq. 4.4.41, we find p_R with the result

$$p_R = p_\infty + \rho\left(R\ddot{R} + \frac{3\dot{R}^2}{2}\right) \tag{4.4.43}$$

Finally (can you believe it?) we find the desired result—the internal bubble pressure—by substituting p_R into Eq. 4.4.39. The result is

$$P(t) = \frac{2\sigma}{R} + \frac{4\mu\dot{R}}{R} + p_\infty + \rho\left(R\ddot{R} + \frac{3\dot{R}^2}{2}\right) \tag{4.4.44}$$

It is very easy to lose sight of the physics of the problem after so much mathematical manipulation. At this point we have a model that relates the pressure $P(t)$ inside a gas bubble to the rate of change in radius of the bubble. There are three contributions to the internal pressure. One arises from surface tension. One accompanies the viscous resistance to deformation. The third contribution (the last term of Eq. 4.4.44) is associated with inertial effects in the liquid. Even in a liquid having a vanishingly small viscosity, a pressure would arise associated with the acceleration of the mass of liquid that surrounds the bubble. What we call "inertia" here is the resistance of the fluid to this acceleration. A force is required to move the fluid out of the way of the growing bubble.

It is interesting to examine the magnitude of some of the pressures and accelerations associated with bubble growth or collapse. We do this in Example 4.4.1.

EXAMPLE 4.4.1 *Growth of a Bubble in a Very Viscous Liquid*

Tiny bubbles are entrained within a highly viscous molten polymer that is extruded from a wire coating die at an absolute pressure (prior to extrusion) of 10^6 Pa (about 10 atm). The bubbles have an initial radius of 100 μm. The liquid has a viscosity at the temperature of the die of 100 Pa·s (1000 poise). Since the pressure within the bubbles exceeds that of the ambient medium after extrusion, the bubbles grow. Assume that the pressure remains constant at 10^6 Pa within the bubbles during growth. (This assumption might hold for bubble growth due to the vapor pressure of a volatile liquid within the bubbles.) How long will the bubbles take to attain a radius of 1 mm?

Equation 4.4.44 is a differential equation for $R(t)$, given $P(t)$. Let's look first at a simplifying assumption. We will neglect the contributions of surface tension and inertia, considering only viscous effects. This approach is suggested by the high viscosity of the liquid. We will examine the restrictions on this approximation shortly. We write Eq. 4.4.44 as

$$\frac{4\mu\dot{R}}{R} = P - p_\infty = \Delta P \tag{4.4.45}$$

For ΔP we have a constant value of $10 - 1 = 9$ atm $= 9 \times 10^5$ Pa. Equation 4.4.45 is easily solved by integration to yield

$$\frac{R}{R_o} = \exp\left(\frac{\Delta P t}{4\mu}\right) \tag{4.4.46}$$

We have used an initial condition in the form

$$R = R_o \quad \text{at} \quad t = 0 \tag{4.4.47}$$

To grow to a radius of 1 mm, or $R/R_o = 10$, the time required is found from

$$\ln\frac{R}{R_o} = \left(\frac{\Delta P t}{4\mu}\right) = \ln 10 = 2.3 \tag{4.4.48}$$

Then the time required for this extent of growth is

$$t = \frac{9.2\mu}{\Delta P} = \frac{9.2 \times 100}{9 \times 10^5} \approx 10^{-3}\,\text{s} \tag{4.4.49}$$

This is an extremely short time! Any result that seems counter intuitive, should start you looking for errors in the arithmetic, or for poor assumptions in the model. You won't find any here. Bubble growth is indeed an extremely rapid process in this case because the driving pressure is large, even though the liquid has a very high viscosity (100 Pa · s is about the viscosity of cold honey).

Now let's look at some of the terms that we neglected in writing Eq. 4.4.45. The surface tension contribution is

$$P_\sigma = \frac{2\sigma}{R} \tag{4.4.50}$$

In the worst case (the smallest bubble), when $R = R_o = 100$ μm, and taking a typical surface tension value of 0.05 N/m (see Table 2.2.1), we find

$$P_\sigma = \frac{2 \times 0.05\,\text{N/m}}{10^{-4}\,\text{m}} = 10^3\,\text{N/m}^2 = 10^3\,\text{Pa} \tag{4.4.51}$$

This pressure is about three orders of magnitude below the driving pressure for this process. Hence we can safely neglect surface tension in this example. (This might not be possible for bubbles growing in low viscosity liquids under very small internal pressures.)

We also neglected inertial terms in writing Eq. 4.4.45. Let's compare the inertial terms to the viscous term. If we use Eq. 4.4.46 for $R(t)$, we find the following results:

$$V \equiv \frac{4\mu\dot{R}}{R} = \Delta P \tag{4.4.52}$$

$$I \equiv \rho\left(R\ddot{R} + \frac{3\dot{R}^2}{2}\right) = \frac{5}{32}\frac{\rho R_o^2(\Delta P)^2}{\mu^2}\left(\frac{R}{R_o}\right)^2 \tag{4.4.53}$$

Then we find the ratio of the inertial to viscous terms, I/V, in the form

$$\frac{I}{V} = \frac{5}{32}\frac{\rho \Delta P R_o^2}{\mu^2}\left(\frac{R}{R_o}\right)^2 \tag{4.4.54}$$

In this example we take

$$\rho = 10^3\,\text{kg/m}^3 \quad \Delta P = 9 \times 10^5\,\text{Pa} \quad R_o = 10^{-4}\,\text{m} \quad \mu = 100\,\text{Pa·s}$$

and find (note the use of consistent SI units)

$$\frac{I}{V} = \frac{5}{32} \frac{10^3\,\text{kg/m}^3 \times 9 \times 10^5\,\text{Pa} \times 10^{-8}\,\text{m}^2}{(100)^2\,\text{Pa}^2 \cdot \text{s}^2} \left(\frac{R}{R_o}\right)^2 \approx 1.4 \times 10^{-4} \left(\frac{R}{R_o}\right)^2 \quad \textbf{(4.4.55)}$$

Unless R/R_o becomes huge, the inertial terms in this viscous liquid will be negligible.

In real engineering systems, bubble growth may not occur in a way that yields spherical bubbles. For example, the bubbles may be deformed if they are close together (as in foam growth) or near a solid boundary—perhaps because they formed on the surface of a containing vessel. Also, in real systems, the behavior of the gas inside the bubble may be complex. The internal gas pressure may not be constant, for example, if the liquid is vaporizing at the bubble surface as the bubble grows. If the amount of gas in the bubble is fixed (because, e.g., it is insoluble in the liquid), the gas pressure may satisfy some thermodynamic constraint. If the bubble growth is as rapid as illustrated in Example 4.4.1, the assumption of *adiabatic* expansion may be valid, and the pressure–volume relationship may have to satisfy $PV^\gamma = $ constant, where V is the bubble volume at any time, the pressure is P, and γ is the ratio of the heat capacities, C_p/C_v.

4.4.4   Viscous Resistance to a Freely Falling Object: A Long Cylinder in a Coaxial, Fluid-Filled Tube

Now we examine the motion of a body falling through a viscous liquid. We will find the velocity field by solving, as before, the differential equations that describe the constraints of conservation of mass and momentum. This time we will find that we must return to those constraints in an *integral* form to find the velocity of the body through the fluid.

We begin with a definition of the problem, which is presented graphically in Fig. 4.4.4. A long solid circular cylinder of length L and radius R_1 falls under the attraction of gravity, coaxially, through a liquid held within a concentric cylindrical vessel of radius R_2, *closed at the bottom*. The axis of the system is vertical and gravity acts downward. We want to find the velocity U at which this cylinder falls along the axis, assuming that somehow the concentric geometry is maintained. We need to find the velocity field in the fluid between the two cylindrical surfaces first, because this velocity field will exert

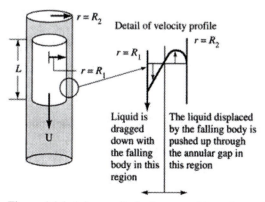

Figure 4.4.4 A long cylinder immersed in a viscous liquid falls under the action of gravity, within a concentric cylinder that is closed at the bottom.

a shear stress on the face of the falling cylinder, thereby controlling the motion of the cylinder. (Notice how thinking about the physics dictates the strategy of analysis.)

We choose a *moving* coordinate system in formulating this problem, fixed in the moving cylinder, with $z = 0$ at the upper surface of the moving cylinder. We make this choice because in the "laboratory" coordinate system, this flow field is not at steady state. If we were to pick a fixed point in the liquid, we would observe the velocity changing with time as the cylinder moved through that region. If the cylinder fell at a steady velocity, however, then in a coordinate system attached to the falling cylinder, the fluid between the two cylinders would appear to be in steady flow. This choice of a moving coordinate system is part of the problem-solving *strategy*.

Assumption	**Consequence**
1. Steady state has been reached.	We set all time derivatives $\partial/\partial t = 0$. U is not a function of time.
2. The liquid is incompressible and Newtonian, and the flow is isothermal.	We can use the Navier–Stokes equations.
3. The flow is fully developed in the region between R_1 and R_2. This assumption, which ignores the end effects as liquid enters and exits the annular region, might be a good approximation if $L \gg R_2 - R_1$. The flow is laminar and axisymmetric.	The only velocity component is u_z and it is a function only of r.

We choose a cylindrical coordinate system, in view of the symmetry and the geometry. Because of symmetry, we can write

$$\frac{\partial}{\partial \theta} = 0 \tag{4.4.56}$$

and

$$u_\theta = 0 \tag{4.4.57}$$

The continuity equation in cylindrical coordinates (Eq. 4.1.5 in the most general case) has the form

$$\frac{\partial \rho}{\partial t} + \frac{1}{r}\frac{\partial}{\partial r}(\rho r u_r) + \frac{1}{r}\frac{\partial}{\partial \theta}(\rho u_\theta) + \frac{\partial}{\partial z}(\rho u_z) = 0 \tag{4.4.58}$$

With the assumptions of steady state, constant density, and axial symmetry, this reduces to

$$\frac{1}{r}\frac{\partial}{\partial r}(r u_r) + \frac{\partial u_z}{\partial z} = 0 \tag{4.4.59}$$

The neglect of end effects means that the velocity profile does not vary in the z direction. Thus

$$\frac{\partial u_z}{\partial z} = 0 \tag{4.4.60}$$

This lets us integrate Eq. 4.4.59 to find

$$r u_r = \text{constant for all } r \tag{4.4.61}$$

But liquid cannot cross the solid boundaries, so

$$ru_r = 0 \quad \text{on} \quad r = R_1 \quad \text{and} \quad R_2 \tag{4.4.62}$$

Hence $u_r = 0$ everywhere. Thus the vanishing of u_r is not a separate assumption, but a consequence of the assumptions of axisymmetric constant density fully developed flow.
  Now let's look at the momentum equations.

Momentum (Navier–Stokes) Equations

r (Eq. 4.3.24d)

$$\left.\begin{array}{l} u_r = 0 \\ u_\theta = 0 \\ g_r = 0 \end{array}\right\} \text{leads to} \frac{\partial p}{\partial r} = 0 \tag{4.4.63}$$

θ (Eq. 4.3.24e)

$$\left.\begin{array}{l} u_\theta = 0 \\ \dfrac{\partial}{\partial\theta} = 0 \\ g_\theta = 0 \end{array}\right\} \text{leads to} \frac{\partial p}{\partial\theta} = 0 \tag{4.4.64}$$

Hence

$$p = p(z) \text{ only} \tag{4.4.65}$$

z (Eq. 4.3.24f)

$$\left.\begin{array}{l} \dfrac{\partial}{\partial t} = 0 \\[4pt] u_r = 0 \\[4pt] u_\theta = 0 \\[4pt] \dfrac{\partial u_z}{\partial z} = 0 \\[4pt] \dfrac{\partial}{\partial\theta} = 0 \end{array}\right\} \text{leads to } 0 = -\frac{\partial p}{\partial z} + \mu\frac{1}{r}\frac{\partial}{\partial r}\left(r\frac{\partial u_z}{\partial r}\right) + \rho g_z \tag{4.4.66}$$

(Compare this to Eq. 3.2.22.) Since the pressure is not a function of r, and the velocity is *only* a function of r, this equation is of the form

$$0 = F(z) + G(r) \tag{4.4.67}$$

or

$$F(z) = \frac{\partial p}{\partial z} - \rho g_z = -G(r) \tag{4.4.68}$$

We cannot write $F(z) = -G(r)$ for any (r, z) unless

$$F = -G = \text{constant} = C \tag{4.4.69}$$

Thus we find two equations:

$$\frac{dp}{dz} - \rho g_z = C \quad \text{and} \quad \mu \frac{1}{r} \frac{d}{dr}\left(r \frac{du_z}{dr}\right) = C \tag{4.4.70}$$

We may integrate the pressure equation immediately and find

$$p - p_o = (C + \rho g_z)z \tag{4.4.71}$$

or

$$p - \rho g z - p_o = Cz \tag{4.4.72}$$

where p_o is a constant of integration. (We drop the subscript on g in the rest of the analysis.) We now define a modified pressure variable $\mathcal{P}$ to include the gravitational contribution to pressure:

$$p - \rho g z = \mathcal{P} \tag{4.4.73}$$

and, setting $p = p_o$ at $z = 0$, which is just an arbitrary datum for pressure, we find

$$p_o = \mathcal{P}_o \tag{4.4.74}$$

This lets us write

$$\mathcal{P} - \mathcal{P}_o = Cz \tag{4.4.75}$$

Recall that the z axis is taken as positive downward, and its origin is the upper end of the falling cylinder. At the lower end of the falling cylinder ($z = L$), we write the pressure as

$$\mathcal{P}_L - \mathcal{P}_o = CL \tag{4.4.76}$$

This permits us to write the constant C as

$$C = \frac{\mathcal{P}_L - \mathcal{P}_o}{L} = -\frac{\Delta\mathcal{P}}{L} \tag{4.4.77}$$

Now we return to the differential equation (Eq. 4.4.70) for the velocity field, using this result for C:

$$\mu \frac{1}{r} \frac{d}{dr}\left(r \frac{du_z}{dr}\right) = -\frac{\Delta\mathcal{P}}{L} \tag{4.4.78}$$

Two successive integrations yield

$$u_z = -\frac{\Delta\mathcal{P}}{4\mu L} r^2 + A \ln r + B \tag{4.4.79}$$

where A and B are the constants of integration. At this point we need a reminder that the constant C in Eq. 4.4.70 is not known (we hid this defect in the course of writing Eq. 4.4.77). Thus it is now the pressure *difference* across the ends of the falling cylinder, $\Delta\mathcal{P}$, that is not known. In summary, we have three unknown constants at this point, A, B, and $\Delta\mathcal{P}$. We are going to need two boundary conditions on u_z along with some boundary condition on the pressure field. In fact, things will get worse in just a moment.

First, let's apply the no-slip boundary condition to the velocity u_z at the surfaces R_1 and R_2. Keep in mind that we have chosen to use a *moving* coordinate system: the inner cylinder is stationary and the outer cylinder appears to have a velocity upward of $-U$. These boundary conditions then take the forms

$$u_z = -U \quad \text{on} \quad r = R_2 \tag{4.4.80}$$

$$u_z = 0 \quad \text{on} \quad r = R_1 = \kappa R_2 \tag{4.4.81}$$

Before the algebra gets out of hand, let's introduce a simpler notation (cf. Eq. 3.3.17). With

$$\frac{u_z}{U} = \phi \tag{4.4.82}$$

$$\frac{\Delta \mathscr{P} R_2^2}{4 \mu L U} = \Phi \tag{4.4.83}$$

$$\frac{r}{R_2} = s \tag{4.4.84}$$

we may write the solution for the velocity field in a nondimensional format[4]:

$$\phi = -\Phi s^2 + \frac{A}{U} \ln s R_2 + \frac{B}{U} \tag{4.4.85}$$

and the boundary conditions are

$$\phi = -1 \quad \text{on} \quad s = 1 \tag{4.4.86}$$

$$\phi = 0 \quad \text{on} \quad s = \kappa \tag{4.4.87}$$

This yields the velocity profile in the form (cf. Eq. 3.3.18)

$$\phi(s) = -1 + \Phi(1 - s^2) + \frac{1 - \Phi(1 - \kappa^2)}{\ln \kappa} \ln s \tag{4.4.88}$$

Keep in mind that Φ is unknown at this point, since the pressure difference generated by this flow is unknown. However, this pressure difference must satisfy a constraint based on the need of the liquid displaced by the falling cylinder to go somewhere. *If the bottom of the surrounding coaxial cylinder is closed, as we have assumed, then the liquid can only go back up the annular region.* The volume flowrate through the annular region must then balance the volumetric displacement due to the falling cylinder.

Soon we will apply this statement in a mathematical form. It is a little simpler first to transform Eq. 4.4.88 back to the *fixed* coordinate system, where the free inner cylinder falls at velocity $+U$ and the outer cylinder is stationary. We just add $+U$ to the velocity field; or, in dimensionless variables, we add $+1$ to ϕ in Eq. 4.4.88, yielding

$$\phi'(s) = \Phi(1 - s^2) + \frac{1 - \Phi(1 - \kappa^2)}{\ln \kappa} \ln s \tag{4.4.89}$$

It is not difficult to see that ϕ' (the dimensionless velocity in *fixed* coordinates) satisfies the expected boundary conditions in the fixed coordinate system. Now we impose the condition that the bottom of the outer cylinder is closed, and the liquid displaced by the falling cylinder must flow through the annular region. This physical statement takes

---

[4] Be careful of the notation here! We have a lowercase ϕ and a capital Φ. When the going gets tough, the tough learn Greek!

the mathematical form (cf. Eq. 3.3.19)[5]

$$\int_{R_1}^{R_2} 2\pi u'_z(r) r \, dr = -\pi R_1^2 U = -\pi \kappa^2 R_2^2 U \tag{4.4.90}$$

or

$$\int_\kappa^1 2\phi' s \, ds = -\kappa^2 \tag{4.4.91}$$

When Eq. 4.4.89 is substituted into this integral and the integration is carried out, the result is an algebraic equation for Φ as a function of κ. This has the form (cf. Eq. 3.3.23)

$$\Phi = \frac{1}{(1 + \kappa^2) \ln \kappa + (1 - \kappa^2)} \tag{4.4.92}$$

(Φ is negative, by the way.) From Eq. 4.4.88, we see that we have the velocity profile $\phi(s)$ in terms of the single parameter κ, since $\Phi(\kappa)$ is given in Eq. 4.4.92. However, ϕ (and ϕ') have been made nondimensional using U, which is still unknown. *In fact it is U that we are after.* Hence we must add an additional *constraint*—an additional equation—to the analysis. The requirement is that the stresses acting on the surface of the falling cylinder retard the fall of the cylinder under the influence of gravity. Then the additional equation is a force balance, namely,

(shear force + pressure force) opposes gravity force

First let's look at the shear *stress*.

$$\tau_{rz}\big|_{R_1} = -\mu \frac{du'_z}{dr}\bigg|_{R_1} = -\frac{\mu U}{R_2} \frac{d\phi'}{ds}\bigg|_{s=\kappa} \tag{4.4.93}$$

To find $d\phi'/ds$ at $s = \kappa$, we must differentiate the velocity profile, given in Eq. 4.4.89. If we do that (without algebraic errors—see Problem 4.35), we find

$$\frac{d\phi'}{ds}\bigg|_{s=\kappa} = \frac{1 - \kappa^2}{1 + \kappa^2} \frac{J}{\kappa} \tag{4.4.94}$$

which we simplify by writing

$$J \equiv \left[\ln \kappa + \frac{(1 - \kappa^2)}{(1 + \kappa^2)} \right]^{-1} \tag{4.4.95}$$

It is not hard to verify that J is a negative number, and so the shear stress, according to the definition in Eq. 4.4.93, is a positive number. Since we are now working in the fixed coordinate system, *a positive stress points down*, in the direction of motion of the falling cylinder. But we expect the shear stress to oppose the motion, hence to point upward! This apparent dilemma is resolved by recalling that what we calculate in Eq. 4.4.93 is the stress exerted *on the fluid by the surface* at $s = \kappa$. In the force balance preceding Eq. 4.4.93, we are writing the forces exerted on the cylinder *by the fluid*. That force is $-\tau_{rz}$ times the area on which it acts.

Next we examine the pressure force acting on the falling cylinder. Because of viscous effects, there is a pressure gradient along the direction of motion. As a consequence,

---

[5] Since the displaced flow is upward through the annular region, Eq. 4.4.90 gives a negative volumetric flowrate.

on the circular faces of the cylinder, at $z = 0$ and $z = L$, there are different pressures, hence a net pressure difference acting on this area. The force due to the pressure difference, *opposing the motion,* is just $(p_L - p_o)$ times the face area of the cylindrical end.

When we put these ideas together, the force balance takes the form

$$+\tau_{rz}A_1 + (p_L - p_o)A_2 = \text{body force} = \rho_{\text{solid}}gLA_2 \tag{4.4.96}$$

We write $\tau_{rz}A_1$ with a plus sign because that is the magnitude of the shear force opposing the motion. The respective areas in Eq. 4.4.96 are simply

$$A_1 = 2\pi R_1 L \quad \text{and} \quad A_2 = \pi R_1^2 \tag{4.4.97}$$

Note that we are using the weight of the cylinder as the body force. Why are we not accounting for buoyancy? The reason is that buoyancy, which arises from the gravitational field, will appear in the pressure terms p_o and p_L if gravity appears in the Navier–Stokes equation, as it does (see Eq. 4.4.66). To see this, let's subtract the weight of the fluid displaced by the cylinder (this is the buoyancy term) from both sides of Eq. 4.4.96:

$$\tau_{rz}A_1 + (p_L - p_o)A_2 - \rho_{\text{fluid}} gLA_2 = \text{net body force}$$
$$= \rho_{\text{solid}} gLA_2 - \rho_{\text{fluid}} gLA_2 \tag{4.4.98}$$

or

$$\tau_{rz}A_1 - \Delta\mathscr{P}A_2 = \Delta\rho gLA_2 \tag{4.4.99}$$

where $\Delta\rho$ is the difference in density between the solid and the fluid. Hence, if we include buoyancy in the body force (as in Eq. 4.4.99), we must use the pressure difference $\Delta\mathscr{P}$ in place of Δp. Equations 4.4.96 and 4.4.99 are different forms of the same physical statement.

In terms of dimensionless variables the left-hand side of Eq. 4.4.99 becomes

$$-\frac{\mu U}{R_2}\left(2\frac{d\phi'}{ds}\bigg|_{s=\kappa} + \frac{4R_1}{R_2}\Phi\right) = \Delta\rho gR_1 \tag{4.4.100}$$

With Eqs. 4.4.92 and 4.4.94 we may replace ϕ' and Φ in terms of κ and solve for a dimensionless terminal velocity as

$$\Theta \equiv \frac{\mu U}{\Delta\rho gR_2^2} = -\frac{\kappa^2}{2J} = \frac{\kappa^2}{2}\left[\ln\frac{1}{\kappa} - \frac{(1-\kappa^2)}{(1+\kappa^2)}\right] \tag{4.4.101}$$

Given a value of κ, we may find Θ from Fig. 4.4.5. With values of $\Delta\rho$ and μ, we may then find the expected terminal velocity U.

This completes the model of the system comprising a freely falling cylinder in a coaxial, fluid-filled tube, and we have achieved our goal of developing a relationship with which to predict the *steady* velocity of fall (the so-called terminal velocity) of the cylinder, given geometrical and physical properties. Since a measurement of the terminal velocity of the inner cylinder would yield a value for the liquid viscosity, if all the other parameters were known, we could design a viscometer based on this model.

A number of the assumptions of the model bear further examination. One is the neglect of entrance and exit effects—the assumption of fully developed flow over the entire length L. We can evaluate this assumption definitively only by solving the full Navier–Stokes equations (retaining the assumptions of steady axisymmetric incompressible flow), a task that can be carried out computationally but not analytically. At this stage we offer the intuitive argument that if $L/(R_2 - R_1)$ is large, we might expect that the entrance and exit regions would be small with respect to the length L.

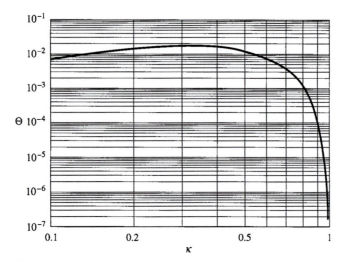

Figure 4.4.5 Terminal velocity function for a falling cylinder in a *closed* concentric tube, from Eq. 4.4.101.

Another potentially dangerous assumption is that once released, the inner cylinder will not wander slightly off axis. If such movement were to occur, the flow would no longer be axisymmetric. Axisymmetry can be maintained for very small gaps ($\kappa \approx 1$), but this is quite difficult for larger gaps, and special care is essential.

We have also assumed that the flow is at steady state and that the inner cylinder falls at its terminal velocity. There certainly is a period of acceleration following the release of the cylinder. At this stage we do not know how long it lasts, but we have obviously assumed that the time of acceleration is short in comparison to the time over which we measure the fall of the cylinder. Later, when we estimate the time scale of such transient viscous flows, we will be able to confirm that this time is short in most applications of interest to us.

We have assumed laminar flow, in the sense that the flow is not turbulent. At a high enough velocity, however, the falling cylinder could induce turbulent flow. Hence this model is restricted to slow fall in a high viscosity liquid. We might expect the restriction to laminar flow to take the form of a limitation on the Reynolds number for this flow. Some experimental studies suggest that departure from laminar flow occurs when a Reynolds number (as defined for this geometry) satisfies

$$\text{Re} = \frac{\rho U (1 - \kappa) R_2}{\mu} \geq 0.1 \qquad (4.4.102)$$

Note that in this definition the length scale chosen is $(1 - \kappa)R_2$, the gap between the cylinders. The value of 0.1 used here for the limiting Reynolds number is probably of the right order of magnitude, but a more precise value should be established experimentally. This could be done by measuring the fall velocity of a cylinder in liquids of known, successively lower viscosity. At some point we should observe a significant deviation between the measured and predicted terminal velocity. If this experiment is carried out with several liquids, using various combinations of radii R_1 and R_2, while keeping $L/(R_2 - R_1)$ large to ensure the elimination of end effects, it should be possible to find a Reynolds number above which the model is no longer adequate.

EXAMPLE 4.4.2   *Design of a Simple Falling Cylinder Viscometer*

We need a viscometer for use with liquids of viscosities in the range of 0.1 to 1 Pa·s. Will a falling cylinder viscometer work well in this application? To answer this question, let's see what design conditions would give us a useful instrument.

The conceptual design is along the lines of Fig. 4.4.4. A long cylinder will be released from rest to fall along the axis of a surrounding concentric tube, closed at its bottom. Our analysis of this system has already implied some design constraints. For example, the cylinder should be long in comparison to the radial gap occupied by the liquid.

Let's begin with an arbitrary but easily manufactured geometry: a stainless steel cylinder of diameter 1 cm, with an axial length of 10 cm. The test liquid will be confined in a tube of diameter 1.2 cm, closed at the bottom. This gives a κ value of 1/1.2 = 0.833. The immediate concern now is whether this cylinder will fall at a speed that lends itself to accurate measurement.

We may calculate U by finding the value of Θ from Eq. 4.4.101 (or with less accuracy from Fig. 4.4.5), using 1 Pa·s for the viscosity, 7800 kg/m^3 for the density of steel, and assuming a liquid density of 1100 kg/m^3. We find $U = 1.34 \times 10^{-3}$ m/s—a very low velocity, but one that is easily measured. For example, in a convenient and easily measured fall time, 100 s, the cylinder would fall 13.4 cm. The design seems all right for use with a 1 Pa·s liquid. For the lower end of the viscosity range, 0.1 Pa·s, the velocity will be too high for convenient and accurate measurement. But if we use a second inner cylinder, of diameter 1.1. cm, we halve the gap and the additional frictional resistance compensates for the lower viscosity, with the result that the slightly larger cylinder in the 0.1 Pa·s liquid has a velocity of 5×10^{-3} m/s. Over a fall time of 40 s, the cylinder will fall 20 cm.

Apparently then, we can build a viscometer consisting of an outer cylindrical container of diameter 1.2 cm and a set of falling cylinders with diameters in the range of 1 to 1.1 cm and axial lengths of 10 cm. The cylindrical container can be 50 cm in height, and to minimize any acceleration and deceleration phenomena at the start and end of the motion of the falling cylinder, we will measure the fall velocity over the central 30 cm of the outer tube.

As suggested by earlier comments, we decide to check the Reynolds numbers that might be observed with this design. Applying Eq. 4.4.102 to the design, we find that the Reynolds number is well within the criterion specified for laminar flow.

EXAMPLE 4.4.3   *Determination of Viscosity from a Falling Cylinder Experiment*

A stainless steel cylinder (diameter, 1 cm; axial length, 10 cm) falls through a test liquid confined in a tube of diameter 1.2 cm, closed at the bottom. The velocity of fall is observed to be $U = 1$ mm/min. The liquid density is 890 kg/m^3. The temperature is 25°C. What is the liquid viscosity?

For this viscometer we have a κ value of 1/1.2 (=0.833). From Fig. 4.4.5 we find $\Theta = 7 \times 10^{-4}$. Taking the density of stainless steel as 7800 kg/m^3 and using the data of this example (converted to consistent units), we invert Eq. 4.4.101 to solve for the viscosity. The result, $\mu = 102$ Pa·s, indicates a very high viscosity liquid.

4.4.5 Slow Flow Around a Solid Sphere: A Two-Dimensional Flow

In Section 4.4.4 we set up a model for the forces acting on a cylindrical object moving coaxially through a surrounding cylinder filled with a viscous liquid. The interaction of the flow with the stationary solid boundaries of the flow exerted a strong influence on the forces acting on the moving cylinder, and therefore on the steady velocity of the cylinder. This example focuses on the modeling of the forces acting on a body moving steadily through an *unbounded* fluid. The simplest geometry we can select, but in fact one of the most important from a practical point of view, is the sphere (Fig. 4.4.6). We will fix the coordinate system at the center of the sphere. The flow can then be regarded as moving relative to the stationary sphere. A point in the fluid, which is outside the sphere, is described by its spherical coordinates (r, θ, ϕ) relative to the center of the sphere.

We will assume that the flow is laminar and symmetric about the z axis. In that case, there is no velocity component in the ϕ direction, but there are components in the r and θ directions. Hence this is a *two-dimensional flow*, the first such flow we have considered. The force of gravity can be introduced after we have developed the velocity profiles; for this part of the analysis, we will ignore it.

We will assume that the flow is isothermal, and at steady state. We will assume that the fluid is Newtonian and incompressible. Then we can reduce the Navier–Stokes equations (remember—we are using spherical coordinates here) to the following forms (see Table 4.3.1):

Eq. 4.3.24g

$$\rho\left[u_r\frac{\partial u_r}{\partial r}+\frac{u_\theta}{r}\frac{\partial u_r}{\partial \theta}-\frac{u_\theta^2}{r}\right]=-\frac{\partial p}{\partial r}+\mu\left[\frac{1}{r^2}\frac{\partial^2}{\partial r^2}(r^2 u_r)+\frac{1}{r^2\sin\theta}\frac{\partial}{\partial \theta}\left(\sin\theta\frac{\partial u_r}{\partial \theta}\right)\right] \quad \textbf{(4.4.103)}$$

Eq. 4.3.24h

$$\rho\left[u_r\frac{\partial u_\theta}{\partial r}+\frac{u_\theta}{r}\frac{\partial u_\theta}{\partial \theta}+\frac{u_r u_\theta}{r}\right]=-\frac{1}{r}\frac{\partial p}{\partial \theta}+\mu\left[\frac{1}{r^2}\frac{\partial}{\partial r}\left(r^2\frac{\partial u_\theta}{\partial r}\right)\right.$$
$$\left.+\frac{1}{r^2}\frac{\partial}{\partial \theta}\left(\frac{1}{\sin\theta}\frac{\partial}{\partial \theta}(u_\theta\sin\theta)\right)+\frac{2}{r^2}\frac{\partial u_r}{\partial \theta}\right] \quad \textbf{(4.4.104)}$$

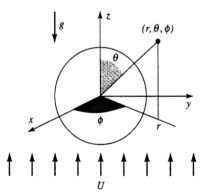

Figure 4.4.6 A uniform flow at velocity U approaches a rigid sphere.

Note that in reducing the Navier–Stokes equations, we have discarded all derivatives $\partial/\partial\phi$ and all terms with u_ϕ because of the assumed symmetry about the z axis.

We must add the continuity equation, (Eq. 4.1.7), which takes the form (Eq. 4.1.8c):

$$0 = \left[\frac{1}{r^2}\frac{\partial}{\partial r}(r^2 u_r) + \frac{1}{r\sin\theta}\frac{\partial}{\partial\theta}(\sin\theta\,u_\theta) \right] \tag{4.4.105}$$

This set of equations is much more complicated than any we have examined previously. We have two velocity components. We have *partial differential equations,* not ordinary differential equations. Most distressing of all, we have *nonlinear* equations. Since the left-hand sides of Eqs. 4.4.103 and 4.4.104 have terms that are *second-order* in velocity components, these equations cannot be solved analytically.

We can introduce an approximation here. It is based on the notion that if the velocity is small enough, second-order terms in velocity should be small compared with first-order terms. This is just the statement that the square of a number less than one is less than the number itself. The primary question is: How small is small? What criterion might we use to argue that the nonlinear terms in the Navier-Stokes equations are negligible?

To answer this question, we first nondimensionalize the equations above, assisted by two natural parameters. We will compare all velocities to the approach velocity U. We do not expect any velocity component to be able to exceed the magnitude of U here. There is also a natural length scale in this problem—the radius of the sphere. We will rescale the space-coordinate variable r by dividing it by R. Hence we now define the following variables:

$$s \equiv \frac{r}{R} \qquad \tilde{u}_r \equiv \frac{u_r}{U} \qquad \tilde{u}_\theta \equiv \frac{u_\theta}{U} \tag{4.4.106}$$

Now the dynamic equations take the form

$$\frac{\rho U^2}{R}\left[\tilde{u}_r\frac{\partial\tilde{u}_r}{\partial s} + \frac{\tilde{u}_\theta}{s}\frac{\partial\tilde{u}_r}{\partial\theta} - \frac{\tilde{u}_\theta^2}{s} \right] = -\frac{1}{R}\frac{\partial p}{\partial s} + \frac{\mu U}{R^2}\left[\frac{1}{s^2}\frac{\partial^2}{\partial s^2}(s^2\tilde{u}_r) + \frac{1}{s^2\sin\theta}\frac{\partial}{\partial\theta}\left(\sin\theta\frac{\partial\tilde{u}_r}{\partial\theta} \right) \right] \tag{4.4.107}$$

and

$$\frac{\rho U^2}{R}\left[\tilde{u}_r\frac{\partial\tilde{u}_\theta}{\partial s} + \frac{\tilde{u}_\theta}{s}\frac{\partial\tilde{u}_\theta}{\partial\theta} + \frac{\tilde{u}_r\tilde{u}_\theta}{s} \right] = -\frac{1}{Rs}\frac{\partial p}{\partial\theta} + \frac{\mu U}{R^2}\left[\frac{1}{s^2}\frac{\partial}{\partial s}\left(s^2\frac{\partial\tilde{u}_\theta}{\partial s} \right) \right.$$
$$\left. + \frac{1}{s^2}\frac{\partial}{\partial\theta}\left(\frac{1}{\sin\theta}\frac{\partial}{\partial\theta}(\tilde{u}_\theta\sin\theta) \right) + \frac{2}{s^2}\frac{\partial\tilde{u}_r}{\partial\theta} \right] \tag{4.4.108}$$

We now have two dimensionless velocity variables that are functions of the two dimensionless space variables s and θ. How do we nondimensionalize the third dependent variable, the pressure p? There is no natural pressure parameter in this flow. But note that $\mu U/R$ has the units of pressure. Let's use this fact to define a dimensionless pressure as

$$\tilde{p} \equiv \frac{p}{\mu U/R} \tag{4.4.109}$$

It is not clear what physical meaning might be attributed to this dimensionless pressure, but with this definition the dynamic equations may now be written as

$$\frac{\rho UR}{\mu}\left[\tilde{u}_r\frac{\partial\tilde{u}_r}{\partial s} + \frac{\tilde{u}_\theta}{s}\frac{\partial\tilde{u}_r}{\partial\theta} - \frac{\tilde{u}_\theta^2}{s} \right] = -\frac{\partial\tilde{p}}{\partial s} + \left[\frac{1}{s^2}\frac{\partial^2}{\partial s^2}(s^2\tilde{u}_r) + \frac{1}{s^2\sin\theta}\frac{\partial}{\partial\theta}\left(\sin\theta\frac{\partial\tilde{u}_r}{\partial\theta} \right) \right] \tag{4.4.110}$$

and

$$\frac{\rho U R}{\mu}\left[\tilde{u}_r\frac{\partial \tilde{u}_\theta}{\partial s}+\frac{\tilde{u}_\theta}{s}\frac{\partial \tilde{u}_\theta}{\partial \theta}+\frac{\tilde{u}_r\tilde{u}_\theta}{s}\right]=-\frac{1}{s}\frac{\partial \tilde{p}}{\partial \theta}+\left[\frac{1}{s^2}\frac{\partial}{\partial s}\left(s^2\frac{\partial \tilde{u}_\theta}{\partial s}\right)\right.$$

$$\left.+\frac{1}{s^2}\frac{\partial}{\partial \theta}\left(\frac{1}{\sin\theta}\frac{\partial}{\partial \theta}(\tilde{u}_\theta\sin\theta)\right)+\frac{2}{s^2}\frac{\partial \tilde{u}_r}{\partial \theta}\right] \qquad \textbf{(4.4.111)}$$

Keep in mind that we are only doing algebra here—no physics. Equations 4.4.110 and 4.4.111 differ from Eqs. 4.4.103 and 4.3.104 only in *format;* no terms have been removed. Hence they are still nonlinear partial differential equations. Then what have we gained by this exercise? Very little, unless we stare at the equations long enough to recognize that the coefficient $(\rho U R/\mu)$ of the left-hand side of both is a Reynolds number for this flow. We can now reach two conclusions:

1. The Reynolds number arises naturally in the course of nondimensionalization of the Navier–Stokes equations.
2. In the limit of very small values of $\rho U R/\mu = $ Re, we may neglect the entire left-hand side of the Navier–Stokes equations for this flow.

In a loose sense, a vanishing Reynolds number implies a very small velocity, and thus we refer to the approximate set of equations that results from setting Re $= 0$ as the "creeping flow" equations. They are also referred to as the "Stokes flow" equations. In what follows, then, we work with the creeping flow equations for flow about a solid sphere.

Because the equations have second-order *derivatives* of both velocity components, we need four boundary conditions. Two of them will express the no-slip condition at the surface of the sphere:

$$\tilde{u}_\theta=\tilde{u}_r=0 \qquad \text{on the surface} \quad r=R \qquad \textbf{(4.4.112)}$$

(Actually, $\tilde{u}_r=0$ is the statement that the fluid does not penetrate the surface $r=R$.)

Where do we impose the other boundary conditions? There is no other boundary to this flow because we have assumed that the sphere is in an unbounded fluid. We expect intuitively that far from the sphere the fluid behaves as if the sphere did not exist to disturb the flow, and that is all we can say. Hence far from the sphere we impose the condition that the radial and angular velocity components correspond to uniform flow along the z axis. Taking note of the geometry in Fig. 4.4.6, we see that in the spherical coordinate system these components must satisfy the following boundary conditions far from the sphere:

$$\left.\begin{array}{l}u_\theta=-U\sin\theta\\[4pt]u_r=U\cos\theta\end{array}\right\} \qquad \text{for} \quad r\rightarrow\infty \qquad \textbf{(4.4.113)}$$

Now we must set about solving these differential equations. This is not a trivial task. Most students in an introductory fluid dynamics course do not have sufficient experience to know how to go about solving such a set of equations. We will illustrate a solution technique here, but little will be lost at this stage if you skip directly to the solutions below. Of course, if you're curious, . . .

We begin by noting that so far, we have made no use of the continuity equation, given as Eq. 4.4.105 for this flow, and that often the form of the boundary conditions is reflected to some extent in the solutions themselves. These observations motivate us to try solutions of the form (see Eqs. 4.4.113)

$$\tilde{u}_\theta=G(s)\sin\theta \qquad \text{and} \qquad \tilde{u}_r=F(s)\cos\theta \qquad \textbf{(4.4.114)}$$

(We are trying a method called "separation of variables," in which it is assumed that in the solution, the two independent variables s and θ appear in functions that are separated as a product of a function of s and a function of θ.) This trial solution is substituted into the continuity equation, and we find

$$\frac{\cos \theta}{s^2}\left[2s\,F(s) + s^2 \frac{dF(s)}{ds} \right] + \frac{2\cos\theta}{s}\,G(s) = 0 \qquad \textbf{(4.4.115)}$$

Note that the $\cos \theta$ term can be divided out of both terms, leaving a differential equation for the functions F and G that contains only the independent variable s. Thus the assumption of separability, and the assumed sine and cosine forms for the velocity components, are compatible with the equation of conservation of mass. Equation 4.4.115 provides a relationship between the unknown functions F and G:

$$\left[F(s) + \frac{s}{2}\frac{dF(s)}{ds} \right] + G(s) = 0 \qquad \textbf{(4.4.116)}$$

Remember that we still have three uknowns: $F(s)$, $G(s)$, and the pressure function $\bar{p}(s, \theta)$, about which we have said nothing. We can eliminate pressure from the dynamic equations (Eqs. 4.4.110 and 4.4.111) in the following manner, though we will not write out the algebraic details. We set the left-hand side of Eq. 4.4.110 to zero (creeping flow) and then differentiate with respect to θ. We introduce the functions F and G. We get a lot of terms, but one stands out: $(\partial^2 \bar{p}/\partial s \partial \theta)$. Next we set the left-hand side of Eq. 4.4.111 to zero (creeping flow), multiply the terms on the right by s, introduce the functions F and G, and then differentiate with respect to s. Again, we get a lot of terms, but there is a term $(\partial^2 \bar{p}/\partial \theta \partial s)$ as in the other equation. (The order of differentiation does not matter, so the two terms are identical.) If we now subtract these two equations, we eliminate this pressure function. We have, at this point, a *single* equation with *third-order* derivatives of F and G. If we use Eq. 4.1.116 to eliminate G, we end up with a single ordinary differential equation for the function $F(s)$. The price we have paid is that we now have a *fourth-order* equation to solve. It takes the form

$$s^4 \frac{d^4 F(s)}{ds^4} + 8s^3 \frac{d^3 F(s)}{ds^3} + 8s^2 \frac{d^2 F(s)}{ds^2} - 8s \frac{dF(s)}{ds} = 0 \qquad \textbf{(4.4.117)}$$

This is called a linear homogeneous equation (an Euler equation) and it is solved by assuming a solution of the form

$$F(s) = s^n \qquad \textbf{(4.4.118)}$$

where n is to be found. Upon substituting this form into Eq. 4.4.117, we obtain a fourth-order polynomial in n. In this case the four roots of the polynomial are -3, -1, 0, and 2. Then the general solution for $F(s)$ is

$$F(s) = as^{-3} + bs^{-1} + cs^0 + ds^2 \qquad \textbf{(4.4.119)}$$

The four coefficients a, b, c, and d are found through application of the four boundary conditions.

Given Eqs. 4.4.112 and 4.4.113 on the radial velocity, we can write the boundary conditions on $F(s)$ as

$$F = 0 \quad \text{on} \quad s = 1$$
$$F = 1 \quad \text{for} \quad s \to \infty \qquad \textbf{(4.4.120)}$$

The boundary conditions on the θ-velocity component require

$$G = 0 \qquad \text{on} \quad s = 1$$
$$G = -1 \qquad \text{for} \quad s \to \infty \tag{4.4.121}$$

If we use Eq. 4.4.116 to relate G to F, we find that by imposing on G the conditions of Eq. 4.4.121, we obtain two more conditions on F:

$$\frac{dF}{ds} = 0 \qquad \text{on} \quad s = 1$$
$$\frac{dF}{ds} = 0 \qquad \text{for} \quad s \to \infty \tag{4.4.122}$$

When these four boundary conditions on F are applied to Eq. 4.4.119 we are able to solve for the integration constants, with the result that we may write the velocity functions as

$$F(s) = 1 - \frac{3}{2s} + \frac{1}{2s^3}$$
$$G(s) = -1 + \frac{3}{4s} + \frac{1}{4s^3} \tag{4.4.123}$$

Next we need to find the pressure distribution. The easiest approach is to substitute the solutions for the two velocity components into Eq. 4.4.111. Then we integrate the pressure term $\partial \bar{p}/\partial \theta$ with respect to θ and find, after some substantial algebra, that

$$\bar{p} = \bar{p}_\infty - \frac{3}{2s^2} \cos \theta \tag{4.4.124}$$

The term $\bar{p}_\infty$ is the (dimensionless) pressure at a large distance from the sphere.

We have now completed the solution to the dynamic equations for the creeping flow of a Newtonian fluid around a sphere. You can accept this solution, or you can carry out the details just outlined. But you cannot expect to "see" where these equations came from without doing the "busy work." It's your choice.

What do we do with this mathematical model? The most sensible thing at this point is to use the solution to model measurable phenomena related to this flow and examine the correspondence of the prediction(s) to the observation(s). We will calculate the steady velocity of a sphere in response to the action of a steady externally applied force, which is easily measured. Since, however, we expect that the motion is determined by the forces exerted by the fluid on the moving sphere, we will have to calculate the stresses acting on the surface of the sphere.

> Look, this is a tough problem. There's lots of detail, and it's easy to get lost in the math. Don't give up. There's some really important stuff to be learned here, if you'll stick with it. But this doesn't mean you have to torture yourself. This is a good time to take a break. Eat some chocolate and watch the Three Stooges for 30 minutes. Then come back refreshed.

Figure 4.4.7 shows the two kinds of stress acting on the surface of the sphere: the fluid exerts a shear stress, tangential to the surface, and a radial normal stress, perpendicular to the surface. In general, at any point on the surface of the sphere, both stresses have components in the direction of motion.

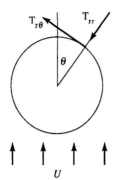

Figure 4.4.7 Normal and shear stresses acting on the surface of a sphere.

We will now find the components of force, in the direction of motion, that arise from these two kinds of stress. At any point on the surface the normal stress has a component $(-T_{rr} \cos \theta)$ in the direction of motion. The total force due to this stress is obtained by integrating it over the entire surface of the sphere:

$$F_P = \int_0^{2\pi} \int_0^{\pi} [-T_{rr}(R, \theta) \cos \theta] R^2 \sin \theta \, d\theta \, d\phi = 2\pi \int_0^{\pi} \left[\left(-p + 2\mu \frac{\partial u_r}{\partial r} \right)_R \cos \theta \right] R^2 \sin \theta \, d\theta$$

$$(4.4.125)$$

where we assume an incompressible Newtonian fluid surrounding the sphere, so $\mathbf{T} = p\boldsymbol{\delta} - \mu\boldsymbol{\Delta}$.

Using Eq. 4.4.124 for $p(r)$, and Eq. 4.4.123 for $F(s)$ so that we can evaluate $\partial u_r/\partial r$, we find, after a lengthy exercise in calculus,

$$F_P = 2\pi\mu RU \qquad (4.4.126)$$

In the same manner, we see that the viscous stress component $(T_{r\theta} \sin \theta)$ may be integrated to give a viscous force acting in the direction of the external flow:

$$F_V = \int_0^{2\pi} \int_0^{\pi} [T_{r\theta}(R, \theta)\sin \theta] R^2 \sin \theta \, d\theta \, d\phi \qquad (4.4.127)$$

To find the shear stress component $T_{r\theta}$ we must go back to Table 4.2.1 and find $\Delta_{r\theta}$. Then

$$T_{r\theta} = -\mu \left[r \frac{\partial}{\partial r} \left(\frac{u_\theta}{r} \right) + \frac{1}{r} \frac{\partial u_r}{\partial \theta} \right] \qquad (4.4.128)$$

Using Eqs. 4.4.123 and 4.4.114 for each velocity component, and converting to the dimensional velocity $\mathbf{u} = U\bar{\mathbf{u}}$, we find that the integration finally yields

$$F_V = 4\pi\mu RU \qquad (4.4.129)$$

We conclude that the total force resisting the flow of fluid relative to the sphere is composed of two parts: a viscous shear term F_V, usually referred to as the *frictional drag*, and a pressure term F_P, usually referred to as the *form drag*. Their sum is the total force arising from the relative motion:

$$F_{SL} = F_P + F_V = 6\pi\mu RU \qquad (4.4.130)$$

$$\text{Stokes' law}$$

This is called Stokes' law. Thus one test of the model's limitations is the correspondence of Stokes' law (Eq. 4.4.130) to observations on the terminal velocity of spherical particles. The simplest test is to measure the terminal velocity of a sphere of known radius, placed in a fluid of known viscosity. Such experiments are easily performed, and the body of observational experience indicates that Stokes' law is a valid model for situations that satisfy a criterion based on the Reynolds number:

$$\text{Re} = \frac{2\rho_f U R}{\mu} \le 0.1 \qquad \text{for Stokes' law to hold} \tag{4.4.131}$$

Note that this Reynolds number is defined with $2R$, or the sphere diameter, as the length scale, and the *fluid* density—not the particle density. We usually refer to motion of a particle relative to a fluid such that $\text{Re} < 0.1$ as creeping flow. We will describe some other limitations to the use of Stokes' law at the conclusion of Example 4.4.4, an application of the model to a problem of practical importance.

EXAMPLE 4.4.4   *Gravitational Sedimentation of Small Particles*

A common application of Stokes' law is in the estimation of the time required for small particles, suspended in a fluid, to fall through the fluid under the influence of gravity. Imagine the situation presented schematically in Fig. 4.4.8. We assume that solid spheres of several different diameters are uniformly distributed in a quiescent fluid at some time, and thereafter the spheres fall under the influence of gravity. We would like to estimate the time required for the spheres to fall to the lower bounding surface.

If we assume that Stokes' law is valid when each sphere is surrounded by a fluid containing other spheres (we will have to discuss some limitations of its application shortly), then at steady state, when the particles have achieved their terminal velocities, the resistive force due to the relative motion of each sphere with respect to the surrounding fluid must be balanced by the gravitational force. The gravitational force on a particle suspended in a fluid is just its weight, less the buoyancy, or

$$F_g = \tfrac{4}{3}\pi R^3 (\rho_s - \rho_f) g \tag{4.4.132}$$

Subscripts on the densities (ρ) refer to the solid and fluid, respectively. When we equate F_g to F_{SL}, we find

$$U = \frac{2R^2}{9\mu}(\rho_s - \rho_f) g \tag{4.4.133}$$

Suppose we have the following data:

$$\mu = 1.8 \times 10^{-5}\,\text{Pa}\cdot\text{s (air at 25°C)} \qquad \rho_s = 1000\,\text{kg/m}^3 \qquad \rho_f \approx 0\,\text{kg/m}^3 \qquad g = 9.8\,\text{m/s}^2$$

The spheres will be assumed to have radii of 0.5, 1, and 2 μm.

Figure 4.4.8 Spherical particles of various sizes falling through a fluid.

Then we easily calculate the terminal velocities as 30 μm/s for the smallest particle, and 121 and 484 μm/s, respectively, for the two larger particles. Except for the 2 μm particle, these are extremely small velocities. Because of the quadratic dependence of velocity on particle radius, there is a significant differential in the terminal velocity as a function of radius. Thus if there is an initially uniform distribution of particle sizes throughout the fluid, we may be able to separate the particles according to particle size effectively by a technique based on the difference in sedimentation velocity.

A simple but crude example of the performance of a particle separator can be worked out along the following lines. We suppose that a fluid initially contains the three particles sizes given above, in *equal* numbers and uniformly distributed spatially throughout a container of height $H = 12$ cm. Settling begins at time $t = 0$. A time of 248 s is required for *all* the largest particles to reach the lower surface, since any 2 μm sphere farthest from the bottom, at a distance of 12 cm, requires 248 s to travel that distance. In that time, the middle-sized particles travel only 25% of that distance, since their terminal velocity is only 25% of that of the largest particles. Hence the lower 25% of all the particles of radius 1 μm will settle out. Only 6.25% of the smallest particles will have settled out in 248 s. Thus, after 248 s, the fluid contains no large particles, and nearly equal numbers of the two smaller particles. (The fluid will actually be composed of 56% in the smallest size (the 0.5 μm particle) and 44% in the middle particle size. These are *number* percentages.)

In this simple example we assume that the particles do not interfere with one another and that it is possible to have an initially uniform spatial distribution of particles. While these assumptions may not hold exactly, the point of the example is still valid: gravitational settling can be used to separate particles according to size, and Stokes' law provides a basis for analysis of the performance of such a system.

If we return to the model for the velocity field relative to the sphere (Eqs. 4.4.123 and 4.4.114) and note the boundary conditions *far* from the sphere (Eqs. 4.4.120 and 4.4.121), we may deal with the following question: How far must a sphere be from another sphere, or from a solid boundary, to ensure the validity of the assumption of this model—that the sphere is surrounded by an *infinite body* of fluid? From Eq. 4.4.123 we see that if s exceeds 15, the velocity in the fluid falls below 10% of U. We could take this as a criterion of "far from the sphere." In other words, a point in the fluid located more than 15 radii from a sphere that is moving relative to the surrounding fluid is hardly "aware" that the sphere is there. Thus we infer that if the surrounding boundaries (the container) of the fluid are no closer than 15 radii to the center of the sphere, or if the spheres in a suspension are no closer than 15 radii to one another, Stokes' law will hold approximately. (Of course we are also assuming that the Reynolds number is small and that the fluid is Newtonian.)

4.4.6   Flow Around a Deformable Particle

Stokes' law gives a good description of the flow around a *rigid* spherical particle as long as the Reynolds number is small, as noted above. The situation is much more complex in the case of deformable particles such as liquid drops and gas bubbles. The viscous stresses can deform a mobile interface, causing a gas bubble or liquid drop to deform into an ellipsoidal, or more complex, shape, thereby altering the flow field. In fact, drops or bubbles that are large enough are not stable and break up into smaller drops or bubbles. This property of particles is familiar to anyone who has watched large air bubbles released at the bottom of a tank of water.

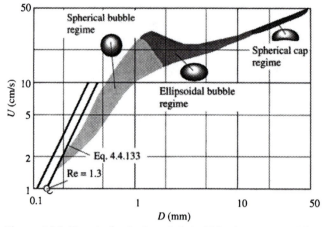

Figure 4.4.9 Terminal velocity of air bubbles in water at 20°C. Lower-line (Eq. 4.4.133) is Stokes' law for a rigid sphere. Line above and parallel is Stokes' law for a gas bubble (Eq. P4.17.2).

For very small drops or bubbles (perhaps a radius of the order of 1 mm), surface tension forces are often strong enough to maintain the spherical shape. We still might see departures from Stokes' law, but these would be due to the failure of the no-slip condition to hold at a mobile interface. (See Problem 4.17.) With gas bubbles, the situation is even more complex because contaminants, either small particles or surfactants, can produce complex alterations in the flow at the bubble/liquid interface. Figure 4.4.9 gives an example of the difficulty in predicting the terminal velocity of gas bubbles. (A similar figure, with additional discussion, can be found in Clift *et al., Bubbles, Drops, and Particles,* Academic Press, San Diego, 1978.) The wide variance in the data reflects uncontrolled contamination of the water. Even in the case of pure water, the terminal velocity is a complex function of the bubble diameter because of the distortion of the bubble shape, and ultimately because of the oscillation of large bubbles as they rise through water.

4.4.7 A Summary of the Modeling Procedure

We have examined a variety of fluid dynamics problems, but one procedure for developing a mathematical model is common to each of our cases (Examples 4.4.1–4.4.4). It is appropriate to review the modeling approach we have been using and will continue to use as we study even more complicated problems in the next section of the text.

Before we can set about simplifying the Navier–Stokes equations, we must *visualize the physics* of the problem of interest. We seek to simplify the geometry and the physics, and at an early stage of the analysis we must recognize (or introduce assumptions about) the boundary conditions that constrain the flow physically and geometrically. Usually this leads us to a set of assumptions regarding the dimensionality of the flow: Is the flow one- or two-dimensional? Is the flow fully developed, or does the velocity vary in the direction of flow? Can we argue that some velocity components vanish by virtue of symmetry of the boundaries? What approximations are we willing to make for the sake of reducing the Navier–Stokes equations to a set of equations that can be solved analytically? Finally, we must be prepared to evaluate the model by comparison to experimental data, or by assessing the relative orders of magnitude of terms we had assumed to be negligible.

4.5 FORMULATION OF THE DYNAMIC EQUATIONS FOR SOME COMPLEX BUT INTERESTING PROBLEMS: ENGINEERING APPROXIMATIONS

Many important and interesting problems are more complex than the ones illustrated in Section 4.4. We offer a few examples here to provide some practice at the activity of simplifying the continuity and Navier–Stokes equations. For each case we introduce, as well, some crude engineering approximations that lead to analytical solutions to the equations. This has the advantage of producing *testable* predictions from which we can learn something of the "art" of engineering analysis.

EXAMPLE 4.5.1 *Steady Radial Flow Between Parallel Disks*

Many commercial plastic items are manufactured by the process of injection molding, in which a molten polymer is pumped into a cavity that has the shape of the desired object. Inexpensive plastic camera lenses are produced in this way. We examine here a flow field that has some of the characteristics of an injection molding process for the manufacture of disk-shaped objects.

The geometry we will study is shown in Fig. 4.5.1. A viscous liquid is pumped continuously through an entrance tube to the axis of a pair of parallel disks. The flow then moves radially outward toward the circumference of the disks. The circumferential region is exposed to atmospheric pressure. While the confined radial flow feature occurs in some injection molding processes, we will not end up with a model of injection molding here because injection molding is a *transient* process in which the liquid is pumped into a *closed* cavity against a steadily increasing pressure until the cavity is filled. Our example is considerably simplified in comparison to a commercial polymer process, but it allows us to practice working with the continuity and Navier–Stokes equations and relating physical concepts to mathematical formalism.

The goal of the analysis that follows is to find a relationship between the flowrate and the pressure forcing the flow. We will make the following assumptions as we attempt to simplify the Navier–Stokes equations:

The disks are rigid, no-slip, surfaces.
The flow is symmetric about the z axis, so $\partial/\partial\theta = 0$, and there is no velocity component u_θ.
The fluid is Newtonian, isothermal, and incompressible.

Although some polymers are nearly Newtonian, and the mold filling process is sometimes close to isothermal, most molten polymers that are injection molded are not Newtonian liquids, and the injection molding process is usually far from isothermal (since the molten polymer must be solidified after the mold has been filled). We will not be deterred by these practicalities. Our intent is not to develop a model of commercial injection molding but to represent a steady radial flow between parallel disks. We simply

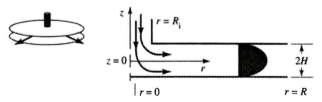

Figure 4.5.1 Radial flow between parallel disks.

point out that when facing a complex problem such as injection molding, it may be useful to have a model of a simpler flow that shares some of the features (radial flow between parallel surfaces) of the more complex process. With these thoughts in mind we add one more assumption:

The flow is steady in time.

Let's look first at the continuity equation, Eq. 4.1.7 in this case. Because of the symmetry about the z axis it appears that it will be simplest to work in cylindrical coordinates. Then the continuity equation becomes (use Table 4.1.1, and let $\mathbf{A} = \mathbf{u}$.)

$$0 = \frac{1}{r}\frac{\partial(ru_r)}{\partial r} + \frac{\partial u_z}{\partial z} \tag{4.5.1}$$

Notice that this is a two-dimensional flow. The velocity vector has two nonzero components.

Now we must go into Table 4.3.1 and simplify the Navier–Stokes equations as much as possible, consistent with the assumptions we have made *so far*. (We reserve the right to introduce subsequent assumptions later.) Hence we find

$$\rho\left[u_r\frac{\partial u_r}{\partial r} + u_z\frac{\partial u_r}{\partial z}\right] = -\frac{\partial p}{\partial r} + \mu\left\{\frac{\partial}{\partial r}\left[\frac{1}{r}\frac{\partial(ru_r)}{\partial r}\right] + \frac{\partial^2 u_r}{\partial z^2}\right\} \tag{4.5.2}$$

$$\rho\left[u_r\frac{\partial u_z}{\partial r} + u_z\frac{\partial u_z}{\partial z}\right] = -\frac{\partial p}{\partial z} + \mu\left\{\frac{1}{r}\frac{\partial}{\partial r}\left[r\frac{\partial u_z}{\partial r}\right] + \frac{\partial^2 u_z}{\partial z^2}\right\} - \rho g \tag{4.5.3}$$

If we make no further assumptions, we are stuck with this set of three simultaneous coupled, nonlinear, partial differential equations. The nonlinearity is a serious problem—it prevents us from finding an analytical solution. It arises from the inertial terms (the left-hand side of the Navier–Stokes equations), and unless we have a basis for neglecting these terms (assuming, e.g., a very low Reynolds number "creeping" flow), the computational problem we face is serious. At this point (and this is all we wanted to accomplish now), we have a well-formulated problem, with three equations in three unknowns: u_z, u_r, and p. Well—not quite! We must include *boundary conditions* in the formulation of the problem. First, we will impose no-slip boundary conditions on the solid surfaces. On the lower disk ($z = -H$) the no-slip condition is simply

$$u_r = u_z = 0 \quad \text{on} \quad z = -H, 0 \le r \le R \tag{4.5.4}$$

On the upper disk ($z = +H$) we have

$$u_r = u_z = 0 \quad \text{on} \quad z = +H, R_i \le r \le R \tag{4.5.5}$$

In the entry tube we have a no-slip condition:

$$u_r = u_z = 0 \quad \text{on} \quad r = R_i, H \le z \le H + L_e \tag{4.5.6}$$

where L_e is the length of the entry tube.

We have assumed that the flow is symmetric, so we write

$$u_r = 0 \quad \text{and} \quad \frac{\partial u_z}{\partial r} = 0 \quad \text{on} \quad r = 0, -H \le z \le H + L_e \tag{4.5.7}$$

A boundary condition on pressure can be written if we assume that the fluid exiting the disks along the circumference $r = R$ is at atmospheric pressure. Note that pressure appears in the Navier–Stokes equations only in derivatives. Hence the pressure can be found only with respect to some arbitrary constant level. We will take atmospheric pressure to have the value $p = 0$ (this is the same as regarding p as "gage pressure"), and so our boundary condition on pressure at the outlet is expressed as

$$p = 0 \quad \text{for} \quad -H \leq z \leq +H, r = R \qquad \textbf{(4.5.8)}$$

We must impose some conditions on the flow into the entrance tube. We have two choices. One is a condition on the entry pressure:

$$p = P_{\text{entr}} \quad \text{at} \quad z = H + L_e, 0 \leq r \leq R_i \qquad \textbf{(4.5.9)}$$

The flow rate through the system would then be determined from the solution to the equations. Alternatively, we could impose a boundary condition on the entry flow:

$$u_z = -U \quad \text{and} \quad u_r = 0 \quad \text{at} \quad z = H + L_e, 0 \leq r \leq R_i \qquad \textbf{(4.5.10)}$$

Then the pressure required to produce this flow would be determined from the solution. Either choice of boundary condition yields a well-defined boundary value problem, and the physics of the case at hand (what we actually impose on the system—the entrance pressure or the flowrate) would determine the choice.

To solve the equations that govern this flow we must now make a decision. We could implement a numerical method of solving Eqs. 4.5.1–4.5.3, subject to the boundary conditions given. (There are commercial programs available for such fluid dymamics problems.) Or we could choose to introduce further assumptions, and simplify the equations to a linear form that would yield an analytical solution. Of course, the key issue would be whether the further simplifications are consistent with the physics. The test of consistency would lie in the extent to which predictions of the resulting simplified mathematical model mimic the observations.

We will take the bold course, introducing additional assumptions (approximations) that permit us to find an analytical solution for the flow field.

The first thing we will do is neglect the entry tube entirely.

We will assume that the entry tube is short, and of a large diameter, with the result that the pressure loss across the length L_e is very small in comparison to the loss across the radial length R of the pair of disks. Presumably there are geometries of interest to us that support this approximation and others that do not. Hence we expect that if the resulting model is of any value, it will be for geometries such that $H \ll R_i \ll R$. In some sense, then, the flow field depicted in Fig. 4.5.1 may be replaced by that of Fig. 4.5.2. The most important thing to note is that we now imply that the flow region of interest is strictly bounded by the shaded region of Fig. 4.5.2: $R_i \leq r \leq R$ and $-H \leq z \leq +H$.

Consistent with having thrown away the entry region, we now introduce the assumption that since the boundaries of the flow are parallel to the z plane, the velocity vector is strictly parallel to that plane, or

$$\mathbf{u} = [u_r(r, z), 0, 0] \qquad \textbf{(4.5.11)}$$

Figure 4.5.2 Flow region in which the entry region is neglected.

This is a statement that the flow between the parallel disks is everywhere unidirectional—there is no entry region where the flow adjusts from the input tube to the disk region. With this assumption, the continuity equation reduces to

$$0 = \frac{1}{r}\frac{\partial(ru_r)}{\partial r} \tag{4.5.12}$$

Since $u_z = 0$ (by assumption), Eq. 4.5.3 reduces to the hydrostatic law. In Eq. 4.5.2, one of the viscous stress terms vanishes by virtue of Eq. 4.5.12. Even so, Eq. 4.5.2 is still nonlinear because of the presence of the term $u_r(\partial u_r/\partial r)$ on the left-hand side. (The other nonlinear term vanishes with the assumption that $u_z = 0$.) Hence, at this stage, we must solve

$$\rho u_r \frac{\partial u_r}{\partial r} = -\frac{\partial p}{\partial r} + \mu \frac{\partial^2 u_r}{\partial z^2} \tag{4.5.13}$$

unless we are able to argue away the nasty term on the left-hand side.

Before we proceed further, let's nondimensionalize Eq. 4.5.13 by defining

$$\frac{u_r}{U} = u^* \qquad \text{and} \qquad \frac{p}{\mu U/H} = p^* \tag{4.5.14}$$

and

$$\frac{r}{H} = r^* \qquad \text{and} \qquad \frac{z}{H} = z^* \tag{4.5.15}$$

For the moment, U is undefined except that it has the units of a velocity. It is *not* the velocity U defined earlier in the boundary condition (Eq. 4.5.10), which is not relevant to the current analysis. With these definitions, Eq. 4.5.13 takes the form

$$\left(\frac{\rho U H}{\mu}\right) u^* \frac{\partial u^*}{\partial r^*} = -\frac{\partial p^*}{\partial r^*} + \frac{\partial^2 u^*}{\partial z^{*2}} \tag{4.5.16}$$

Although U is undefined, the coefficient of the left-hand side of Eq. 4.5.16 has the form of a Reynolds number. We can now choose a sensible definition of a characteristic velocity U. We suppose that the volume flowrate through the system is denoted by Q, which might be specified as a constraint on the dynamics of the system. If that were the case, we would be seeking the pressure required to produce this value of Q. On the other hand, if the pressure driving the flow were specified, Q would be unknown, and it would be found as part of the solution to the dynamic equations. But regardless of whether we know Q, there certainly is a well-defined Q for this steady state system, and we can use it to define a velocity scale by writing

$$Q = 4\pi R_i H U \tag{4.5.17}$$

Note that since $4\pi R_i H$ is the cross-sectional area of the entrance to the flow region, U is the average velocity at the entrance. This is a perfectly reasonable choice of a velocity scale to use for nondimensionalization. Keep in mind, however, that Eq. 4.5.17 is only a *definition* of U in terms of Q. Since we don't know Q yet, neither do we know U.

Do we have any basis on which to argue that a flow of this type would exhibit low Reynolds numbers under conditions of engineering interest? One way to answer this question is to calculate the Reynolds number for an arbitrary set of specified conditions. We choose the following values:

$$R = 0.05 \text{ m} \qquad R_i = 0.005 \text{ m} \qquad H = 0.002 \text{ cm}$$
$$Q = 10^{-4} \text{ m}^3/\text{s} \qquad \mu = 100 \text{ Pa} \cdot \text{s} \qquad \rho = 1000 \text{ kg/m}^3$$

Calculation yields $U = 0.8$ m/s and Re $= 0.016$, a small Reynolds number. We have chosen a high viscosity for this estimate—a 100 Pa · s liquid is more viscous than cold honey—but even if the viscosity were smaller by a factor of 10, we would still have a small Reynolds number. The value selected for Q is large. At this Q we are pumping 6 L/min through a region of very small volume. Even at this high flowrate, the Reynolds number is small. On the basis of this calculation, we conclude that there are flows of engineering interest, in this kind of geometry, involving high viscosity liquids, in which the Reynolds number would be small. With that argument in mind, we assume "creeping flow," neglect the left-hand side of Eq. 4.5.16, and seek a solution to the linearized dynamic equation, which we write as

$$0 = -\frac{\partial p^*}{\partial r^*} + \frac{\partial^2 u^*}{\partial z^{*2}} \tag{4.5.18}$$

We can begin the solution process by noting that the continuity equation (Eq. 4.5.12) has the solution (in dimensionless variables)

$$u^* = \frac{1}{r^*} C(z^*) \tag{4.5.19}$$

Although C is not a function of r^*, it could be a function of z^*. If we introduce this result into Eq. 4.5.18, we find

$$0 = -\frac{\partial p^*}{\partial r^*} + \frac{1}{r^*} \frac{d^2 C}{dz^{*2}} \tag{4.5.20}$$

which we rearrange to the form

$$r^* \frac{\partial p^*}{\partial r^*} = \frac{d^2 C}{dz^{*2}} = A \tag{4.5.21}$$

Since C is not a function of r^*, its second derivative (denoted by A) is not a function of r^*. Since there is no flow in the z direction (Eq. 4.5.11), the only variation of p^* in the z direction is hydrostatic. We see this simply by setting $u_z = 0$ in Eq. 4.5.3. It follows then that the solution for p^* must lead to

$$p = -\rho g z + fn(r) \tag{4.5.22}$$

where $fn(r)$ is an unknown function of r alone. Then it follows that $\partial p / \partial r$ is not a function of z, so the first term in Eq. 4.5.21 is not a function of z^*. We conclude, finally, that since A is not a function of either r^* or z^*, it must be a constant. This permits us to integrate Eq. 4.5.21 immediately to yield

$$C = -\frac{A}{2}[1 - (z^*)^2] \tag{4.5.23}$$

(In the course of the integration we find two constants that are determined by imposing the no-slip condition on u^* at $z^* = \pm 1$.) The pressure distribution also follows from integration of Eq. 4.5.21, while incorporating Eq. 4.5.22, with the result

$$p^* = A \ln r^* + P^* - \frac{\rho g H^2}{\mu U} z^* \tag{4.5.24}$$

where P^* is a constant of integration. For a boundary condition on pressure, we set

$$p^* = 0 \quad \text{at} \quad r^* = \frac{R}{H} \tag{4.5.25}$$

but in doing so we have dropped the (small) effect of hydrostatic pressure on the flow, from which it follows that

$$P^* = -A \ln \frac{R}{H} \tag{4.5.26}$$

Note that the constant A is still undetermined. We may now find A by writing the volume flowrate in the form

$$Q = 2 \int_0^H 2\pi(ru_r)\, dz = 2H^2 U \int_0^1 2\pi C\, dz^* \tag{4.5.27}$$

After introducing Eq. 4.5.23 for C into the integral, and using Eq. 4.5.17 for U, we find

$$A = -\frac{3R_i}{H} \tag{4.5.28}$$

The algebra is flying about like angry hornets. Before we get stung, let's get back to the physics. We may now write the velocity field in the form

$$u_r(r, z) = \left(\frac{3Q}{8\pi H}\right)\frac{1}{r}\left[1 - \left(\frac{z}{H}\right)^2\right] \tag{4.5.29}$$

The details of the velocity field may be of less interest to us than the pressure drop–flowrate relationship. Hence we need to find the pressure at the entrance to the disk region (i.e., p at $r = R_i$). From Eqs. 4.5.24, 4.5.26, and 4.5.28, we find

$$p^* = \frac{pH}{\mu U} = -\frac{3R_i}{H} \ln \frac{r}{R} \tag{4.5.30}$$

If we define the pressure drop as

$$\Delta P = p(R_i) - p(R) = p(R_i) \tag{4.5.31}$$

we may rearrange Eq. 4.5.30, after setting $r = R_i$, to

$$\Delta P = -\frac{3\mu Q}{4\pi H^3} \ln \frac{R_i}{R} \tag{4.5.32}$$

This is a testable prediction. We need only obtain data for Q and ΔP in several geometries (i.e., for different values of H, R, and R_i) and over a range of viscosities. The results should suggest the limitations on using this model for engineering design. Some data for a related flow, presented in Problem 4.49, indicate the extent to which this type of analysis is useful.

EXAMPLE 4.5.2 *A Biomedical Flow Device*

Many commercial systems for analyzing blood and urine samples are based on the principles of wicking and steady planar flow. Figure 4.5.3 will help to define the fluid dynamics of wicking. In a typical flow device used in biomedical analysis, a drop of the sample to be tested is brought into contact with an opening in one end of the test cell. Capillary action is sufficient to draw the sample into the test chamber because the height of the opening is small, of the order of a millimeter. An air vent at the other end of the cell permits displaced air to escape and maintains atmospheric pressure at the advancing surface of the sample. The test chamber is essentially a region bounded

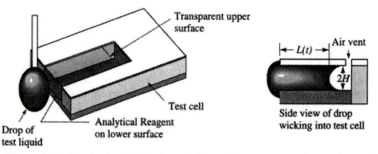

Figure 4.5.3 Device in which a drop is drawn into a narrow planar channel by capillary action (wicking).

by a pair of parallel planes. A pair of parallel sharp rails (Fig. 4.5.4) controls the position at which the drop stops. One of the planes is coated with an analytical reagent that dissolves into the sample and causes its color to change. An optical system "reads" the color of the sample through a transparent region of the second (upper) plane. The fluid dynamic issue in the design of this system is the rate at which the sample liquid is drawn into the test cell. We want to model this process so we can estimate that rate.

Our first step is to think about the physics of this system. Assuming that the plane of the device is horizontal, the only force available to "drive" the liquid into the cell arises from surface tension. The inflow is retarded by the viscous resistance of the liquid to deformation. If we sketched the flow region and labeled the various forces acting on the liquid, we would come out with a drawing like Fig. 4.5.4.

This is a very complex flow, largely because of the free surfaces at the two ends. In addition, it is a *transient* flow, because as the liquid intrudes into the channel, the area over which the viscous stresses are exerted increases. We might anticipate that the liquid will enter rapidly but slow down as viscous effects grow larger.

How does the surface tension force manifest itself, and "draw" the liquid into the channel? With reference to Fig. 4.5.4, we see that there is a force acting to the right, with a component in the plane of the channel given by

$$\frac{F_\sigma}{W} = 2\sigma \cos \theta \tag{4.5.33}$$

The factor of 2 accounts for the forces on the top and bottom surfaces. We write the force per unit width by using the width W of the channel in the direction normal to the page. We will assume that $W \gg H$.

Opposing this force is the viscous resistance, but how are we to determine this term without first solving for the flow field? We will assume that we can approximate the flow field by taking the solution to a simpler problem that has some of the key features of this unsteady channel flow. The problem we select is that of *fully developed steady*

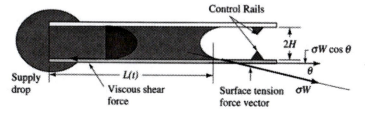

Figure 4.5.4 Force diagram for wicking of a drop into the channel.

flow of a viscous liquid through a parallel plate channel. The channel is completely filled with the liquid. Our thinking is that when the penetration length L has become large compared to H, the end effects may not be very important and the steady flow model may provide at least an approximation to the viscous resistance of the unsteady case. At least it's worth a try.

Our immediate goal, then, is to derive an expression for the steady flow field within a parallel plate channel, caused by a pressure difference ΔP across the length L. (Later we will have to relate the pressure difference to the surface tension.)

Note what we are doing here. We have replaced a difficult problem with a simpler, solvable one, hoping that the solution to the simple problem will shed some light on the behavior of the more complex flow that is of interest to us. We begin by assuming that the velocity vector, in Cartesian coordinates, has the form

$$\mathbf{u} = (u_z(y), 0, 0) \tag{4.5.34}$$

The geometry is given in Fig. 4.5.5.

With the assumed fully developed velocity profile (Eq. 4.5.34) the continuity equation (Table 4.1.1, Eq. 4.1.8a) is already satisfied, and yields no information. From the dynamic equations (Table 4.3.1), we find the following simplifications. Equation 4.3.24a is not used, on the grounds that there is no flow in the x direction. Equation 4.3.24b yields the hydrostatic law, and will not be useful to us. Equation 4.3.24c takes the form

$$\frac{\partial p}{\partial z} = \mu \frac{\partial^2 u_z}{\partial y^2} \tag{4.5.35}$$

By arguments that should be familiar at this point, we note that since the left-hand side of Eq. 4.5.35 is not a function of y, and the right-hand side is not a function of z, each term is a constant C. Thus we are quickly led to integrate

$$\mu \frac{d^2 u_z}{dy^2} = C \tag{4.5.36}$$

twice with respect to y and to use as boundary conditions the no-slip assumption:

$$u_z = 0 \quad \text{on} \quad y = \pm H \tag{4.5.37}$$

The result is

$$u_z = -\frac{CH^2}{2\mu}\left[1 - \frac{y^2}{H^2}\right] \tag{4.5.38}$$

We now go back to Eq. 4.5.35 and integrate

$$\frac{\partial p}{\partial z} = C \tag{4.5.39}$$

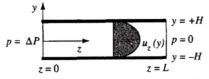

Figure 4.5.5 Geometry for analysis of steady planar flow.

with the boundary condition

$$p = \Delta P \quad \text{at} \quad z = 0 \tag{4.5.40}$$

and obtain the result

$$p = \Delta P + Cz \tag{4.5.41}$$

All that remains is the determination of the constant C, which can be found in terms of the volume flowrate (per unit width) by writing

$$\frac{Q}{W} = 2 \int_0^H u_z \, dy = -\frac{2CH^3}{3\mu} \tag{4.5.42}$$

Hence

$$-C = \frac{3\mu Q}{2WH^3} \tag{4.5.43}$$

But we have a second relationship for C, since we have taken the pressure at the downstream end of the channel (at $z = L$) to be $p = 0$. From Eq. 4.5.41, then, we find

$$-C = \frac{\Delta P}{L} = \frac{3\mu Q}{2WH^3} \tag{4.5.44}$$

Thus we not only eliminate C from the analysis, we also find the pressure drop–flowrate relationship, which we rewrite in the form

$$Q = \frac{2WH^3 \, \Delta P}{3\mu L} \tag{4.5.45}$$

We may refer to this as Poiseuille's law for a wide channel (*cf.* Eq. 3.2.34). It is the steady flow solution for the *geometry* of interest to us here. However, it is based on the idea that the channel is *completely wetted with liquid* over the entire length L. This is distinct from the *dynamics* of interest to us, where the length of the wetted region is increasing with time. In fact, our goal is to develop a model of the function $L(t)$. To what degree, then, is Eq. 4.5.45 at all useful, since it applies to conditions different from those of Fig. 4.5.4? In other words, how can we use Eq. 4.5.45?

We will make the following assumption:

> *If a liquid is moving into a parallel plate channel under a constant pressure differential, Eq. 4.5.45 holds in this transient case and we simply replace* L *by* L(t).

Under what conditions, if any, might this statement be true? We can think of one condition in which it would *not* be true. In the early stage of penetration, when $L(t)$ is smaller than or simply comparable to H, the assumption of fully developed flow is not likely to hold. On the other hand, fully developed flow might exist once $L(t)$ has grown to be several times larger than H. Should we build a theory on "might" or "maybe?" Why not? The worst that can happen is that when the model is tested in the laboratory, it fails to describe the phenomenon of interest to a satisfactory degree.

In the absence of any better suggestion, we move ahead with the assumption above.

Our next step is to relate ΔP, the pressure drop driving the flow, to the surface tension effect at the meniscus. We already have such a relationship. It is the Young–Laplace equation—Eq. 2.2.7. For the planar geometry, one of the radii of curvature is infinite. (A plane corresponds to a sphere of infinite radius.) The other radius depends on H, and on the contact angle θ that the meniscus makes with the solid boundary. If we regard the meniscus as having a cylindrical surface of radius R, it is not difficult to find that R and H are related by a simple geometrical relationship:

$$R = \frac{H}{\cos \theta} \tag{4.5.46}$$

Then, according to the Young–Laplace equation, the driving pressure is given by

$$\Delta P = \frac{\sigma \cos \theta}{H} \tag{4.5.47}$$

We introduce this expression into Eq. 4.5.45, with the result

$$Q = \frac{2WH^2 \sigma \cos \theta}{3\mu L(t)} \tag{4.5.48}$$

From the geometry, we can write the volume flowrate as

$$Q = 2HW \frac{dL}{dt} \tag{4.5.49}$$

Eliminating Q between these last two equations, we obtain a differential equation for $L(t)$:

$$L \frac{dL}{dt} = \frac{H\sigma \cos \theta}{3\mu} \tag{4.5.50}$$

The solution, satisfying the initial condition $L = 0$ at $t = 0$, is

$$L = \left[\frac{2H\sigma \cos \theta}{3\mu} t \right]^{1/2} \tag{4.5.51}$$

If the required extent of penetration is L^*, the time to reach that length is given by

$$t^* = \left(\frac{3\mu}{2H\sigma \cos \theta} \right) (L^*)^2 \tag{4.5.52}$$

With this model we can now estimate the time required for a liquid to penetrate a channel of given geometry. For example, suppose we take the following physical properties:

$$\mu = 0.003 \text{ Pa} \cdot \text{s} \qquad \sigma = 0.050 \text{ N/m} \qquad \theta = 85°$$

and choose

$$H = 0.0005 \text{ m} \quad \text{and} \quad L^* = 0.03 \text{ m}$$

Then we find, from Eq. 4.5.52,

$$t^* = 1.9 \text{ s}$$

Whether this is a realistic estimate depends of course on the degree to which the model applies to unsteady penetration. An experimental study should be designed at this point, to shed some light on the newly raised question. This mathematical model can serve as an aid in the design of the experimental study. This is an important use of an approximate flow analysis.

EXAMPLE 4.5.3 *Film Thickness on a Center-Fed Rotating Disk*

A lubricating film of liquid is fed to the axis of a rotating disk from a small tube centered at the axis of rotation, as shown in Fig. 4.5.6. The liquid film flows out radially under the action of centrifugal acceleration. We are interested in the mean film thickness across the radius and the degree of nonuniformity of the film thickness, and we want to determine how these features depend on the feed rate, the rotation rate, and the liquid properties. We are able to derive a model of this system using the tools developed so far.

Thinking back to our discussion of Example 4.5.1, we must wonder about the dynamics of the flow near the axis of rotation. While Fig. 4.5.6 suggests what specific shape we might expect for the free surface near the axis, there is no guarantee that the surface profile will in fact resemble that of the sketch. In the exit region, where the flow is really very complex, the nonlinear terms in the Navier–Stokes equations will be important under most conditions of flow. We would expect the distance L of the tube exit from the disk surface to affect the shape of the liquid surface, at least near the exit region. We are not going to try to model the flow in the exit region. Instead, we will assume that beyond some radial position r^* the film profile $h(r)$ is smooth in some sense, and we seek a model for $h(r)$ in that region.

Let's begin with the continuity equation, which we take as Eq. 4.1.8b for an incompressible liquid. Replacing the vector $\mathbf{A}$ by the velocity vector $\mathbf{u}$, we have

$$0 = \frac{1}{r}\frac{\partial(ru_r)}{\partial r} + \frac{\partial u_z}{\partial z} \tag{4.5.53}$$

(We have assumed that the flow is axisymmetric, and so $\partial u_\theta/\partial\theta = 0$.)

We first introduce the approximation that the flow is largely radial, with only a small u_z component. Then Eq. 4.5.53 yields the approximation

$$0 \approx \frac{1}{r}\frac{\partial(ru_r)}{\partial r} \tag{4.5.54}$$

If this held exactly, we would conclude that

$$ru_r \neq f(r) \tag{4.5.55}$$

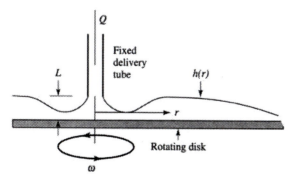

Figure 4.5.6 Liquid is fed onto the surface of a rotating disk.

Next we simplify the radial component of the momentum equation, in this case Eq. 4.3.24d. In doing so we assume steady state flow and, as before, axisymmetry. We neglect the term multiplying u_z on the left-hand side, since we have assumed that the flow is largely radial. From Eq. 4.5.55, the first of the viscous terms on the right-hand side vanishes. The next two vanish because all $\partial/\partial\theta = 0$ for axisymmetric flow. There is no gravitational force in the r direction, so $g_r = 0$. The result is

$$\rho u_r \frac{\partial u_r}{\partial r} - \rho \frac{u_\theta^2}{r} = -\frac{\partial p}{\partial r} + \mu \frac{\partial^2 u_r}{\partial z^2} \tag{4.5.56}$$

Our next assumption is that while the liquid flows out radially across the disk, its angular velocity matches that of the disk. That is, the liquid moves as a rigid body in the angular direction, and so

$$u_\theta = r\omega \tag{4.5.57}$$

where ω is the rotational speed of the disk in radians (not revolutions) per unit of time. Then Eq. 4.5.56 becomes

$$\rho u_r \frac{\partial u_r}{\partial r} - \rho r\omega^2 = -\frac{\partial p}{\partial r} + \mu \frac{\partial^2 u_r}{\partial z^2} \tag{4.5.58}$$

Now let's look at the pressure distribution function, $p(r, z)$. To begin, consider the z component of the momentum equation, Eq. 4.3.24f. If we again ignore any effect of the velocity component u_z (which is assumed to be small), we may write the following immediately:

$$0 \approx \frac{\partial p}{\partial z} + \rho g_z \tag{4.5.59}$$

We expect that for *thin* films the contribution of gravity to the pressure distribution is small. Then a good approximation is

$$\frac{\partial p}{\partial z} \approx 0 \tag{4.5.60}$$

Hence p is approximately uniform across the thickness of the liquid film. Along the free surface, the pressure must be equal to the external pressure, except for any contribution arising from surface tension and the curvature of the free surface that arises from the fact that $h = h(r)$. We will assume that the shape of the free surface is smooth enough to guarantee that surface tension does not give rise to an appreciable contribution to the radial pressure profile. Then, since the external pressure is uniform (it is simply the ambient atmospheric pressure), we conclude that

$$\frac{\partial p}{\partial r} \approx 0 \tag{4.5.61}$$

everywhere within the liquid film. As a consequence, Eq. 4.5.58 becomes

$$\rho u_r \frac{\partial u_r}{\partial r} - \rho r\omega^2 = \mu \frac{\partial^2 u_r}{\partial z^2} \tag{4.5.62}$$

Let's pause for a moment and review the ideas we are promoting here. As usual, we are developing a mathematical model by the process of simplifying the equations for the conservation of mass and momentum. Some assumptions seem quite reasonable. It is not difficult to imagine that we can create the flow suggested by the sketch in Fig. 4.5.6 in such a way that the flow is

steady state
axisymmetric
moving rigidly in the angular direction with the disk rotation

It may be less obvious that we can create a flow for which the film thickness is

thin
fairly uniform

In fact, it is not even obvious what we mean by "thin" and "fairly uniform" *in a quantitative and testable sense.* But we make the assumptions anyway, because if they *are* true they will move us rapidly along a path toward a mathematical model that can be solved analytically. Still, this does not relieve us of the responsibility, ultimately, for checking the assumptions by direct observation of the flow. We often find that while some of these assumptions are not valid under any arbitrary set of conditions that would produce the flow sketched in Fig. 4.5.6, there may be a set of operating conditions that does produce a flow satisfying all the simplifying assumptions. Then the question remaining is whether this set of operating conditions is so restrictive that it does not produce a flow of any practical interest.

Now let's get back to model building. We cannot devise a simple analytical solution for Eq. 4.5.62 for the radial velocity profile because of the nonlinear term (the first term on the left-hand side). (We could solve the equation numerically, but this is not a trivial procedure.) Do we have any basis for assuming that the nonlinear term is small compared with the other contributions to the momentum balance? The answer may not be obvious, and we may be reluctant to introduce an assumption without a strong physical argument to support it. On the other hand, the temptation is great, since the removal of this term will lead us almost immediately to a simple analytical model. Do we yield to temptation?

Yes! The decision is simple:

Our goal is to obtain a simple analytical model of this flow.
Our path is blocked by this nonlinear term.
We cannot achieve our goal easily unless . . .
We throw away the offensive term.

Then we can achieve our goal. At that point the issue becomes the following:

> *Can we confirm the validity of the model through experimental observations?*

This could include the determination that the range of experimental conditions over which the model does mimic the observations is too narrow to be of any use. (Remember, we are engineers. We *use* models. We don't simply display and admire them.)

> *In the absence of an experimental program, is there any rational basis for using the simple model to determine its own range of validity?*

Let's not get sidetracked here. Continue modeling. Solve the following differential equation:

$$-\rho r \omega^2 = \mu \frac{\partial^2 u_r}{\partial z^2} \qquad (4.5.63)$$

with the boundary conditions

$$\frac{\partial u_r}{\partial z} = 0 \quad \text{on} \quad z = h(r) \tag{4.5.64}$$

which expresses the assumption that there is no shear stress on the free surface, and

$$u_r = 0 \quad \text{on} \quad z = 0 \tag{4.5.65}$$

which is simply the no-slip assumption on the solid surface.

We integrate Eq. 4.5.63 twice with respect to z, impose these conditions, and find[6]

$$u_r = \frac{\rho r \omega^2 h^2}{\mu} \left[\frac{z}{h} - \frac{1}{2} \left(\frac{z}{h} \right)^2 \right] \tag{4.5.66}$$

Does anyone remember our original goal? It was to find the mean film thickness, and the degree of uniformity of the film across the surface of the disk. The one piece of information we have not used so far is this: the film is created by delivering liquid at the axis of rotation from a tube, at a constant volumetric flowrate Q. Then the film thickness profile must be determined by this delivery rate, and we simply write a mass balance in the form

$$Q \equiv 2\pi r \int_0^{h(r)} u_r dz = \frac{2\pi \rho r^2 \omega^2 h^3}{3\mu} \tag{4.5.67}$$

or

$$h(r) = \left(\frac{3\mu Q}{2\pi \rho r^2 \omega^2} \right)^{1/3} \tag{4.5.68}$$

From this result we may find that the mean film thickness is

$$\bar{H} \equiv \frac{1}{\pi R^2} \int_0^R 2\pi r h(r) \, dr = \left(\frac{81 \mu Q}{16\pi \rho R^2 \omega^2} \right)^{1/3} \tag{4.5.69}$$

For a measure of the uniformity we can calculate the root-mean-square deviation of the film thickness from its average value (averaged across the surface of the disk). Specifically, we define a uniformity measure σ as

$$\sigma \equiv \left\{ \frac{1}{\pi R^2} \int_0^R 2\pi r \left[\frac{h(r) - \bar{H}}{\bar{H}} \right]^2 dr \right\}^{1/2} \tag{4.5.70}$$

By this definition, σ is essentially the *variance* of the film thickness distribution—the square root of the average squared deviation of the film thickness—from its mean value. From Eqs. 4.5.68 and 4.5.69 we may write

$$\frac{h(r)}{\bar{H}} = \frac{2}{3} \left(\frac{r}{R} \right)^{-2/3} \tag{4.5.71}$$

From Eq. 4.5.70 we then perform the indicated integration and find

$$\sigma = 0.5774 \tag{4.5.72}$$

---

[6] Note that $r u_r$ is a function of r. This violates Eq. 4.5.54! At this point we have some uncertainty as to the value of this solution. See Problem 4.87.

Two conclusions are apparent. The degree of nonuniformity of the film thickness, as defined by σ, is independent of all parameters in the problem. This implies that we have no control over uniformity in this flow. The second conclusion is that the film thickness is very nonuniform, since this value of σ essentially states that the nonuniformity of the film thickness is about 58% of the mean value.

Let's look a little more closely at the film thickness profile (Fig. 4.5.7): the film thickness deviates dramatically from the mean as the axis of rotation is approached. In fact, inspection of Eq. 4.5.71 shows us that the model predicts that the film thickness will go to infinity as r approaches the axis. The reason for this is obvious: since the area for flow gets small like r as r approaches zero, the film thickness must blow up in such a way that (according to Eq. 4.5.67) r^2h^3 remains constant. Of course, this does not happen in the real flow, as Fig. 4.5.6 suggests. We don't expect this model to work well near the axis, and we should not draw any conclusions from the behavior of the model in that region.

With these ideas in mind, and recalling that the value calculated from Eq. 4.5.70 includes the region near the axis, we reconsider our evaluation of the nonuniformity. For example, suppose we examine the nonuniformity over the outer half of the disk, which we define as the region from $R/2$ to R. (This is not half the area of the disk, but in fact three-fourths of the area.) Then we define and calculate

$$\sigma_{1/2} \equiv \left\{ \frac{4}{3\pi R^2} \int_{R/2}^{R} 2\pi r \left[\frac{h(r) - \overline{H}}{\overline{H}} \right]^2 dr \right\}^{1/2} \tag{4.5.73}$$

with the result that

$$\sigma_{1/2} = 0.2232 \tag{4.5.74}$$

As expected, the film shows a much smaller variance over the outer region of the disk.

Other features of this model can be explored, and some are suggested as homework problems. At this stage we must not forget that we obtained this simple analytical model as a consequence of imposing a number of approximations (assumptions) that permitted us to simplify the Navier–Stokes equations. Now we have the responsibility for evaluating some of these assumptions. *This is an essential feature of the model-building exercise.*

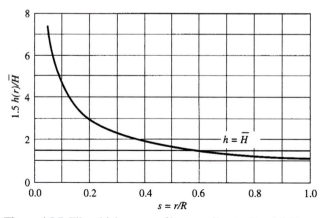

Figure 4.5.7 Film thickness profile according to Eq. 4.5.71.

EXAMPLE 4.5.4 *Film Thickness on a Center-Fed Rotating Disk: Evaluation of Approximations*

Let's begin with a key assumption that led to the simplification of Eq. 4.5.62, which we rewrite as follows:

$$\rho u_r \frac{\partial u_r}{\partial r} - \rho r \omega^2 = \mu \frac{\partial^2 u_r}{\partial z^2} \tag{4.5.75}$$

In Eq. 4.5.63, we dropped the underlined term without any strong physical justification, simply as a matter of expediency. Dropping that term permitted us to obtain an analytical solution for the velocity profile, and subsequently for the film thickness.

Now that we have a solution for u_r, Eq. 4.5.66, we can calculate the nonlinear term and compare it with the terms retained. In doing this we will find that whether the dropped term is negligible is determined by the values of certain dimensionless groups. Now we have a quantitative test that we can use to ascertain whether our simple model can be applied with confidence to a given problem.

Our first task, then, is to calculate the so-called inertial term

$$I \equiv \rho u_r \frac{\partial u_r}{\partial r} \tag{4.5.76}$$

and compare it with the centrifugal term

$$C \equiv \rho r \omega^2 \tag{4.5.77}$$

Our question is, Under what conditions is $I/C \ll 1$?

This is a very straightforward but tedious exercise in differential calculus and algebra. (See Problem 4.48.) The result is

$$I = \frac{\rho^3 \omega^4 q^{4/3}}{3\mu^2} F\left(\frac{z}{h}\right) r^{-5/3} \tag{4.5.78}$$

where

$$F\left(\frac{z}{h}\right) \equiv \left(\frac{z}{h}\right)^2 - 2\left(\frac{z}{h}\right)^3 + \frac{3}{4}\left(\frac{z}{h}\right)^4 \tag{4.5.79}$$

and

$$q \equiv \frac{3\mu Q}{2\pi\rho\,\omega^2} = r^2 h^3 \tag{4.5.80}$$

The function $F(z/h)$ is largest (in absolute value) at $z = h$, where

$$\left|F\left(\frac{z}{h}\right)\right| = 0.25 \tag{4.5.81}$$

Hence we find that

$$\left|\frac{I}{C}\right| = \frac{1}{12}\left(\frac{3}{2\pi}\right)^{4/3}\left(\frac{\rho Q^2}{\mu\omega R^4}\right)^{2/3}\left(\frac{R}{r}\right)^{8/3} \tag{4.5.82}$$

The importance of the inertial term grows larger, relative to the centrifugal term, as the origin $r = 0$ is approached. As already noted, we should not try to apply this model too close to the axis of the disk. If we pick a value of $r/R = 0.25$, evaluate the numerical coefficient in Eq. 4.5.82, and round off the numbers, we find that

$$\left|\frac{I}{C}\right| \leq \left(\frac{\rho Q^2}{\mu \omega R^4}\right)^{2/3} \tag{4.5.83}$$

over most of the surface of the disk. Thus we discover that the relative role of inertial versus centrifugal forces is determined by the dimensionless grouping of parameters on the right-hand side of Eq. 4.5.83.

Let's look at some numbers. Consider a disk of radius $R = 5$ cm and a liquid of viscosity $\mu = 0.2$ poise and density $\rho = 1.5$ g/cm^3. The rotational speed will be 100 rad/s (≈ 1000 rpm). We require a film thickness greater than 100 μm over the entire disk. From Eq. 4.5.68, if we set $h = 100$ μm at $r = 5$ cm, we find that the required feed rate of liquid is $Q = 3.9$ cm^3/s. Equation 4.5.83 then indicates that $|I/C| < 0.015$. Hence the approximation in question is quite good for this case. For thinner films and higher rotation rates (more typical, e.g., of lubrication of a magnetic recording disk), the approximation is even better.

By the same procedure, we can confirm that the other approximations introduced in the derivation of Eq. 4.5.66 are valid under a range of conditions of practical interest.

EXAMPLE 4.5.4 *A Model of Torsional Flow Between Parallel Disks*

Imagine a horizontal disk-shaped surface onto which we carefully place a drop of liquid. Next we take a second disk, identical to the first and parallel to it, and lower it toward the first disk until the drop begins to be squeezed. We squeeze until the drop flattens and completely fills the space between the two disks. Now we rotate the upper disk about its axis, at a constant rotational speed Ω. As a consequence, the liquid between the two surfaces is in a rotational or torsional flow. Figure 4.5.8 shows a schematic of this system.

Our goal is to develop a model from which the flow field may be found, leading to an expression for the relationship of the torque required to keep the upper disk in steady motion as a function of the geometry (disk radius and the spacing between the disks) and the viscosity of the liquid. If we can achieve this, we will have a model that can be used to design a viscometer suitable to the study of very small samples of liquid. Such devices are especially important in the study of biological liquids such as blood and mucus, since knowledge of their viscosity is useful in the diagnosis of certain pathological conditions.

If we assume laminar Newtonian incompressible isothermal axisymmetric flow at rotational speeds low enough to prevent radial (centrifugal) flow, the θ component of

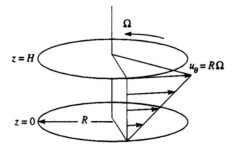

Figure 4.5.8 Torsional flow between parallel disks.

the Navier–Stokes equations reduces to the form

$$0 = \mu \left(\frac{\partial}{\partial r} \left[\frac{1}{r} \frac{\partial}{\partial r} (r u_\theta) \right] + \frac{\partial^2 u_\theta}{\partial z^2} \right) \tag{4.5.84}$$

Notice that we have here a case of a two-dimensional flow: the velocity vector **u** has only a single component u_θ, but u_θ is a function of both r and z. This is a linear partial differential equation, but you may not be too familiar with methods for solving such equations. Let's think about the physics and geometry of this flow for a moment, in search of clues to the nature of the velocity profile. In particular, let's look at the boundary conditions on the flow field. Of course, on the lower, stationary, boundary, we have a no-slip condition, and so we must write

$$u_\theta = 0 \quad \text{on} \quad z = 0 \tag{4.5.85}$$

We also require a no-slip condition on the upper, rotating, surface, which leads us to

$$u_\theta = r\Omega \quad \text{on} \quad z = H \tag{4.5.86}$$

This boundary condition suggests the possibility that the velocity might have the functional form

$$u_\theta = r\Omega F(z) \tag{4.5.87}$$

with $F(z)$ taking on the values

$$F(z) = 0 \quad \text{on} \quad z = 0$$
$$\tag{4.5.88}$$
$$F(z) = 1 \quad \text{on} \quad z = H$$

The simplest function $F(z)$ that behaves this way is

$$F(z) = \frac{z}{H} \tag{4.5.89}$$

Hence we ask the question: Can the relation

$$u_\theta = \frac{rz\Omega}{H} \tag{4.5.90}$$

satisfy the momentum equation (Eq. 4.5.84)? It is easy to confirm that it can. Hence we have "found" a solution to Eq. 4.5.84 that also satisfies the boundary conditions on the velocity. We conclude that Eq. 4.5.90 is the velocity field we were seeking.

With this result we can find the shear stress on the rotating surface. It is (see Table 4.2.1 for $\Delta_{\theta z}$)

$$T_{\theta z} = -\mu \Delta_{\theta z} = -\mu \frac{\partial u_\theta}{\partial z} = -\frac{r\Omega\mu}{H} \tag{4.5.91}$$

The force on that surface is obtained by integrating the shear stress across the disk surface:

$$F = \int_0^R dF = \int_0^R -T_{\theta z} dA = \int_0^R \frac{r\Omega\mu}{H} 2\pi r \, dr = \frac{2\pi\Omega\mu R^3}{3H} \tag{4.5.92}$$

The sign becomes positive because we are finding the force exerted by the liquid on the surface. Of more practical interest would be the torque required to turn the rotating disk. This follows from

$$T = \int_0^R -rT_{\theta z}dA = \int_0^R \frac{r^2\Omega\mu}{H} 2\pi r\, dr = \frac{\pi\Omega\mu R^4}{2H} \tag{4.5.93}$$

The model is complete, and we can find the viscosity of a liquid from a measurement of the torque required to maintain the rotational speed Ω for a given set of values of R and H.

EXAMPLE 4.5.6 *Evaluation of the Design of a Rotational Disk Viscometer*

A viscometer is designed as shown in Fig. 4.5.9. The torque is fixed by hanging a known weight W on a frictionless pulley attached to the rim of the upper disk. The rotational speed is measured. Derive an expression for the expected rotational speed as a function of the weight W, with the fluid viscosity as a parameter. What sample volume is required?

For a selected weight W the torque exerted will be

$$T = RW \tag{4.5.94}$$

The sample volume is

$$V = \pi R^2 H \tag{4.5.95}$$

For the geometrical parameters $H = 10^{-4}$ m and $R = 0.03$ m, we find the required sample volume to be

$$V = \pi(0.03)^2 \times 10^{-4} = 2.8 \times 10^{-7} \mathrm{m}^3 = 0.28\, \mathrm{mL} \tag{4.5.96}$$

This sample volume is small enough to make the viscometer suitable for use with biological fluids.

From Eqs. 4.5.93 and 4.5.94 we find

$$\Omega = \frac{2HW}{\pi\mu R^3} = \frac{2.36W}{\mu} \tag{4.5.97}$$

For a liquid of viscosity $\mu = 6$ m Pa·s, a 1-gram "weight" ($W = 9.8 \times 10^{-3}$ N) will produce a rotational speed of

$$\Omega = \frac{2.36\,(9.8 \times 10^{-3})}{0.006} = 3.85\, \mathrm{rad/s} = 0.61\, \mathrm{rev/s} \tag{4.5.98}$$

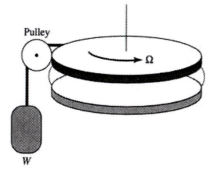

Figure 4.5.9 Rotational disk viscometer.

SUMMARY

Often the flows of interest to us are so complex that the simple one-dimensional force–balance method of Chapter 3 cannot be used. In this chapter we learn how to describe three-dimensional flows, including unsteady flows. For a Newtonian fluid the fluid dynamics are completely described by the solutions (for the pressure p and the components of velocity $\mathbf{u}$) of the equation that expresses conservation of mass (the continuity equation)

$$\frac{\partial \rho}{\partial t} = - \nabla \cdot \rho \mathbf{u}$$

and the equations that describe conservation of momentum (the Navier–Stokes equations). There are three Navier–Stokes equations for a three-dimensional flow, but we can express all three in a vector format as

$$\rho \left[\frac{\partial \mathbf{u}}{\partial t} + \mathbf{u} \cdot \nabla \mathbf{u} \right]_i = - [\nabla p]_i + \mu \, [\nabla^2 \mathbf{u}]_i + \rho f_i$$

Table 4.3.1 presents these equations in component forms for Cartesian, cylindrical, and spherical coordinate systems.

An essential part of the model-building (problem-solving) process is the establishment of *boundary conditions* that the solutions for p and $\mathbf{u}$ must satisfy. This is where we introduce simplifications, assumptions, and approximations into the analysis. This stage is guided by our understanding, observation, and intuition regarding the physics, and driven by our motivation to create the least complex mathematical model that will lead to testable predictions. In the end, reality (correspondence to observations) is the test of the success of a model-building exercise.

The remainder of Chapter 4 is devoted to presentation of a variety of examples of the development of mathematical models of flows of interest to engineers. The testing of these and other models is left to the homework exercises that follow.

PROBLEMS

4.1 Since the density ρ appears outside the derivatives when we pass from Eq. 4.3.16 to Eq. 4.3.17, it might appear that we have assumed that the fluid is of constant density. Give the derivation of Eq. 4.3.17 and show that this is not the case.

4.2 Return to Section 4.4.1, where we calculated the torque driving the inner cylinder (Eq. 4.4.13). Calculate the torque required to hold the outer cylinder in place. Is the result as you expected?

4.3 In Eq. 4.4.10 we have the velocity profile due to angular drag flow between concentric cylinders. Derive expressions for the shear rate $\Delta_{\theta r}$ and the shear stress $T_{\theta r}$ as functions of radial position r/R. Plot the ratio of the shear rates at the inner and outer surfaces as a function of the geometrical parameter κ. What conclusion can you draw from your results?

4.4 The following data have been obtained in a coaxial cylinder viscometer for which $L = 0.1$ m, $\kappa R = 0.01$ m, and $\kappa = 0.9$. Note the units given for Ω.

Ω (rpm)	3	5	10	30	50	100
$10^6 T$ (N·m)	5.2	9	17	53	87	180

What is the viscosity of this fluid?

4.5 The viscosity of blood is being determined in a coaxial cylinder viscometer. Equation 4.4.13 is the basic equation for converting data on torque versus rotational speed into viscosity data. Assume that blood is a Newtonian fluid with a viscosity of 0.005 Pa·s at 25°C. For the viscometer being used, the outer cylinder has $R = 2$ cm and the rotating cylinder has $R = 1.8$ cm. If the limitation is the maintenance of laminar flow, what is the highest shear rate at which the viscosity can be measured? (See Eq. 4.4.15.)

4.6 Red blood cells are fragile and can be destroyed by large shear stresses. Suppose the viscometer of Problem 4.5 is in use and that the shear stress must be kept below 200 Pa. What is the highest shear rate at which you could safely measure the viscosity of blood in this instrument?

4.7 Derive Eq. 4.4.26 from the stated boundary conditions.

4.8 Pure glycerin at 25°C flows over a vertical plane surface to form a laminar liquid film, as in Fig. 4.4.2. The density of the liquid is 1250 kg/m³. Plot the film thickness as a function of the surface velocity of the film. Use Eq. 4.4.28 to set the upper limit on the film thickness that you will plot.

4.9 A small gas bubble is formed within a viscous liquid. The bubble has an initial diameter of 4 mm, and the liquid is at a uniform pressure of 10^5 Pa. Gravity does not act on the liquid. The gas is not soluble in the liquid. At time $t = 0$ the liquid is suddenly subjected to a uniform pressure of 2×10^5 Pa, and the bubble begins to collapse. Assume that the collapse occurs isothermally, so that the gas and liquid remain at constant temperature. Find the "history" of the collapse: that is, find the function $R(t)$. Take the liquid viscosity to be 100 Pa·s. Keep in mind that the *gas* pressure must change as the volume of the bubble changes, and assume that the gas obeys the ideal gas law. Introduce any assumptions that permit you to obtain an analytical solution for $R(t)$, but justify the assumptions.

4.10 Replace the assumption of isothermal collapse, in Problem 4.9, with the assumption of *adiabatic* collapse. Use a value of $\gamma = 1.4$ in the adiabatic relationship $PV^\gamma = $ constant.

4.11 Work through the derivation of the terminal velocity of a falling cylinder within a fluid-filled coaxial tube, but let the tube be *open* at both ends and immersed within a large body of the same fluid. Since the bottom is open, the displaced fluid does not have to move counter to the falling cylinder, and there is no pressure other than hydrostatic. Derive an equation analogous to Eq. 4.4.101.

4.12 Design a falling cylinder viscometer for use with Newtonian liquids of viscosities in the range 10 to 100 Pa·s.

4.13 A Newtonian fluid is in laminar isothermal flow through a channel of rectangular cross section, bounded by two pairs of parallel planes. One pair is separated by the distance $2B$, and the other by the distance W, such that $W \gg B$. With such a geometry we might neglect the effect of the walls of the channel that are separated by the distance W and assume that the flow does not vary in the z direction. (See Fig. P4.13.1.)

a. Derive the analog to Poiseuille's law for this geometry. That is, find the $Q(\Delta P)$ relationship analogous to Eq. 4.5.45, but do so for low speed compressible viscous flow of an ideal gas, air at 20°C.

b. Compare the model derived in part a to the data given in Fig. P4.13.2. The data for the flow of air at 20°C correspond to the following conditions:

$$L = 4 \text{ cm} \qquad 2B = 0.03 \text{ cm}$$

$$W = 5 \text{ cm} \qquad P_L \ll P_o$$

4.14 Calculate and plot the mean free path λ for air as a function of pressure at 20°C. Plot λ/B versus ΔP and comment on the results, with reference to Problem 4.13.

4.15 In the analysis of steady flow about a solid sphere, we made the pressure nondimensional by defining

$$\tilde{p} \equiv \frac{p}{\mu U/R} \qquad \textbf{(P4.15.1)}$$

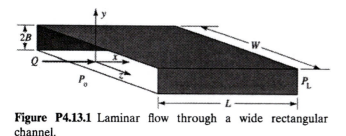

Figure P4.13.1 Laminar flow through a wide rectangular channel.

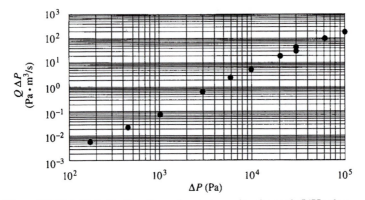

Figure P4.13.2 Data for flow through a rectangular channel. O'Hanlon, *J. Vac. Sci. Technol.*, **A5,** 98 (1987).

There seems to be no physical basis for this particular choice. We apparently chose to normalize p with a group of characteristic parameters of the flow that had the units of pressure. Suppose, instead, we had made another choice, such as

$$\tilde{p} \equiv \frac{p}{\rho U^2} \qquad \text{(P4.15.2)}$$

How does this change the analysis that follows? Can you draw any conclusions from our observations?

4.16 In deriving Stokes' law we assumed that gravity did not act on the fluid and did not in any way alter the flow field. When the derivation was complete, we put gravity into the force balance through the introduction of the buoyancy term. How do we know that gravity does not alter the flow around the sphere? Return to the starting point, Eqs. 4.4.103 and .104, and put the r and θ components of gravity into the equations. Then define a pressure

$$\mathscr{P} = p - \rho g r \cos \theta \qquad \text{(P4.16)}$$

Carry through the analysis with this pressure term, and show that we still recover Stokes' law.

4.17 In the development of Stokes' law, we assumed that the fluid satisfies a no-slip boundary condition at the surface of the sphere. Suppose the sphere is a gas bubble that rises in a viscous liquid. Because the gas is not a viscous fluid—its viscosity is several orders of magnitude below that of any liquid—it does not seem appropriate to impose the no-slip condition on the tangential velocity. Instead, we would assume that no shear stress can be transmitted across the interface, or

$$T_{r\theta} = -\mu \left[r \frac{\partial}{\partial r} \left(\frac{u_\theta}{r} \right) + \frac{1}{r} \frac{\partial u_r}{\partial \theta} \right] = 0 \qquad \text{at} \quad r = R$$

$$\text{(P4.17.1)}$$

a. Show how the analysis based on the application of this boundary condition changes the boundary conditions given in Eqs. 4.4.120 and 4.4.121.

b. Find the velocity components for this case.

c. Show that Stokes' law for a gas bubble changes from Eq. 4.4.133 to

$$U = \frac{R^2}{3\mu} (\rho_s - \rho_f) g \qquad \text{(P4.17.2)}$$

4.18 Consider steady, fully developed, laminar isothermal Newtonian flow through a long capillary. In Chapter 3 we derived Poiseuille's law for the $Q(\Delta P)$ relationship (Eq. 3.2.34). We also considered the case of capillary that had a thin annular coating of a liquid with a viscosity μ' different from that of the primary flow. We supposed that the coating thickness did not change in time and was uniform along the capillary axis. This permitted us to derive an analog to Poiseuille's law for this case (Eq. 3.2.57).

If we were not aware of the presence of this thin annular layer of liquid, we might apply Poiseuille's law and calculate the viscosity from the $Q(\Delta P)$ relationship. Thus we would write an "apparent viscosity" as

$$\mu_{\text{app}} = \frac{\pi R^4 \Delta P}{8QL} \qquad \text{(P4.18)}$$

Using this definition of apparent viscosity, and the results given in Example 3.2.3, find an expres-

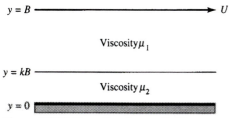

$y = B$ ———————→ U

Viscosity μ_1

$y = kB$ ———————————

Viscosity μ_2

$y = 0$

Figure P4.20 Drag flow in parallel layers.

sion for the error made in using Eq. P4.18 for calculation of the viscosity of the primary fluid in the case of an annular film of thickness $(1 - \kappa)R$, where R is the radius of the capillary. Prepare a plot of the error as a function of the viscosity ratio μ/μ', for several values of κ.

4.19 Give the full details that lead to Stokes' law, beginning with Eq. 4.4.125 and using the results developed prior to that for the velocity and pressure terms.

4.20 A pair of viscous liquids is initially layered between two large parallel planes, as shown in Fig. P4.20. The lower surface is stationary, and the upper surface moves in its own plane at uniform velocity U. Find the steady state velocity profiles in the two fluids.

4.21 Prove that in the limit as κ approaches 1, Eq. 4.4.101 takes the form

$$\Theta = \frac{(1 - \kappa)^3}{6} \quad \text{for} \quad \kappa \to 1 \quad \textbf{(P4.21)}$$

4.22 Give the components of the Δ tensor for the growth of a bubble, as described in Section 4.4.3.
4.23 Using Eq. 4.5.29, write out the components, in matrix form, of the rate of deformation tensor for steady radial flow between parallel disks.

4.24 A system is designed for supporting a disk on an oil film, making use of the principles of radial flow described in Example 4.5.1. The system is as described in Fig. 4.5.1, except that the entry tube is into the lower disk. Find the weight that can be supported on the upper disk (i.e., the disk opposite the entry tube) for the following conditions:

$R = 0.05$ m $R_i = 0.002$ m $H = 0.001$ cm
$Q = 10^{-8}$ m^3/s $\mu = 1$ Pa·s $\rho = 800$ kg/m^3

What ΔP is required to produce this flow?
4.25 In developing Eq. 4.5.52, no account was taken of the viscous resistance to flow of the gas displaced by the intruding liquid. Modify the analysis of the problem of estimating the time required for a liquid to penetrate a channel of given geometry to take this into account, and offer an estimate of the importance of the gas flow in retarding the liquid flow.
4.26 Derive a model for the time required for a liquid to completely "wick" into a long horizontal capillary of circular cross section. Present the result in the form of Eq. 4.5.52.

Apply the model to the case that the capillary has length $L = 0.2$ m and inside diameter $D = 2$ mm. The liquid has a surface tension of 0.02 N/m and a viscosity of 0.1 Pa·s. The contact angle is $0°$.
4.27 Find the terminal velocity of the body falling through the liquid contained in the four-sided rectangular guide, as shown in Fig. P4.27. You must also present a simple approximate mathematical model that relates the terminal velocity U to the geometry, and to appropriate physical properties.

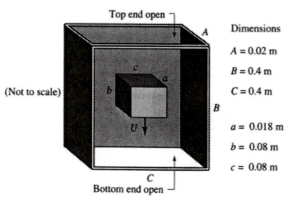

Dimensions

$A = 0.02$ m
$B = 0.4$ m
$C = 0.4$ m

$a = 0.018$ m
$b = 0.08$ m
$c = 0.08$ m

Figure P4.27 A rectangular solid falls through a four-sided guide.

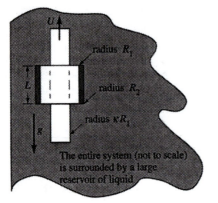

Figure P4.29 A long cylinder moves coaxially through a collar.

The falling body is a solid with the dimensions a, b, c shown in Fig. P4.27. The solid has a density of 5000 kg/m³. The Newtonian liquid that surrounds the body has a viscosity of 20 Pa · s, and a density of 1000 kg/m³.

The rectangular guide has rigid vertical sides with the dimensions A, B, C shown in Fig. P4.27. Open at the top and bottom ends, the guide is completely immersed in and surrounded by a large body of the liquid described above. Hence, as the inner body falls, the fluid it displaces does not move counter to the body, as would occur if the bottom end of the guide were closed.

The entire system is in Earth's gravitational field, and a means is provided to maintain the position of the falling body along the central axis of the guide, with its surfaces parallel to those of the guide.

State clearly any simplifying assumptions that you make, and offer brief supporting arguments where appropriate. It is not necessary to derive the analytical model from first principles if you can present and support a valid argument for the flow field for this problem.

4.28 In the derivation of Eq. 4.4.13 we neglected any end effect that could arise if the inner cylinder had a flat bottom. To shed some light on this feature of the design of a concentric cylinder viscometer, analyze the flow between parallel concentric disks, one of which is rotating about the common axis of the pair of disks. Begin with a set of simplifying assumptions, simplify the Navier–Stokes equations, and find an expression for the torque associated with this end of the

viscometer. Give a geometrical criterion for neglecting the end torque in comparison to that on the cylindrical face of the viscometer.

4.29 An infinitely long cylinder of circular cross section is continuously moving vertically upward through a large body of a viscous liquid, as in Fig. P4.29. This cylinder is surrounded by a short annular cylindrical collar, open at both ends, and a means is provided to maintain the collar coaxial with the moving cylinder. If there were no motion of the cylinder, the collar would fall because of its weight. The cylinder velocity U is adjusted until the collar is fixed in space (does not fall).

a. Write the basic force balance that describes this system when it is in equilibrium.

b. Derive a model for the viscosity of the liquid, in terms of the geometry, the speed U, and the densities of the liquid and collar.

c. What is the viscosity μ (Pa · s) if the speed U that halts the fall of the collar is found to be 0.01 m/s? The following data are available:

$R_1 = 3.00$ mm　　　$R_2 = 4.00$ mm　　　$\kappa = 0.99$
$L = 20$ mm　　　$\rho_{\text{liquid}} = 1000$ kg/m³
　　　　　　　　$\rho_{\text{collar}} = 5000$ kg/m³

4.30 A sphere of radius R rotates steadily about its vertical axis with an angular velocity Ω (rad/s) in an infinite Newtonian liquid of viscosity μ (Fig. P4.30). Assume that the liquid is quiescent far from the sphere. Beginning with the Navier–Stokes equations, derive, but do not attempt to solve, a differential equation for the velocity profile $u_\phi(r, \theta)$ in the liquid when the rotation is slow enough to permit the neglect of centrifugal forces. Write a set of boundary conditions for the differential equation.

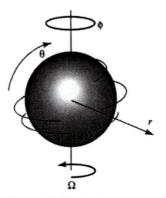

Figure P4.30 A sphere rotates about a vertical axis.

4.31 A viscous liquid is pumped into the region between parallel planes, as shown in Fig. P4.31. The upper assembly that forms the upper surface, and through which the liquid is pumped at steady flowrate Q, "floats" on a liquid film that is maintained at a uniform thickness $2H$. The lower assembly confines the liquid to flow in a rectilinear manner. The boundaries between the upper and lower assemblies are machined with such close tolerances that the liquid does not leak out of the system except through the regions $2H \times W$ at each end.

Derive and present an explicit expression for the vertical force $\mathbf{F}$ required to hold the plates a distance $2H$ apart, in terms of the geometry and the flowrate Q, and the properties of the liquid.

4.32 In Section 4.4.2 we presented an analysis of fully developed flow in a laminar film flowing down a vertical *planar* surface. Derive the analogous model for the case of a fully developed laminar film flowing down a vertical *cylindrical* surface. In particular, do the following.

a. Show that the velocity profile is given by

$$u(r) = \frac{gr_\delta^2}{2\nu}\left[\ln\left(\frac{r}{r_0}\right) - \frac{1}{2}\left(\frac{r^2 - r_o^2}{r_\delta^2}\right)\right]$$

(P4.32.1)

where r_o is the radius of the cylinder, and the film thickness δ is given by

$$\delta \equiv r_\delta - r_o \qquad \text{(P4.32.2)}$$

b. Show that the film thickness may be expressed as a function of the flowrate in a dimensionless format given by

$$\beta = \left(\tfrac{3}{4}\,\mathrm{Re}\right)^{1/3} \qquad \text{(P4.32.3)}$$

where the Reynolds number for the falling film

is defined as

$$\mathrm{Re} \equiv \frac{4Q}{2\pi r_o \nu} = \frac{4\Gamma}{\nu} \qquad \text{(P4.32.4)}$$

and Γ is the mass flowrate per unit length normal to the flow—in this case the appropriate length is the perimeter of the cylindrical surface. We define β, a complicated function of the film thickness, by

$$\beta^3 \equiv \frac{3}{4}\frac{r_\delta^4 g}{r_o \nu^2}\left\{-\ln\left(\frac{r_o}{r_\delta}\right) - \frac{1}{4}\left[3 + \left(\frac{r_o}{r_\delta}\right)^4\right] + \left(\frac{r_o}{r_\delta}\right)^2\right\}$$

(P4.32.5)

Experimental data consistent with Eq. P4.32.3 are presented by Mudawar and Houpt [*Int. J. Heat Mass Transfer*, **36**, 3437 (1993)].

c. Show that in the limit of $\delta/r_o \ll 1$, Eq. P4.32.3 reduces to Eq. 4.4.27. You will need to use approximation formulas for $\ln(1 + \delta/r_o)$, and for $(1 + \delta/r_o)^n$. You will find that to get the desired result, you must expand these expressions out to cubic terms in δ/r_o.

4.33 Refer to Problem 4.32. Panelo obtained the data shown in Fig. P4.33 for the film thickness on a vertical cylinder. How well do these data agree with the model given by Eq. P4.32.3? To what degree does Eq. 4.4.27 provide an accurate model of the data? Cylinder radii were varied in the range $0.733 < r_o < 6.35$ mm.

4.34 Derive Eq. 4.4.92.

4.35 Derive Eq. 4.4.94.

4.36 Derive Eq. 4.4.101.

4.37 Wehbeh et al. [*Phys. Fluids*, **A5**, 25 (1993)] present data obtained in a "falling cylinder viscometer" of the type described in Example 4.4.2. Figure P4.37 shows their data, plotted as a dimensionless force as a function of the geometri-

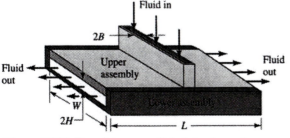

Figure P4.31 Flow into a parallel plate channel.

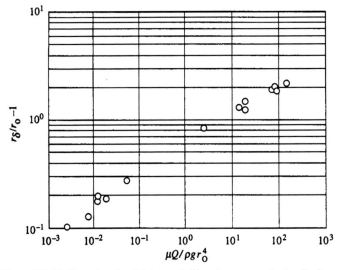

Figure P4.33 Data for film thickness falling down a vertical cylinder. (Panelo, unpublished work.)

cal parameter κ. We presume that the investigators calculated the force from the "net" weight (weight less buoyancy) of the falling cylinder. They refer to this as a "drag force," but it is clearly the sum of the drag and pressure forces on the cylinder. (Figure 11 of Wehbeh *et al.* has a factor of 4π on the y axis—it should be 2π.)

Begin with Eq. 4.4.101 and develop an expression for the function plotted in Fig. P4.37, and plot it on the figure. Do the data agree well with the model?

Your final result should be

$$\frac{force}{2\pi\mu UL} = \left(\ln\frac{1}{\kappa} - \frac{1 - \kappa^2}{1 + \kappa^2} \right)^{-1} \quad \textbf{(P4.37)}$$

4.38 a. Repeat the development given in Section 4.4.4, but take the case of a cylinder falling through a concentric tube that is completely immersed in the test fluid but *open* at both ends.
b. For this case, calculate the dimensionless drag force defined in Fig. P4.37 and plot it against κ

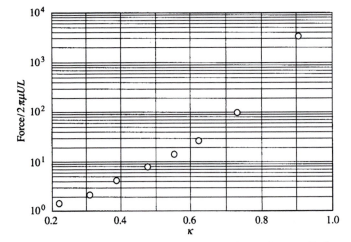

Figure P4.37 Data for dimensionless drag force for a cylinder falling in a closed tube. (Replotted from the data of Wehbeh et al. (1993)).

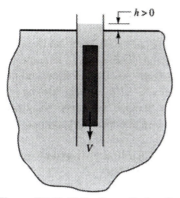

Figure P4.40 Fall of a cylinder through an open tube.

on Fig. P4.37. Does the presence of a closed lower end strongly influence the terminal velocity of the cylinder? Is this true regardless of the value of κ?

4.39 A stainless steel solid cylinder 1 cm in diameter and 10 cm long falls coaxially through a viscous liquid ($\rho = 1200$ kg/m^3, $\mu = 10$ Pa·s) that fills a concentric tube of inside diameter D. In one case the tube is closed at the bottom; in the other it is open and immersed within the viscous liquid. Find the terminal velocity in both cases, for $D = 1.1$ cm, and $D = 5$ cm.

4.40 A concentric falling cylinder viscometer is designed as shown in Fig. P4.40. After the cylinder reaches its terminal velocity, do you expect the free surface within the tube to exhibit an h value that is positive (as shown), negative, or zero? Explain your reasoning and support it by a model of this flow.

Your result for the absolute value of h

should be

$$|h| = \left(\frac{4\mu VL}{g\rho R_1^2}\right) \frac{\kappa^2}{-(1 + \kappa^2)\ln \kappa - (1 - \kappa^2)}$$
(P4.40)

where $\kappa = R_1/R_2 < 1$.

4.41 We wish to study the terminal velocity of a disk falling under the influence of gravity through a large (nearly unbounded) body of a viscous liquid. To maintain the orientation of the disk such that it moves in the direction of the normal vector to its circular face, the design of Fig. P4.41 is suggested. Design the "guide" in such a way that its associated frictional force is less than 3% of the force arising on the face of the disk. It is known that the force on the face of a very thin disk, moving as shown, is given by

$$F_z = -16\mu a U \qquad (P4.41)$$

where U is the steady velocity and a is the disk radius.

4.42 Go back to Eq. 4.5.61. Doesn't this seem to be a strange result? We know that centrifugal forces can give rise to large radial pressure distributions. Think of the spin cycle of a washing machine. Examine the following problem. Liquid is confined to a disk-shaped region, as shown in Fig. P4.42. Find the radial pressure distribution at a rotational speed of 100 rad/s, for $R = 0.05$ m, in a liquid of density $\rho = 1500$ kg/m^3. What feature of the physics distinguishes this flow from that of Example 4.5.3?

4.43 There is another form of simplification of Eq. 4.5.62. Assume that the viscous terms are unimportant, and solve Eq. 4.5.62 with the right-

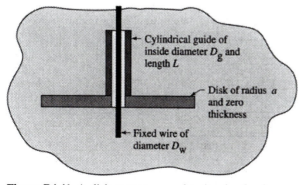

Figure P4.41 A disk moves normal to its circular face.

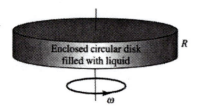

Figure P4.42 A rotating liquid fills a disk-shaped container.

hand side set equal to zero. Find an expression for $h(r)$ for this case, and plot $h(r)$ for the case of a rotational speed of 100 rad/s, for $R = 0.05$ m, in a liquid of density $\rho = 1500$ kg/m³. Under what physical conditions might this be a useful mathematical model?

4.44 Return to Example 4.5.3. At very high rotational speeds, say $\omega > 500$ rad/s, a rotating disk acts as a centrifugal pump, and its motion induces a strong flow of air across the surface of the liquid on the disk. This airflow is strong enough to exert a shear stress on the liquid. Hence Eq. 4.5.64 must be replaced. Under some conditions the shear stress is known to take the following form:

$$T_{rz} = \tfrac{1}{2} \omega^{3/2} \nu_{air}^{-1/2} \mu_{air} r \qquad \textbf{(P4.44)}$$

where ν_{air} is the kinematic viscosity of the surrounding air.

Derive the equation that replaces Eq. 4.5.67, in this case.

4.45 Derive Eq. 4.5.66.

4.46 We wish to achieve a mean film thickness of 10 μm over a rotating disk for the case of a rotational speed of 100 rad/s, for $R = 0.05$ m, in a liquid of density $\rho = 1500$ kg/m³ and viscosity $\mu = 0.1$ Pa·s. What feed rate Q is required? What is the fractional change in film thickness between the radii $r = 0.02$ m and $r = 0.04$ m? (Take "fractional change" to mean the difference Δh divided by the mean film thickness over that region.)

4.47 In simplifying Eq. 4.3.24d in order to obtain Eq. 4.5.66, we assumed that

$$\frac{\partial}{\partial r}\left[\frac{1}{r}\frac{\partial}{\partial r}(ru_r)\right] \ll \frac{\partial^2 u_r}{\partial z^2} \qquad \textbf{(P4.47)}$$

Use our analytical solution (Eq. 4.5.66) to find a dimensionless group that determines conditions under which this is a good approximation. Evaluate the validity of the approximation for the case of a disk of radius $R = 5$ cm, and a liquid of viscosity $\mu = 0.2$ poise and density $\rho = 1.5$ g/cm³. The rotational speed will be 100 rad/s. We require a film thickness greater than 100 μm over the entire disk.

4.48 Derive Eq. 4.5.78. Keep in mind from Eq. 4.5.67 that r^2h^3 is a constant; therefore, you should look for this group to appear in the algebra. Also remember that

$$\frac{\partial}{\partial r}\left(\frac{z}{h(r)}\right) = -\frac{z}{h^2}\frac{dh}{dr} \qquad \textbf{(P4.48)}$$

4.49 The data shown in Fig. P4.49 were obtained by Chen for radial flow between parallel disks.

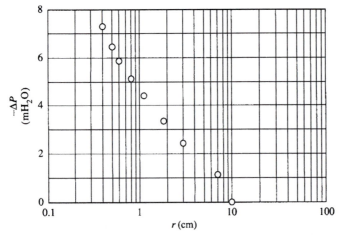

Figure P4.49 Data for inward flow between parallel disks. Chen, *J. Mec.*, **5**, 245 (1996).

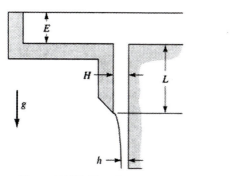

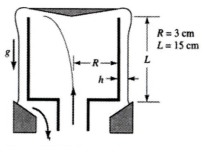

Figure P4.51 System for producing a thin film on the outside of a cylinder.

Figure P4.50 Flow from a reservoir through a planar slot.

However, the flow was radially *inward,* rather than outward as in Example 4.5.1. Modify the model given in that example so that it applies to inward flow. Compare the model with Chen's data.

For this experiment, the flowrate was $Q = 10.65 \text{ cm}^3/\text{s}$, and the liquid was water with a measured *kinematic* viscosity given as 1.06 cSt (centistokes). The disk separation was $2H = 0.104 \text{ mm}$, and $R_i = 0.3 \text{ cm}$ and $R = 10.2 \text{ cm}$. The pressure given in the figure is *negative*; suction is required to draw the water through the system. The pressure at $r = R$ was taken arbitrarily as $p = 0$. Chen gives these pressures in units of meters of water, as labeled on Fig. P4.49.

4.50 A viscous liquid flows from a reservoir through a wide planar slot as shown in Fig. P4.50. The liquid level in the reservoir, E, is constant. When steady flow is achieved, what is the film thickness h?

$E = 5 \text{ cm}$ $H = 0.3 \text{ cm}$ $L = 5 \text{ cm}$
$\mu = 4 \text{ poise}$ $\rho = 1.2 \text{ g/cm}^3$

4.51 Part of the design of an oxygenator for synthetic blood involves the system sketched in Fig.

P4.51. A fluorocarbon liquid is pumped upward through a tube of radius R. The liquid then overflows and falls, by gravity, as a thin film attached to the outside of the cylindrical surface. The liquid has a viscosity of $\mu = 2 \text{ Pa} \cdot \text{s}$, and a density of $\rho = 950 \text{ kg/m}^3$. What is the maximum volumetric flowrate that can be pumped through the system if h is not to exceed 3 mm?

4.52 A long cylindrical rod is continuously drawn through a viscous liquid that initially fills the concentric tube that surrounds the rod. The system (Fig. P4.52) is leak-tight, except at the downstream end, which is penetrated by a capillary tube that permits liquid to leave the system. At the upstream end, where the rod enters the system, a small vent tube permits outside air to enter as the liquid is pumped out of the system. Under the following conditions, what is the initial leakage rate through the capillary tube?

$L = 1 \text{ m}$ $L_c = 1 \text{ cm}$ $\mu = 1 \text{ Pa} \cdot \text{s}$ $R = 1 \text{ cm}$
$d = 1 \text{ mm}$ $V = 2 \text{ m/s}$ $\kappa = 0.5$

4.53 A long cylinder is to be "floated" in a cylindrical trough by the shear and pressure forces exerted by the flow of a high viscosity liquid. For the system sketched in Fig. P4.53, develop

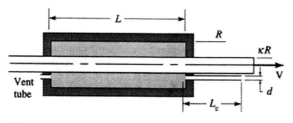

Figure P4.52 Rod drawn through a concentric tube, with leakage through a small capillary.

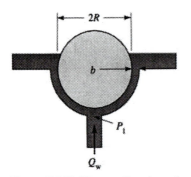

Figure P4.53 Viscous flotation of a long cylinder.

a mathematical model for the pressure P_1 and the volumetric flowrate per unit of axial length Q_w in terms of geometrical and liquid properties, and the weight of the cylinder, per unit of axial length. Assume that the radial separation b is maintained constant.

4.54 A "drag coefficient" for the steady motion of particles through a fluid is often defined as

$$C_D \equiv \frac{8gR_e}{3U^2} \qquad \text{(P4.54)}$$

where R_e is an "equivalent radius" defined as the radius of a sphere having the same volume V as the particle. The data shown in Fig. P4.54 were obtained for the terminal velocity of gas bubbles rising through water purposely contaminated with surface-active solutes. Are these data

well described by Stokes' law? How about Eq. P4.17.2?

4.55 An extrusion coating die is designed as shown in Fig. P4.55. Fluid is extruded from the slit and is dragged through the "land" region by the lower moving substrate. The fluid exits the land region and creates a film of thickness H on the moving substrate, as shown. Because the flow exits the die under pressure, it is possible for some of the fluid to move opposite the motion of the substrate, into the "backflow" region. Develop a model for the length of the backflow region L_2, in terms of the design and operating variables of this system. Do the case of isothermal Newtonian flow. Plot L_2/L_1 as a function of the coating thickness ratio H/H_o. Assume that the flow is two-dimensional and that the slit is very thin compared with the lengths L_2 and L_1. Provide a relationship between the coating thickness ratio H/H_o and the pressure at the exit of the slit, P_{slit}.

4.56 Design a viscometer suitable for measuring viscosities in the range 10 to 1000 poise. The design is to be based on the concept sketched in Fig. P4.56, where D is a cylindrical vessel containing a viscous fluid A and B is a cylindrical "bob" that is dragged up through the fluid by a frictionless pulley system (C) utilizing a weight W.

A "design" requires specification of all geometrical dimensions of the system. Present a de-

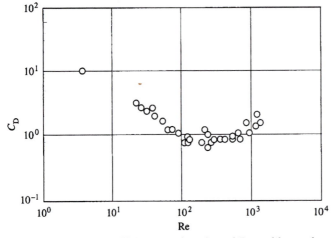

Figure P4.54 Drag coefficient as a function of Reynolds number for bubbles rising through water with surfactants. Hartunian and Sears, *J. Fluid Mech.*, **3**, 27 (1957).

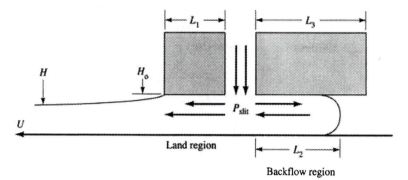

Figure P4.55 Schematic of flow from a slit die.

sign equation that relates viscosity to the geometry of the system, and list the measurements you would make of the behavior of the system.

What range of W values (in grams mass) do you require?

4.57 An infinitely long cylinder of circular cross section is drawn continuously and at constant speed through a concentric tube, as shown in Fig. P4.57.1. If the initial state of the system, prior to starting the movement of the cylinder, is as shown in Fig. P4.57.2a, where the reservoir fluid does not initially fill the annular region surrounding the cylinder, what is the steady state value of the depth h of the liquid surface in the annular region surrounding the moving cylinder? Suppose the initial state is as shown in Fig. P4.57.2b. What will the final state look like? Support your answers with appropriate models for the steady state flows. Assume that gravity acts opposite the direction of the motion of the cylinder.

4.58 A viscous liquid is supplied from a vertical capillary tube to a cylinder, and the resulting film of liquid flows radially across the cylinder to the edge. Beyond the edge, the liquid flows down the side of the cylinder under gravity. See Fig. P4.58 for the details of the geometry and flow. The following data have been obtained for H_g, the film thickness along the side, at various

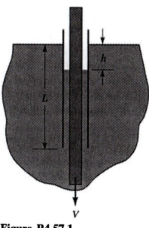

Figure P4.57.1

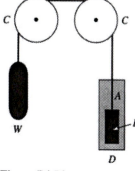

Figure P4.56

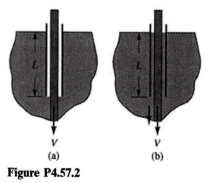

Figure P4.57.2

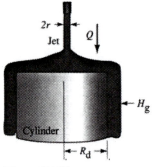

Figure P4.58

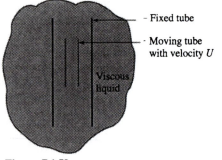

- Fixed tube

- Moving tube with velocity U

Viscous liquid

Figure P4.59

ing conditions:

flowrates of the liquid. What is the viscosity of the liquid? The liquid density is $\rho = 1019$ kg/m³.

R_d (m)	Q (m³/s)	H_g (m)
6.35×10^{-3}	5.2×10^{-7}	1.8×10^{-3}
6.35×10^{-3}	1.3×10^{-6}	2.5×10^{-3}
1.27×10^{-2}	2.5×10^{-6}	2.5×10^{-3}
2.50×10^{-2}	2.5×10^{-6}	1.9×10^{-3}

4.59 A thick-walled cylindrical tube, open at both ends, is moving at a steady velocity U within and coaxial to a larger but fixed cylindrical tube. Both tubes are immersed in a large body of a viscous liquid. Figure P4.59 shows the geometry. Does the liquid within the inner tube move relative to the inner tube? Answer for two cases: (a) the outer fixed tube has a closed bottom and (b) the outer fixed tube is open at the bottom. In both cases, develop a model for the flow fields inside and outside the inner tube, and relate the volume flowrates inside and out to the parameters of the system. Define the geometrical parameters carefully, and distinguish inner and outer radii of the tubes in your notation.

4.60 With reference to the geometry shown in Fig. P4.59, develop a model for the terminal velocity of the inner tube, falling under the acceleration of gravity, for the case in which the outer fixed tube is closed at the bottom. Find the terminal velocity in the case described for the follow-

density of the inner tube: $\rho = 2400$ kg/m³
inside diameter of the outer fixed tube: 0.02 m
inside diameter of the inner tube: 0.012 m
outside diameter of the inner tube: 0.016 m
density of the liquid: $\rho = 1400$ kg/m³
length of the outer fixed tube: 0.2 m
length of the inner tube: 0.1 m

4.61 Using Eqs. 4.4.1, 4.4.2, and 4.4.3, find $p(r, z)$. What boundary conditions do you impose on p?

4.62 We are interested in an expression for the velocity profile $u_z(r)$ for the film flow depicted in Fig. P4.58, in the region along the side of the cylinder where H_g is constant.
a. By simplifying the continuity and Navier–Stokes equations, present a differential equation for $u_z(r)$.
b. Solve this differential equation and present the solution for $u_z(r)$.
    State clearly the assumptions that permit you to remove various terms from the continuity and Navier–Stokes equations. Do not use a planar approximation to the geometry.

4.63 Portalski [*Chem. Eng. Sci.*, **18**, 787 (1963)] presents the following data for film thickness on a smooth vertical plate. Plot the data in a form suitable for comparison to the simple model developed in Section 4.4.2. Portalski's Reynolds number is defined as in Eq. 4.4.28.

Re	177	212	283	318	388	424	494	706	848	990
100δ (cm)	1.99	2.11	2.43	2.55	2.80	2.87	3.08	3.53	3.87	4.11

Re	1130	1414	1554	1838	1979	2261	2403	2544	2826	2968
100δ (cm)	4.48	4.97	5.22	5.67	5.94	6.37	6.58	6.91	7.43	7.62

Re	3250	3393	3534	3674	3816
100δ (cm)	8.13	8.26	8.52	8.61	8.87

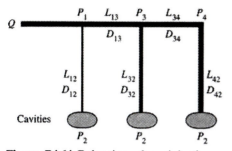

Figure P4.64 Balancing of an injection molding system.

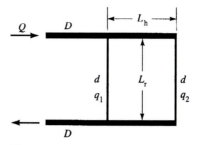

Figure P4.66 Flow distribution system for a solar hot water heater.

4.64 Many plastic items are manufactured by the process of injection molding. Molten polymer is pumped to a network of tubes, each of which terminates at the entrance to a "cavity" shaped like the article to be molded. Flow balancing is necessary to ensure that all cavities fill simultaneously. Balance the network shown in Fig. P4.64. Assume that the cavities are identical, and that the flow is Newtonian and isothermal. Neglect any pressure losses associated with entrance and exit effects from one element of the network to another. Assume that all the pressures labeled P_2 are atmospheric.

Do the special case that all the L_{ij} are equal, and $D_{13} = D_{34} = D_{42} = D$. Specifically, find values of D_{12}/D and D_{32}/D that yield a balanced network.

4.65 Repeat Problem 4.64, but allow the lengths L_{12} and L_{32} to be variable. Now find values of D_{12}/D and D_{32}/D that yield a balanced network, but with the constraint that the volume within the tubes of lengths L_{12} and L_{32} be minimized. Hence you will also have to provide values for L_{12}/L and L_{32}/L, where $L = L_{13} = L_{34} = L_{42}$.

4.66 A simple flow distribution system for a solar hot water heater is sketched in Fig. P4.66. Neglecting entrance and exit effects, and pressure losses at pipe bends, find the extent to which the system is unbalanced, as defined by the ratio q_1/q_2. Prepare a plot of q_1/q_2 as a function of the ratio d/D, in the range $0.1 < d/D < 1$, all other geometrical parameters being fixed. Do two cases: $L_h/L_r = 0.1$ and 1.0. Assume laminar flow.

4.67 Repeat Problem 4.66, but for the plumbing configuration shown in Fig. P4.67.

4.68 A viscometer can be designed, based on the flow described in Section 4.4, where a long cylinder falls under gravity through a concentric tube closed at its bottom. Some features of this so-called falling cylinder viscometer are discussed by Gui and Irvine [*Int. J. Heat Mass Transfer*, **37**, suppl. 1, 41, (1994)]. In particular, the authors investigate, numerically, the end effects we ignored in developing the model that results in Eq. 4.4.101. In effect, they write our equation in a form explicit in viscosity:

$$\mu = \frac{\Delta\rho g R_2^2}{U\Theta(\kappa)} \frac{1}{\varphi(\kappa,\lambda)} \qquad \text{(P6.68.1)}$$

In this expression, $\Theta(\kappa)$ is the function defined in Eq. 4.4.101, and plotted in Fig. 4.4.5. The function $\varphi(\kappa,\lambda)$ is an end correction factor and is determined from the numerical solutions that account for end effects. The parameter λ is simply

$$\lambda = \frac{L}{R_2} \qquad \text{(P4.68.2)}$$

Gui and Irvine present a statistical curve fit to their numerical results, which leads to

$$\varphi(\kappa, \lambda) = 1 + \frac{1}{10\lambda}$$
$$(0.746 + 48.9\kappa - 140\kappa^2 + 148\kappa^3 - 57.7\kappa^4)$$
$$- 0.003\left(1 - \frac{10}{\lambda}\right) \qquad \text{(P4.68.3)}$$

for $\lambda < 30$.

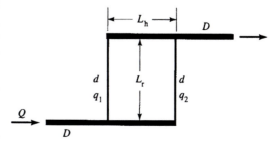

Figure P4.67 Flow distribution system for a second solar hot water heater.

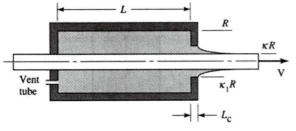

Figure P4.69

Plot $\varphi(\kappa)$ for $\lambda = 4$ and 10. For each value of λ, what is the maximum error in calculated viscosity due to failure to account for end effects, and for what κ is the maximum observed?

4.69 A long cylindrical rod is continuously drawn through a viscous liquid that initially fills the large concentric tube that surrounds the rod. See Fig. P4.69. The system is not leak-tight at the downstream end, where the rod exits the system through a concentric orifice of radius $\kappa_1 R$. At the upstream end, where the rod enters the system, a small vent tube permits outside air to enter as the liquid leaks from the system. Give the initial leakage rate through the orifice under the following conditions: $L = 1$ m, $L_c = 1$ cm, $\mu = 1$ Pa·s, $R = 1$ cm, $V = 2$ m/s, $\kappa = 0.5$, $\kappa_1 = 0.51$.

Someone suggests that if a more viscous liquid is used, the leakage rate will be reduced. Is that true?

4.70 In the process of performing the procedure called coronary angioplasty, a surgeon places a cylindrical catheter of diameter D_i inside a coronary blood vessel. It is important to understand the degree to which the presence of the catheter reduces blood flow within the vessel. We can develop a simple model of this flow if we begin

with the geometry sketched in Fig. P4.70. A resistance factor R can be defined by the equation

$$R = \frac{\Delta P}{LQ} = \frac{128\,\mu}{\pi D_o^4}F \qquad \text{(P4.70)}$$

a. Derive a model for the factor F that appears in this equation and plot F versus κ.

b. Will the shear stress acting along the arterial surface be increased or decreased by the presence of the catheter? Assume that $\Delta P/L$ remains constant despite the presence of the catheter. Find and plot the ratio of the shear stress with the catheter in place to that in the absence of the catheter, as a function of $\kappa = D_i/D_o$.

4.71 We have considered annular axial pressure flow when the cylinders are concentric. If they are eccentric (the inner cylinder off-center: see Fig. P4.71), there is an altered resistance to flow. A resistance factor R_e may be defined as

$$R_e = \frac{\Delta P}{LQ} = \frac{128\mu}{\pi D_o^4}FG \qquad \text{(P4.71)}$$

where F is the function defined in Eq. P4.70 and G accounts for the effect of eccentricity. Theoretical solutions for eccentric annular flow at low Reynolds number yield Table P4.71, listing values of the G function.

Table P4.71 Eccentricity Factor G

κ	0	0.25	$\dfrac{2e}{D_o - D_i}$ 0.5	0.75	1.0
0.1	1	0.94	0.82	0.70	0.61
0.25	1	0.93	0.77	0.62	0.50
0.50	1	0.92	0.74	0.57	0.44
0.75	1	0.915	0.73	0.54	0.42

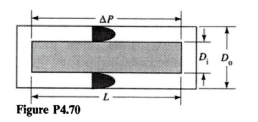

Figure P4.70

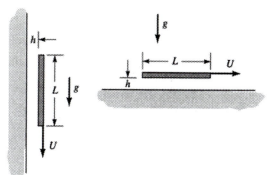

Figure P4.72

For a given pressure drop, by what factor is the volumetric flowrate increased by an inner cylinder that just touches the surrounding annular surface (instead of being concentric), as a function of $\kappa = D_i/D_o$?

4.72 Consider two flows, as sketched in Fig. P4.72. In both cases a planar solid of length (in the direction of motion) L, width W, and thickness t is being dragged at constant speed U parallel to a large planar surface, at a constant distance h from the surface. An unbounded Newtonian liquid surrounds the solid surfaces. The geometry is such that $L \gg h$, $W \gg L$, $t \ll h$. Students often respond to this pair of flows with the intuitive reaction that the flows are different, because gravity—in some sense—aids the vertical flow and does not affect the horizontal flow. Prove that in fact the flows are identical by showing that the velocity profiles in the region bounded by the two planar surfaces are identical in both cases. Assume that for very small Reynolds numbers, and for $L \gg h$, entrance and exit effects are negligible, since this allows you to solve the problem analytically.

Does the "equality" of the two flows depend on this assumption regarding entrance and exit effects?

4.73 An open-ended, thin-walled cylindrical tube (tube A in Fig. P4.73) surrounds a long cylinder (B). The entire system is filled with a Newtonian oil. Tube A is released, and falls

through the oil under the action of gravity. Assume that tubes A and B remain coaxial. Find the steady terminal velocity of tube A for the following conditions:

$L = 10$ cm	$H = 100$ cm	$R_1 = 1$ cm
$R_2 = 1.1$ cm	$R_3 = 1.2$ cm	$R_4 = 20$ cm
$\mu_{oil} = 10$ Poise	$\rho_{oil} = 1$ g/cm^3	
$\rho_{tube} = 6$ g/cm^3		

4.74 In Section 4.4.2 we examined the problem of laminar film flow down a vertical planar surface. Because of the simplicity of the geometry, especially the (assumed) planar free surface, it was possible to obtain an analytical model for the flow. Such film flows are often used to promote heat or mass transfer across a liquid/solid or liquid/gas interface. A common strategy is to use a corrugated surface instead of a planar surface because of the increased surface area that can be achieved, per unit of surface width.

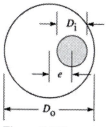

Figure P4.71

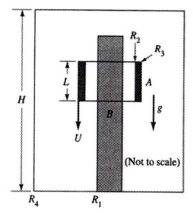

Figure P4.73

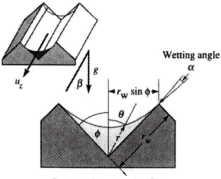

Wetting angle
α

Cross section normal to flow

Figure P4.74.1 Gravity-driven flow in a triangular groove.

However, the more complex solid boundary has a large effect on the film flow, slowing it relative to the planar film because there is more wetted perimeter at which the liquid must satisfy a no-slip boundary condition. To assess the utility of such a system, we need a model for the extent to which flow is retarded by the increased surface. No "closed-form" analytical solution is possible to this problem, but an infinite series solution has been worked out by Ayyaswamy et al. [*Trans. ASME, J. Appl. Mech.*, **41**, 332 (1974)].

a. Assuming steady, fully developed, gravity-driven isothermal Newtonian flow of an incom-

pressible liquid, write the equation the velocity field $u_z(r, \theta)$ must satisfy when r and θ are cylindrical coordinates, as shown in Fig. P4.74.1. Write the boundary conditions that the liquid must satisfy on the wetted surfaces $\theta = \pm\phi$. To ensure the absence of parameters, nondimensionalize the equation.

b. What *physical* boundary condition must be satisfied on the free surface?

For various wetting angles α (see Fig. P4.74.1). Ayyaswamy *et al.* calculate a dimensionless flowrate, which we plot in Fig. P4.74.2 as a function of the half-angle ϕ. In this plot Q^* is defined as

$$Q^* = \frac{Q\nu}{r_w^4 g \cos\beta} \qquad \text{(P4.74.1)}$$

One measure of the effect of the bounding surfaces on retarding the film flow would be the ratio

$$\chi \equiv \frac{Q_{\text{groove}}}{Q_{\text{plane}}} \qquad \text{(P4.74.2)}$$

but we need to select a basis for comparison. We can use as a "unit width" for the planar film flow a width

$$W = 2r_w \sin\phi \qquad \text{(P4.74.3)}$$

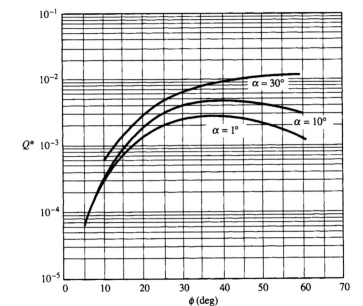

Figure P4.74.2 Dimensionless flowrate down a triangular groove.

How do we compare film thicknesses for the two cases? The maximum utilization of the triangular geometry occurs when the liquid just contacts the bounding surface at the tip $r = r_w$. Use $\delta = r_w$, and find the ratio χ defined in Eq. P4.74.2. Plot χ versus ϕ, and discuss the result. Is it what you would expect?

Suppose, for δ, you used the distance between the vertex of the triangle and the position of the liquid interface, along the line $\theta = 0$. Plot the resulting χ' based on this choice and compare it with χ.

4.75 With reference to Problem 4.74, a vertical grooved surface is to be used to promote mass transfer from a liquid stream under the following conditions: kinematic viscosity of the liquid $\nu = 10^{-4}$ m^2/s, $\alpha = 10°$, $\phi = 30°$, $r_w = 0.01$ m.
a. Find the volumetric flowrate required to ensure the achievement of the required value of r_w.
b. If there is a planar (ungrooved) section just before the grooved region, how thick will the film be in the planar region? Is the film flow laminar in this planar region?

4.76 A Newtonian liquid flows slowly into a planar channel, as shown in Fig. P4.76. The planes that confine the flow are parallel, and very wide in the direction perpendicular to the page. Hence the flow is two-dimensional. The upper plane "floats" freely in such a way that it remains parallel to the lower, fixed plane. A *weight w* sits on the upper plane. The upper plane itself may be considered weightless. The volume flowrate per width W entering the system is Q/W, and the entrance region is a plane parallel region of length L_e and thickness $2B$. It width is the same as that of the upper region. The pressure (relative to the atmosphere) at the entrance plane **A** is P_o. Derive an expression for the plane half-

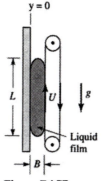

Figure P4.77

separation H, in terms of w and the other parameters.

4.77 A Newtonian viscous liquid is flowing between two wide planar surfaces, as shown in Fig. P4.77. The surface along $y = 0$ is fixed in space, and the surface along $y = B$ has a steady speed U parallel to the fixed plane. If U were zero, the liquid film would flow downward under the action of gravity. Derive an expression for the speed U such that the liquid neither falls nor rises. What is the shear force (per unit width normal to the page) on the fixed surface under these conditions?

4.78 Two immiscible viscous liquids are delivered to a vertical surface in such a way that the two liquids flow downward, under gravity, in the two adjacent planar parallel layers as shown in Fig. P4.78.
a. For the case $Q_1/Q_2 = 1$, $\rho_1/\rho_2 = 1$, and $\mu_1/\mu_2 = 1$, do you expect δ_1/δ_2 to be greater than, equal to, or less than unity? Explain your

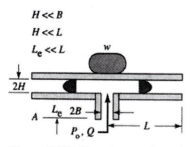

Figure P4.76 Flow into a planar two-dimensional channel.

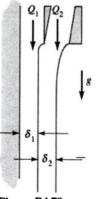

Figure P4.78

reasoning, but make no quantitative calculations.
b. Assuming that the liquids have the same density but different viscosities, derive an expression for the film thickness ratio δ_1/δ_2 in terms of the volume flowrate ratio Q_1/Q_2 and the viscosity ratio μ_1/μ_2.
c. Give the numerical value of δ_1/δ_2 for the case $Q_1/Q_2 = 1$ and $\mu_1/\mu_2 = 1$.

4.79 Two immiscible viscous liquids are delivered to a vertical surface in such a way that the two liquids flow downward, under gravity, in the two adjacent planar parallel layers as shown in Fig. P4.78. For the case $\rho_1/\rho_2 = 1$ and $\mu_1/\mu_2 = M$, derive an expression for the ratio $Q_1/Q_2 = q(M)$ that will yield a film thickness ratio $\delta_1/\delta_2 = 1$.

4.80 A device consists of a pair of parallel, closely spaced solid disks of radius R. The separation between the disk surfaces is $2B$. The entire device is submerged in a viscous liquid, as shown in Fig. P4.80. Develop a model for the time required for the liquid to completely fill the region between the two disk surfaces. Assume that the region is initially empty of liquid. Do not account for the flow into the air vent, which is the vertical tube of radius $D_R/2$. Present the result as a plot of the dimensionless time (give the definition of this time) as a function of the ratio $\alpha = D_R/2R$, for α in the range $[0.01, 0.1]$. Assume that $B \ll R$, $D_R \ll R$, and $B \ll H$. Account for both hydrostatic and surface tension effects.

4.81 With reference to Problem 4.80, find the time to fill the system shown in Fig. P4.80 for the case $R = 5.08$ cm, $2B = 0.00254$ cm, $D_R = 0.1$ cm, $H = 1$ cm, $\mu = 0.032$ poise, and $\sigma = 18$ dyn/cm.

4.82 Tsay et al. [*Int. J. Multiphase Flow*, **16**, 853 (1990)] present the experimental data tabulated below for the liquid film thickness for ethanol

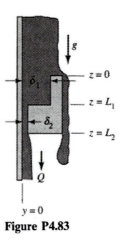

Figure P4.83

at 30°C, falling under gravity down a smooth vertical plane. Use the analysis given in Section 4.4.2 to predict these results, and compare your predictions to these data, where $\dot{m}$ is the mass flowrate per unit width of the plane. The density of ethanol at this temperature is 788 kg/m³. The viscosity is 0.93×10^{-3} Pa·s.

$\dot{m}$ (kg/m·s)	δ ($\times 10^{-4}$ m)
0.01	1.67
0.02	2.11
0.04	2.66

4.83 A viscous liquid flows steadily, under the action of gravity, as a thick laminar film down a vertical planar surface that occupies the plane $y = 0$, as in Fig. P4.83. At some point $z = 0$ the liquid encounters the first of a pair of planar narrow restrictions in the region $0 < z < L_2$ that are parallel to the surface at $y = 0$. We want to develop a model for the volume flowrate through the lower narrow restriction as a function of the separation between the planar surfaces, δ_2.

Write, but do not attempt to solve, the differential equations that the velocity component u_z and the pressure p must satisfy in each of the regions $0 < z < L_1$ and $L_1 < z < L_2$. Write the boundary conditions that u_z and p must satisfy in each region. Assume that $L_1 \gg \delta_1$, $L_2 - L_1 \gg \delta_2$. You may not assume that $\delta_1 \gg \delta_2$.

4.84 A viscous Newtonian fluid is confined between large parallel plates, as shown in Fig. P4.84. The motion is a combination of drag and pressure flow. Derive an expression for the velocity profile. Find an expression for the value

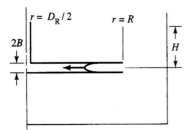

Figure P4.80

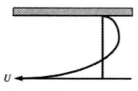

Figure P4.84

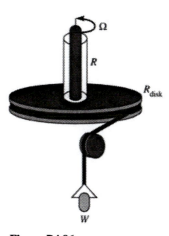

Figure P4.86

of U that yields a zero shear stress at the fixed plate. Draw the velocity profile under these conditions. Draw the velocity profile if U is half this value.

4.85 Begin with the solution for the velocity profile in axial annular pressure flow (Eq. 3.3.26) and prove that the solution satisfies a force balance of the form

$$\Delta P A_c = 2\pi R L \tau_R - 2\pi \kappa R L \tau_{\kappa R} \quad \textbf{(P4.85)}$$

Give the physical meaning of each term of this balance. Why aren't the two (shear) force terms on the right-hand side additive, instead of subtractive?

4.86 A rotational viscometer is designed according to the sketch given in Fig. P4.86. A viscous liquid is set in annular angular motion in the region between two coaxial cylinders, the inner one of which rotates at angular speed Ω. The outer cup of radius R is rigidly attached to a disk, which is free to rotate on a frictionless bearing. If no force opposes the rotation of the disk, the disk and the *inner* cylinder will rotate at the same speed Ω (rad/s). If, as shown in the sketch, a string is wrapped around the circumference of the disk and strung over a frictionless pulley, and a weight is hung at the end of the string, the motion of the system will be arrested when the weight is sufficient to balance the torque exerted by the motion of the viscous liquid in the annular space between the two cylin-

ders. For the case that $R_{\text{disk}} = 5$ cm, $R = 1$ cm, $\kappa = 0.85$, and $L = 15$ cm, derive an expression for the viscosity of the liquid as a function of rotational speed Ω and the weight W that just stops rotation. (L is the wetted length of the inner rotating cylinder.)

If the system is limited to rotational speeds up to 4 rad/s, what weights must be provided if the liquids might have viscosities up to 1 Pa·s?

4.87 In Example 4.5.3 we introduced an approximation (Eq. 4.5.54) that permitted us to find the radial velocity field in the form given in Eq. 4.5.66. But this result for u_r does not satisfy Eq. 4.5.54!

a. Obtain an expression for $(1/r)[\partial(ru_r)/\partial r]$.

b. Use the continuity equation to show that $u_z \neq 0$, and that $\partial u_z/\partial z \neq 0$, and in fact that

$$u_z = -\frac{1}{3}\frac{\rho\omega^2 h^2 z}{\mu} \quad \textbf{(P4.87.1)}$$

c. Find an expression for u_z/u_r and present a criterion for the assumption that the flow is almost strictly radial.

Chapter 5

Dimensional Analysis and Dynamic Similarity

In this chapter we return to the use of dimensional analysis as an aid to understanding the dynamics of complex systems. We begin with the principle of dynamic similarity and its implications for the design of properly scaled experiments. Dimensional analysis also provides an important guide to the correlation of experimental data.

In the preceding chapters we applied the principles of conservation of mass and momentum, and we derived equations that permitted us to obtain *analytical* solutions for the flow fields in several *simple* problems. Several features of these problems made them simple: for example, the velocity vector often had only one nonzero component, and that velocity component often was a function of only one space variable. (The analysis of Stokes flow about a rigid sphere was an exception, as was torsional flow between parallel disks.) A second simplification was allowed because the solid boundaries of the systems studied were always the coordinate surfaces of a simple coordinate system—planar, cylindrical, or spherical.

These simple analyses, though often useful to us, are also very limited. In real engineering systems we are faced with flow fields bounded by surfaces of complex shape. The velocity field is usually a function of at least two space variables, as well as of time. The velocity vector usually has more than a single component. The nonlinear terms of the Navier–Stokes equations (the left-hand side arising from the $[\mathbf{u} \cdot \nabla \mathbf{u}]$) terms are often not negligible. We need to examine, now, a methodology appropriate to the study of these more complex flows.

We are not going to deal with the question of how we *solve* these multivariable nonlinear partial differential equations. Our goal is to show that even when we cannot solve the equations, we can use rational analysis of the equations to guide our further understanding of the fluid dynamics they model. To a large degree we are talking about *dimensional analysis,* a topic introduced in Chapter 1 through the Buckingham pi theorem, but now in a form guided more closely by the physics.

We have already introduced some other aspects of dimensional analysis. As a matter of routine, we put our mathematical models into nondimensional formats by defining nondimensional dependent and independent variables. In the course of doing this we "discovered" the occurrence of dimensionless groups, such as the Reynolds number and the Weber number. This routine changed the "format," the appearance of the equations. It did not change the mathematics or the physics of the problem.

Our goal, as engineers, is to be able to predict the performance of fluid dynamic systems over a range of potential operating conditions. A mathematical model serves that purpose. But in the absence of such a model, it is still possible to develop predictive capability from an efficiently designed experimental program. Dimensional analysis, based on the principle of dynamic similarity, teaches us how to design the experiments and how to manipulate and examine the resulting data, to yield reliable predictions.

5.1   THE PRINCIPLE OF DYNAMIC SIMILARITY

We begin by considering an arbitrary flow field for which we can identify a characteristic velocity and a characteristic length scale. By "characteristic" we mean that this choice of velocity and length scales has some simple relationship to the fluid motion. For example, for a stirred tank we might choose the tip velocity of the stirrer and the diameter of the tank. For a liquid through which gas is being bubbled we might choose the ratio of volume flowrate of gas to cross-sectional area of the tank as a velocity scale, and we might choose the diameter of the tube from which the bubbles emerge as a length scale. As we shall see, these choices are arbitrary. Some may seem more natural—more sensible—than others, but we really are free to select a velocity and a length that characterize the flow field that is of interest.

The one thing we know about the velocity vector **u** and the pressure p in some complex flow is that they are constrained to satisfy the Navier–Stokes equations, the continuity equation, and a set of initial and boundary conditions. (We are assuming Newtonian incompressible flow.) Hence we begin with the equations in a vector format:

$$\nabla \cdot \mathbf{u} = 0 \tag{5.1.1}$$

$$\rho \left(\frac{\partial \mathbf{u}}{\partial t} + \mathbf{u} \cdot \nabla \mathbf{u} \right) = -\nabla p + \mu \nabla^2 \mathbf{u} + \rho \mathbf{g} \tag{5.1.2}$$

(We will look at boundary conditions later.)

We are now going to make these equations dimensionless by introducing a characteristic velocity U and a length scale L. We define a new set of space coordinates first. We choose Cartesian coordinates when we write out the component forms, but the arguments we are going to present would hold in any coordinate system. The dimensionless coordinates are labeled as a vector with components

$$\mathbf{x}^* = (x^*, y^*, z^*) = \left(\frac{x}{L}, \frac{y}{L}, \frac{z}{L} \right) \tag{5.1.3}$$

We define a new velocity vector

$$\mathbf{u}^* = \frac{1}{U} \mathbf{u} \tag{5.1.4}$$

(We imply that each component of **u** is divided by the *same* velocity scale U.)

We will define dimensionless differential operators. Since the ∇ operator has components

$$\nabla \equiv \frac{\partial}{\partial \mathbf{x}} = \frac{1}{L} \frac{\partial}{\partial \mathbf{x}^*} \tag{5.1.5}$$

it follows that

$$\nabla^* = L\nabla \quad \text{and} \quad \nabla^{*2} = L^2 \nabla^2 \tag{5.1.6}$$

Here we note that ∇^2 has dimensions of $1/L^2$, and ∇ has dimensions of $1/L$, regardless of the choice of coordinate system.

At this stage we can rewrite Eqs. 5.1.1 and 5.1.2 as

$$\nabla^* \cdot \mathbf{u}^* = 0 \tag{5.1.7}$$

$$\rho U \left(\frac{\partial \mathbf{u}^*}{\partial t} + \frac{U}{L} \mathbf{u}^* \cdot \nabla^* \mathbf{u}^* \right) = -\frac{1}{L} \nabla^* p + \frac{\mu U}{L^2} \nabla^{*2} \mathbf{u}^* + \rho \mathbf{g} \tag{5.1.8}$$

Now we divide both sides of Eq. 5.1.8 by $\rho U^2/L$ to yield

$$\frac{L}{U} \frac{\partial \mathbf{u}^*}{\partial t} + \mathbf{u}^* \cdot \nabla^* \mathbf{u}^* = -\frac{\nabla^* p}{\rho U^2} + \frac{\mu}{\rho UL} \nabla^{*2} \mathbf{u}^* + \frac{gL}{U^2} \tag{5.1.9}$$

Looking at the time-dependent term, we see that we can define a dimensionless time as

$$t^* = \frac{Ut}{L} \tag{5.1.10}$$

Let's define (again by inspection of Eq. 5.1.9) a dimensionless pressure as

$$p^* = \frac{p}{\rho U^2} \tag{5.1.11}$$

Then Eq. 5.1.9 becomes

$$\frac{\partial \mathbf{u}^*}{\partial t^*} + \mathbf{u}^* \cdot \nabla^* \mathbf{u}^* = -\nabla^* p^* + \frac{1}{\text{Re}} \nabla^{*2} \mathbf{u}^* + \frac{1}{\text{Fr}} \frac{\mathbf{g}}{g} \tag{5.1.12}$$

(Here we denote the magnitude $|\mathbf{g}|$ by g.)

At this stage, two dimensionless groups have emerged in a natural way from this analysis (really—manipulation—not "analysis"). The Reynolds number is defined generally by

$$\text{Re} = \frac{\rho UL}{\mu} \tag{5.1.13}$$

A new group we haven't seen before, called the Froude number,[1] is defined by

$$\text{Fr} = \frac{U^2}{gL} \tag{5.1.14}$$

Before we proceed further we must return to a consideration of the *boundary* conditions on the Navier–Stokes equations.

Boundary conditions can be expressed in a very simple format if the boundaries of the system are coordinate planes. Suppose, for example, that the flow is confined in the y direction by a pair of rigid parallel planes with y coordinates $y = 0$ and $y = B$. When we nondimensionalize a no-slip condition in such a geometry, we go from

$$u_x = 0 \quad \text{on} \quad y = 0 \tag{5.1.15}$$

$$u_x = U \quad \text{on} \quad y = B \tag{5.1.16}$$

to

$$u_x^* = 0 \quad \text{on} \quad y^* = 0 \tag{5.1.17}$$

---

[1] Many fluid dynamics books define the Froude number (Fr) as the square root of our definition, which seems to be more commonly found in chemical engineering literature. The reader must be wary of the potential confusion arising from this difference.

$$u_x^* = 1 \quad \text{on} \quad y^* = \frac{B}{L} \tag{5.1.18}$$

with similar equations for the velocity components u_y and u_z. Note Eqs. 5.1.17 and 5.1.18 contain a new dimensionless group, B/L. If the two systems are geometrically similar, then B/L will be identical in the two systems. Most often, the boundary conditions give rise to a set of dimensionless geometrical factors, called "scale factors," or "shape factors," and these are automatically identical in two geometrically similar systems, since, by definition, the systems have the same shape but are of differing size (scale).

A much more complex boundary condition could arise in a system in which one boundary of the flow is a curved interface with an immiscible fluid. For example, suppose we were considering the dynamics of spreading of a drop on a rigid surface, as in Fig. 5.1.1. The external fluid is taken to be essentially stationary and stress free. This might be a good approximation for a drop spreading on a surface in air, where the viscosity of the surrounding gas is so low that the motion of the drop cannot induce any motion in the gas strong enough to give rise to measurable stresses. We take the pressure on the gas side of the interface to be constant, and we set that constant to zero, since the absolute value of pressure has no effect on the motion of an incompressible medium (the drop).

Then we must solve Eqs. 5.1.12 and Eq. 5.1.7 within the domain defined by the surface of the drop. This problem has some elements similar to the bubble growth problem in Section 4.4.3. There we wrote a boundary condition on the *radial* stress across the bubble interface. Because the surface of the drop in Fig. 5.1.1 is not a section of a sphere, the stress normal to the surface, T_{nn}, is not radial. A boundary condition on the "continuity" of the normal stress across the surface would include a pressure from the Young–Laplace equation. In the terms of the physical statement, the normal stress at the surface, which is due to the deformation of the fluid within plus the contribution from surface tension, must equal the external normal stress on the surface. The latter is just the external pressure if, as indicated, we ignore dynamic effects in the external fluid. Then a normal stress boundary condition for this type of system could be written as

$$-\sigma\left(\frac{1}{R_1} + \frac{1}{R_2}\right) + T_{nn,\text{inside}} = -\sigma\left(\frac{1}{R_1} + \frac{1}{R_2}\right) + p - \mu\,\Delta_{nn} = T_{nn,\text{outside}} = 0 \tag{5.1.19a}$$

or

$$p - \mu\,\Delta_{nn} - \sigma\left(\frac{1}{R_1} + \frac{1}{R_2}\right) = 0 \tag{5.1.19b}$$

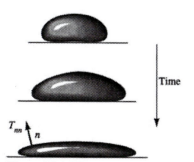

Figure 5.1.1 Three "snapshots" in the history of a liquid drop spreading on a surface.

where R_1 and R_2 are the radii of curvature of the surface at any position and at any instant of time. In general, the surface shape is not known (it could be what we are solving for), so

$$R_1 = R_1(x, y, z, t) \tag{5.1.20}$$

and the same for R_2. When we nondimensionalize Eq. 5.1.19b, we find[2]

$$\frac{p}{\rho U^2} - \frac{\mu}{\rho U L} \Delta^*_{nn} - \frac{\sigma}{\rho U^2 L}\left(\frac{L}{R_1} + \frac{L}{R_2}\right) = 0 \tag{5.1.21}$$

or

$$p^* - \frac{1}{\mathrm{Re}}\Delta^*_{nn} - \frac{1}{\mathrm{We}}\left(\frac{1}{R_1^*} + \frac{1}{R_2^*}\right) = 0 \tag{5.1.22}$$

We have defined a dimensionless radius of curvature function as

$$R_1^* \equiv \frac{R_1}{L} = R_1^*(x^*, y^*, z^*, t^*) \tag{5.1.23}$$

and the same for R_2^*.

Now we observe that in nondimensionalizing a dynamic boundary condition that involves an interface across which surface tension acts, there appears a new dimensionless group:

$$\mathrm{We} = \frac{\rho U^2 L}{\sigma} \tag{5.1.24}$$

This group is the Weber number, which we met earlier (Fig. 1.1.3.) It plays an important role in problems in which surface tension exerts a strong influence on the flow. Notice that the Weber number has come into the nondimensional formulation in a natural way. We did not have to stop and formulate a dimensionless group involving surface tension, by the methods illustrated in Chapter 1.

At this stage we see that the nondimensionalization of the dynamic equations and a boundary condition on stress at the free surface yields three *kinematic* dimensionless groups. (By "kinematic" we mean that all three groups involve the *motion*, since each has the velocity scale U in it.) In general, consideration of the boundary conditions will also lead to the appearance of some dimensionless geometrical factors (shape factors) that characterize the shape of the boundaries that confine the flow. We do not find any shape factors in the specific example of the spreading of a drop on a plane surface because the solid boundary has no shape! A plane has no characteristic length scale. The drop itself has a shape, but the shape of the drop as it spreads is unknown, and must be obtained as part of the solution of the dynamic equations.

One additional boundary condition could be imposed in this specific example. We might assume that the contact angle made by the drop along the perimeter of its contact with the solid surface must remain equal to the equilibrium contact angle θ_c that characterizes the *static* contact of this liquid with this solid in air. This is an assumption that *might* be valid, but its validity is not dictated by any independent argument we can make. If it is valid, then an additional dimensionless group is the contact angle θ_c

---

[2] Regardless of the complexity of the geometry, the deformation gradient term Δ_{nn} in Eq. 5.1.21 has dimensions of inverse time. Hence the dimensionless deformation gradient Δ^{**}_{nn} is just defined as $(L/U)\,\Delta_{nn}$. Be alert! ∇ has dimensions of $1/L$; Δ has dimensions of $1/t$.

itself, which is a physical property of the materials under study. Thus we make an important point here, which is that since the boundary conditions reflect our assumptions about the physics at the boundaries of the flow, additional dimensionless groups may arise from the boundary conditions, but we cannot state a priori what these groups will be. They will depend on the specific physics of the flow problem being studied. We can, however, say that the Navier–Stokes equations alone give rise to two groups—those of the Reynolds number and the Froude number, when they are nondimensionalized as above. (But see Problem 5.4.)

Where is all this leading? We now make a trivial but fundamental statement:

> *If two flows occur in geometrically similar systems in which viscous effects and interfacial phenomena occur, and if Re, We, and Fr are the same in both systems, the two systems are dynamically similar.*

If we photograph each of two *geometrically similar* systems (e.g., two stirred tanks, two piping geometries, two spreading drops), we cannot distinguish one from the other without knowing the length scales in the photographs. The two systems are identical except for the length scale L. All geometrical features have the same shapes.

Two systems that are *dynamically similar* are identical in fluid dynamics in the sense that $\mathbf{u}^*(x^*, y^*, z^*, t^*)$ and $p^*(x^*, y^*, z^*, t^*)$ are *identical functions* of the dimensionless time and space coordinates in the two systems. In the case of the spreading drop, the shape of the drop would be the same in the two systems, at the same value of t^*. (Keep in mind that the shape of the drop is part of the *dynamics* of the problem.)

For cases of free boundary flow with surface tension, as found in the spreading drop problem, the principle of dynamic similarity requires that the groups Re, Fr, and We be identical in two geometrically similar systems if the two flows are to be *dynamically* similar.

Let's look at an application of the *principle of dynamic similarity,* which is useful in the design and interpretation of experimental observations.

EXAMPLE 5.1.1   *Design of an Experimental Study of Ink-Jet Printing*

Ink-jet printing involves the collision of a very small drop of ink with a planar surface. In some cases, the surface is smooth. Solvent in the ink may dissolve in the surface, but the surface is not permeable to the liquid. (By contrast, printing on paper involves capillary flow of the ink *into* the porous paper matrix—another interesting problem. See Chapter 10, Section 10.3.) We want to know how much a drop "spreads" on impact with a surface. Our expectation is illustrated in Fig. 5.1.2.

In the printer of interest the drop volume is $V = 8$ nanoliters (nL) $= 8 \times 10^{-12}$ m^3. A drop in this microscopically small system has a viscosity of 5 centipoise (cP: 5 cP =

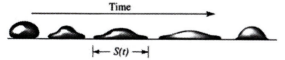

Time

$\leftarrow S(t) \rightarrow$

Figure 5.1.2 Sequence of events as a drop hits a surface, spreads radially, and finally returns to an equilibrium shape; $S(t)$ is the diameter of the contact boundary of the drop on the surface.

0.005 Pa·s) and a surface tension in air of 40 dyn/cm (0.04 N/m). The drop velocity just before impact is $U = 1$ m/s. We want to design a scaled-up version of this process to facilitate observation and measurement of the drop dynamics.

First, we define a length scale and a velocity scale. An arbitrary but rational choice is to pick a length scale equal to the cube root of the drop volume:

$$L = V^{1/3} = (8 \times 10^{-12})^{1/3} \text{ m} = 2 \times 10^{-4} \text{ m} \tag{5.1.25}$$

For a velocity scale an obvious choice is the drop velocity just before impact:

$$U = 1 \text{ m/s} \tag{5.1.26}$$

We will assume that the drops in the application are so small that gravity is not a significant factor. This may not be the case in our experimental system, but for the moment, let's say that the experiment is performed with drops so small that gravitational effects are unimportant. Then we ignore the Froude number, since terms involving **g** would be dropped from the Navier–Stokes equations.

We find

$$\text{Re} = \frac{\rho U L}{\mu} = \frac{1000(1)0.0002}{0.005} = 40 \tag{5.1.27}$$

We certainly expect surface tension to play an important role in drop dynamics. Hence we want to note that the Weber number in the application is

$$\text{We} = \frac{\rho U^2 L}{\sigma} = \frac{1000(1)^2(0.0002)}{0.04} = 5 \tag{5.1.28}$$

According to the principal of dynamic similarity, we must design an experiment such that Re = 40 and We = 5. Hence

$$\text{Re} = 40 = \left(\frac{\rho U L}{\mu} \right)_{\text{exp}} \tag{5.1.29}$$

and

$$\text{We} = 5 = \left(\frac{\rho U^2 L}{\sigma} \right)_{\text{exp}} \tag{5.1.30}$$

Notice especially that since we claim that the role of gravity can be ignored, we do not have to consider the Froude number in designing a dynamically similar system. This claim may be incorrect! If it is, we will be misled when we attempt to predict the behavior of the small-scale system from the observations on the larger scale system.

To review for a moment, we are claiming that if we perform an experiment under conditions that are dynamically similar to those of the ink-jet printer, the behavior of the drop dynamics in the microscopic system can be inferred from *observations* of the behavior in the larger scale system. Our goal is to apply this principle to the experimental design and find a set of appropriate operating conditions and physical properties for the test fluids.

We have five parameters at our disposal (ρ, U, L, μ, and σ), but they must satisfy two constraints (Eqs. 5.1.29 and 5.1.30). Hence we must select (i.e., we are *free* to select) three (five parameters minus two constraints) of these parameters.

We choose σ, ρ, and L because we think it will be easy to satisfy whatever requirements come out of Eqs. 5.1.29 and 5.1.30 on μ and U. On the other hand, the surface tension and density of liquids lie in a very narrow range, and we have less control over these properties when we select test liquids. Let's pick values for these physical properties that we know to be typical of liquids:

$$\sigma = 0.05 \text{ N/m}$$
$$\rho = 1000 \text{ kg/m}^3$$

and let's pick a convenient drop size, say

$$L = 0.002 \text{ m}$$

Notice that we have increased the *linear scale* in the model by a factor of 10 over that in the system of interest.

Then we impose the constraints

$$\text{Re} = 40 = \left(\frac{1000U \times 0.002}{\mu} \right)_{\text{exp}} \tag{5.1.31}$$

and

$$\text{We} = 5 = \left(\frac{1000U^2 \times 0.002}{0.05} \right)_{\text{exp}} \tag{5.1.32}$$

These two equations yield the solutions

$$U = 0.354 \text{ m/s} \tag{5.1.33}$$

and

$$\mu = 0.018 \text{ Pa} \cdot \text{s} \tag{5.1.34}$$

The experimental protocol requires the use of a liquid with the selected properties, drops of diameter approximately[3] 2 mm (probably small enough to validate our assumption that gravity is not a factor), and an impact velocity of about 35 cm/s. These are the conditions for performing a study with larger drops, while maintaining dynamic similarity to the case of interest. They appear to be easily achievable.

But once we have performed the experiment, how do we translate the observations into predictions of the behavior of the microscopic system?

We will measure the rate of spread of the drop from high speed photographs or video imaging, which will give us some function $S(t)$—the diameter of contact of the drop as a function of time. We could plot this as S versus t; but if we plot it as S/L versus Ut/L, we are really plotting S^* versus t^*, and the principle of dynamic similarity asserts that $S^*(t^*)$ will be the identical function of *dimensionless* time in the real (small ink drop) system. Once we have measured S^* versus t^* for the large system, we convert to a prediction of $S(t)$ for the *small* system through the use of the length and velocity scales of the *small* system.

We can apply this idea to the following hypothetical set of data. Figure 5.1.3 shows a plot of the contact diameter S of a drop as a function of time, obtained under the following conditions as specified above for dynamic similarity:

drop volume $V = 8 \times 10^{-9} \text{ m}^3 \, (L = 0.002 \text{ m})$ impact velocity $U = 35.4 \text{ cm/s}$
density $\rho = 1000 \text{ kg/m}^3$ surface tension $\sigma = 0.05 \text{ N/m}$
viscosity $\mu = 18 \text{ mPa} \cdot \text{s}$

The data are replotted in Fig. 5.1.4 by defining

$$S^* = S/L \tag{5.1.35}$$

---

[3] L is not the diameter (see Eq. 5.1.25). The actual diameter would be 1.6 mm.

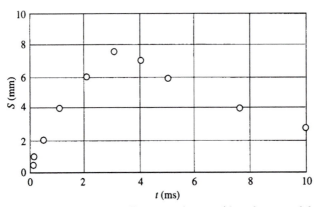

Figure 5.1.3 Drop spreading upon impact (data from model system).

and

$$t^* = Ut/L \tag{5.1.36}$$

With Fig. 5.1.4 available, we can predict the behavior of a drop spreading under the originally stated conditions of the ink-jet printer. The results are shown in Fig. 5.1.5. Note that the time scale is in units of *micro*seconds (μs).

In Example 5.1.1 our concern was with the design of an experiment to model one specific situation. The physical properties and drop speed on impact are specified for the ink-jet printer. The principle of dynamic similarity tells us how to design *the one* experiment that mimics the application of interest. In this example, it has been *assumed* that gravity plays no role in affecting the spreading of such small drops. The dynamics are *assumed* to be controlled by the balance between inertial forces, surface tension, and viscous effects. Our modeling involves only the Reynolds and Weber numbers. If we are incorrect, and gravity does play a role, the observed dynamics will not agree with the predictions.

In systems where gravity, viscosity, and surface tension *all* play a role, it is necessary to scale up (or down) while keeping all three dynamic groups—Re, We, and Fr—the

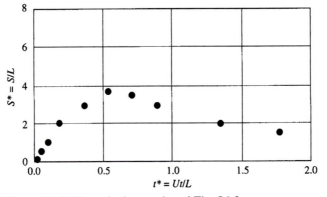

Figure 5.1.4 Dimensionless replot of Fig. 5.1.3.

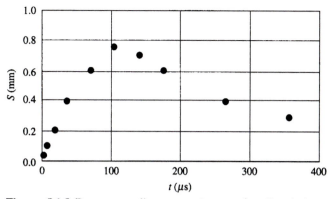

Figure 5.1.5 Drop spreading upon impact (predicted from model system).

same. This leads to severe constraints on the design of a dynamically similar experiment, as Example 5.1.2 illustrates.

EXAMPLE 5.1.2   *Design of an Experimental Study of Oil Film Entrainment*

Imagine that an oil spill has occurred on the surface of a body of water. The spill is of a finite volume, and the oil floats on the water and has spread to cover a large area of the surface. One technique for removing the oil from the water surface is "skimming," using a technology described in more detail later (Chapter 6, Section 6.6). We want to design an experiment that models some of the features of operation of a commercial oil skimmer. Figure 5.1.6 defines the characteristics of the system, in which a moving belt of material is continuously withdrawn through the water/oil interface. The belt is made from a hydro*phobic* material—it is preferentially wet by the oil phase. As a consequence, in the neighborhood of the interfaces, water is rejected by the belt and oil is entrained and lifted out of the surface. Of course gravity exerts a pull on the oil and limits the amount of oil that can be removed. The observation is that somewhere on the belt, upstream of the entrainment region, the oil film reaches an equilibrium thickness. The fluid dynamics of such a system is quite complex, and it is appropriate to design an experimental model to learn more about how the rate of oil entrainment is related to operating parameters such as the belt speed U and physical properties such

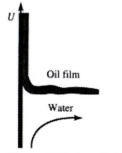

Figure 5.1.6 An oil skimmer.

as the oil viscosity and density, and the interfacial tension of the oil with air. For simplicity we will neglect any potential influence of the interfacial tension of oil in contact with water. This turns out to be of minor importance. However, we must consider the surface tension of the oil film in contact with the air.

In the absence of any further information, it seems reasonable to suppose that the dynamics are dependent on the balance between inertial, viscous, surface, and gravitational forces. If that is the case, then we must model the skimmer so that all three groups (Re, We, and Fr) are the same in the experimental model and in the full-scale real system. A set of constraints on the experimental design then is

$$\text{Re} = \left(\frac{\rho U L}{\mu}\right)_{\text{model}} = \left(\frac{\rho U L}{\mu}\right)_{\text{system}} \tag{5.1.37}$$

$$\text{We} = \left(\frac{\rho U^2 L}{\sigma}\right)_{\text{model}} = \left(\frac{\rho U^2 L}{\sigma}\right)_{\text{system}} \tag{5.1.38}$$

$$\text{Fr} = \left(\frac{U^2}{gL}\right)_{\text{model}} = \left(\frac{U^2}{gL}\right)_{\text{system}} \tag{5.1.39}$$

Thus we have three equations among six parameters. If we take g as a given fixed constant, we have only five parameters and two degrees of freedom (number of parameters minus number of equations). Hence we can solve these equations after selecting values for two of the parameters. Normally we might select the scale factor for the experimental model, since we would like to control the degree of scale-up or scale-down in the geometry. Hence we choose a specific value for k in the expression

$$L_{\text{model}} = kL_{\text{system}} \tag{5.1.40}$$

Since the density of liquids varies over such a narrow range, we might simply choose to use liquids of nearly equal densities:

$$\rho_{\text{model}} = \rho_{\text{system}} \tag{5.1.41}$$

We have no more degrees of freedom at this point, so we must solve Eqs. 5.1.37 through 5.1.39 to find the following results:

$$U_{\text{model}} = k^{1/2} U_{\text{system}} \tag{5.1.42}$$

$$\mu_{\text{model}} = k^{3/2} \mu_{\text{system}} \tag{5.1.43}$$

$$\sigma_{\text{model}} = k^2 \sigma_{\text{system}} \tag{5.1.44}$$

While this is a simple solution to the mathematical problem posed, the solution has some practical implications that make the result nearly useless. Normally, we rescale an experimental system as a matter of convenience because the system of interest is either too large or too small to permit convenient, practical study. Hence k is normally quite different from unity. This creates a problem when we examine Eq. 5.1.44, which says that to design a properly scaled experiment we must use a fluid with a surface tension that differs from that of the real system by a factor of k^2. This constraint creates a problem because surface tensions of liquids with respect to air lie in a fairly narrow range, typically about 0.015–0.075 N/m. Since this is at most a factor of 5, the linear scale factor would be restricted to, at most, a factor of $5^{1/2} = 2.24$. This is seldom enough of a difference in linear scale to justify a model experiment.

Thus we conclude that when all three dynamic groups (Re, We, Fr) play an important role in controlling the physics of the flow, it is extremely difficult to design an experimental model *significantly different in length scale from the system of interest* that satisfies the requirements of dynamic similarity.

One of the lessons of this section is that while dimensional analysis is a very helpful tool, its proper use requires that we know enough about the physics to make knowledgeable evaluations of the relative importance of the three dynamic groups, Re, We, and Fr. Dimensional analysis does not itself provide that information. In a more general sense, dimensional analysis tells us how to organize or correlate experimental data so that the data will have *broader* predictive value. Let's turn then to a discussion of rational principles that guide the *correlation* of data.

5.2   CORRELATION OF DATA

Again, we select a flow problem in which surface tension acts across a free surface. We make this choice because free surface problems are complex, and we want to illustrate the power of dimensional analysis on complex, nontrivial problems. In addition, however, there is a wide range of technologically important fluid dynamics problems that involve two fluid phases so disposed that surface tension plays an important role. In our model system (Fig. 5.2.1), a cylindrical steel roll rotates about a horizontal axis while partially submerged in a bath of a Newtonian liquid. A liquid film is entrained on the surface of the roll rotating out of the bath. That liquid film is then transferred to a second moving surface, as shown in the figure. The roll and the second surface may move independently. The purpose of the system is to coat a thin uniform film of the liquid onto the second surface.

The observation that motivates the problem presented here is that if the roll speed is too high, the plunging surface of the roll causes air to be entrained in the liquid bath. The presence of air bubbles then disrupts the uniformity of the coating process. We would like to be able to predict the critical roll speed at which air entrainment occurs, and it would be useful to have the prediction in a form that can be applied with some confidence to a wide range of liquid properties. A solution of this problem by analytical methods, via the Navier–Stokes equations, is impossibly difficult. A numerical solution of this complex free boundary problem would be very difficult to achieve, and a numerical approach would be quite expensive and time-consuming. In contrast, we could rather quickly develop an experimental program and produce a database of observations on entrainment over a range of experimental conditions. Throughout this discussion, we assume that an experimental study is to be carried out. Dimensional analysis can aid the design of the experiments as well as the correlation of the data.

We will denote the linear velocity of the roll surface by U and note that in terms of rotational speed

$$U = R\Omega \qquad (5.2.1)$$

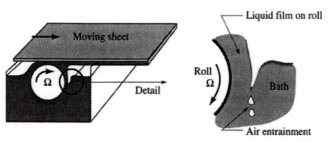

Figure 5.2.1 A roll coating system. If the roll speed is too high, air bubbles become entrained into the liquid when the roll plunges back into the bath.

with Ω in radians per second. We will measure the critical speed for entrainment, U^*, and we would like to convert this value into a dimensionless kinematic group. From our discussion in Section 5.1 we have three kinematic groups: Re, We, and Fr. Intuitively, we expect that surface tension plays an important role in resisting air entrainment. Hence we might choose the Weber number to characterize the critical roll speed:

$$\text{We}^* = \frac{\rho U^{*2}L}{\sigma} \tag{5.2.2}$$

This would be a poor choice, however, because we have no natural length scale L in this problem. We cannot use the film thickness, H, because it is unknown *a priori* and must itself be found experimentally or from theory. We need a length scale independent of the behavior of the system. While the roll radius might seem to be a likely candidate as a scaling factor for lengths, observation of the physics suggests otherwise. In most applications of interest, the film thickness is very small in comparison to the roll radius. Typically, H/R is of the order of 0.01. Under these conditions the roll appears to be a plane because its radius of curvature is so large compared with the film thickness. Moreover, data on the critical velocity of entrainment are found to be independent of the roll radius. Hence the roll radius R is not a suitable scaling factor in this problem. We conclude that there is no natural length scale in this problem.

It seems, then, that the Weber number is not a good choice for a dimensionless critical speed, except that it is the only dimensionless group that has surface tension in it. But we can eliminate L among the three dimensionless groups we think are important. For example, the grouping

$$\frac{\text{We}}{\text{Re}} = \frac{\mu U}{\sigma} = \text{Ca} \tag{5.2.3}$$

defines the so-called capillary number. This group has the attributes that we want—it is a dimensionless velocity that has surface tension in it, but it does not include a length scale. Note that since the capillary number is formed from the Reynolds and Weber numbers, it is not an *additional* dimensionless group. Among Re, We, Fr, and Ca, only three are independent. For another such example, recall that we introduced a dimensionless group called the Bond number (see Eq. 2.1.10) in a discussion of drops falling from capillaries. It is not difficult to see that

$$\frac{\text{We}}{\text{Fr}} = \frac{\rho g L^2}{\sigma} = \text{Bo} \tag{5.2.4}$$

We have decided to use the capillary number for the dimensionless critical entrainment velocity.[4] We will denote the critical value by Ca.*.

We would like to be able to predict Ca* from a knowledge of the physical properties of the liquid that is being coated. According to the principle of dynamic similarity, the critical value Ca* depends on Re and Fr. According to our argument that the radius of the roll is not important, the critical value Ca* depends on no parameters other than Re and Fr. But both Re and Fr are *kinematic* parameters. We want a relationship between Ca* and physical properties alone. If such a relationship exists, it must come out of the dimensionless groups (Re, Fr, We) already accounted for. Is this possible?

---

[4] We don't use the Bond number here because it is not a *kinematic* group—it does not have velocity in it. In addition, it does contain L.

Yes, although not by trivial inspection. We can, however, confirm three characteristics of the grouping

$$\frac{\text{Re}^{4/3}\text{Fr}^{1/3}}{\text{We}} = \sigma\left(\frac{\rho}{\mu^4 g}\right)^{1/3} \equiv \gamma \tag{5.2.5}$$

It is dimensionless (it must be, since it is a ratio of dimensionless numbers); it clearly contains only physical properties of the system; and it contains neither a velocity scale nor a length scale. Therefore, γ is sometimes called the property number.

We can also "discover" γ by analysis, in the following manner. We will require that some dimensionless number N—a combination of Re, Fr, and We—in the form

$$N = \text{Re}^a\text{Fr}^b\text{We}^c \tag{5.2.6}$$

be independent of velocity U and length L. Thus we must choose coefficients a, b, and c such that

$$\left(\frac{\rho UL}{\mu}\right)^a\left(\frac{U^2}{gL}\right)^b\left(\frac{\rho U^2 L}{\sigma}\right)^c = \alpha U^0 L^0 \tag{5.2.7}$$

(The zero exponents on U and L imply that U and L do not actually appear on the right-hand side of Eq. 5.2.7.) By equating powers of U and L on the two sides of the equation, we obtain two algebraic equations for the three constants:

$$a + 2b + 2c = 0$$
$$a - b + c = 0 \tag{5.2.8}$$

We can pick an arbitrary value of a, say $a = 1$, and then solve for b and c. The result is $b = \frac{1}{4}$ and $c = -\frac{3}{4}$, or

$$N = \frac{\text{Re Fr}^{1/4}}{\text{We}^{3/4}} = \left(\frac{\sigma^3 \rho}{\mu^4 g}\right)^{1/4} \tag{5.2.9}$$

This group serves our purpose. It is sometimes common to take the $\frac{4}{3}$ power of N, which is then simply proportional to σ, and this is exactly γ defined above.

Let's pause for a brief summary of our status. Dimensional analysis of the Navier–Stokes equations, including a boundary condition at an interface where surface tension is important, leads to the prediction that the dynamic behavior depends on three independent kinematic dimensionless groups: Re, We, and Fr. Other combinations of these groups give rise to additional dimensionless groups: Bo, Ca, and γ. However, these latter groups are not independent of the first three mentioned. At this stage of our discussion, then, we are led to speculate (supported by the principle of dynamic similarity) that air entrainment occurs at a critical linear speed U^* such that data obtained over a range of physical properties can be correlated in the form

$$\text{Ca}^* = \Re(\gamma) \tag{5.2.10}$$

By $\Re(\gamma)$ we imply an unknown functional relationship between Ca* and γ. If the relationship exists, we may determine it from experimental data. An example of such data is given in Fig. 5.2.2. Over a very wide range of the property number γ, the data seem to define a *single* curve, and this implies that there is a single unique relationship between Ca* and γ. The advantage of this dimensionless correlation of data is that if we need to *predict* the performance of a system operating under a set of conditions for which there are no data, we simply use the correlation. If the correlation is applicable to the conditions under which the system will be operated, it is not necessary to perform additional experiments. Indeed, if we had absolute faith that no other dimensionless

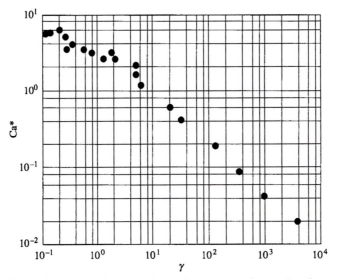

Figure 5.2.2 Data for critical entrainment speed, correlated as capillary number versus property number. Bolton and Middleman, *Chem. Eng. Sci.*, **35,** 597 (1980).

groups affected the performance of this coating system, we could produce a useful correlation with a very small number of carefully selected experiments. (See Problem 5.27.) For example, six experiments performed by choosing liquids such that their properties yielded γ values of 0.1, 1, 10, . . . up to $\gamma = 10^4$ would produce a correlation permitting accurate interpolation between the data points.

5.3   INSPECTIONAL ANALYSIS

At this stage it is natural to ask why this complicated analysis of the Navier–Stokes equations is necessary. If we simply know, or guess, that the critical speed will depend on μ, ρ, σ, and g, and not on a length scale, can't we arrive at Eq. 5.2.10 by classical dimensional analysis? The answer, of course, is Yes. The five parameters (U^*, μ, ρ, σ, g) contain three fundamental dimensions (mass m, length L, and time t). Arguments based on dimensional homogeneity imply that only *two* independent dimensionless groupings can be formed from these five parameters. (The number is the difference between the number of parameters and the number of dimensions: 5 minus 3 in this case. This is the Buckingham pi theorem.)

In many cases it is possible and even easier to use classical dimensional analysis, rather than manipulating the Navier–Stokes equations. However, we illustrate now a form of manipulation of the Navier–Stokes equations that leads to much stronger results than simple dimensional analysis. We consider the problem of flow of an incompressible Newtonian fluid through a "tube" of odd cross-sectional shape. The cross section, however, is uniform in size and shape along the axis of the tube (see Fig. 5.3.1).

Our goal is to infer something about the relationship between the pressure drop ΔP and the flowrate Q in this tube, and from that relationship to arrive at an efficient way to obtain and correlate data for such a system. We imply that the cross-sectional shape is so complex that a theoretical (analytical) solution for the flow field cannot be obtained.

It should be clear that according to our earlier analysis of the Navier–Stokes equations we would expect the pressure drop ΔP across a length L of tube to depend on the

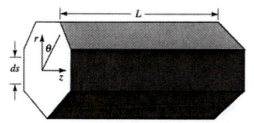

Figure 5.3.1 Flow along the z axis through a tube of odd cross section.

Reynolds number, the Froude number, and geometrical shape factors, in an expression of the form:

$$\frac{\Delta P}{\rho U^2} = \Psi[\text{Re, Fr, shape factors}] \qquad (5.3.1)$$

In this kind of problem we would normally introduce the average velocity $U = Q/A_c$ (where A_c is the cross-sectional area of the tube) for defining a Reynolds number and a Froude number. No Weber number dependence is expected, since there is no interface between two fluids in this system. Of course the function Ψ is not given by any dimensional analysis.

We assume steady state, isothermal Newtonian flow, and we write the axial momentum equation in cylindrical coordinates. We ignore the effect of gravity. (Then the Froude number will disappear from the dimensional analysis.) In the z direction we have

$$\rho\left(u_r\frac{\partial u_z}{\partial r} + \frac{u_\theta}{r}\frac{\partial u_z}{\partial \theta} + u_z\frac{\partial u_z}{\partial z}\right) = -\frac{\partial p}{\partial z} + \mu\left[\frac{1}{r}\frac{\partial}{\partial r}\left(r\frac{\partial u_z}{\partial r}\right) + \frac{\partial^2 u_z}{\partial z^2} + \frac{1}{r^2}\frac{\partial^2 u_z}{\partial \theta^2}\right] \qquad (5.3.2)$$

Note that we could have just as easily chosen Cartesian coordinates for this problem. Neither coordinate system fits the boundaries of the flow.

We will now make this equation dimensionless by introducing the velocity U and a length scale D. We are free[5] to select a definition of D. For example, we could define D as $\sqrt{A_c}$. All velocity components are normalized to U, and the coordinates r and z are normalized to D. Pressure is normalized as in Eq. 5.1.11. The result is

$$\left(u_r^*\frac{\partial u_z^*}{\partial r^*} + \frac{u_\theta^*}{r^*}\frac{\partial u_z^*}{\partial \theta} + u_z^*\frac{\partial u_z^*}{\partial z^*}\right) = -\frac{\partial p^*}{\partial z^*} + \frac{1}{\text{Re}}\left[\frac{1}{r^*}\frac{\partial}{\partial r^*}\left(r^*\frac{\partial u_z^*}{\partial r^*}\right) + \frac{\partial^2 u_z^*}{\partial z^{*2}} + \frac{1}{r^{*2}}\frac{\partial^2 u_z^*}{\partial \theta^2}\right]$$
$$(5.3.3)$$

5.3.1 Fully Developed Laminar Flow

Let's consider the case of *fully developed laminar flow,* for which the velocity vector has only the single component u_z, which may be a function of r and θ, but not of z, since the flow is assumed to be fully developed. Then Eq. 5.3.3 simplifies considerably, since the entire left-hand side vanishes, and we are left with

$$0 = -\frac{\partial p^*}{\partial z^*} + \frac{1}{\text{Re}}\left[\frac{1}{r^*}\frac{\partial}{\partial r^*}\left(r^*\frac{\partial u_z^*}{\partial r^*}\right) + \frac{1}{r^{*2}}\frac{\partial^2 u_z^*}{\partial \theta^2}\right] \qquad (5.3.4)$$

---

[5] But some choices make more sense than others! For example, for fully developed flow the length L of the "tube" would not be a logical choice of a scaling factor because it does not affect the velocity profile.

We need a second equation, since we have two unknowns here: u_z^*, and p^*. For fully developed flow the radial component of the momentum equation simplifies drastically to the form

$$0 = -\frac{\partial p^*}{\partial r^*} \qquad (5.3.5)$$

We need boundary conditions that reflect the physics of the flow. For example, we impose the no-slip condition in the form

$$u_z^* = 0 \qquad \text{on} \quad r^* = R^*(\theta) \qquad (5.3.6)$$

Because we have not specified the shape of the cross section, the mathematical description of the surface is given in its most general form in Eq. 5.3.6. It is easier to look at a specific example if we wish to know more explicitly what the boundary condition would look like. For example, suppose the tube were rectangular and its boundaries were given by the planes

$$y = \pm H \qquad \text{and} \qquad x = \pm B \qquad (5.3.7)$$

with respect to a Cartesian coordinate system with origin at the center of the rectangle. In cylindrical coordinates (not the natural choice for this geometry, but we can live with it), the boundaries are at

$$r = \frac{B}{\cos \theta} \qquad \text{for} \quad 0 \le \theta \le \arctan H/B$$

$$r = \frac{H}{\sin \theta} \qquad \text{for} \quad \arctan H/B \le \theta \le \frac{\pi}{2} \qquad (5.3.8)$$

in the first quadrant, with similar expressions in the other three quadrants. In nondimensional form we would write the no-slip condition as

$$u_z^* = 0 \qquad \text{on} \qquad \begin{array}{ll} r^* = \dfrac{B/D}{\cos \theta} & \text{for } 0 \le \theta \le \arctan H/B \\[2mm] r^* = \dfrac{H/D}{\sin \theta} & \text{for } \arctan H/B \le \theta \le \dfrac{\pi}{2} \end{array} \qquad (5.3.9)$$

If we choose the length scale D to be H or B, we see that Eqs. 5.3.9 introduce one dimensionless shape factor, H/B. In the more general case of Eq. 5.3.6, the function $R^*(\theta)$ includes any relevant shape factors.

Returning now to the general geometry, we write the boundary condition on pressure in the form of a specification of zero pressure at the inlet of the tube, so that

$$p^* = 0 \qquad \text{at } z^* = 0 \qquad (5.3.10)$$

(With this choice of inlet pressure, the outlet pressure is just the negative of the pressure drop ΔP.) The solutions to Eqs. 5.3.4 and 5.3.5 are of the form

$$u_z^* = u_z^*(r^*, \theta, \text{Re}, \text{shape factors})$$
$$p^* = p^*(z^*, \text{Re}, \text{shape factors}) \qquad (5.3.11)$$

From Eq. 5.3.5 we can state that p^* does not depend on r^*. (We have not written the θ component of the momentum equation, but it would simply give $\partial p^*/\partial \theta = 0$. Hence p^* does not depend on θ either.) Thus Eq. 5.3.4 is of the form

$$C(z^*) = -\frac{dp^*}{dz^*} = F(r^*, \theta) \qquad (5.3.12)$$

The left-hand side, which is a function *only* of z^*, can equal the right-hand side, which is *not* a function of z^*, only if $C(z^*)$ is a constant. C is simply the dimensionless pressure gradient, and when the definitions of p^* and z^* are introduced, we find that

$$C(z^*) = \frac{\Delta P\, D}{\rho U^2 L} \tag{5.3.13}$$

At this stage we can make the following statement: since C is a constant, it can depend only on the *constants* that appear in the differential equations and the boundary conditions that define the flow. Thus we can state that

$$C = f'(\text{Re, shape factors}) \tag{5.3.14}$$

Of course the functional dependence f' is not known. But classical dimensional analysis would have given us this same result, without the need to manipulate the Navier–Stokes equations. We show now that if we manipulate the equations a bit more cleverly, we can extract conclusions that go beyond what we can get from classical dimensional analysis.

To begin, we state a guiding principle:

In nondimensionalizing the constraining equations for a system, we should go as far as possible to remove all parameters from explicit appearance in the resulting nondimensional equations.

Let's apply that principle, and then see what it does for us.

Look back at Eqs. 5.3.4 and 5.3.5, which are the constraining equations for u_z^* and p^*. One parameter, the Reynolds number, appears (in Eq. 5.3.4). No Reynolds number appears in the boundary condition on velocity (Eq. 5.3.6) or pressure (Eq. 5.3.10). Can we remove the Reynolds number from Eq. 5.3.4? Yes, by defining a new dimensionless axial coordinate as

$$z^{**} = \frac{z^*}{\text{Re}} \tag{5.3.15}$$

Equation 5.3.4 now can be replaced by

$$0 = -\frac{\partial p^*}{\partial z^{**}} + \left[\frac{1}{r^*}\frac{\partial}{\partial r^*}\left(r^*\frac{\partial u_z^*}{\partial r^*}\right) + \frac{1}{r^{*2}}\frac{\partial^2 u_z^*}{\partial\theta^2}\right] \tag{5.3.16}$$

Equation 5.3.5 is unchanged. The boundary condition on velocity (Eq. 5.3.6) is unchanged. The boundary condition on pressure becomes

$$p^* = 0 \qquad \text{at } z^{**} = 0 \tag{5.3.17}$$

Equations 5.3.11 become

$$u_z^{**} = u_z^{**}(r^*, \theta, \text{shape factors})$$
$$p^{**} = p^{**}(z^{**}, \text{shape factors}) \tag{5.3.18}$$

The double asterisks (u^{**} and p^{**}) are just a reminder that these symbols come from Eqs. 5.3.16 and 5.3.17. We have not redefined u^* or p^*.

Before proceeding further, let's make a force balance in the axial direction. We apply the force balance to the solid surfaces of the duct, along which viscous shear stresses act, and to the open surfaces at $z = 0$ and $z = L$, across which pressures act. We will write the shear stress as $\tau_{zn} = \tau_{nz}$, by which we mean the shear stress, which exerts a force in the z direction, acting in the plane of the bounding surfaces. The normal vector **n** will vary in the θ direction. The force balance is simply a statement that the pressure drop across the length L arises from the wall shear exerted on the fluid:

$$\Delta P A_c = \int_0^L \left[\int_s - \tau_{zn} ds \right] dz \tag{5.3.19}$$

The bracketed integral is a contour integral along the cross-sectional perimeter of the tube, and ds is the differential length along the perimeter (see Fig. 5.3.1). The cross-sectional area of the tube at any point along the tube axis is given by A_c.

For a Newtonian fluid, we can write

$$\tau_{zn} = -\mu \Delta_{zn} \tag{5.3.20}$$

For a specific shape function $r = R(\theta)$ for the perimeter, we could, at least in principle, write the components of Δ_{zn} in cylindrical coordinates. It is not necessary to do so in the argument we are developing here.

Now we will nondimensionalize Eq. 5.3.19 using U and D as the velocity and length scales. We make the arbitrary but specific choice $D = \sqrt{A_c}$. Equation 5.3.19 takes the form

$$\frac{\Delta P}{\rho U^2} = \frac{1}{\text{Re}} \int_0^{L/D} \left[\int_{s^*} \Delta_{zn}^* \, ds^* \right] dz^* \tag{5.3.21}$$

Equation 5.3.11 holds, since we are using z^*, not z^{**}, at this point. Then it follows that Δ_{zn}^* is a function of θ, Re, and shape factors, and so the contour integral is just some constant K' that depends on Re and shape factors. Taking the integral of K' over z^*, we find

$$\frac{\Delta P}{\rho U^2} = \frac{K'(\text{Re, shape factors})}{\text{Re } D/L} \tag{5.3.22}$$

We look back at Eqs. 5.3.13 and 5.3.14: we have made no progress. We have simply replaced the unknown function f' by the unknown function K'.

Now let's see what happens when we use z^{**} in the force balance—Eq. 5.3.19. Equation 5.3.21 changes to a form in which Re does not appear except in the upper limit on the z^{**} integral:

$$\frac{\Delta P}{\rho U^2} = \int_0^{(L/D\,\text{Re})} \left[\int_{s^*} \Delta_{zn}^{**} \, ds^* \right] dz^{**} \tag{5.3.23}$$

We note the fact that Δ_{zn}^{**} depends on θ, and shape factors, and so the contour integral is just some constant K that depends on the shape factors. Thus we now write the integral as

$$\frac{\Delta P}{\rho U^2} = \frac{K(\text{shape factors})}{\text{Re } D/L} \tag{5.3.24}$$

or more simply

$$\frac{D\Delta P}{\rho L U^2} \text{Re} = K(\text{shape factors}) \tag{5.3.25}$$

Compare this result to Eq. 5.3.22: it is a much stronger result than what we had before. Equation 5.3.25 states that for fully developed laminar flow of a Newtonian fluid through a tube of arbitrary, but axially uniform, cross section, this particular combination of parameters on the left-hand side of the equation is a constant, at any flow rate—as long as the flow remains laminar. (Keep in mind our assumption of only a single component u_z of velocity implies laminar fully developed flow.) Our result implies that if we take a single measurement of the flow rate and corresponding pressure drop for the flow of any Newtonian fluid through a tube *of the shape of interest,* and if the flow is laminar and fully developed, that single measurement will give us the constant K that can be used for any other flow rates, other fluids, and tubes of varying sizes (but the same shape!) *as long as we do not violate the assumption of laminar fully developed flow.*

We may summarize the point of this analysis very simply.

1. An exercise in classical dimensional analysis would begin with the statement that the pressure drop would depend on the following parameters:

$$\Delta P = \Delta P(U, L, D, \mu, \rho, \text{ and other length scales}) \tag{5.3.26}$$

Ignoring "other length scales" for a moment, we have six parameters (counting ΔP), which involve three fundamental dimensions (mass, length, and time), and so we expect that there are three independent dimensionless groups (6 parameters minus 3 dimensions)—for example, Re, L/D, and $\Delta P/\rho U^2$. Dimensional analysis would simply require that

$$\frac{\Delta P}{\rho U^2} = fn(\text{Re}, L/D, \text{other shape factors}) \tag{5.3.27}$$

This result is valid for any flow of a Newtonian fluid, whether laminar and fully developed or not.

2. For the case of laminar fully developed flow through a tube of uniform cross section, we may carry out a procedure called "inspectional analysis," the main technique of which is the removal, as far as possible, of any dimensionless groups from explicit appearance in the governing equations. This is achieved through appropriate choices of dimensionless variables. (In the case studied, the key is the choice of z^{**} over z^*.) As a consequence, we obtain a much "stronger" result: Eq. 5.3.25 above.

We can offer a useful interpretation of the factor K here, by returning to the force balance given in Eq. 5.3.19. For fully developed flow, the z integration is trivial, and we may write the integral of the shear stress around the perimeter of the tube in terms of the average stress:

$$\Delta P A_c = \left[\int_s -\tau_{zn}\, ds \right] L = \bar{\tau} S L \tag{5.3.28}$$

where $\bar{\tau}$ is the magnitude of the *wall* shear stress averaged *along the perimeter* of the cross section—not the internal shear stress averaged over the cross section of the flow—and S is the total perimeter on which the shear stress acts (usually called the "wetted perimeter"). We now define the "friction factor," denoted f, as

$$f = \frac{\bar{\tau}}{\frac{1}{2}\rho U^2} \tag{5.3.29}$$

(The factor of $\frac{1}{2}$ in the denominator is arbitrary, but customary.) Using Eq. 5.3.28, we may write f as

$$f = \frac{\Delta P A_c/SL}{\frac{1}{2}\rho U^2} \tag{5.3.30}$$

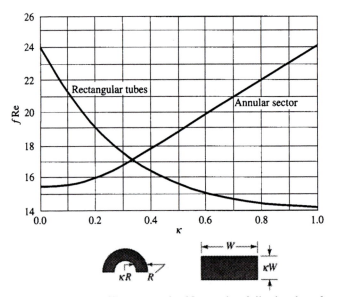

Figure 5.3.2 Some $f\,\mathrm{Re}$ curves for Newtonian fully developed laminar flow.

The ratio of the cross-sectional area to the wetted perimeter is a length scale called the "hydraulic radius," denoted r_h. This definition allows us to write the friction factor in the form

$$f = \frac{(\Delta P/L)r_h}{\frac{1}{2}\rho U^2} \tag{5.3.31}$$

Earlier we indicated that the choice of length scale D used in the nondimensionalization of the Navier–Stokes equations, hence in the definition of the Reynolds number, is an arbitrary choice. We state now that a common convention is to use

$$D_h = 4r_h \tag{5.3.32}$$

so that

$$\mathrm{Re} = \frac{4r_h U\rho}{\mu} \tag{5.3.33}$$

Had we made that choice earlier, we would have still obtained Eq. 5.3.25, but in the form

$$f\,\mathrm{Re} = F(\text{shape factors}) \tag{5.3.34}$$

All the information we need to describe laminar fully developed flow of an isothermal incompressible Newtonian fluid through a tube or duct of axially uniform cross section is the $f\,\mathrm{Re}$ product for a given cross-sectional geometry. Figure 5.3.2 shows some examples of $f\,\mathrm{Re}$ products calculated theoretically for flows through a few relatively simple geometries.

EXAMPLE 5.3.1   *Laminar Flow Through a Semicircular Pipe*

Let's consider the laminar flow of a viscous liquid through a long tube or pipe whose cross section normal to the flow is that of a semicircle of radius R. According to Eq.

5.3.34, the f Re product is some constant for this shape. In Figure 5.3.2 we have an f Re curve for annular sectors as a function of κ. A semicircle corresponds to the case $\kappa = 0$. From the graph we find

$$f \, \text{Re} = 15.5 \qquad (5.3.35)$$

From Eqs. 5.3.31 and 5.3.33, we may write this as

$$f \, \text{Re} = \frac{\Delta P}{L} \frac{r_h}{\frac{1}{2}\rho U^2} \frac{4r_h U \rho}{\mu} = \frac{8\Delta P \, r_h^2}{\mu U L} = 15.5 \qquad (5.3.36)$$

The hydraulic radius for this shape is

$$r_h = \frac{A_c}{S} = \frac{\frac{1}{2}\pi R^2}{\pi R + 2R} = \frac{\pi}{4 + 2\pi} R \qquad (5.3.37)$$

This permits us to write Eq. 5.3.36 in the form

$$\frac{\Delta P}{L} = 20.8 \frac{\mu U}{R^2} = 13.2 \frac{\mu Q}{R^4} \qquad (5.3.38)$$

From Eq. 3.2.34, Poiseuille's law for a circular cross section gives a coefficient of $8/\pi = 2.55$ instead of 13.2. For a given volumetric flowrate, the pressure drop is considerably greater in the semicircular tube because the effective radius is smaller, at fixed R, and the pressure drop–flowrate relationship is very strongly dependent on the tube radius. It is interesting to note that the factor by which the pressure drop is increased, namely $13.2/2.55 = 5.18$, is of the same order as the fourth power of the ratio of the diameter of a circular tube to 4 times the hydraulic radius of the semicircular tube:

$$\frac{D}{4r_h} = \frac{2R}{4\pi R/(4 + 2\pi)} = 1.64 \quad \text{and} \quad (1.64)^4 = 7.18 \qquad (5.3.39)$$

5.3.2 Entry Region Flow

If we relax the assumption of fully developed flow, the analysis that leads to Eq. 5.3.24 must be modified, and we find that the f Re product depends on more than just geometry. We go back to Eq. 5.3.23 and note that if the flow is not fully developed, the bracketed term in the integral is still a function of z^{**}. If we introduce the definition of the friction factor, we may write Eq. 5.3.23 as

$$f \, \text{Re} = \frac{\text{Re} \, D}{2L} \int_0^{(L/D\text{Re})} \left[\int_{s^*} \Delta_{zn}^{**} \, ds^* \right] dz^{**} \qquad (5.3.40)$$

The integration over the perimeter (the ds^* integration) cannot be carried out unless we know the flow field. But whatever the flow field, the integral of Δ_{zn}^{**} along the perimeter (i.e., along the path ds^*) will depend only on the axial position z^{**}. Because of the way we have nondimensionalized the equations, we can say that Δ_{zn}^{**} is not a function of Re. If we name the ds^* integral $h(z^{**})$, then Eq. 5.3.40 yields

$$f \, \text{Re} = \frac{\text{Re} \, D}{2L} \int_0^{(L/D\text{Re})} [h(z^{**})] \, dz^{**} = \frac{\text{Re} \, D}{2L} G(L/D \, \text{Re}) \qquad (5.3.41)$$

where we have written the integral over dz^{**} as

$$\int_0^{(L/D\text{Re})} [h(z^{**})] \, dz^{**} = G(L/D \, \text{Re}) \qquad (5.3.42)$$

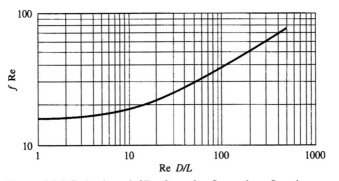

Figure 5.3.3 Behavior of f Re for tube flow when flow is not fully developed.

This integral is unknown also, but it is a definite integral over the limits $[0, L/D \text{ Re}]$, and so the integral depends on (is a function only of) the upper limit L/D Re. Thus we conclude from Eq. 5.3.41 that if the flow is not fully developed, the f Re product is not a constant, as we found earlier for fully developed flow. We imply, but didn't write, that there is a dependence on shape factors as well. Then Eq. 5.3.41 has the form

$$f \text{ Re} = F(\text{Re } D/L, \text{ shape factors}) \tag{5.3.43}$$

Before, for fully developed flow, a *single experiment* would yield the f Re product for a tube or duct of a given shape. Now, when the flow is not fully developed, we conclude that a *single curve* of f Re versus Re D/L will provide all the information we need. This fact (or prediction, until we examine reality, i.e., experimental data) provides a very efficient guideline to our investigation of the pressure–flow relationship in a duct of odd cross section. Without this prediction, we would expect to have to obtain data for a variety of fluids, flowrates, and tube sizes before being able to produce a useful correlation of the data. Inspectional analysis simplifies this task enormously. Unfortunately it cannot be applied to all flow problems with as much efficiency and simplicity as in the example illustrated here.

Figure 5.3.3 shows the f Re behavior for flow in the entry region of circular tubes, based on theoretical calculations. The nature of the theory is such that these calculated curves should not be used for tubes for which $L/D < 1$. Indeed, even when the L/D is of the order of 100, for Reynolds numbers of the order of 1000 (but still laminar flow), excess pressure losses associated with the flow upstream of (i.e., just before) the entrance give rise to significant departures of the f Re product from the line shown in Fig. 5.3.3. The entrance geometry of the tube becomes important in those cases. (See Problem 5.11.)

5.4 EXPERIMENTAL DESIGN

Sometimes we have a very limited goal in studying a specific flow—for example, we might be designing a process for a reservoir that is to be filled initially with a Newtonian liquid but which will be pumped out until it is nearly empty, through an open pipe that enters the reservoir bottom as shown in Fig. 5.4.1. Studies with similar systems indicate that at some critical height H the free surface forms a vortex that is sucked into the tube, entraining air in the liquid. It is this "event" that we wish to avoid. We want to design a scale model of the system, determine the critical height in the scale model, and then infer the critical height in the real system.

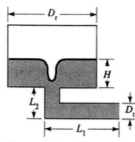

Figure 5.4.1 Air entrainment during drainage of a reservoir.

Why do we not simply build, and make the measurement in, a real system? The answer may be that the real system does not yet exist, or performing experiments on it may be prohibitively expensive. Suppose, for example, that the real system has dimensions

$$D_r = 1 \text{ m} \quad D_t = 0.03 \text{ m} \quad L_1 = 0.5 \text{ m} \quad L_2 = 0.3 \text{ m}$$

and that the liquid is a molten ceramic at a temperature of 1000 K. Obviously experiments on this system would entail considerable expense. Hence we want to design a "model experiment" that mimics the behavior of the real system.

The molten ceramic system must operate at a withdrawal rate of $Q = 12$ L/s = 1.2×10^{-2} m^3/s.

Now suppose that the viscosity, density, and surface tension of the molten ceramic are

$$\mu = 2.4 \text{ Pa} \cdot \text{s} \quad \rho = 2400 \text{ kg/m}^3 \quad \sigma = 0.25 \text{ N/m}$$

Based on our earlier discussions, we expect that in a properly designed model experiment, the model and the real system will be geometrically and dynamically similar. Hence we would require that

$$\text{Re}_{\text{model}} = \text{Re}_{\text{real}} \tag{5.4.1}$$

$$\text{Fr}_{\text{model}} = \text{Fr}_{\text{real}} \tag{5.4.2}$$

$$\text{We}_{\text{model}} = \text{We}_{\text{real}} \tag{5.4.3}$$

We include the Weber number because there is a free surface that could play an important role in this problem. If the length scale is large enough, however, the radius of curvature of the vortex may be so large that surface tension phenomena will be rendered secondary to viscous and gravitational phenomena. Another way to argue this point is to examine the ratio of surface tension effects to gravitational effects. That is, we examine the ratio

$$\frac{\text{We}}{\text{Fr}} = \frac{\rho g L^2}{\sigma} = \text{Bo} \tag{5.4.4}$$

which is just the Bond number. We might expect that a length scale characteristic of the radius of curvature would be of the same order as the tube diameter $D_t = 0.03$ m. Setting $L = 0.03$ m and using the other parameters already given, we find

$$\text{Bo} = \frac{\text{We}}{\text{Fr}} = \frac{2400 \times 9.8 \times 0.03^2}{0.25} = 84.7 \tag{5.4.5}$$

The simplest way to interpret the significance of a large Bond number is by inspection of Eq. 5.4.4. We may think of the Bond number as the ratio of a pressure characteristic of hydrostatic effects ($\rho g L$) to a pressure characteristic of surface tension effects (σ/L). A large Bond number implies that surface tension effects are secondary to gravitational effects in the real system. Hence we will ignore Eq. 5.4.3. That is, we will assume that surface tension effects are unimportant and need not be considered in designing the experiment.

Next we will calculate values of Re and Fr characteristic of the real system. This requires a specification of the flow rate Q, given above as 1.2×10^{-2} m³/s.

The Reynolds number is then found from

$$\text{Re} = \frac{\rho U D_t}{\mu} = \frac{4\rho Q}{\pi D_t \mu} = 509 \tag{5.4.6}$$

The Froude number for this system can now be calculated, and we find

$$\text{Fr} = \frac{U^2}{g D_t} = \frac{16 Q^2}{\pi^2 g D_t^5} = 980 \tag{5.4.7}$$

Our goal is to design a model experiment. We must select a length scale D_t for the model, and of course all other lengths must scale to maintain geometrical similarity. We must select values of liquid properties ρ and μ. However, these two properties appear only in the Reynolds number, and as the ratio μ/ρ (the "kinematic viscosity"). Hence we regard this pair as a single parameter. Finally, we have the velocity U (or the flowrate Q). Thus we have three parameters, but only two constraints that they must satisfy—Eqs. 5.4.1 and 5.4.2. This means that we are free to choose one of the parameters. We choose the length scale so that we can control the size of the model. We decide to scale down by one order of magnitude, so that D_t in the model is 0.3 cm.

Equation 5.4.1 requires that

$$\left[\frac{Q}{\nu D_t}\right]_{\text{model}} = \left[\frac{Q}{\nu D_t}\right]_{\text{real}} \tag{5.4.8}$$

where we let $\nu = \mu/\rho$. Since the scale factor has been selected as $D_{t\,\text{model}} = 0.1 D_{t\,\text{real}}$, we now write

$$\left[\frac{Q}{\nu}\right]_{\text{model}} = \left[\frac{0.1Q}{\nu}\right]_{\text{real}} \tag{5.4.9}$$

Equation 5.4.2 requires that

$$\left[\frac{Q^2}{D_t^5}\right]_{\text{model}} = \left[\frac{Q^2}{D_t^5}\right]_{\text{real}} \tag{5.4.10}$$

or

$$[Q^2]_{\text{model}} = 10^{-5}[Q^2]_{\text{real}} \tag{5.4.11}$$

Thus (using $Q_{\text{real}} = 0.012$ m/s) we find that the flowrate in the model system must be maintained at

$$Q_{\text{model}} = 3.8 \times 10^{-5} \text{ m}^3\text{/s} = 38 \text{ cm}^3\text{/s} \tag{5.4.12}$$

The required value of kinematic viscosity for the model flow then follows from Eq. 5.4.9 and is

$$\nu_{\text{model}} = 3.16 \times 10^{-5} \text{ m}^2\text{/s} \tag{5.4.13}$$

If we select a test liquid with a typical liquid density of 1000 kg/m^3, we find that the test liquid must have a viscosity of

$$\mu_{\text{model}} = 3.16 \times 10^{-2} \text{ Pa} \cdot \text{s} \qquad \textbf{(5.4.14)}$$

Such a liquid could be made from an aqueous solution of corn syrup or glycerol.

We have now selected a scale factor (the model is one-tenth full scale in all linear dimensions) and a test liquid (defined by the kinematic viscosity given in Eq. 5.4.13). We must withdraw liquid from the model reservoir at a rate of 38 cm^3/s. Then the model will be geometrically and dynamically similar to the real system. By performing a series of experiments, we can observe the critical height H at which air is entrained. (In principle, a single experiment will suffice. A series of replicate experiments permits us to get an average value for H that will be more reliable, statistically, than a single measurement.) Since the observed H is just a characteristic length for the dynamic system, the corresponding value in the real system will be obtained through the scale factor, and so

$$H_{\text{real}} = 10 \times H_{\text{model}} \qquad \textbf{(5.4.15)}$$

We have reduced the length scales in designing the model system as outlined here. Do we still satisfy the assertion made earlier that surface tension is unimportant? Look at the Bond number in the model system:

$$\text{Bo} = \frac{1000(9.8)(0.003)^2}{0.06} = 1.5 \qquad \textbf{(5.4.16)}$$

Surface tension may be important in the model system, since the Bond number is now two orders of magnitude smaller than its value (Eq. 5.4.5) in the large system. This diminution occurred because we went to a much smaller length scale for the model, and surface curvature effects (remember the Young–Laplace equation?) may affect the physics. We could check on this point by doing our model experiments with liquids of several surface tensions and looking for any influence of surface tension on the results. For example, we could easily add small amounts of a liquid soap to the aqueous corn syrup or glycerol solution used in the model experiments.

With this design, we can estimate other features of the experimental design prior to building and operating the system. For example, we have specified a very high viscosity test liquid, and we should check to see what kind of pressure requirement is imposed if we are to maintain the specified flowrate. (See Problem 5.19.)

EXAMPLE 5.4.1 *A Model for the Flow Field in a CVD Reactor*

We want to study the flow field in a CVD reactor such as the one described in Section 1.1.3. In the application of interest, the gas consists of a mixture of 1% silane in helium. At a gas temperature of 700 K and a pressure of 13.5 Pa, the flowrate is 10^4 L/min. The central inlet tube diameter is 1 cm. We wish to model this flow using a cold liquid instead of a hot reactive gas. A scale model of the CVD reactor, smaller than the real reactor by a *linear* scale factor of 3, is available. If we choose a liquid of viscosity 10 cP, at what volume flowrate should we operate? Assume that a properly scaled experiment can be based just on equal Reynolds numbers.

First, we need the viscosity and density of helium. We will ignore the effect of the small fraction of silane on the gas properties needed here. From Table 3.4.3 we find the parameters necessary for use in Eq. 3.4.8:

$$M = 4 \qquad \sigma = 2.6 \text{ Å} \qquad \varepsilon/k = 10.2 \text{ K} \qquad kT/\varepsilon = 700/10.2 = 68.61$$

$$\Omega_\mu \text{ (Fig. 3.4.2)} = 0.62$$

$$\mu(\text{poise}) = 2.7 \times 10^{-5} \frac{(MT)^{1/2}}{\sigma^2 \Omega_\mu} = 3.4 \times 10^{-4} \text{ poise} \tag{5.4.17}$$

For the gas density we use the ideal gas law:

$$\rho = \frac{PM}{R_G T} = \frac{(13.5 \times 10^{-5} \text{ atm})(4)}{82(700)} = 9.4 \times 10^{-9} \text{ g/cm}^3 \tag{5.4.18}$$

The volume flowrate is

$$Q = 10^4 \text{ L/min} = 1.67 \times 10^5 \text{ cm}^3/\text{s} \tag{5.4.19}$$

Using the inlet diameter as the length scale for the Reynolds number, we find

$$\text{Re} = \frac{4Q\rho}{\pi D \mu} = \frac{4(1.67 \times 10^5)\, 9.4 \times 10^{-9}}{\pi(1)\, 3.4 \times 10^{-4}} = 5.8 \tag{5.4.20}$$

We must use the same Reynolds number in the model, for which $D = 0.33$ cm and $\mu = 0.1$ poise. Assume, in the absence of information, that for a liquid at room temperature, $\rho = 1$ g/cm^3. Then

$$Q = \frac{\pi D \mu \text{Re}}{4\rho} = \frac{\pi(0.33)(0.1)5.8}{4(1)} = 0.15 \text{ cm}^3/\text{s} \tag{5.4.21}$$

is the flowrate that provides the same Reynolds number at which the CVD reactor operates.

In assuming that the Reynolds number is the only dimensionless group we must match in designing a properly scaled experiment, we are neglecting the potential role of gravity. If there are significant temperature gradients in the CVD reactor, and this is usually the case in commercial reactors, the flow field will be strongly affected by buoyancy-driven flows. Several additional dimensionless groups enter into a proper analysis of such a system, and it is no simple matter to design a scaled experiment that accounts for these thermal effects. Hence we can regard the answer given for Q as an estimate that is valid only in the limit of near-isothermal behavior.

SUMMARY

The ideas of dimensional analysis, introduced in Chapter 1, are extended to give us a principle of dynamic similarity. This principle both permits us to design and interpret properly scaled experiments and guides the correlation of data. We find that nondimensionalization of the Navier–Stokes equations introduces two dimensionless groups, the Reynolds number and the Froude number. The latter is unimportant in flows where gravity plays no role. If we are considering flows with a phase boundary (e.g., flows involving drops or bubbles, or liquid films or jets in air), then interfacial tension may give rise to significant forces that must be accounted for *in the boundary conditions*. A consequence of this requirement is the introduction of an additional dimensionless group involving interfacial tension. The principle of dynamic similarity states that (at least for isothermal, incompressible, low Mach number flows) the fluid dynamics are completely controlled by the values of *no more than* three dynamic dimensionless groups: the Reynolds number (Re $= \rho U L/\mu$), which expresses the relative importance of inertial forces to viscous forces, the Froude number (U^2/gL), which expresses the relative importance of inertial forces to gravitational forces, and the Weber number ($\rho U^2 L/\sigma$), which expresses the relative importance of inertial forces to interfacial forces. All other dimensionless groups that might come out of application of the Buckingham pi theorem can be shown to be expressible in terms of these three groups, hence are not independent of them.

PROBLEMS

5.1 Suppose, instead of using Eq. 5.1.11, we had chosen to define $p*$ as

$$p* = \frac{p}{\mu U/L} \qquad \text{(P5.1)}$$

How, if at all, would Eq. 5.1.12 be changed?

5.2 An incompressible Newtonian liquid moves relative to a solid disk oriented as shown in Fig. P5.2. Write the nondimensional formulation of the continuity and dynamic equations for this flow. Write the no-slip boundary condition in nondimensional form. Display the dimensionless geometrical or shape factors that fall naturally out of this formulation. Do the case of steady flow.

5.3 Suppose we had nondimensionalized the pressure p in Eqs. 5.1.19 with Eq. P5.1 in Problem 5.1. What form would Eq. 5.1.22 have taken? Would this change any conclusions?

5.4 Suppose that instead of using Eq. 5.1.11, we had chosen to normalize the pressure using surface tension:

$$p* = \frac{pL}{\sigma} \qquad \text{(P5.4)}$$

What form would Eq. 5.1.12 have taken? Would this change any conclusions?

5.5 A coating roll rotates in an ink bath and deposits the ink on a moving sheet, as shown in Fig. 5.2.1. The liquid ink has a viscosity of 0.01 Pa · s, and a surface tension of 0.04 N/m. The ink density is 1200 kg/m^3, and the liquid is Newtonian. The coating roll has a radius of 5 cm. What is the maximum rotational speed of the roll, in units of radians per second, if air entrainment is to be avoided?

5.6 In Section 5.2 we introduced a number of dimensionless groups relevant to flows with a free surface across which surface tension acts. A dimensionless group not mentioned in that section is the Ohnesorge number

$$Z \equiv \frac{\mu}{(\rho D \sigma)^{1/2}} \qquad \text{(P5.6)}$$

Show that Z is not independent of We and Re.

5.7 In writing Eq. 5.3.2 we did not account for the possible influence of gravity. Suppose that gravity acted in the $+z$ direction. Show that if we define a new pressure $\mathcal{P} = p - \rho g z$, none of the conclusions drawn in Section 5.3 would be

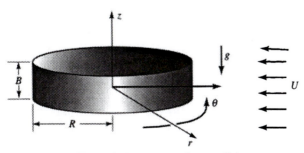

Figure P5.2 Flow relative to a stationary disk.

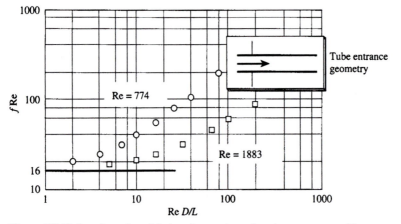

Figure P5.12 Laminar flow friction factor data showing entrance effects.

changed, and we would still obtain Eq. 5.3.25, except that in place of ΔP we would have $\Delta \mathscr{P}$.

5.8 Begin with Eq. 3.3.30 for axial annular pressure flow. Rearrange it to the form of f Re as a function of the shape factor κ. Define r_h for this geometry. Plot f Re versus κ.

5.9 Rework Problem 3.2 of Chapter 3, but consider tubes whose cross sections are rectangular or annular segments. (See Fig. 5.3.2 for the geometrical parameters.) For the rectangular tube, take $W = 0.004$ m and $\kappa = 0.5$. For the annular segment, take $R = 0.002$ m and $\kappa = 0.5$.

5.10 Rework the air entrainment example in Section 5.4 to include surface tension as an important parameter.

5.11 Shah [*J. Fluids Eng.*, **100**, 177 (1978)] presents an equation for correlating data on the friction factor for laminar Newtonian flow through tubes of circular cross section. His correlation takes the form

$$f\,\text{Re} = \frac{16 - 3.44/\zeta^{1/2} + 0.31/\zeta}{1 + 0.00021/\zeta^2} + \frac{3.44}{\zeta^{1/2}}$$

$$\text{(P5.11.1)}$$

where

$$\zeta = \frac{L}{D\,\text{Re}} \qquad \text{(P5.11.2)}$$

It is easy to see that f approaches the expected value of 16/Re as ζ grows large. Define an "entry length" as the tube length L_e such that f has approached within 1% of the fully developed value [i.e., $f = 1.01(16)/\text{Re}$] and use Eq. P5.11.1 to give the value of ζ_e corresponding to L_e.

5.12 Ghajar and Madon [*Exp. Thermal Fluid Sci.*, **5**, 129 (1992)] present the data shown in Fig. P5.12, obtained for laminar flow in tubes of circular cross section. Do the data agree with Shah's correlation (see Eq. P5.11.1)?

5.13 A molten wax is pumped through a long capillary of diameter 0.3 cm at a volume flowrate of 10 mL/s. The jet that issues from the capillary breaks down into a stream of droplets, as shown in Fig. P5.13. The wax properties are:

$$\rho = 0.9 \text{ g/cm}^3$$
$$\mu = 5 \text{ cP}$$
$$\sigma = 45 \text{ dyn/cm}$$

We wish to model the drop breakup characteristics of this system using an aqueous solution of corn syrup. Assume that the aqueous solution

Figure P5.13

has a density of $\rho = 1.2$ g/cm^3 and a surface tension $\sigma = 65$ dyn/cm, independent of concentration.

Specify the viscosity, capillary diameter, and volume flowrate needed to model the external wax flow.

5.14 Under gravity-free conditions, a Newtonian liquid jet is projected at a rigid smooth surface, and we are interested in observing the details of the impact of the jet with the surface. For example, we are interested in the fraction of the mass of the liquid jet that is converted to droplets.

Preliminary observations indicate that a very high speed video recording system would serve to resolve the details of this flow. We cannot afford such a system. Instead, we believe that by choosing a more viscous liquid we can "slow down" the dynamics to a degree that will allow us to use a much slower, and less expensive, recording system.

Since the density and the surface tension of common liquids lie in a narrow range, we will select for the "slow" jet a liquid having the same density and surface tension as the "fast jet."

We want to slow the dynamics by a factor of 1000. This means that a sequence of events that would require a time interval t_1 in the real system would need a time interval of 1000 t_1 in the model system. By what factor do we have to increase the viscosity of the "slow" fluid, relative to that of the "fast" fluid, to achieve this?

By what factor do we have to change the length scales in the "slow" system? For example, if we use the jet diameter D as a characteristic length, what is the required value of D_{slow}/D_{fast}?

5.15 In an application of interest to us, a static drop sits on a smooth plane horizontal surface, and we must predict the shape of the drop. The drop is so small, however, that we cannot measure its shape precisely enough to verify our prediction. Someone suggests that we use a system that will simulate reduced gravity to study a larger drop. If we measure the shape of the larger drop at reduced gravity, we will be able to infer the shape of the smaller drop at normal gravity.

We decide to use a drop that is a thousand times larger in volume than the small drop of interest. We will use the same liquid, and it will sit on the same surface. At what value of reduced gravity will the shapes of the two drops (the large, scaled-up drop and the small drop of concern in this problem) be identical?

5.16 The time t_e to empty a bottle filled with a Newtonian liquid is expected to depend on the volume of the bottle V (we assume that it is initially filled to capacity), the viscosity μ and density ρ of the liquid, and the gravitational constant g. Assume that surface tension does not play a role. Thus we have five parameters (t_e, V, μ, ρ, g) that involve the three fundamental dimensions (mass m, length L, and time t). The Buckingham pi theorem tells us that there will be a relationship among these five variables that involves only two dimensionless groups. But our analysis of the Navier–Stokes equations leads us to expect the dimensionless emptying time (call it t_e^*) to be a function of the Reynolds number and the Froude number: $t_e^* = f(\text{Re}, \text{Fr})$—which implies three dimensionless groups t_e^*, Re, Fr. We need to resolve this apparent dilemma.

Show that the resolution lies in the absence from this problem of any given external velocity imposed on the fluid dynamics. Begin by defining a length scale L and a time scale T. Choose a time scale that does not involve viscosity. What are your choices of L and T?

Define a velocity scale U as L/T. Now (here's the important part!) show that with these definitions, we do not have *both* a Reynolds number *and* a Froude number in this problem. Hence there is only one dynamic dimensionless number G (a dimensionless group involving parameters related to the viscous and gravitational forces) that a dimensionless emptying time ($t_e^* = t_e/T$) depends on.

Give the forms for t_e^* and G.

5.17 A continuously moving section of a cylindrical filament plunges vertically into a bath of liquid at a speed U (see Fig. P5.17). We want to design an experimental program and develop

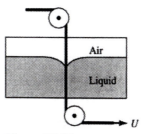

Figure P5.17

an empirical model of the critical speed of the filament at which air is entrained with the filament as it passes into the liquid. We expect that there will be some dimensionless dynamic group G^*, which will include the critical speed U^*, and that G^* will depend on other dimensionless groups $G_1, G_2, \ldots$ characteristic of this system. Preliminary data suggest that the results depend on the filament radius R, so we cannot ignore this parameter in designing an experiment.

Define a group G^*, and define any other dimensionless groups, appropriate to the physics of this problem, for correlation with your data on U^*. How many independent dimensionless groups does G^* depend on? Define those groups.

5.18 A pasta-producing machine extrudes raw pasta dough through an annular semicircular die of length L as in Fig. P5.18. Calculate the pressure, in units of psig, required to produce 10 kg/h of pasta under the following conditions: $\mu = 10$ Pa·s, $\rho = 1200$ kg/m³, $L = 5$ cm, $R_1 = 0.5$ cm, $R_2 = 1$ cm.

5.19 Estimate the pressure drop required to drive the flow in the model experiment designed in Section 5.4.

5.20 Consider a gas bubble rising slowly through a viscous liquid. When the bubble reaches the surface it will burst and collapse, and under some conditions the collapse will create liquid droplets that are projected above the surface. A study of such phenomena by Takahashi et al. (*Int. Chem. Eng.*, **21**, 251 (1981)] led to a dimensionless correlation of the critical (minimum) bubble diameter that will yield drops on bubble collapse. Assume that the critical bubble diameter D_B depends on the viscosity μ and density ρ of the liquid, on the interfacial tension σ between the liquid and the gas bubble, and on the gravitational constant g. Define a dimensionless group proportional to D_B, and state what other

group(s) this dimensionless critical bubble diameter could depend on. These other groups should *not* include D_B.

5.21 With reference to Problem 5.20, Takahashi et al. found that the minimum bubble diameter that would yield a droplet upon bubble collapse corresponded to a constant value of the Bond number, defined as (cf. Eq. 5.2.4)

$$\text{Bo} \equiv \frac{\rho D_B^2 g}{\sigma} = 1.43 \qquad \textbf{(P5.21)}$$

a. Is this observation compatible with our expectations based on dimensional analysis?

b. A bubble of diameter $D_B = 1$ mm rises through a liquid of viscosity 1 Pa·s, density 1200 kg/m³, and interfacial tension 0.05 N/m. Will a liquid drop be ejected when the bubble bursts at the surface? What is the smallest bubble that will yield a drop?

5.22 Bhunia and Lienhard [*J. Fluids Eng.*, **116**, 338 (1994)] studied the splattering that occurs when a jet of a low viscosity liquid strikes a rigid surface. They found the critical conditions for splattering, which they define as the conversion of more than 5% of the mass flow of the jet into a spray upon impact of the jet, to depend on the distance L of the nozzle exit from the surface, the nozzle diameter D, and the velocity U and surface tension σ of the jet.

a. From principles of dimensional analysis, what dimensionless groups determine the critical conditions for splattering? For example, on what dimensionless groups would a dimensionless critical distance depend?

b. Bhunia and Lienhard correlate their data with the expression

$$\frac{L_{\text{crit}}}{D} = \frac{130}{1 + (5 \times 10^{-7}\,\text{We}^2)} \qquad \textbf{(P5.22.1)}$$

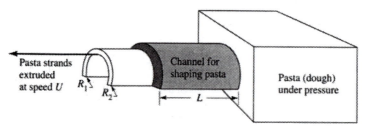

Figure P5.18 Pasta is extruded through the channel of length L.

where the Weber number is defined as

$$\text{We} = \frac{\rho U^2 D}{\sigma} \qquad \text{(P5.22.2)}$$

Is this observation consistent with your answer to part a? Explain your response.

5.23 In a cleaning system you are designing for the removal of small particles from semiconductor wafers, a liquid jet of an isopropanol/water solution will be directed at the wafers at a 90° angle. The jet is to be created by flow of the solution through a capillary of inside diameter $D_i = 1$ mm. Other considerations indicate that the jet velocity should be 10 m/s. The surface tension of the solution is $\sigma = 0.042$ N/m at the temperature (300 K) of application. The viscosity is that of water. Using the correlation of Bhunia and Lienhard (Eq. P5.22.1 in Problem 5.22), specify the distance that the nozzle exit should be from the "target." As a criterion, use a separation distance that is *half* the distance that would produce splattering, when "splattering" is defined as in Problem 5.22.

5.24 A classroom demonstration on the time to drain water from a set of initially filled inverted bottles yielded the following results:

where V is the filled bottle volume (L), D is the internal diameter of the bottle neck (m), g is the acceleration of gravity (9.8 m^2/s), and t is the time to empty (s). The constant is different for each bottle shape.

a. For our experiments, how should t depend on V, according to Whalley's correlation? (Is the relationship linear, or does some other power relationship apply? If the latter, give the power n in $t = KV^n$.)

b. For *geometrically similar* bottles, how should t depend on V, according to Whalley's correlation? (Is the relationship linear, or does some other power relationship apply? If the latter, give the power n in $t = KV^n$.)

c. Plot our data as a test of Whalley's correlation. Do the data agree with the correlation?

5.25 At the end of Section 5.1 we got in trouble with surface tension when we chose to model a system by keeping density constant and arbitrarily fixing the scale factor k. Suppose, instead, we keep density and surface tension the same (as well as g). What scale factor, viscosity, and characteristic velocity must we use in the properly scaled model experiment? Is this a better

Bottle volume, V (liters)	2	1.5	0.7	0.4	0.36	0.18
Time to drain, t (s)	18.6	17.3	9.5	4.2	4.8	2.9

The bottles were not geometrically similar, but the same cap size fit each bottle. Hence the neck diameter was the same in each case.

In a study of bottle emptying, Whalley [*Int. J. Multiphase Flow*, **17**, 145 (1991)] presents a correlation of data of the form

$$\frac{4V}{g^{1/2}D^{5/2}t} = \text{constant}$$

method of experimental design? Explain your answer.

5.26 The following bubble sizes were observed in various liquids for very slow gas flow through two different capillaries of diameters D_0. Plot the data in some dimensionless form. Describe the reason for your choice of dimensionless parameters.

Fluid	Liquid Viscosity (cP)	Liquid Density (g/cm^3)	Surface Tension (dyn/cm)	Bubble Diameter (cm) $D_0 = 0.293$ cm	Bubble Diameter (cm) $D_0 = 0.438$ cm
Water/air	0.81	0.996	70.2	0.514	0.612
Water/H$_2$	0.81	0.996	70.2	0.516	0.630
7.5% Ethanol/air	1.07	0.982	51.4	0.480	0.583
45% Sugar/air	52	1.2	58.1	0.486	0.585
Wesson oil/air	57	0.92	36.5	0.411	0.491

5.27 In Section 5.2 we argue, with no proof, that the critical entrainment speed depends on physical properties through a relationship between Ca and γ. If the roll radius R were large enough, and/or the rotational speed Ω large enough, we might expect centrifugal acceleration to give rise to a pressure that could be comparable to the pressure associated with surface tension in the neighborhood of the entrainment region. Derive a dimensionless group that represents the ratio of these two pressures. From Section 2.3 in Chapter 2 we expect a centrifugal pressure to be of the order of $\rho R^2 \Omega^2$. From Section 2.2 we expect surface tension to give rise to a pressure of the order of σ/L, where L is some length scale characteristic of the radius of curvature in the region of entrainment. Here is the hard part of the problem: How do we characterize L in terms of the other parameters of this system? Make some suggestions, come up with some dimensionless groups, and show how each group can be related to the Reynolds, Froude, and/or Weber numbers.

5.28 An f Re plot for laminar Newtonian flow through long straight ducts whose cross sections are isoceles triangles is shown in Fig. P5.28.
a. What is the ratio of cross-sectional areas of a duct of circular cross section to one whose cross section is an equilateral triangle, under conditions that equal pressure gradients produce equal volumetric flowrates?
b. What error would we make if we predicted the volumetric flowrate through the equilateral triangular duct by using Poiseuille's law with the diameter replaced by four times the hydraulic radius of the equilateral triangular duct?

5.29 Given the pressure drop–flowrate data below, obtained for flow through a duct of triangular cross section, plot the expected curve for pressure drop versus flowrate for flow through a duct of the following characteristics, under the implied conditions:

s = length of side of (equilateral) triangle		10 cm
L = axial length of duct		250 cm
ρ = fluid density		1 g/cm^3
μ = fluid viscosity		0.1 poise
expected range of flow (Q)		100–1000 cm^3/s

Data for Flow in Equilateral Triangular Duct

$s = 1$	2	10 cm
$L = 10$	2	10 cm
$\rho = 1$	1	1 g/cm^3
$\mu = 1$	0.01	0.1 poise
$Q = 60$	10	60 cm^3/s
$\Delta P = 1.2 \times 10^4$	5	18 dyn/cm^2

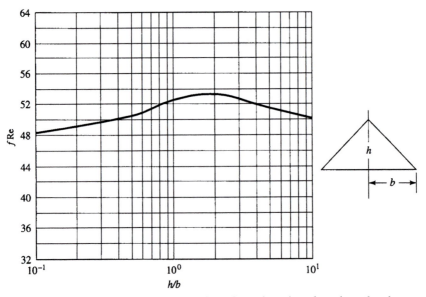

Figure P5.28 f Re product for laminar flow through an isoceles triangular duct.

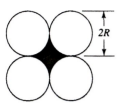

Figure P5.30 The four-cusp geometry.

5.30 Figure P5.30 shows the cross section of a "four-cusp channel" that runs along the axis of a set of parallel solid cylindrical rods. Such a flow geometry is of interest in considering the dynamics of cooling of fuel rods in a nuclear reactor. A numerical solution for laminar fully developed flow through a four-cusp channel leads to the result

$$f = \frac{26.3}{\text{Re}} \quad (P5.30.1)$$

In this expression, the Reynolds number is defined as

$$\text{Re} = \frac{\rho U D_h}{\mu} \quad (P5.30.2)$$

where U is the average velocity through the channel and D_h is the hydraulic diameter. In this case, the friction factor has been defined as

$$f = \frac{2(\Delta P/L)D_h}{\rho U^2} \quad (P5.30.3)$$

which differs from our definition, given in Eq. 5.3.31, by a factor of 4. (But $D_h = 4r_h$, as in Eq. 5.3.32.)

a. Show that the hydraulic diameter is

$$D_h = 2R\frac{4 - \pi}{\pi} \quad (P5.30.4)$$

b. Consider the case of a laminar fully developed flow through a four-cusp channel. By what factor is the volumetric flow reduced in a four-cusp channel, relative to that through a circular tube of the same open cross-sectional area, at a given pressure gradient?

c. Find the radius R_c of a circular tube that has the same open area as the four-cusp channel. Give your answer in two forms: R_c/R and R_c/D_h.

5.31 Find the pressure drop required to pump water at 20°C through a horizontal four-cusp channel (see Problem 5.30) of axial length $L = 1$ m at a Reynolds number of 1000, for the case $R = 1$ cm.

5.32 Parrott et al. [*Int. J. Multiphase Flow*, **17**, 119 (1991)] reported experiments on the onset of gas entrainment into the discharge of a tank (see Fig. P5.32). They found the critical flowrate

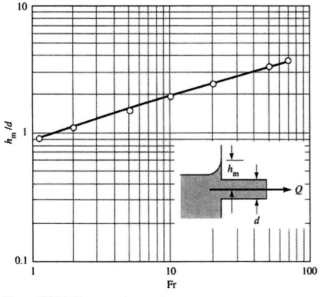

Figure P5.32 Data on air entrainment.

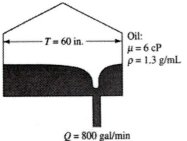

$T = 60$ in.

Oil:
$\mu = 6$ cP
$\rho = 1.3$ g/mL

$Q = 800$ gal/min

Figure P5.33 A draining tank.

(in terms of a Froude number) at which gas is first entrained. This flowrate depends on the height of the surface of the liquid (actually, the height of the meniscus h_m, as shown in Fig. P5.32) relative to the axis of the discharge pipe, of diameter d.

A tank of water is to be discharged under the following conditions: $d = 5$ mm, $Q = 1$ gal/min, and $g = 1$ m/s². Give a value expected for the meniscus height, h_m, just before air entrainment into the outlet pipe occurs. How does this differ from the value to be expected at normal gravity, $g = 9.8$ m/s²?

Parrott et al. define the Froude number as

$$\text{Fr} = \frac{U}{(gd)^{1/2}} \qquad \text{(P5.32.1)}$$

which is the square root of the definition we use in this text.

5.33 A large tank of vegetable oil drains through a pipe in the bottom of the tank, as shown in Fig. P5.33. We wish to design an experiment that will simulate the draining process, with particular concern for the conditions under which air will be entrained in the outflow as the head in the tank gets small. Design a scale model experiment using water as the test fluid. In particular, specify the flowrate that simulates the conditions given on Fig. P5.33, and give the scaling factor, defined as the ratio of the diameter of the model tank to that of the real tank.

5.34 Suppose that the system described in Problem P5.33 is to be operated at a lower temperature, such that the viscosity of the oil increases to 10 cP. Given the model tank size found in solving Problem P5.33, what flowrate Q must be used to provide a dynamically similar system?

5.35 Scheller and Bousfield [*AIChE J.*, **41,** 1357 (1995)] present data on the maximum degree of radial spread of liquid drops that impact onto a planar surface. A selection of their data for high viscosity liquids is tabulated below. The "dimensionless maximum spreading radius" $R*$ is defined as the measured maximum spreading radius divided by the diameter of the drop prior to impact. Test these data against the following speculations:

a. Viscous effects dominate the spreading dynamics, and surface tension forces are negligible. Subsequent to impact, gravitational effects are negligible.

b. Surface tension forces are important, in addition to viscous effects. Subsequent to impact, gravitational effects are negligible.

What do you conclude?

ρ (kg/m³)	μ (mPa·s)	σ (N/m)	U (m/s)	D (m)	$R*$
1236	296	0.065	1.44	0.00319	1.47
			2.52		1.73
			3.25		1.88
			3.70		1.91
1154	32.0	0.0405	1.42	0.00285	2.28
			2.39		2.52
			3.20		2.67
			3.93		2.80
			4.96		2.97
1121	12.77	0.0415	1.37	0.00287	2.45
			2.36		2.99
			3.17		3.26
			3.93		3.38
			4.96		3.78

5.36 Additional data from Scheller and Bousfield (see Problem 5.35), obtained with low viscosity liquids, are presented below.

a. Examine the speculation that for these data, spreading is controlled by the balance between surface tension and the inertia of the drop impact, and viscosity is not relevant. What do you conclude?

b. Examine the speculation that surface tension forces are important, in addition to viscous effects. Postimpact gravitational effects are negligible. What do you conclude?

ρ (kg/m³)	μ (mPa·s)	σ (N/m)	U (m/s)	D (m)	R^*
999	1.05	0.0731	1.34	0.00365	3.11
			2.44		4.28
			3.18		4.44
			3.85		4.99
			4.74		5.39
1029	2.35	0.0474	1.95	0.00308	3.72
			2.87		4.16
			3.58		4.47
			4.27		4.78
			4.88		4.90

5.37 A liquid jet issues from a circular orifice of diameter 0.01 cm at a mean velocity of 30 cm/s. The liquid has a density $\rho = 0.9$ g/cm³, a viscosity $\mu = 0.05$ poise, and a surface tension $\sigma = 35$ dyn/cm. Drops formed from this jet strike a fixed planar surface normal to the axis of the jet, at a distance of 10 cm from the orifice. We wish to model the dynamics of the impact of the drops with the surface, using a larger system to facilitate obtaining high quality video images of the impact region. For that purpose, an orifice of diameter 0.1 cm is available, as is a liquid of density $\rho = 1.2$ g/cm³ and surface tension $\sigma = 60$ dyn/cm. The viscosity of this test liquid can be adjusted without changing the surface tension and density. In a properly scaled experiment,

what viscosity liquid would you use, and what jet velocity is required?

5.38 A drop of molten metal falls under gravity through a vacuum onto a rigid planar surface. Isothermal conditions are maintained. The metal has the following properties: density $\rho = 6$ g/cm³, viscosity $\mu = 1.0$ poise, surface tension $\sigma = 100$ dyn/cm. The drop diameter is $D = 1$ cm, and the drop speed on impact is $U = 25$ cm/s. We want to model this system in the laboratory using an ordinary liquid of density $\rho = 1.0$ g/cm³ and a surface tension $\sigma = 50$ dyn/cm. What drop size, impact speed, and liquid viscosity are required for the model experiment?

How would your model experiment change if this system operated in a gravity-free environment? Recommend an experimental design (i.e., values for D, U, ρ, μ, σ) for this case.

5.39 Burley and Jolly [*Chem. Eng. Sci.*, **39**, 1357 (1984)] present data, tabulated below, on the critical speed U^* at which air entrainment occurs when a continuous dry tape enters a liquid bath, as shown in Fig. P5.39. Would you expect the correlation of data shown in Fig. 5.2.2 to apply to this set of data? Why? Plot the data according to the correlation of Fig. 5.2.2, and comment on the result.

Liquid	σ (N/m)	ρ (kg/m³)	μ (Pa·s)	U^* (m/s)
Ethylene	0.05	1110	0.22	0.70
Propylene glycol	0.04	1040	0.85	0.21
Polyethylene glycol 300	0.049	1120	1.32	0.20
Polyethylene glycol 400	0.05	1120	2.0	0.145
Glycerol/water 1	0.065	1230	3.3	0.105
Glycerol/water 2	0.064	1210	1.6	0.215
Glycerol/water 3	0.065	1190	0.64	0.487

Figure P5.39 A tape plunges vertically into a liquid bath and entrains air at the critical speed U^*.

5.40 Pamperin and Rath [*Chem. Eng. Sci.,* **19,** 3009 (1995)] present data on the bubble diameter D_b formed in clean water above a submerged capillary of diameter D_c through which air is flowing. In addition to measurement of the bubble diameter, these investigators indicate the average velocity U of the gas flow as it crosses the outlet of the capillary. Do these data follow the model presented in Chapter 2, and plotted there in Fig. 2.1.10? Discuss and explain any differences that you observe.

D_c (mm)	D_b/D_c	U (m/s)
0.39	7.5	9.55
0.39	9.5	17.4
0.39	10.5	21.4
0.39	11.3	24.7
0.80	7.7	10.5
0.80	7.9	12.2
0.80	8.7	20.0

5.41 In a manufacturing process under development, a molten metal droplet will strike a ceramic surface. Since the surface is at the same temperature as the metal, no droplet solidification will occur. Preliminary evidence indicates that for a given droplet diameter there is a critical relative velocity between the droplet and the surface at which splashing occurs, and smaller drops are created from the "parent" droplet. The droplet velocity is normal to the surface in this process. Because it is important to operate under conditions that avoid splashing, we will carry out an extensive experimental study to determine the critical splash velocity U_{crit} as a function of drop-

let diameter D. It is too difficult to work with the molten metal system in the laboratory, and so we are obliged to use a model liquid instead. The properties of the molten metal, at the required temperature of operation, are as follows: $\rho = 7000$ kg/m^3, $\sigma = 0.15$ N/m, $\mu = 3$ mPa·s.

A model liquid is available with a density $\rho = 10{,}000$ kg/m^3 and a surface tension $\sigma = 0.25$ N/m. The viscosity of the liquid is a strong function of temperature, and over the range of temperatures acceptable for the experimental program, the viscosity–temperature data are shown in Fig. P5.41.

The recommended manufacturing conditions involve a drop diameter of $D = 3$ mm striking the surface at a relative velocity of 1m/s. Under conditions of *full* dynamic similarity, assuming that viscous, gravitational, and interfacial effects are *all* important, specify the experimental conditions (U, D, and μ) that will simulate the recommended manufacturing conditions. At what temperature will you do this experiment?

5.42 In Chapter 9 we will consider the laminar boundary layer equations that describe flow near the wall when a turbulent flow moves parallel to a planar wall or plate. For now, take these equations as appropriate for the description of the velocity field (u, v) near the wall:

$$u\frac{\partial u}{\partial x} + v\frac{\partial u}{\partial y} = \frac{\mu}{\rho}\frac{\partial^2 u}{\partial y^2} \qquad \textbf{(P5.42.1)}$$

$$\frac{\partial u}{\partial x} + \frac{\partial v}{\partial y} = 0 \qquad \textbf{(P5.42.2)}$$

Using the method of inspectional analysis, prove that the mean friction factor for this flow, defined as

$$f \equiv \frac{\bar{\tau}}{\frac{1}{2}\rho U^2} \qquad \textbf{(P5.42.3)}$$

satisfies

$$f = \frac{2C}{\mathrm{Re}_L^{1/2}} \qquad \textbf{(P5.42.4)}$$

where the Reynolds number is based on the length of the plate:

$$\mathrm{Re}_L = \frac{\rho UL}{\mu} \qquad \textbf{(P5.42.5)}$$

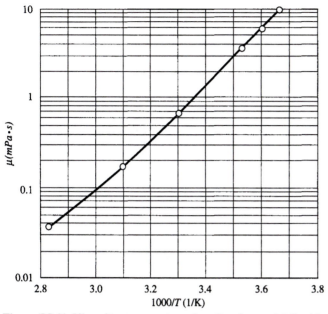

Figure P5.41 Viscosity–temperature data for the model liquid.

The mean shear stress at the plate surface is defined as

$$\bar{\tau} = \frac{1}{L}\int_0^L \tau\, dx = \frac{1}{L}\int_0^L \mu \left(\frac{\partial u}{\partial y}\right)_{y=0} dx$$

$$(P5.42.6)$$

5.43 Repeat Example 5.4.1, but use air at 300 K and 1 atm total pressure as the model fluid.

5.44 At the end of Example 5.4.1 it is noted that in the presence of significant temperature gradients, buoyancy-driven flows will affect the performance of the reactor. Assume that in addition to the parameters considered in that exam-

ple, the following physical parameters are important under *nonisothermal* conditions, and find the additional dimensionless groups that must be considered in a properly scaled experiment.

ΔT—a characteristic temperature difference between the incoming gas and the reactive surface

$\beta = (1/\rho)\partial\rho/\partial T$—the temperature dependence of the gas density (β in reciprocal kelvins)

$k[W/(m \cdot K)]$—the thermal conductivity of the gas

$C_p[(W \cdot s)/kg \cdot K]$—the heat capacity of the gas

$g[m/s^2]$—the acceleration of gravity

Chapter 6

Nearly Parallel Flows

The flows we consider in this chapter are a little more complex than those in Chapter 4 for which we have obtained analytical solutions. We will examine two-dimensional nonturbulent flows that are "nearly parallel." The classical example of such a flow occurs in lubrication phenomena, and these flows are often called "lubrication flows" even when the application has nothing to do with lubrication per se. Hence we begin with a study of a classic lubrication problem: the slider bearing. This is followed by a series of examples drawn from a wide range of flow fields that yield analytical solutions when this type of "lubrication" approximation is imposed on the flow.

6.1 THE SLIDER BEARING

A wide planar surface moves steadily, in its own plane, at a velocity U. In the region $0 < x < L$ there is a fixed solid plane maintained at an angle to the moving surface. The geometry is shown in Fig. 6.1.1. The angle θ is assumed small, in the neighborhood of $10°$ or less. The surfaces are completely surrounded by a liquid. We will restrict our attention to steady state, isothermal, incompressible Newtonian flow. We will find that such converging flows can produce very high pressures in the region between the two surfaces, and as a consequence a large force exists normal to the surfaces. If the "fixed surface" were free to move in the y direction it would do so. Alternatively, if it were loaded with an opposing force, this surface could be held stationary (i.e., truly "fixed"). Hence the stationary surface is capable of supporting or bearing a load, and this flow configuration is often referred to as a "slider bearing." It plays an important role in the design of lubricated surfaces in machinery with surfaces in relative motion.

We begin an analysis of the slider bearing by noting that there are two characteristic length scales to consider: the x component of the length of the fixed surface L, and the gap spacing H_1. We will take θ as the third geometrical parameter, in which case H_o follows from H_1 and θ, since

$$\tan \theta = \frac{H_1 - H_o}{L} \tag{6.1.1}$$

We assume that in comparison to the length scale L, the surfaces are very wide in the z direction. Since all length scales are normally quite small in such systems, we ignore the effect of gravity. Then we may take the flow to be two-dimensional with velocity components u_x and u_y, each of which depends on both x and y. In addition, the pressure distribution $p(x, y)$ is an unknown function.

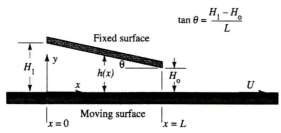

Figure 6.1.1 Geometry of the slider bearing. The motion of the lower surface drags fluid from left to right.

We begin with the continuity and Navier–Stokes equations, for steady two-dimensional flow:

$$\frac{\partial u_x}{\partial x} + \frac{\partial u_y}{\partial y} = 0 \tag{6.1.2}$$

$$\rho\left(u_x\frac{\partial u_x}{\partial x} + u_y\frac{\partial u_x}{\partial y}\right) = -\frac{\partial p}{\partial x} + \mu\left(\frac{\partial^2 u_x}{\partial x^2} + \frac{\partial^2 u_x}{\partial y^2}\right) \tag{6.1.3}$$

$$\rho\left(u_x\frac{\partial u_y}{\partial x} + u_y\frac{\partial u_y}{\partial y}\right) = -\frac{\partial p}{\partial y} + \mu\left(\frac{\partial^2 u_y}{\partial x^2} + \frac{\partial^2 u_y}{\partial y^2}\right) \tag{6.1.4}$$

Suppose we nondimensionalize these equations as we did in Chapter 5. We pick some length scale, say H_1, and a velocity scale U. We define

$$\tilde{x} = \frac{x}{H_1} \qquad \tilde{y} = \frac{y}{H_1} \tag{6.1.5}$$

$$\tilde{u}_x = \frac{u_x}{U} \qquad \tilde{u}_y = \frac{u_y}{U} \qquad \tilde{p} = \frac{pH_1}{\mu U} \tag{6.1.6}$$

We will end up doing an analysis valid for low Reynolds number (viscous) flows. In anticipation of this, we use a viscous pressure scale, $\mu U/H_1$, rather than an inertial pressure ρU^2, in making p dimensionless. Then we write Eqs. 6.1.2 through 6.1.4 as follows:

$$\frac{\partial \tilde{u}_x}{\partial \tilde{x}} + \frac{\partial \tilde{u}_y}{\partial \tilde{y}} = 0 \tag{6.1.7}$$

$$\frac{\rho U H_1}{\mu}\left(\tilde{u}_x\frac{\partial \tilde{u}_x}{\partial \tilde{x}} + \tilde{u}_y\frac{\partial \tilde{u}_x}{\partial \tilde{y}}\right) = -\frac{\partial \tilde{p}}{\partial \tilde{x}} + \left(\frac{\partial^2 \tilde{u}_x}{\partial \tilde{x}^2} + \frac{\partial^2 \tilde{u}_x}{\partial \tilde{y}^2}\right) \tag{6.1.8}$$

$$\frac{\rho U H_1}{\mu}\left(\tilde{u}_x\frac{\partial \tilde{u}_y}{\partial \tilde{x}} + \tilde{u}_y\frac{\partial \tilde{u}_y}{\partial \tilde{y}}\right) = -\frac{\partial \tilde{p}}{\partial \tilde{y}} + \left(\frac{\partial^2 \tilde{u}_y}{\partial \tilde{x}^2} + \frac{\partial^2 \tilde{u}_y}{\partial \tilde{y}^2}\right) \tag{6.1.9}$$

We see that the left-hand side has as a coefficient a Reynolds number, based on the length scale H_1. Hereafter we will restrict the analysis to flows for which this Reynolds number is so small that the terms on the left-hand side may be neglected. Physically this approximation corresponds to flows such that viscous effects completely dominate inertial effects. This is not true for many flows of interest, but there are equally many flows in which this is a good approximation. Hence the neglect of the inertial terms is a restriction, and it defines a particular class of flows that is important enough to warrant specific study.

Then at vanishing Reynolds number the equations that constrain the motion take the form

$$\frac{\partial \tilde{u}_x}{\partial \tilde{x}} + \frac{\partial \tilde{u}_y}{\partial \tilde{y}} = 0 \tag{6.1.10}$$

$$0 = -\frac{\partial \tilde{p}}{\partial \tilde{x}} + \left(\frac{\partial^2 \tilde{u}_x}{\partial \tilde{x}^2} + \frac{\partial^2 \tilde{u}_x}{\partial \tilde{y}^2}\right) \tag{6.1.11}$$

$$0 = -\frac{\partial \tilde{p}}{\partial \tilde{y}} + \left(\frac{\partial^2 \tilde{u}_y}{\partial \tilde{x}^2} + \frac{\partial^2 \tilde{u}_y}{\partial \tilde{y}^2}\right) \tag{6.1.12}$$

We will state appropriate boundary conditions shortly.

We now introduce the idea that the flow is *nearly* parallel. If θ were exactly zero the flow would be exactly parallel, except for some end effects as the fluid enters and leaves the region $0 < x < L$. In the parallel region of flow there would be no y component of velocity. The continuity equation would give

$$\frac{\partial \tilde{u}_x}{\partial \tilde{x}} = 0 \quad \text{for} \quad \text{parallel flow} \tag{6.1.13}$$

[handwritten: & from continuity eqn]

and it would follow then that, for $\tilde{u}_y = 0$, Eq. 6.1.12 would give

$$0 = -\frac{\partial \tilde{p}}{\partial \tilde{y}} \tag{6.1.14}$$

For a truly parallel flow $\partial \tilde{u}_x / \partial \tilde{x} = 0$, so Eq. 6.1.11 yields

$$0 = -\frac{d\tilde{p}}{d\tilde{x}} + \frac{d^2 \tilde{u}_x}{d\tilde{y}^2} \tag{6.1.15}$$

By arguments similar to those we have used before, we find that each term in Eq. 6.1.15 is a constant *in the case of parallel flow*. With the boundary condition that the pressure be uniform outside the region $0 < x < L$, it follows that there is no pressure variation over the region $0 < x < L$ for parallel flow, and the velocity profile is simply

$$\tilde{u}_x = 1 - \tilde{y} \quad \text{for} \quad \text{parallel flow} \tag{6.1.16}$$

(We have assumed no-slip conditions on the two solid surfaces.) This is not a very interesting flow field, in the sense that it is so simple. The velocity profile is linear, and there is no pressure variation within the parallel region. This flow field is called plane Couette flow, or simply plane parallel drag flow.

Now we introduce the idea that the flow is *nearly* parallel to the lower boundary. By this assumption we imply that the velocity u_y is so small that

$$0 = -\frac{\partial \tilde{p}}{\partial \tilde{y}} \tag{6.1.17}$$

still holds. Furthermore, we assume that if the flow is nearly parallel, then the velocity gradient in the x direction is much smaller than that in the y direction. (When the flow is *exactly* parallel there is no gradient of velocity in the x-direction.) This permits us to write Eq. 6.1.11 in the form

$$0 = -\frac{d\tilde{p}}{d\tilde{x}} + \frac{\partial^2 \tilde{u}_x}{\partial \tilde{y}^2} \tag{6.1.18}$$

Although the forms are the same, there is a difference between this equation and Eq. 6.1.15. We are not assuming that u_x is independent of the x coordinate. (That is why we use $\partial \tilde{y}$ rather than $d\tilde{y}$ in the second derivative term.) Rather, we assume that

$\partial^2 \tilde{u}_x/\partial \tilde{x}^2 \ll \partial^2 \tilde{u}_x/\partial \tilde{y}^2$. We will examine this approximation later, because its validity is not obvious at this point.

We now integrate Eq. 6.1.18 twice with respect to $\tilde{y}$ and find

$$\tilde{u}_x = \left(1 - \frac{\tilde{y}}{\eta}\right) - \frac{\eta^2}{2}\frac{d\tilde{p}}{d\tilde{x}}\left[\frac{\tilde{y}}{\eta} - \left(\frac{\tilde{y}}{\eta}\right)^2\right] \tag{6.1.19}$$

The plane of the upper boundary is denoted by

$$\eta = \frac{h(x)}{H_1} = 1 - \tilde{x}\tan\theta \tag{6.1.20}$$

In writing Eq. 6.1.19 we have used as boundary conditions on velocity that

$$\tilde{u}_x = 0 \quad \text{on} \quad \tilde{y} = \eta(\tilde{x}) \tag{6.1.21a}$$

and

$$\tilde{u}_x = 1 \quad \text{on} \quad \tilde{y} = 0 \tag{6.1.21b}$$

In Eq. 6.1.19 the velocity is an explicit function of $\tilde{y}$ and an implicit function of $\tilde{x}$, first through the dependence of η on $\tilde{x}$ and as well through the (as yet unknown) dependence of $\tilde{p}$ on $\tilde{x}$. We must now find this pressure distribution function.

We begin by writing the volumetric flowrate of fluid (per unit width in the z direction) across any plane normal to the x direction, in dimensionless terms:

$$\frac{Q}{W} = \int_0^{h(x)} u_x(x, y)\,dy = UH_1 \int_0^\eta \tilde{u}_x(\tilde{x}, \tilde{y})\,d\tilde{y} \tag{6.1.22}$$

or

$$\lambda \equiv \frac{Q}{WUH_1} = \int_0^\eta \tilde{u}_x(\tilde{x}, \tilde{y})\,d\tilde{y} = \frac{\eta}{2} - \frac{\eta^3}{12}\frac{d\tilde{p}}{d\tilde{x}} \tag{6.1.23}$$

Keep in mind that at this point we do not know λ; it is some dimensionless constant, however. We may now integrate Eq. 6.1.23 to find the pressure function. First we rewrite Eq. 6.1.23 to give the pressure gradient as

$$\frac{d\tilde{p}}{d\tilde{x}} = 12\left[\frac{1}{2\eta^2} - \frac{\lambda}{\eta^3}\right] \tag{6.1.24}$$

We then integrate term by term to find

$$\tilde{p}(\tilde{x}) = \tilde{p}(0) + 6\int_0^{\tilde{x}}\frac{d\tilde{x}}{\eta^2} - 12\lambda\int_0^{\tilde{x}}\frac{d\tilde{x}}{\eta^3} \tag{6.1.25}$$

We now impose boundary conditions on the pressure at the ends of the region $0 < x < L$:

$$\tilde{p}(\tilde{x}) = 0 \quad \text{at } \tilde{x} = 0 \quad \text{and} \quad \tilde{x} = \frac{L}{H_1} \equiv \Lambda \tag{6.1.26}$$

Note that we are *asserting* that the pressure is atmospheric at both ends of the bearing. Note, as well, the appearance of a new dimensionless geometrical parameter, Λ. For the geometry shown in Fig. 6.1.1,

$$\Lambda = \frac{1 - \kappa}{\tan\theta} \tag{6.1.27}$$

where $\kappa \equiv H_0/H_1$. Hence Λ is not independent of geometrical parameters already introduced.

We set $\tilde{p}(\Lambda) = 0$ in Eq. 6.1.25. [Remember to put Λ in the upper limit of the integrals when we set $\tilde{p}(\Lambda) = 0$.] This permits us to write λ as

$$\lambda = \frac{1}{2}\left[\frac{\int_0^\Lambda \frac{d\tilde{x}}{\eta^2}}{\int_0^\Lambda \frac{d\tilde{x}}{\eta^3}}\right] \tag{6.1.28}$$

When the indicated integrations have been performed, using Eq. 6.1.20 for η, we find a very simple result for λ:

$$\lambda = \frac{\kappa}{1 + \kappa} \tag{6.1.29}$$

[handwritten: $K = H_2/H_1$]

We see that λ depends strictly on the geometry. Now the pressure distribution may be written by performing the indicated integrations in Eq. 6.1.25. The result is found to be

[handwritten: $P^*(x)=$] $\tilde{p} = \frac{6\tilde{x}(\eta - \kappa)}{\eta^2(1 + \kappa)} \tag{6.1.30}$

Inspection of the pressure gradient (Eq. 6.1.24) indicates that there is a maximum pressure at a position defined by

$$\frac{d\tilde{p}}{d\tilde{x}} = 0 = 12\left[\frac{1}{2\eta^2} - \frac{\lambda}{\eta^3}\right] \tag{6.1.31}$$

[handwritten across page: For final velocity profile $\frac{dp^*}{dx^*} \Rightarrow U_x$ $Q = WUH_1\left(\frac{\eta}{2} - \frac{\eta^3}{12}\frac{dp^*}{dx^*}\right)$ Force $\int_0^1 dF = \int_0^1 W dx \, P$ plug in]

or

[handwritten: $[dA = Wdx]$]

$$\eta = 2\lambda \tag{6.1.32}$$

This corresponds to a position

$$\xi \equiv \frac{x}{L} = \frac{1}{1 + \kappa} \tag{6.1.33}$$

With these results we may show that the maximum pressure is

$$\tilde{p}^{\text{max}} = \frac{3\Lambda}{2}\frac{1 - \kappa}{\kappa(1 + \kappa)} \tag{6.1.34}$$

This completes the model for this planar lubrication flow. Let's review what we have found.

1. This nonparallel drag flow generates a pressure distribution under the fixed surface. If the flow were parallel, there would be no pressure variation.
2. The pressure reaches a maximum value somewhere between the inlet and outlet. Equation 6.1.33 gives the position of the maximum pressure. For small values of κ, the maximum is very close to the downstream end of the gap.
3. The maximum pressure increases with increasing viscosity, speed, and bearing length.

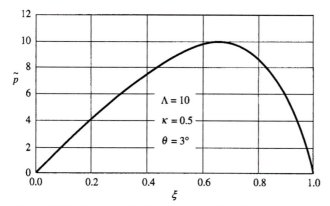

Figure 6.1.2 Pressure distribution in a slider bearing.

An example of a pressure distribution is shown in Fig. 6.1.2. While this flow field has the appearance of a drag flow, since no external pressure differential is applied across the ends of the system, the converging character of the flow generates a pressure gradient, and the resulting pressure distribution affects the velocity field. Thus this flow is a combination of drag flow and pressure flow. The velocity profiles at several positions in the bearing region are shown in Fig. 6.1.3, for the specific case of $\kappa = 0.5$. Notice that before the maximum in the pressure ($\xi < 0.67$ for the case $\kappa = 0.5$) the pressure acts *against* the flow, pushing it back in some loose sense. After this point ($\xi > 0.67$) the pressure gradient is in the direction of flow and aids the flow. Keep in mind that the flowrate is constant through the bearing region, so although the velocity profiles are distorted by the pressure field, the volumetric flowrate is not altered.

At one point along the bearing, the velocity profile is exactly linear. It is not difficult to show that this occurs at the point where the pressure is a maximum, hence where the pressure *gradient* vanishes. This is the point given by Eq. 6.1.33 above. To the left or right of this point the flow is against or with the pressure gradient, respectively, and the velocity profile is distorted accordingly (see Problem 6.78).

This pressure distribution exerts a vertical force against the stationary bearing surface, and thus the flow field is capable of supporting a load. The load per unit width is found from the vertical component of the pressure force, and so is given by

$$\overbrace{\Pi}^{\text{load per unit width}} = \int_0^L p(x) \cos \theta \, dx = \frac{6\mu U \cos \theta}{\tan^2 \theta} \left(-\ln \kappa - 2 \frac{1 - \kappa}{1 + \kappa} \right) \tag{6.1.35}$$

Lubrication flows are capable of supporting substantial loads, as Example 6.1.1 indicates.

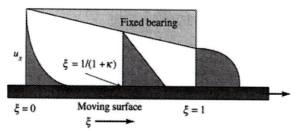

Figure 6.1.3 Velocity profiles in the bearing region, for the case $\kappa = 0.5$.

EXAMPLE 6.1.1   *Load Capacity of a Slider Bearing*

A slider bearing is lubricated with a liquid of viscosity 1.0 Pa·s. The relative motion of the surfaces is $U = 1.0$ m/s. The geometry is such that $L = 5$ mm, $H_o = 0.025$ mm, and $\theta = 3°$. We want to find the load-bearing capacity of this system. What is the Reynolds number for this flow?

Equation 6.1.27 may be rearranged to the form

$$\frac{1 - \kappa}{\kappa} = \frac{L}{H_o} \tan \theta \qquad \text{(6.1.36)}$$

from which we find

$$\kappa = 0.087 \qquad \text{(6.1.37)}$$

For an angle of $\theta = 3°$ we find $\cos \theta \approx 1$ and $\tan^2 \theta = 0.00275$. From Eq. 6.1.35 the load per unit width is

$$\Pi = 1650 \text{ N/m} \qquad \text{(6.1.38)}$$

(1650 newtons corresponds to the weight, on Earth, of 168 kg or 370 pounds. Thus a bearing with a width of 0.1 m would support a load of 165 N or 37 pounds.)

The Reynolds number has been assumed small in this lubrication analysis, to permit the neglect of the nonlinear (inertial) terms of Eqs. 6.1.8 and 6.1.9. For this case we find

$$\text{Re} = \frac{\rho U H_1}{\mu} = \frac{\rho U(L \tan \theta + H_o)}{\mu} = \frac{1000 \times 1 \times [0.005 \tan 3° + (2.5 \times 10^{-5})]}{1.0} = 0.29$$
$$\text{(6.1.39)}$$

While this value is less than unity, it is not small, and the neglect of the inertial terms is marginal in this case.

6.1.1   Examination of the Nearly Parallel Assumption

Equation 6.1.18 is based on the *assumption* that

$$\frac{\partial^2 \bar{u}_x}{\partial \bar{x}^2} \ll \frac{\partial^2 \bar{u}_x}{\partial \bar{y}^2} \qquad \text{(6.1.40)}$$

Let's assess the validity of this assumption. One way to do this is by direct calculation of the ratio

$$\Re \equiv \frac{\partial^2 \bar{u}_x}{\partial \bar{x}^2} \Big/ \frac{\partial^2 \bar{u}_x}{\partial \bar{y}^2} \qquad \text{(6.1.41)}$$

We begin with Eq. 6.1.19 for $\bar{u}_x$ and perform the indicated differentiations. Keep in mind that η is a function of $\bar{x}$ (Eq. 6.1.20). The denominator of $\Re$ above is simplified by recognizing that, from Eqs. 6.1.18 and 6.1.24,

$$\frac{\partial^2 \bar{u}_x}{\partial \bar{y}^2} = \frac{d\bar{p}}{d\bar{x}} = 12 \left[\frac{1}{2\eta^2} - \frac{\lambda}{\eta^3} \right] \qquad \text{(6.1.42)}$$

Hence our intermediate result is

$$\Re \equiv \frac{\partial^2 \bar{u}_x}{\partial \bar{x}^2} \Big/ \frac{\partial^2 \bar{u}_x}{\partial \bar{y}^2} = \frac{\left(\dfrac{10}{\eta^3} - \dfrac{36\lambda}{\eta^4} \right) \bar{y} \tan^2 \theta}{\dfrac{6}{\eta^2} - \dfrac{12\lambda}{\eta^3}} \qquad \text{(6.1.43)}$$

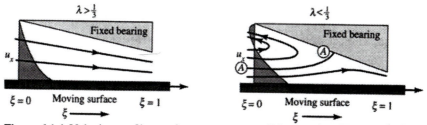

Figure 6.1.4 Velocity profiles at the entrance to a slider bearing. For $\lambda < \frac{1}{3}$, there is a line AA above which fluid recirculates.

This is a function of position, and we will evaluate $\Re$ at the downstream end of the bearing, where $\bar{y} = \eta = \kappa$. The result is

$$\Re = \frac{(5\kappa - 18\lambda) \tan^2 \theta}{[3\kappa(\kappa - 1)]/(\kappa + 1)} \tag{6.1.44}$$

It is not difficult to confirm that for small angles, typical of bearing surfaces, $\Re$ is indeed small compared to unity. For example, for $\kappa = 0.2$ we find $\Re = 0.038$ for $\theta = 5°$ and $\Re = 0.156$ for $\theta = 10°$. We conclude that the validity of the assumption in question (Eq. 6.1.40) depends primarily on how small the angle θ is. Of course this is the expected conclusion, but Eq. 6.1.44 provides a quantitative measure of the value of $\Re$.

6.1.2 Effect of Geometry on the Character of the Flow

Further analysis demonstrates that the character of the flow field depends on the parameter κ. The simplest way to demonstrate this is to examine the slope of the velocity profile at the upper surface, at the entrance to the bearing region, where $\eta = 1$. From Eq. 6.1.19 we find

$$\frac{\partial \bar{u}_x}{\partial \bar{y}} = -\frac{1}{\eta} - \frac{\eta^2}{2} \frac{d\bar{p}}{d\bar{x}} \left(\frac{1}{\eta} - \frac{2\bar{y}}{\eta^2} \right) \tag{6.1.45}$$

With Eq. 6.1.24 this becomes

$$\frac{\partial \bar{u}_x}{\partial \bar{y}} \bigg|_{\eta=1} = 2 - 6\lambda \tag{6.1.46}$$

From Eq. 6.1.29, λ can vary over the range $[0, 1]$. Hence the velocity gradient can be positive or negative, depending on whether λ is less than or greater than $\frac{1}{3}$. Further analysis along this line leads to the conclusion that the velocity field may include a recirculation region under some circumstances, depending on the value of κ. This is suggested in Fig. 6.1.4. A quantitative analysis of these profiles is left as a problem at the end of this chapter (Problem 6.78).

In most real bearings the loaded bearing "floats" on the lubricant film and takes a position ensuring that the load is balanced by the pressure distribution. Determination of the floating position is easily carried out, as Example 6.1.2 illustrates.

EXAMPLE 6.1.2   *Floating Position of a Slider Bearing*

A gross load of 10 kg is supported on a planar bearing with a length $L = 0.1$ m. The bearing angle is maintained at 3°. The width of the bearing is 0.1 m. The viscosity of the lubricant is 10 mPa · s. Our concern is that when the system is powered down, the linear speed U will fall to a level that will cause the bearing separation H_o to fall below 50 μm, at which point the lubricated surfaces could be damaged. Our goal is to determine the linear speed at which this would occur, with a view to designing a mechanical system for preventing near contact.

The simplest way to handle this computationally is to invert Eq. 6.1.35 to a form explicit in U:

$$U = \Pi \frac{\tan^2 \theta}{6\mu \cos \theta} \left(-\ln \kappa - 2\frac{1 - \kappa}{1 + \kappa} \right)^{-1} \tag{6.1.47}$$

From Eq. 6.1.36 we may solve for H_o in the form

$$H_o = \frac{L \kappa \tan \theta}{1 - \kappa} \tag{6.1.48}$$

With the parameters specified already, a choice of a particular value for κ yields a pair of values, from Eqs. 6.1.47 and 6.1.48, of $H_o(U)$. The load per unit width Π is found from

$$\Pi = \frac{10 \text{ kg} \times 9.8 \text{ m/s}^2}{0.1 \text{ m}} = 980 \text{ N/m} \tag{6.1.49}$$

Results are shown in Fig. 6.1.5. Note that although we desire a set of values of H_o as a function of U, the computation is made much easier by picking κ values (equivalent to picking H_o values) and solving directly for U. This avoids trial-and-error solution of κ from Eq. 6.1.47. It appears that $U = 16$ m/s is the critical speed below which the bearing separation H_o will fall below 50 μm.

Since the load becomes infinite as κ goes to zero, according to Eq. 6.1.35, this simple lubrication theory predicts that *any* load could be supported by a sufficiently thin film, at some fixed bearing angle. In reality, when the gap becomes too small, surface roughness causes solid–solid contact, which leads to wear and damage to the surfaces that are supposed to be lubricated. In addition, as the gap gets very small, the nature

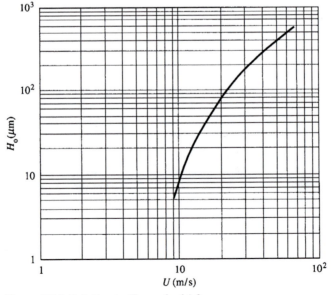

Figure 6.1.5 Solution to Example 6.1.2.

of the flow may change from laminar to turbulent, in which case our simple lubrication model will no longer hold.

For very small separations, pressures may get so high that the viscosity of the lubricant is altered, and the bearing surfaces may distort elastically in the region of the maximum pressure. In addition, frictional heating of the lubricant may occur, resulting in a reduction in viscosity, and possibly in thermal degradation of the fluid. A more complex form of lubrication theory is required to treat such complex, but important, problems. We do not treat such difficult problems in this introductory text.

We turn now to another nearly parallel flow.

6.2 LAMINAR FLOW THROUGH A LEAKY TUBE

There are a number of examples of technologically important flows in which a liquid passes down the axis of a tube of circular cross section under conditions such that the wall of the tube is permeable to the liquid. In Fig. 6.2.1, which shows the geometry and the anticipated flow field, the tube has a uniform inside radius R, and the length of the permeable section is L. Beyond an axial position $z = 0$, the tube wall is to be regarded as *uniformly* permeable, so that there is a continuous *volumetric flux j* (with units of $m^3/m^2 \cdot s$) of liquid through the wall. To be able to neglect the inertial terms in the Navier–Stokes equations, we assume that the flow along the tube axis is characterized by a low Reynolds number. We also assume that the system is at steady state. We assume the flow is axisymmetric: $\mathbf{u} = [u_r(r, z), 0, u_z(r, z)]; p = p(r, z)$.

Our goal is to develop a model with which to determine the fractional leakage associated with the flow, and the dependence of the leakage on design and operating parameters.

We begin with Table 4.3.1 (cylindrical coordinates) and write the r and z components as

$$0 = -\frac{\partial p}{\partial r} + \mu \left[\frac{\partial}{\partial r} \left(\frac{1}{r} \frac{\partial}{\partial r} (r u_r) \right) + \frac{\partial^2 u_r}{\partial z^2} \right] \tag{6.2.1}$$

and

$$0 = -\frac{\partial p}{\partial z} + \mu \left[\frac{1}{r} \frac{\partial}{\partial r} \left(r \frac{\partial u_z}{\partial r} \right) + \frac{\partial^2 u_z}{\partial z^2} \right] \tag{6.2.2}$$

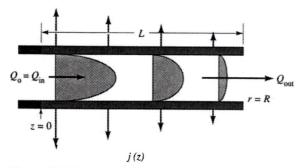

Figure 6.2.1 Laminar flow through a leaky tube. Because the pressure falls in the direction of flow, the leakage decreases along the axis.

We need a third equation (we have three unknowns—count them), and that is provided by the continuity equation:

$$0 = \frac{1}{r} \frac{\partial}{\partial r} (ru_r) + \frac{\partial u_z}{\partial z} \tag{6.2.3}$$

We need one additional equation that characterizes the leakage flow as a function of the internal pressure in the tube. We assume a linear relationship, and write

$$j(z) = K[p(z) - p_a] \tag{6.2.4}$$

where p_a is the pressure in the region outside the tube. This equation defines a *permeability coefficient K* of the tube wall. We may take $p_a = 0$, or we may regard p as the "gage" pressure, the pressure relative to the external pressure. We will assume here that the pressure outside the leaky tube is atmospheric.

Because there is a leakage flow across the tube wall, there is a *radial* velocity function. Right at the tube wall, this velocity is identical to the volumetric flux, so

$$j(z) = u_r(R, z) = u_R(z) \tag{6.2.5}$$

(This relationship will be useful as a boundary condition later.)

We will now introduce an important simplifying constraint or assumption. (Keep in mind that *assumptions are constraints.* The generality and utility of a solution based on a particular assumption is limited or constrained because of that assumption.) We will assume that the leakage flow is small in some sense that we will quantify later. Then the flow is almost parallel, except very near the wall, where the *axial* velocity gets small (to satisfy no-slip in the axial direction) and the *radial* velocity remains finite.

We now ask a question: If the leakage is small enough, could the axial velocity possibly have a functional dependence on radial position nearly the same as in the case of no leakage, but with a minor correction that accounts for leakage? For example, with Eqs. 3.2.33, 3.2.34, and 3.2.26 as a basis, we may propose to write the velocity profile for "leaky" Hagen–Poiseuille flow in the form

$$u_z = 2U(z) \left[1 - \left(\frac{r}{R} \right)^2 \right] = 2U_o \left[1 - \left(\frac{r}{R} \right)^2 \right] f(z) \tag{6.2.6}$$

Here we use the average velocity at the inlet of the tube, $U_o = Q_o / \pi R^2$ and introduce an unknown function $f(z)$. We may use the continuity equation to write u_r in terms of u_z:

$$u_r = -\frac{1}{r} \int_0^r r \frac{\partial u_z}{\partial z} \, dr \tag{6.2.7}$$

We now differentiate Eq. 6.2.6 with respect to z, put the result into the integral above, and perform the integration with respect to r. The result is

$$u_r = -RU_o f'(z) \left[\left(\frac{r}{R} \right)^2 - \frac{1}{2} \left(\frac{r}{R} \right)^4 \right] \tag{6.2.8}$$

where $f' = \partial f / \partial z$.

At the tube surface ($r = R$), we find

$$u_r(R, z) \equiv u_R(z) = -\frac{RU_o}{2} f' \tag{6.2.9}$$

Note what we have accomplished here: we have replaced the two unknowns u_r and u_z by the single unknown $f(z)$. But we still have to answer the question: Is this assumed form for u_z given in Eq. 6.2.6 compatible with the principle of conservation of momentum? To find out, we must substitute u_z in Eq. 6.2.2 and see if it is possible to find a solution for the function $f(z)$. (A key point to remember is that we are assuming that f is a function only of z.) Equation 6.2.2 gives us

$$\frac{\partial p}{\partial z} = \mu \left[-\frac{8U_o}{R^2} f + 2U_o \left(1 - \left(\frac{r}{R}\right)^2 \right) f'' \right] \tag{6.2.10}$$

From Eqs. 6.2.4, 6.2.5, and 6.2.9, we see that

$$\frac{\partial p}{\partial z} = \frac{1}{K} \frac{du_R}{dz} = -\frac{RU_o}{2K} f'' \tag{6.2.11}$$

Upon substituting this result into Eq. 6.2.10, we obtain the following differential equation for the function f:

$$f'' = \frac{16K\mu}{R^3} f - \frac{4K\mu}{R} \left[1 - \left(\frac{r}{R}\right)^2 \right] f'' \tag{6.2.12}$$

Since the radial variable r appears in this differential equation for f, it follows that f is a function of *both* z and r, which violates our original assumption that $f = f(z)$ only. Thus the assumed form of the axial velocity distribution, Eq. 6.2.6, does not satisfy conservation of z momentum. *Note, however, that if the dimensionless group $4K\mu/R$ is small compared to unity, the "offending" term (the term with the r dependence) can be neglected relative to the second-derivative term on the left-hand side of the equation.* We claim (or at least hope), then, that Eq. 6.2.6 is a good approximation if the leakage is small enough, and specifically if

$$\frac{4K\mu}{R} \ll 1 \tag{6.2.13}$$

The function $f(z)$ must then be found as the solution to the following ordinary differential equation (Eq. 6.2.12 with the approximation from Eq. 6.2.13):

$$f'' - \frac{16K\mu}{R^3} f = 0 \tag{6.2.14}$$

We require two boundary conditions. One states that at the entrance to the leaky section (i.e., at $z = 0$), the flow is fully developed and given by the classical Hagen–Poiseuille solution. This is equivalent to the condition

$$f = 1 \quad \text{at} \quad z = 0 \tag{6.2.15}$$

The second condition is simply that

$$f \text{ remains finite} \quad \text{for} \quad z \to \infty \tag{6.2.16}$$

Equation 6.2.14 is easily solved then, and the solution that satisfies these boundary conditions may be written as

$$f = \exp\left[-4\beta \frac{z}{R} \right] \tag{6.2.17}$$

where a dimensionless permeability parameter is defined as

$$\beta = \left(\frac{K\mu}{R} \right)^{1/2} \tag{6.2.18}$$

Our next task is to find the axial profile of pressure, $p(z)$.

From Eqs. 6.2.11 and 6.2.14 we find

$$\frac{dp}{dz} = -\frac{8\mu U_o}{R^2} f = -\frac{8\mu U_o}{R^2} \exp\left(-4\beta \frac{z}{R}\right) \tag{6.2.19}$$

When we integrate this differential equation with the boundary condition

$$p = p_o \quad \text{at} \quad z = 0 \tag{6.2.20}$$

we find that the pressure distribution down the tube axis is

$$p(z) = p_o + \frac{2\mu U_o}{\beta R}\left[\exp\left(-4\beta \frac{z}{R}\right) - 1\right] \tag{6.2.21}$$

We have defined the pressure p_o relative to atmospheric pressure. If we assume that at the end of the tube section, right at $z = L$, the pressure is atmospheric (this is equivalent to neglecting any pressure losses downstream of the tube end), then upon setting $p = 0$ at $z = L$, we find

$$p_o = \Delta P = \frac{2\mu U_o}{\beta R}\left[1 - \exp\left(-4\beta \frac{L}{R}\right)\right] \tag{6.2.22}$$

This permits us to write the pressure distribution in the form

$$p(z) = \frac{2\mu U_o}{\beta R}\left[\exp\left(-4\beta \frac{z}{R}\right) - \exp\left(-4\beta \frac{L}{R}\right)\right] \tag{6.2.23}$$

We may now find the local leakage flux from Eq. 6.2.4, with the result

$$j = Kp(z) = 2\beta U_o\left[\exp\left(-4\beta \frac{z}{R}\right) - \exp\left(-4\beta \frac{L}{R}\right)\right] \tag{6.2.24}$$

Over the length L the *total* leakage flow is given by

$$Q_L = \int_0^L 2\pi R j(z)\, dz = \pi R^2 U_o\left[1 - \left(1 + 4\beta \frac{L}{R}\right)\exp\left(-4\beta \frac{L}{R}\right)\right] \tag{6.2.25}$$

The fractional loss of fluid over the length L is

$$\phi_L = \frac{Q_L}{Q_o} = \left[1 - \left(1 + 4\beta \frac{L}{R}\right)\exp\left(-4\beta \frac{L}{R}\right)\right] \tag{6.2.26}$$

The model developed here is based on the approximation that permitted us to simplify Eq. 6.2.12. It turns out that if Eq. 6.2.13 holds, the results that follow yield good approximations to the dynamics of this flow. If we examine the model more carefully, however, we find some discrepancies. Problem 6.9 allows us to explore one of these issues and see that there is something fundamentally wrong with the analysis, even though it provides a useful approximation to the behavior of the system. In Section 6.9 we approach this problem again from a more formal and careful direction, and obtain an improved approximate solution.

With the foregoing reservation in mind, we may now examine some practical consequences of our model of leaky flow.

EXAMPLE 6.2.1   A Hollow Fiber Water Purifier

A water purification system is designed along the lines suggested in Fig. 6.2.2. Feed water containing a contaminant enters the module at a flowrate Q_{feed} and a pressure

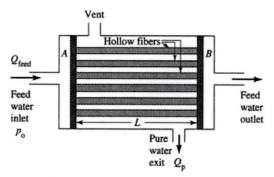

Figure 6.2.2 A hollow fiber membrane module.

p_o relative to atmospheric pressure. From a common header (A), the feed water is distributed uniformly to a set of parallel hollow fiber membranes. The membrane material is permeable to water, but impermeable to the specific contaminant. Hence pure water permeates (passes across the wall of) the hollow fibers and collects at the bottom of the module. The module is vented to atmospheric pressure. The flowrate of the permeate is denoted by Q_p. At the right-hand end the feed water (now slightly more concentrated in contaminant) enters a header (B) and goes to a wastewater drain. The header at B is at atmospheric pressure. There are no significant pressure losses in the header section. The following design and operating data are available to us.

$$p_o = 4 \times 10^5 \text{ N/m}^2 \qquad L = 10 \text{ cm} \qquad R = 50 \text{ } \mu\text{m}$$
$$L/R = 2000 \qquad K = 2 \times 10^{-12} \text{ (m}^3\text{/s} \cdot \text{m}^2\text{)/(N/m}^2\text{)}$$

The module contains 10,000 fibers.

We want to calculate the rate of production of pure water. We assume that there are no osmotic phenomena, and no fouling or plugging of the membranes.

Using a nominal viscosity of 10^{-3} Pa·s for water we find

$$\beta = \left(\frac{K\mu}{R}\right)^{1/2} = \left[\frac{(2 \times 10^{-12}) \times 10^{-3}}{50 \times 10^{-6}}\right]^{1/2} = 6.32 \times 10^{-6} \qquad \textbf{(6.2.27)}$$

This is small enough to satisfy the criterion given in Eq. 6.2.13.

From Eq. 6.2.26 we find the fraction of the feed water that is produced as permeate:

$$\phi_L = 1.24 \times 10^{-3} \qquad \textbf{(6.2.28)}$$

The feed water flow has not been specified. The feed pressure is given, and the feed water flow must be consistent with the available pressure drop of $p_o = 4 \times 10^5$ N/m². We must return to the mathematical model to find the Q_o-versus-ΔP relationship. Setting $Q_o = \pi R^2 U_o$ in Eq. 6.2.22, we find the desired model for $Q_o(\Delta P)$:

$$Q_o = \frac{\pi \beta R^3 \Delta P}{2\mu} \frac{1}{1 - \exp\left(-4\beta \dfrac{L}{R}\right)} \qquad \textbf{(6.2.29)}$$

This is the volumetric flowrate in a single tube of the module. Since the tubes are in parallel, connected to a common header at the pressure p_o, the flowrate is the same in each tube. Returning now to the specific example at hand, we find from Eq. 6.2.29 that

$$Q_o = 10^{-8} \text{ m}^3\text{/s} \qquad \textbf{(6.2.30)}$$

Since the module consists of 10,000 fibers in parallel, the total feed rate is

$$Q_{feed} = 10^{-4} \text{ m}^3/\text{s} \tag{6.2.31}$$

and the permeate rate is

$$Q_P = \phi_L Q_{feed} = 1.24 \times 10^{-7} \text{ m}^3/\text{s} = 0.124 \text{ cm}^3/\text{s} \tag{6.2.32}$$

This unit can produce pure water at the rate of approximately 0.45 L/h.

The model just developed for the performance of a hollow fiber water purification unit is quite useful for design purposes. We can explore the effect of fiber radius on performance, and in general examine the effects of various changes in design and operating conditions. Some problems at the end of this chapter address these issues.

It is interesting at this stage to demonstrate the use of a very similar mathematical model in a second application—one that arises in the field of biomedical technology.

6.3   A DEVICE FOR THE TREATMENT OF HYDROCEPHALUS

Hydrocephalus is a pathological condition that occurs (most often in children) when an obstruction blocks the normal drainage of cerebrospinal fluid (CSF) from the lateral ventricle of the brain. Appropriate treatment is directed toward providing a shunt to relieve the pressure by enabling the CSF to bypass the blockage. In this section we apply hydrodynamic analysis to the design of a proposed tube shunt. Figure 6.3.1 shows the shunt in place and a detail of its design. The shunt is manufactured from a hypodermic tube of inside diameter 0.5 mm and wall thickness 60 μm. In one section, the wall is perforated by a large number of cylindrical holes of diameter 18 μm, and axial length (just the wall thickness) 60 μm.

A typical pressure in the lateral ventricle is 100 mmHg above the pressure at the downstream end of the shunt. This is the pressure that forces CSF into the tube through the holes and thence down the tube axis.

Our goal is to develop a model for the capacity of the shunt to relieve the buildup of pressure. The first step toward this goal is to obtain a $Q(\Delta P)$ relationship.

At any axial position z_1 (see Fig. 6.3.2), the pressure inside the large tube will be p_1, and the pressure drop $P_v - p_1$ arises from viscous losses through the holes at z_1. The same is true at the axial position z_2. At the same time, the pressure drop $p_1 - p_2$ will arise to balance the flow within the large tube over the section $z_2 - z_1$. Thus the interior flow is coupled to the flow through the holes. We are going to develop a mathematical model for this system by a method that is very commonly used in problem solving. We will

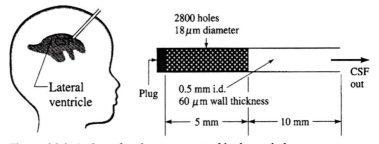

Figure 6.3.1 A shunt for the treatment of hydrocephalus.

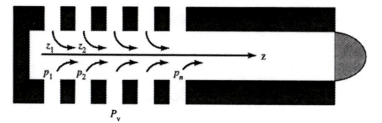

Figure 6.3.2 Detail of the perforated region of the shunt.

look first at isolated pieces of the problem, searching for clues to useful but reasonable approximations that may simplify the development of the model. Then we will apply existing models of simple flows to this more complex flow.

Let's begin the process by looking at Fig. 6.3.2. The figure is not drawn to scale, but we know that the inside diameter of the shunt (0.5 mm) is many times greater than the inside diameter of the perforations (18 μm). We know from the Hagen–Poiseuille law that the pressure drop across a tube depends very strongly on the tube diameter. This suggests the possibility that the interior flow occurs with a pressure loss that is very small in comparison to that across the perforations. In the extreme case, if we were to neglect the pressure loss in the 5 mm length of perforated section, the pressures p_1, p_2, ... would be identical. Let's make that assumption, not because we can argue its validity here, but because we want to see what it leads to.

The pressure drop across each small hole could be calculated from the Hagen–Poiseuille law as

$$P_v - p_i = \frac{128\mu L_{\text{hole}}}{\pi D_{\text{hole}}^4} q_i \tag{6.3.1}$$

If we completely neglect the pressure loss associated with the interior flow, we are assuming that $p_i = 0$ all along the axis of the perforated section. This may not be a good assumption, but we cannot assess it without modeling its features on a conditional basis and then examining the consequences. This will let us solve for q_i for each and every one of the 2800 holes in the perforated section, since all the holes are identical.

The viscosity of CSF at 38°C is 0.00103 Pa·s. The external pressure P_v in units of millimeters of mercury must be converted to pascals (= N/m^2). We find

$$P_v = 100 \text{ mmHg} \times 133.3 \frac{\text{N/m}^2}{\text{mmHg}} = 1.33 \times 10^4 \text{ N/m}^2 \tag{6.3.2}$$

and it follows from Eq. 6.3.1 (and the data in Fig. 6.3.1) that for each hole

$$q_i = 5.54 \times 10^{-10} \text{ m}^3/\text{s} \tag{6.3.3}$$

Since there are 2800 such holes, the total flow that will be delivered to the nonperforated section is

$$Q_o = 1.55 \times 10^{-6} \text{ m}^3/\text{s} \tag{6.3.4}$$

Now let's see what kind of pressure loss is to be expected from such a flowrate in the portion of the 0.5 mm i.d. tube that has no perforations. From Eq. 6.3.1 (but replace q_i with Q_o as the total internal flow), we find (using $L = 10$ mm and $D = 0.5$ mm for the nonperforated section of the tube)

$$\Delta p = 1.04 \times 10^4 \text{ N/m}^2 \tag{6.3.5}$$

This pressure drop is not negligible compared to that (see Eq. 6.3.2) across the perforations. Hence the assumption of negligible pressure loss along the tube axis is not valid.

What we need, then, is a model for flow down the axis of a perforated tube that couples the flow across the wall with the flow down the axis. But we have just such a model in Section 6.2. There *are* some differences—the flow is *into* a tube *closed at one end,* in this case. Still, we should be able to make use of our understanding of the leaky tube problem in developing the model for the perforated shunt.

Our first step is to assign a value to the permeability K, in terms of the discrete geometry of the perforated section. We go back to the definition of K in Eq. 6.2.4, where j is the volumetric flux—a volume flowrate per unit area. For an individual hole in the wall of the tube, this is just the average velocity as given, for example, by the Hagen–Poiseuille law:

$$u_i = j = \frac{D_{hole}^2}{32\mu L_{hole}} \Delta p \tag{6.3.6}$$

or

$$K' = \frac{D_{hole}^2}{32\mu L_{hole}} \tag{6.3.7}$$

However, this is for the isolated hole. It does not account for the fact that the holes occupy only a fraction of the area of the tube. If we define α as the fraction of open (perforated area) in the perforated section of the tube, then

$$K = \frac{\alpha D_{hole}^2}{32\mu L_{hole}} \tag{6.3.8}$$

The perforated section contains 2800 holes in a section of length 5 mm, so

$$\alpha = \frac{2800(\pi D_{hole}^2/4)}{\pi D_{tube} L_{tube}} = \frac{700(18 \times 10^{-6})^2}{0.0005 \times 0.005} = 0.091 \tag{6.3.9}$$

and we find

$$K = (1.49 \times 10^{-5})(m^3/s \cdot m^2)/(N/m^2) \tag{6.3.10}$$

Now that we have all the physical properties in hand, we must go back to our leaky tube model and alter it for the case of the tube *closed* at one end, with the flow *entering* the tube along just one section of the tube. Figure 6.3.3 shows this situation. There is no flow across the boundary at $z = 0$; it is a closed boundary. In the region $0 < z < L_p$ we have the leaky tube. In the region $L_p < z < L_p + L_n$ there is no further leak into the tube, and the flowrate does not vary in the z direction.

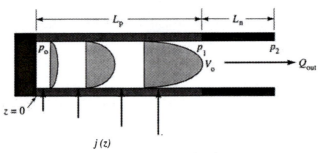

$j(z)$

Figure 6.3.3 Schematic of flow into the shunt.

If we go back over the analysis of flow out of a leaky tube (Section 6.2) we can identify certain equations that hold in the case we are now considering. For example, Eqs. 6.2.6 through 6.2.14 still apply. The first change that we have to make is in the boundary conditions of $f(z)$. Since there is no flow across $z = 0$, we write

$$f = 0 \quad \text{at} \quad z = 0 \tag{6.3.11}$$

We do not know the flow at the boundary $z = L_p$, but we will denote the value of f at that position by f_p and write a second boundary condition as

$$f = f_p \quad \text{at} \quad z = L_p \tag{6.3.12}$$

Then the general solution to Eq. 6.2.14 is written as

$$f = a \exp\left(-8\beta \frac{z}{D}\right) + b \exp\left(+8\beta \frac{z}{D}\right) \tag{6.3.13}$$

and the integration constants a and b are found from the boundary conditions above. This lets us write $f(z)$ (after converting exponential functions to hyperbolic sine functions) as

$$f = f_p \frac{\sinh[8\beta(z/D)]}{\sinh[8\beta(L_p/D)]} = f_p \frac{\sinh(\lambda z/L_p)}{\sinh \lambda} \tag{6.3.14}$$

where $\lambda \equiv 8\beta(L_p/D)$ and β is still defined as

$$\beta = \left(\frac{K\mu}{R}\right)^{1/2} \tag{6.3.15}$$

Note that we have chosen to replace the exponential functions by hyperbolic functions, after using the relationship

$$\sinh u = \frac{e^u - e^{-u}}{2} \tag{6.3.16}$$

To evaluate the second integration constant, or equivalently, to find f_p, we must consider the flow in the nonperforated section. We assume that the Hagen–Poiseuille law holds in that section, and that the velocity field is given by Eq. 6.2.6, with $f = 1$. (But we do not yet know U_o!) What we can say is that the flow leaving the end of the perforated section is just the flow entering the nonperforated section, so

$$f_p = 1 \tag{6.3.17}$$

This is equivalent to defining U_o to be the average velocity *leaving* the downstream end of the shunt. Our task now is to find U_o. We will accomplish this by considering the pressure distribution along the tube axis. Equation 6.2.11 is still valid, but we must apply it to the $f(z)$ function that holds for the present case (i.e., Eq. 6.3.14). When we do this, and the dust from the algebra clears away, we find the pressure profile in the perforated section in the form

$$p(z) = p_o - \frac{4\mu U_o}{\beta D \sinh \lambda}\left(\cosh \frac{\lambda z}{L_p} - 1\right) \tag{6.3.18}$$

This equation contains two unknowns: U_o and p_o. At the end of the perforated section, at $z = L_p$, the pressure is p_1, and from Eq. 6.3.18 we find

$$p_1 = p_o - \frac{4\mu U_o}{\beta D \sinh \lambda}(\cosh \lambda - 1) \tag{6.3.19}$$

We have introduced a third unknown: p_1. We need more equations. We have a second equation for p_1, since we expect (or assume, anyway) that the Hagen–Poiseuille law holds in the nonperforated section. Hence

$$p_1 = p_2 + \frac{32\mu U_o L_n}{D^2} \tag{6.3.20}$$

where p_2 is the pressure at the downstream end, which we assume is known to us from independent physiological measurements, just as is the case with the pressure P_v. At this point we have two equations with the three unknowns. *Don't forget what we are after—we want the velocity U_o since this will give us the drainage rate Q_{out}.*

The final equation comes from the permeability law. If we assume a linear permeability relationship, as we did in writing Eq. 6.2.4 for the leaky tube problem earlier, then the flux is

$$j = K[P_v - p(z)] \quad \text{in the region} \quad 0 \le z \le L_p \tag{6.3.21}$$

With this assumption, and the pressure function given in Eq. 6.3.18, we may write a mass balance that equates the flux across the permeable section of the shunt to the flow that leaves the downstream end of the shunt:

$$Q_{out} = \pi R^2 U_o = \int_0^{L_p} \pi R K(P_v - p(z)) \, dz \tag{6.3.22}$$

This integration yields

$$P_v - p_o = \frac{4\mu U_o}{\beta D \sinh \lambda} \tag{6.3.23}$$

With Eqs. 6.3.19 and 6.3.20, we may eliminate the pressure p_o and solve for U_o. The result may be written in the form

$$\frac{32\mu U_o}{D^2} = \frac{(P_v - p_2)/L_n}{1 + (L_p/L_n)(\coth \lambda)/\lambda} \tag{6.3.24}$$

The volumetric flowrate then follows from

$$Q_{out} = \pi R^2 U_o = \frac{\pi D^4}{128\mu} \frac{\Delta P}{L_n + L_p} \frac{1 + L_p/L_n}{1 + (L_p/L_n)(\coth \lambda)/\lambda} \tag{6.3.25}$$

When we write the solution in the form of Eq. 6.3.24 we see that there are two dimensionless parameters that control the performance of this system. One is the geometrical ratio L_p/L_n. The other is λ. Keep in mind that λ, defined following Eq. 6.3.14, includes the parameter β.

We have now completed the model for a hydrocephalus shunt. What do we do with it? We test it against data. Fortunately, such data are available. This is not always the case when we are "building" models, and we often have to couple an experimental program with model building.

EXAMPLE 6.3.1 **In Vitro *Test of a Hydrocephalus Shunt***

Experiments have been performed on a perforated microtubule designed as a hydrocephalus shunt. The original reference should be consulted for details of the experimental design [Cho and Back, *J. Biomech.* **22,** 335 (1989)]. Data are shown in Fig. 6.3.4 for *square* perforations, 18 μm on a side. The perforations were on only one side of the

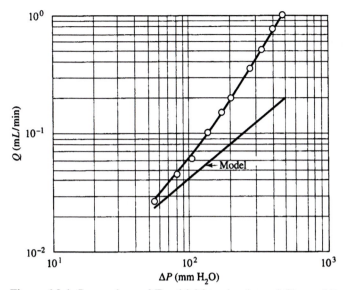

Figure 6.3.4 Comparison of Eq. 6.3.26 to the data of Cho and Back.

tube. Hence the experimental flow was not axisymmetric, as assumed in our model. Nevertheless it is interesting to use our model to calculate a pressure drop–flowrate curve and compare our simple model to these data.

The dimensions given in Fig. 6.3.1 are essentially those of the microtubule used in the experiments described by Cho and Back. (They give $L_p = 4.9$ mm and $L_n = 9$ mm; D is stated to be in the range 0.460–0.580 mm. Because of the strong dependence of flow on the tube diameter D, our predictions would vary considerably with our choice of a single value for D. We have chosen $D = 0.5$ mm as a nominal value.) We will ignore the difference between a square and a circular perforation in this example. The effect is fairly small, and we leave the details to a problem at the end of the chapter. Equations 6.3.24 and 6.3.25 lead to the following model:

$$Q_{out} = \frac{\pi D^4}{128\mu} \frac{\Delta P/L_n}{1 + (L_p/L_n)(\coth \lambda)/\lambda} \tag{6.3.26}$$

where ΔP is the difference in pressure between the fluid outside the shunt and the drainage end of the shunt.

When the various geometrical parameters are introduced, and we use a viscosity for water of 0.001 Pa · s, we predict

$$Q_{out} = 6.5 \times 10^{-11} \, \Delta P \, \text{m}^3/\text{s} \tag{6.3.27}$$

when ΔP is in pascals. This line is shown on Fig. 6.3.4, but in the units used by the original investigators.

Inspection of Fig. 6.3.4 reveals several points of departure between the model and the data. Foremost, we see that the slope of the data indicates that Q is roughly proportional to the square of ΔP, while the model predicts a liner dependence. This nonlinearity cannot be ascribed to the difference between square and circular holes, since low Reynolds number flow through a tube of *any* cross section would be linear in the pressure drop. (This assertion is essentially proven in Chapter 5, Section 5.3.)

We do not know why the data depart so dramatically from this model. Close reading of the original reference offers a possible clue. If the test liquid contained particulate matter of a size of the order of the hole size—18 μm—then the holes could become plugged. Hence all the holes might not be open and available to pass fluid. Higher pressures could force particles through the holes, thereby increasing the number of active holes. This would give rise to a flowrate that increased more than linearly with the pressure drop. We must emphasize, however, that this is only a speculation. The idea is consistent with the observation of Cho and Back that in a second experiment in which the holes were much larger—40 μm—the data for Q versus ΔP were observed to be linear.

6.4 SPREADING OF A VERY VISCOUS DROP

The spreading of a drop of a highly viscous liquid across a surface is a common occurrence in some problems of technological interest to us. When drops of liquid are delivered to a surface, for example, we may want to determine how rapidly the liquid spreads across the surface. Another (though much more complex) example is the spread of oil that has been released into the sea from a broken pipe, cracked reservoir, or damaged ship. In the latter stages of spreading, the liquid film is often thin by comparison to its radius, and the flow is in nearly parallel planes. We illustrate here the development of a model for the spreading of a viscous liquid on a *solid* surface, when the spreading is driven by gravity and opposed by viscous effects. We neglect surface tension forces, an assumption that is reasonable for large drops of high viscosity. We neglect inertial effects, consistent with our statement that we are considering the spreading of a *highly viscous* liquid. We imply that the viscosity is so high that the flow is very slow. Although this is an unsteady state flow that we are considering (the radius of the drop increases with time), we will neglect the acceleration term on the left-hand side of the Navier–Stokes equations, consistent with the notion that this flow is "slow" in some sense.

Figure 6.4.1 shows the geometry for the analysis. An axisymmetric drop of volume V is instantly created at time $t = 0$. Our goal is to develop a model for the radius of the drop as a function of time: $R(t)$. We begin with the assumptions that the flow is quasi-steady, gravity/viscous-dominated, and nearly parallel. Inspection of Eq. 4.3.24d leads us to begin with the following equation for the radial velocity:

$$\frac{1}{\rho}\frac{\partial p}{\partial r} = \nu\frac{\partial^2 u_r}{\partial z^2} \tag{6.4.1}$$

where $\nu = \mu/\rho$ is the kinematic viscosity. This equation reflects the statement, implied above, that the primary forces in balance are those of gravity, through a hydrostatic pressure term, and viscous shear. Notice that since we are examining the *radial* component of the momentum balance, the gravitational term appears only through its influence on the radial pressure gradient. Because of the assumption of "nearly parallel flow," we have neglected the set of terms $\partial[(1/r)\partial(ru_r)/\partial r]/\partial r$ compared to $\partial^2 u_r/\partial z^2$.

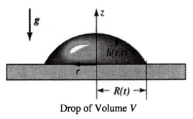

Drop of Volume V

Figure 6.4.1 An axisymmetric drop spreads across the plane $z = 0$.

There is a pressure distribution within the drop that arises solely from gravity if we neglect the effects of surface tension. We may write this hydrostatic pressure as

$$p = p_o + \rho g(h - z) \qquad (6.4.2)$$

Then Eq. 6.4.1 becomes (using Eq. 6.4.2 to eliminate p)

$$g\frac{\partial h}{\partial r} = \nu\frac{\partial^2 u_r}{\partial z^2} \qquad (6.4.3)$$

The continuity equation for this incompressible axisymmetric flow is simply

$$\frac{\partial u_z}{\partial z} + \frac{1}{r}\frac{\partial}{\partial r} ru_r = 0 \qquad (6.4.4)$$

If we integrate Eq. 6.4.4 over the definite limits $z = 0$ to $z = h(r, t)$, we find

$$[u_z(h) - u_z(0)] + \frac{1}{r}\frac{\partial}{\partial r} r \int_0^h u_r\, dz = 0 \qquad (6.4.5)$$

Since the lower surface is assumed impermeable, $u_z(0) = 0$. We may replace $u_z(h)$ with $\partial h/\partial t$. Hence we find an expression for conservation of mass of the drop in the form

$$\frac{\partial h}{\partial t} + \frac{1}{r}\frac{\partial}{\partial r}\left[r \int_0^h u_r\, dz \right] = 0 \qquad (6.4.6)$$

There is another statement of conservation of mass, which simply takes note of the following property: whatever the shape of the drop, its volume remains fixed, as indicated in Eq. 6.4.7.

$$2\pi \int_0^{R(t)} rh\, dr = V \qquad (6.4.7)$$

Now we may integrate Eq. 6.4.3 twice with respect to z and impose the boundary conditions

$$u_r = 0 \quad \text{on} \quad z = 0 \quad \text{(no slip on the solid surface)} \qquad (6.4.8)$$

and

$$\tau_{rz} = -\mu \left.\frac{\partial u_r}{\partial z}\right|_{z=h} = 0 \quad \text{[no shear at the upper surface, } h(r, t)] \qquad (6.4.9)$$

Upon completion of the integration of Eq. 6.4.3, we find the radial velocity in the form

$$u_r = -\frac{g}{2\nu}\frac{\partial h}{\partial r} z(2h - z) \qquad (6.4.10)$$

When this is introduced into Eq. 6.4.6 and the indicated integration is performed, we find a differential equation for $h(r, t)$:

$$\frac{\partial h}{\partial t} - \frac{g}{3\nu}\frac{1}{r}\frac{\partial}{\partial r}\left(rh^3\frac{\partial h}{\partial r} \right) = 0 \qquad (6.4.11)$$

This is a nonlinear partial differential equation in $h(r, t)$. It turns out that it can be solved analytically, with boundary conditions $h = 0$ at $r = R$ and $\partial h/\partial r = 0$ at $r = 0$, and we present the solution (not the procedure) here:

$$h(t) = 0.531\left(\frac{3\nu V}{gt}\right)^{1/4}\left[1 - \left(\frac{r}{R}\right)^2\right]^{1/3} \qquad (6.4.12)$$

After this result has been introduced into Eq. 6.4.7, integration and some algebraic manipulation yields an expression for $R(t)$ in the form

$$\left(\frac{3\nu}{gV^3}\right)^{1/8} R(t) = 0.894 \, t^{1/8} \tag{6.4.13}$$

Thus we predict that the drop will spread slowly, in the sense that R is proportional to the one-eighth power of time.

Experimental data on the spreading of silicone oil drops across a flat horizontal acrylic sheet were presented by Huppert [*J. Fluid Mech.*, **121**, 43 (1982)], who also offered the theoretical analysis outlined here. If you want to follow the details of the solution of Eq. 6.4.11 and the subsequent derivation of Eq. 6.4.13, you will need to put some effort into a study of Huppert's paper. The data are shown in Fig. 6.4.2, and the agreement with the theory is excellent.

Viscous drop spreading is a good example of a complex flow (unsteady, two-dimensional, with an unknown free surface) that can be modeled if one is willing to make rational assumptions about the physics. As always, the bottom line is the test of the correspondence of the model to the observations. For example, the set of data shown in Fig. 6.4.3 is not well described by Eq. 6.4.13. The short time data are very far removed from the predicted behavior. As time increases and the drop spreads further, the data approach the theoretical model.

It is not hard to see that the model might be very poor at short times, but the foregoing statement leaves open the definition of "short time." Let's return to Eq. 6.4.12 for a moment and note the behavior of $h(t)$ as t gets very small. In the limit of vanishingly small time, we see that h becomes unbounded, which certainly violates good physics. Yet the evidence of Fig. 6.4.2 suggests that the model is sound over some range of times. It must be that Eq. 6.4.12 is limited to "large" times in some sense. Since we have assumed from the outset that the flow in the spreading drop is almost parallel, we can expect that the model will make sense only when the height of the drop is very small compared to its radial extent. A simple criterion would be

$$\frac{h(t, r = 0)}{R(t)} < 0.1 \tag{6.4.14}$$

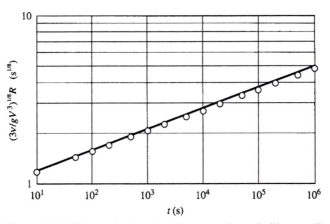

Figure 6.4.2 Huppert's data for the spreading of silicone oil drops. The line is Eq. 6.4.13.

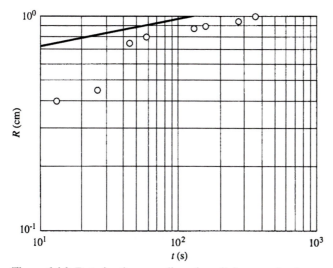

Figure 6.4.3 Data for the spreading of an oil drop on aluminum: $V = 0.024$ cm^3, $\mu = 0.225$ poise, $\rho = 0.841$ g/cm^3. (Mickailey, unpublished data.)

Using Eqs. 6.4.12 and 6.4.13, it is not difficult to show (see Problem 6.23) that a criterion for applying this model, in terms of the time t, may be written in the form

$$\frac{V^{1/3}gt}{\nu} \geq 350 \qquad (6.4.15)$$

Although this criterion is satisfied by all the data shown in Fig. 6.4.3, we find that Eq. 6.4.14 is not satisfied. If we take the shape of the spreading drop to be disklike, with a uniform height $\bar{h}$, then the height of the drop is given approximately by

$$\bar{h} = \frac{V}{\pi R^2} = \frac{0.024}{\pi R^2} \qquad (6.4.16)$$

At the shortest time, where $R = 0.4$ cm, we find $\bar{h}/R = 0.12$, which is close to, but does not quite meet, the criterion given in Eq. 6.4.14. Hence this second set of data (Fig. 6.4.3) provides an appropriate precaution: when a simple model is used to describe a complex flow field, it is important to test the model against data obtained over a broad range of parameters.

6.5   THE HYDRODYNAMIC ENTRY LENGTH FOR A FILM FLOWING ALONG A SURFACE: AN EXAMPLE OF AN INTEGRAL ANALYSIS

In Chapter 4 we considered the gravity-driven flow of a liquid in a thin film attached to one side of a planar surface. We derived a model for the velocity profile, Eq. 4.4.26, and the relationship of the film thickness to the flowrate, Eq. 4.4.27. In that model a key assumption was that the film thickness was independent of the position x in the direction of the flow. It is instructive to develop a model now that gives us some insight into the sort of "entry length" required from the point of origin of the free surface to the region in which the constant film thickness assumption is valid. Figure 6.5.1 shows the geometry. For generality we permit the plane to be at some angle α to the horizontal.

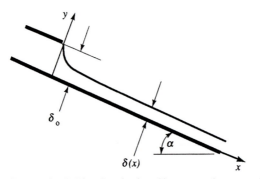

Figure 6.5.1 The developing film on a planar surface.

We assume that the liquid is introduced through a slot at $x = 0$, so that the initial film thickness δ_0 is well defined and known.

The analysis proceeds using a so-called integral method, whereby the Navier–Stokes equations are essentially *averaged* across the liquid film. Then, instead of solving for the axial velocity profile $u_x(x, y)$, we do something quite odd—*we assume a form for the velocity profile*. That being the case, what are we solving for? The answer is that we manipulate the equations into a form such that the only unknown variable is the film thickness as a function of position, $\delta(x)$.

Let's begin by writing the continuity equation, and the momentum equation in the direction of flow, for steady Newtonian laminar flow:

$$\frac{\partial u_x}{\partial x} + \frac{\partial u_y}{\partial y} = 0 \tag{6.5.1}$$

$$\rho\left(u_x \frac{\partial u_x}{\partial x} + u_y \frac{\partial u_x}{\partial y}\right) = -\frac{\partial p}{\partial x} + \mu\left(\frac{\partial^2 u_x}{\partial x^2} + \frac{\partial^2 u_x}{\partial y^2}\right) + \rho g \sin \alpha \tag{6.5.2}$$

Note that we retain the inertial terms on the left-hand side of the x-momentum equation. These terms vanish identically for uniform film flow (review Section 4.4.2), but not for the case of nonuniform film thickness.

Since we have three unknowns (u_x, u_y, and p), we should write a third equation (the y-momentum equation); however, we are going to eliminate pressure from the analysis by the same arguments used in Section 4.4.2 when we developed the model for uniform film thickness. This is our first mention of the notion that the flow is nearly parallel. In effect, we are stating that momentum transfer in the y direction is incidental to the dynamics of this flow as long as the flow is largely in the x direction—nearly parallel to the planar surface. Then $\partial p/\partial y$ is neglected, and if we ignore surface tension effects that might arise from the curvature of the free surface near the exit region, the internal pressure must match the external pressure, which is uniform along the free surface. Hence we make the assumption $\partial p/\partial x = 0$.

For nearly parallel flow, the viscous momentum flow term $\partial^2 u_x/\partial x^2$ may be neglected by comparison to the momentum flow that arises from the shear stress term ($\partial^2 u_x/\partial y^2$); as a result, the mathematical model of this flow takes the form of Eq. 6.5.1 coupled with

$$\rho\left(u_x \frac{\partial u_x}{\partial x} + u_y \frac{\partial u_x}{\partial y}\right) = \mu \frac{\partial^2 u_x}{\partial y^2} + \rho g_\alpha \tag{6.5.3}$$

where we simplify the notation by writing

$$g_\alpha \equiv g \sin \alpha \tag{6.5.4}$$

Now we can begin the "integral analysis." First, we multiply the continuity equation by ρu_x and add the whole equation to the left-hand side of the dynamic equation—Eq. 6.5.3. Keep in mind that the sum of the terms in the continuity equation is zero, so we are adding zero on that side. This gives us

$$\rho\left(u_x\frac{\partial u_x}{\partial x} + u_y\frac{\partial u_x}{\partial y}\right) + \rho\left(u_x\frac{\partial u_x}{\partial x} + u_x\frac{\partial u_y}{\partial y}\right) = \rho\left(\frac{\partial u_x^2}{\partial x} + \frac{\partial u_x u_y}{\partial y}\right) = \mu\frac{\partial^2 u_x}{\partial y^2} + \rho g_\alpha \quad \textbf{(6.5.5)}$$

The "integral analysis" takes its name from the process of integrating Eq. 6.5.5 (and Eq. 6.5.1, as well) across the liquid layer. This is equivalent to averaging the momentum and mass conservation equations across the film. To begin, we multiply each term in the momentum equation above by dy, then integrate term by term from the surface $y = 0$ to the edge of the film, $y = \delta(x)$, where $u_x = U_\delta$. The intermediate result is expressed in Eqs. 6.5.6.[1]

$$\int_0^{\delta(x)}\frac{\partial u_x^2}{\partial x}\,dy + [u_x u_y]_{y=0}^{y=\delta} = -\nu\left[\frac{\partial u_x}{\partial y}\right]_{y=0} + g_\alpha\delta(x) \quad \textbf{(6.5.6)}$$

We impose the boundary conditions:

$$u_x = 0 \quad \text{for} \quad y = 0 \quad \textbf{(6.5.7)}$$

and

$$u_x = U_\delta \quad \text{for} \quad y = \delta \quad \textbf{(6.5.8)}$$

and find

$$\int_0^{\delta(x)}\frac{\partial u_x^2}{\partial x}\,dy + U_\delta u_{y_{(y=\delta)}} = -\nu\left[\frac{\partial u_x}{\partial y}\right]_{y=0} + g_\alpha\delta(x) \quad \textbf{(6.5.9)}$$

If we also average the continuity equation across the film, we find

$$\int_0^{\delta(x)}\left(\frac{\partial u_x}{\partial x} + \frac{\partial u_y}{\partial y}\right)dy = \int_0^{\delta(x)}\frac{\partial u_x}{\partial x}\,dy + u_{y_{(y=\delta)}} = 0 \quad \textbf{(6.5.10)}$$

where we have used the boundary condition

$$u_y = 0 \quad \text{for} \quad y = 0 \quad \textbf{(6.5.11)}$$

Now we have to manipulate the x derivatives inside the integrals of Eqs. 6.5.9 and 6.5.10. It is tempting to bring the derivative outside the integral, but

$$\int_0^{\delta(x)}\frac{\partial u_x}{\partial x}\,dy \neq \frac{d}{dx}\int_0^{\delta(x)}u_x\,dy \quad \textbf{(6.5.12)}$$

(since the upper limit of this integral is a function of x). When this is the case, we must use the Leibniz rule for differentiating integrals. (Buy back your calculus text—second

---

[1] For example, starting on the right-hand side of Eq. 6.5.5, we find the following two terms:

$$\int_0^{\delta(x)}\rho g_\alpha\,dy = \rho g_\alpha\delta(x) \quad \textbf{(6.5.6$'$)}$$

$$\int_0^{\delta(x)}\mu\frac{\partial^2 u_x}{\partial y^2}\,dy = \int_0^{\delta(x)}\mu\frac{\partial}{\partial y}\left(\frac{\partial u_x}{\partial y}\right)dy = \mu\left(\frac{\partial u_x}{\partial y}\right)_0^{\delta(x)} = -\mu\left(\frac{\partial u_x}{\partial y}\right)_0 \quad \textbf{(6.5.6$''$)}$$

The assumption that there is no shear stress along the free surface has been imposed here. This would be a good assumption for liquid films in a gas, but not for films falling through an immiscible liquid phase.

notice.) This states that

$$\frac{\partial}{\partial x}\int_{a(x)}^{b(x)} f(x,y)\,dy = \int_{a(x)}^{b(x)}\frac{\partial f(x,y)}{\partial x}\,dy + f[x,b(x)]\frac{db}{dx} - f[x,a(x)]\frac{da}{dx} \qquad \textbf{(6.5.13)}$$

Now we can write the integrals in Eqs. 6.5.9 and 6.5.10 as

$$\int_0^{\delta(x)}\frac{\partial u_x^2}{\partial x}\,dy = \frac{d}{dx}\int_0^{\delta(x)} u_x^2\,dy - u_{x_{(y=\delta)}}^2\frac{d\delta}{dx} = \frac{d}{dx}\int_0^{\delta(x)} u_x^2\,dy - U_\delta^2\frac{d\delta}{dx} \qquad \textbf{(6.5.14)}$$

and

$$\int_0^{\delta(x)}\frac{\partial u_x}{\partial x}\,dy = -u_{y_{(y=\delta)}} = \frac{d}{dx}\int_0^{\delta(x)} u_x\,dy - U_\delta\frac{d\delta}{dx} = -U_\delta\frac{d\delta}{dx} \qquad \textbf{(6.5.15)}$$

In putting Eq. 6.5.15 into this form we have used Eq. 6.5.10, and we have also imposed a statement of global conservation of mass in the form

$$\frac{d}{dx}\int_0^{\delta(x)} u_x\,dy = 0 \qquad \textbf{(6.5.16)}$$

which follows because the integral above is just the volume flowrate, which cannot vary along the direction of flow.

We now introduce the shear stress at the solid surface:

$$\tau_0 = -\mu\left[\frac{\partial u_x}{\partial y}\right]_{y=0} \qquad \textbf{(6.5.17)}$$

This is used to rewrite the bracketed term on the right-hand side of Eq. 6.5.9. (See the comment in note 1 regarding Eq. 6.5.6″.)

We can finally put all this algebra and calculus together and write Eq. 6.5.9 as

$$\frac{d}{dx}\int_0^{\delta(x)} u_x^2\,dy = \frac{\tau_0}{\rho} + g_\alpha\delta(x) \qquad \textbf{(6.5.18)}$$

Equation 6.5.18 is an averaged version of the conservation of momentum equation. It has two unknowns in it: u_x and $\delta(x)$. (Of course τ_0 is also unknown, but it is related to u_x through Eq. 6.5.17.)

Now we take the next step in the integral method. We assume a specific form for the velocity profile in the film, and we substitute that form into the averaged momentum equation (Eq. 6.5.18). This will lead to a differential equation for $\delta(x)$, from which we may infer the desired information about the entry length for this flow. The main issue is this: How do we make a rational choice (assumption) of a velocity profile? What kinds of general constraint can we impose on the velocity profile?

We expect the velocity profile to satisfy the following conditions:

1. No slip at $y = 0$.
2. No shear at $y = \delta(x)$.
3. The surface velocity U_s must be a function of x.

In addition to these physical requirements, we should choose some function that permits us to perform the integration on the left-hand side of Eq. 6.5.18 analytically.

One possible function to try is the following, a type of profile that satisfies what we call a condition of "similarity."[2]

$$u_x(x, y) = U_s(x) \sin\left[\frac{\pi y}{2\delta(x)}\right] \tag{6.5.19}$$

There are other possibilities as well (see Problem 6.26). With this particular choice we find

$$\int_0^{\delta(x)} u_x^2 \, dy = \frac{\pi^2 (Q/W)^2}{8\delta(x)} \tag{6.5.20}$$

and

$$\tau_0 = -\mu \left[\frac{\partial u_x}{\partial y}\right]_{y=0} = -\frac{\pi^2 \mu (Q/W)}{4\delta^2(x)} \tag{6.5.21}$$

where Q/W is the volumetric flowrate per unit width in the z direction. For this choice of velocity profile Q/W is found to be

$$\frac{Q}{W} \equiv \int_0^{\delta(x)} u_x \, dy = \frac{2\delta(x)U_s(x)}{\pi} \tag{6.5.22}$$

Then Eq. 6.5.18 may be written in the form

$$\frac{d\delta}{dx} = A(\delta_\infty^3 - \delta^3) \tag{6.5.23}$$

where the coefficients A and δ_∞^3 are defined as

$$A \equiv \frac{8g_\alpha}{\pi^2 (Q/W)^2} \tag{6.5.24a}$$

$$\delta_\infty^3 \equiv \frac{\pi^2 \mu (Q/W)}{4\rho g_\alpha} \tag{6.5.24b}$$

The solution $\delta(x)$ to Eq. 6.5.23 has the character that $\delta(x)$ approaches the uniform value δ_∞, since $d\delta/dx$ goes to zero as $\delta(x)$ approaches δ_∞. (That explains the choice of notation for the constant δ_∞.)

In Chapter 4 we obtained the analytical solution for this falling film problem, and we found that the uniform film thickness was (Eq. 4.4.27)

$$\delta_\infty^3 \equiv \frac{3\mu (Q/W)}{\rho g_\alpha} \tag{6.5.25}$$

Equation 6.5.24b predicts an equilibrium film thickness only 6% different from the exact value. (We are comparing the constant $\pi^2/4$ in Eq. 6.5.24b to the constant 3 in Eq. 6.5.25. Note that we take a cube root to get δ_∞.)

Equation 6.5.23 may be integrated analytically, and after some algebraic manipulation the solution may be written in the dimensionless format

$$X = \frac{1}{12}[F(\Delta) - F(\Delta_o)] \tag{6.5.26}$$

---

[2] By this term we mean that for a constant value of the relative distance from the solid surface to the free surface [i.e., for $y/\delta(x)$ = constant], the ratio of the velocity at that value of x and y to the free surface velocity at that value of x is a constant. For example, if we examine the velocity along the curve $y/\delta(x) = 0.5$, we find that regardless of the x value, $u_x(x, y)/U_s(x)$ = constant. For the profile assumed in Eq. 6.5.19, $u_x(x, y)/U_s(x)$ has the value 0.707 along the curve $y/\delta(x) = 0.5$.

where a dimensionless distance from the origin of the free film is defined as

$$X = \frac{\mu}{\delta_\infty \rho (Q/W)} x \qquad (6.5.27)$$

A dimensionless film thickness is defined as

$$\Delta \equiv \frac{\delta(x)}{\delta_\infty} \qquad (6.5.28)$$

and Δ_o is defined as

$$\Delta_o \equiv \frac{\delta_o}{\delta_\infty} \qquad (6.5.29)$$

The initial film thickness δ_o at $x = 0$ is taken to be a known quantity. The function $F(\Delta)$ is given by

$$F(\Delta) \equiv \ln \frac{1 + \Delta + \Delta^2}{(1 - \Delta)^2} + 2\sqrt{3} \tan^{-1}\left(\frac{1 + 2\Delta}{\sqrt{3}}\right) \qquad (6.5.30)$$

and $F(\Delta_o)$ is the same function, but with Δ replaced by Δ_o.

For given liquid, flowrate, and initial film thickness, we may calculate Δ_o. Then Eq. 6.5.26 may be regarded as a relationship for $\Delta(x)$, from which we may find the distance downstream (measured by the dimensionless X) required for the film thickness to (nearly) achieve equilibrium. For example, we may choose $\Delta = 1.1$ for the cases that $\Delta_o > 1$, and $\Delta = 0.90$ for $\Delta_o < 1$. The dependence of X on Δ_o is shown in Fig. 6.5.2, for some values of $\Delta_o > 1$.

The dependence of the (dimensionless) film thickness Δ on the (dimensionless) downstream distance X is shown in Fig. 6.5.3 for the case $\Delta_o = 4$. A comparison of this simple theoretical model to experimental data is available, and the model is found to

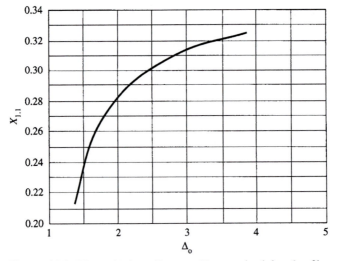

Figure 6.5.2 Dimensionless distance $X_{1.1}$ required for the film thickness to approach within 10 percent of its uniform downstream value.

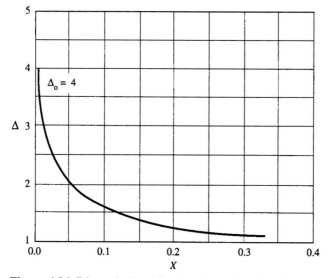

Figure 6.5.3 Dimensionless film thickness as a function of distance X downstream from the origin of the free surface, for $\Delta_o = 4$.

be quite accurate for the purpose of predicting film thickness along the direction of flow. An example of such data (and the appropriate reference) is given in Problem 6.30.

It is instructive to take a moment and contrast this model with the first model of this chapter. In particular, since both involve flows that are (assumed to be) nearly parallel, and both are laminar, steady, two-dimensional flows, why are the analyses so different? Why do we drop the inertial terms in the slider bearing problem, but leave them in the falling film flow development problem? Why do we solve a simplified form of the Navier–Stokes equations in the first case, and use an averaging (integral) method in the second? Why do we assume pressure is uniform in the film problem, eliminating it thereby as a variable, but leave pressure in the analysis of the slider bearing? Wasn't pressure eliminated from the film problem with the argument that the flow was nearly parallel, and isn't the flow also assumed nearly parallel in the slider bearing case? Why don't we start answering some of these questions?

Let's begin with the slider bearing problem. Can we eliminate pressure? Definitely not! The argument used to eliminate pressure in the liquid film problem is based on the *presence of a free surface that is exposed to a region of uniform pressure.* Since the pressures (actually the normal stresses) must be continuous across the free surface, it follows that the pressure does not vary within the liquid. In the slider bearing problem, the upper bounding surface constrains the fluid and is capable of exerting a normal stress (a pressure) that may vary in the direction of flow. There is simply no physical basis for arguing that the pressure is uniform in the x direction in the bearing problem. In fact, we do not want to eliminate pressure in the bearing problem precisely because the pressure distribution is the heart of the problem. It is what gives rise to the load-bearing capacity of this flow.

Then why don't we use an integral method on the slider bearing problem? Simply put, there is no need. We can get an analytical solution for both components of velocity and for pressure from the resulting "lubrication" equations. Of course, we have consid-

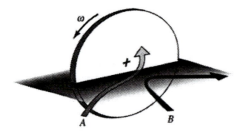

Figure 6.6.1 Operation of the rotating disk skimmer.

ered only the low Reynolds number case of the slider bearing problem. If inertial terms are included in the formulation, an analytical solution is not possible.[3]

Then why don't we drop the inertial terms in the film flow development problem? The answer is that this would leave us without a problem to solve. If there were no inertial terms in Eq. 6.5.3, the only solution compatible with the boundary conditions would be the fully developed flow—constant film thickness—solution. The dynamics of flow development is inseparable from the inertial terms of the dynamic equations. Hence these two problems really are quite different, although they share a few common features.

6.6 RECOVERY FROM AN OIL SPILL: THE ROTATING DISK SKIMMER

When large amounts of oil are released from accidents involving tankers or ocean-floor pipelines, environmental damage can be minimized by means of an efficient method of recovery of the oil floating on the ocean surface. One device used in the event of such a spill is the rotating disk skimmer (Fig. 6.6.1). The rotation of the disk induces a flow toward the disk in one region (path A) and away from the disk in another (path B). Some fluid is entrained by the face of the disk and is lifted above the free surface, where it can be collected by a suitable skimming device (not shown in Fig. 6.6.1).

Even if the surface were steady and planar, which is never the case in ocean spill incidents, the flow field generated by the action of the skimmer would be quite complex. The full fluid dynamics cannot be modeled by the simple methods described in this text. However, if we set our goal at a lower level, we can model an important feature of this system, using the ideas presented in this and the preceding chapters. We begin, then, by defining the problem:

> *Our goal is to develop a simple analytical model that predicts the volumetric flow-rate of oil skimmed from the surface and how that flowrate depends on physical properties of the oil (viscosity, density, surface tension) and the speed of rotation and radius of the disk.*

Before looking at the rotating disk skimmer, let's examine an even simpler problem that has some of the features of this more complex flow. In Fig. 6.6.2 we show a planar

---

[3] A method of solving the slider bearing problem with the inclusion of inertial terms is given in a paper by Tichy and Chen [*Trans. ASME, J. Tribol.,* **107,** 32 (1985)]. These investigators do not use an integral method.

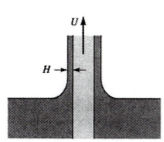

Figure 6.6.2 Withdrawal of a film of liquid by a rising planar sheet.

sheet that is continuously withdrawn upward through the surface of a single phase liquid. We raise the analogous question about the behavior of this system:

> *How does the entrained flowrate depend on physical properties of the liquid (viscosity, density, surface tension) and the removal (rise) speed of the sheet?*

It turns out that even with this geometrical simplification, the solution to this problem cannot be obtained by simple analytical methods appropriate to an introductory course in fluid dynamics. This may seem odd, since this problem is very similar to the falling film problem described in Chapter 4, Section 4.4.2. If we look at a section of the sheet well above the entrainment region, where the oil film thickness is constant at a value H (see Fig. 6.6.3), we may reduce the Navier–Stokes equations to the form

$$0 = \mu \left(\frac{d^2 u_x}{dy^2} \right) - \rho g_x \tag{6.6.1}$$

using the same arguments applied in the derivation of Eq. 4.4.22. (The coordinate system is redefined so that u_x is positive *upward*.) Hence Eq. 4.4.23 is still valid (but with a change in the sign on g):

$$u_x = \frac{\rho g_x}{2\mu} y^2 + ay + b \tag{6.6.2}$$

We still impose a no-slip boundary condition at the solid surface, but since that surface is moving at the speed U, we replace Eq. 4.4.24 with

$$u_x = U \quad \text{at} \quad y = 0 \tag{6.6.3}$$

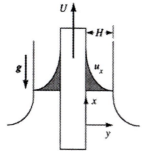

Figure 6.6.3 Liquid film is entrained by a moving sheet.

Continuing to neglect the shear stress along the liquid/gas interface, we may again write the second boundary condition as

$$\tau_{xy} = -\mu \frac{du_x}{dy} = 0 \quad \text{at} \quad y = H \tag{6.6.4}$$

With these boundary conditions, we find that the velocity profile is

$$u_x = U - \frac{\rho g_x H^2}{2\mu} \left[\frac{2y}{H} - \left(\frac{y}{H}\right)^2 \right] \tag{6.6.5}$$

Equation 4.4.27 is now changed to the form

$$Q_W = UH - \frac{\rho g_x H^3}{3\mu} \tag{6.6.6}$$

where Q_W is the flowrate per unit width.

In the falling film problem of Chapter 4, Q (which was the notation for flowrate per unit width in that chapter) was specified and H followed from Eq. 4.4.27, in terms of physical properties. In the current problem, we are after Q_W, and H cannot be specified arbitrarily—it is part of the solution to the dynamics problem itself.

The dependence of H on U and physical properties cannot be found from any simple model of the flow. The dynamics are complicated by the meniscus in the entrainment region. Hence surface tension is going to affect the entrainment of the film. Through a method of analysis more sophisticated than we can illustrate here,[4] it is possible to find an approximate solution for H, which we can write in the following nondimensional format:

$$\left(\frac{\rho g}{\mu U}\right)^{1/2} H = 0.946 \, \text{Ca}^{1/6} - 0.107 \, \text{Ca}^{1/2} + \cdots \tag{6.6.7}$$

It is not hard to confirm that the left-hand side of this equation is nondimensional. On the right-hand side we find the capillary number, a dimensionless group we first met in Chapter 5, defined here as

$$\text{Ca} \equiv \frac{\mu U}{\sigma} \tag{6.6.8}$$

The surface tension that appears in this model is that of the oil with respect to the surrounding air.

Only the two lowest powers in the capillary number are retained in the derivation that leads to Eq. 6.6.7, and so the model is restricted to small values of Ca. Experiments indicate that Eq. 6.6.7 describes the data on entrained film thickness quite well as long as the capillary number is small. A reasonable criterion is Ca ≤ O(1).

The flow field about the rotating disk is much more complex than that for a simple planar film moving vertically upward through the liquid interface. Nevertheless, we are going to attempt to use the planar model for the disk. In the simplest case, when the disk is half-submerged (see Fig. 6.6.4), the velocity of the disk surface is vertically upward right at the interface.

We proceed as follows. We take Eq. 6.6.6 for the flowrate Q_W (per unit width), and we use Eq. 6.6.7 to eliminate the unknown H. We regard this equation to be valid at any radial position r, where the spatially uniform U of the original model is then replaced

---

[4] The details are given in a paper by S. D. R. Wilson, *J. Eng. Math.*, **16**, 209 (1982).

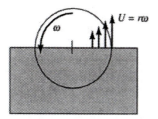

Figure 6.6.4 Schematic of the flow at the disk/liquid interface. Liquid is entrained on the circular face of the disk—not on the cylindrical surface of the rim.

by the radially increasing $U = r\omega$. Then we integrate this Q_W across the "width" 0 to R to find the total flowrate Q entrained by the rotating disk. The final result is

$$\frac{Q}{\omega R^2 (\sigma/\rho g)^{1/2}} = 0.355\, Ca_R^{2/3} - 0.13\, Ca_R \tag{6.6.9}$$

where

$$Ca_R \equiv \frac{\mu R \omega}{\sigma} \tag{6.6.10}$$

In developing this result, we keep only the two lowest powers in the capillary number, consistent with the approximation that led to Eq. 6.6.7. Equation 6.6.9 is tested in Fig. 6.6.5 against data obtained with a viscous oil entrained by a 15 cm radius disk. The data agree quite well with the model, over the range tested. The original reference (cited in the caption) should be studied for further details regarding the experimental procedure, and for a discussion of some factors not accounted for in this simple model that are, however, significant under some conditions of practice.

This is an example of developing a model of a complex system by utilizing and modifying an already existing model of a related system. With our current level of problem-solving skills, we were able to derive Eq. 6.6.6, but some further thought indicated that this was not a useful model because both H and Q_W were unknown. The

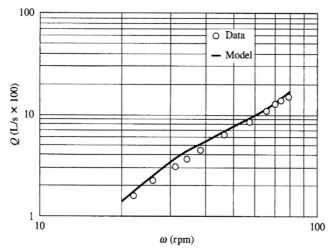

Figure 6.6.5 Experimental data for $R = 15$ cm. Christodoulou et al. [*J. Fluids Eng.,* **112**, 476 (1990)]

key to "solving" this problem was finding Eq. 6.6.7 in the existing fluid dynamics literature. Hence there is a significant advantage to developing the habit of reading the ongoing research literature in the fields that impact the areas of your technological responsibilities. It is even more advantageous to learn how to search the voluminous earlier literature of the fields of interest. (See Problem 6.41.)

EXAMPLE 6.6.1   *Yield from a Rotating Disk Oil Skimmer*

A barge tows an array of disk skimmers through the oily surface that has resulted from an offshore pipe leak. Each skimmer behaves in a manner that is well approximated by the model of Eq. 6.6.9. There are 100 disks, each of radius $R = 18$ cm, rotating at 30 rpm. The oil has a viscosity of 0.2 Pa·s at the ambient temperature of operation. The interfacial tension (oil/air) is 0.032 N/m. The oil density is 900 kg/m^3.

The system recovers only 50% of the oil that is skimmed from the surface. At what rate (kg/h) can the oil be recovered?

We will use Eq. 6.6.9. The capillary number is found to be

$$\text{Ca}_R = \frac{\mu R \omega}{\sigma} = \frac{0.2 \times 0.18 \times (30/2\pi)(1/60)}{0.032} = 0.09 \tag{6.6.11}$$

From Eq. 6.6.9 we find

$$\frac{Q}{\omega R^2 (\sigma/\rho g)^{1/2}} = 0.355(0.09)^{2/3} - 0.13(0.09) = 0.06 \tag{6.6.12}$$

Hence

$$Q = 0.06 \times \omega R^2 \left(\frac{\sigma}{\rho g}\right)^{1/2} = 0.06 \times \frac{30}{2\pi \times 60}(0.18)^2 \left(\frac{0.032}{900 \times 9.8}\right)^{1/2} \tag{6.6.13}$$

$$= 2.9 \times 10^{-7} \text{ m}^3/\text{s} = 0.29 \text{ cm}^3/\text{s}$$

This skimming rate corresponds to 2.6×10^{-4} kg/s or 1 kg/h. Only 50% of that, 0.5 kg/h, is actually recovered. However, this calculation is for one side of the disk, so on the assumption that entrainment occurs on both sides, the projected rate for a single disk is 1 kg/h. For the array of 100 disks, the recovery rate would be 100 kg/h, which corresponds to about 28 gallons of oil per hour.

EXAMPLE 6.6.2   *Dependence of the Recovery Rate on R*

The model given in Eq. 6.6.9 permits us to estimate the sensitivity of the recovery rate to design parameters. In particular, we can enhance the recovery by increasing the disk radius R. For small values of Ca_R we may write Eq. 6.6.9 in the form

$$Q = 0.355\omega R^2 \left(\frac{\sigma}{\rho g}\right)^{1/2} \left(\frac{\mu R \omega}{\sigma}\right)^{2/3} = A R^{8/3} \tag{6.6.14}$$

Hence an increase in the disk radius by a factor of 3 can increase the skimming rate by a factor of nearly 20. One of the important uses of mathematical models is as an aid to determine the sensitivity of the performance of a system to the design and operating parameters that define the system. Even if the model is not exact, and models of interesting systems seldom are, we often find that the parametric sensitivity predicted by the model is in good agreement with observations.

With this idea in mind, we can also observe that Q is predicted to depend on the rotational speed ω to a power just less than quadratic. Hence another way to enhance the skimming rate significantly is to increase the rotational speed.

6.7   FLOATING A DISK ON AN AIR TABLE

In Chapter 4, Example 4.5.1, we considered the flow between parallel disks, with the fluid being introduced through a small hole at the center of the disks. We look now at a related problem. Figure 6.7.1 defines the flow. A thin solid circular disk sits on a porous surface, and air is blown upward through the porous surface against the face of the disk. At some critical point the pressure beneath the disk overcomes the weight of the disk, and the disk lifts off the surface and floats at a height H. We assume that such a configuration is stable and that a uniform air film thickness can be maintained. Our goal is to develop a model for the relationship of the film thickness (the floating height H) as a function of parameters that define this system.

We begin with the following assumptions:

1. The flow is steady.
2. The flow is laminar, and at such a small Reynolds number that inertial effects are unimportant.
3. The flow is axisymmetric.
4. The porous surface is homogeneous in its properties. In particular, we will assume that the flow of air through the porous boundary is linearly proportional to the pressure difference $p_r - p(r)$, where p_r is the reservoir pressure under the porous surface, and $p(r)$ is the radial pressure distribution under the disk. We assume that p_r is independent of radial position.

Then the velocity of the airflow normal to the porous surface can be written in the form

$$u_z^o = \frac{k}{\mu}[p_r - p(r)] \tag{6.7.1}$$

where k is the permeability coefficient of the porous material. Equation 6.7.1 is an *assumed model* for the behavior of the porous medium. It is based on the observation that in many cases of interest, the flow of a fluid through a porous medium is proportional to the pressure difference across the medium and inversely proportional to the viscosity of the fluid. Equation 6.7.1 may be taken as the definition of the permeability coefficient k. Note how this definition differs from that used in Eq. 6.2.4.

For steady, axisymmetric, isothermal Newtonian flow in the absence of inertial effects, the radial component of the Navier–Stokes equations reduces to

$$\frac{\partial p}{\partial r} = \mu \left\{ \frac{\partial^2 u_r}{\partial z^2} + \frac{\partial}{\partial r}\left[\frac{1}{r}\frac{\partial(ru_r)}{\partial r} \right] \right\} \tag{6.7.2}$$

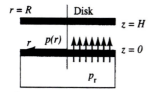

Figure 6.7.1 Flow between a disk and a porous surface.

If we now assume that the flow is nearly parallel, the lubrication approximation suggests that we regard the pressure as a function only of r, and in particular independent of the coordinate z. Consistent with the lubrication approximations, we also neglect the second viscous term, underlined in Eq. 6.7.2. (See Problem 6.36.) As a consequence we may integrate Eq. 6.7.2 twice with respect to the z coordinate. The result is

$$u_r = \frac{1}{2\mu} \frac{\partial p}{\partial r} (z^2 - Hz) \qquad (6.7.3)$$

where we have imposed the no-slip boundary conditions on u_r at the upper and lower surfaces.

Now we make a mass balance on this flow, using Fig. 6.7.2 as a guide. In differential form we write the following statement, which equates the difference in volume flowrates across the boundaries at r and $r + dr$ to the flow that enters the control volume through the porous surface:

$$2\pi (r + dr) \int_0^H (u_r)_{r+dr}\, dz - 2\pi r \int_0^H (u_r)_r\, dz = \frac{k}{\mu}[p_r - p(r)]2\pi r\, dr \qquad (6.7.4)$$

Using Eq. 6.7.3 for u_r, and recalling that we are assuming that $\partial p/\partial r$ is independent of z, we may write this as

$$\frac{H^3}{12\mu} \left\{ \left[r\frac{dp}{dr} \right]_{r+dr} - \left[r\frac{dp}{dr} \right]_r \right\} = -\frac{k}{\mu} r[p_r - p(r)]\, dr \qquad (6.7.5)$$

or

$$\frac{H^3}{12} \frac{1}{r} \frac{d}{dr} \left(r\frac{dp}{dr} \right) = -k[p_r - p(r)] \qquad (6.7.6)$$

This ordinary differential equation may be solved in terms of Bessel functions (see Problem 6.34), but an additional approximation lets us obtain a solution in simpler form. Under some conditions the reservoir pressure may be much greater than the pressure $p(r)$. Then the right-hand side of Eq. 6.7.6 becomes approximately a constant, and we may solve for the pressure distribution by two successive integrations and find

$$p = \frac{3kp_r}{H^3} (R^2 - r^2) \qquad (6.7.7)$$

The boundary conditions imposed on the pressure distribution were atmospheric pressure at the outer edge of the disks,

$$p = 0 \quad \text{at} \quad r = R \qquad (6.7.8)$$

and symmetry at the axis,

$$\frac{\partial p}{\partial r} = 0 \quad \text{at} \quad r = 0 \qquad (6.7.9)$$

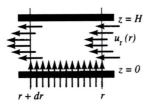

Figure 6.7.2 Schematic for mass balance.

Recall that our goal was a model for the floating height H. The additional physical statement required is the balance of forces acting on the disk. The weight W of the floating disk is supported by the total force exerted on the lower surface of that disk by the pressure distribution $p(r)$:

$$\int_0^R 2\pi r p(r)\, dr = W \tag{6.7.10}$$

Using Eq. 6.7.7 for $p(r)$, we may perform the indicated integration and solve for H. The result is

$$H = \left[\frac{3}{2}\left(\frac{p_r}{W/\pi R^2}\right) kR^2\right]^{1/3} \tag{6.7.11}$$

Assuming that the permeability coefficient of the porous medium is known, this is a testable model, and some data [Hinch and Lemaitre, *J. Fluid Mech.*, **273**, 313 (1994)] are shown in Fig. 6.7.3. The data are nondimensionalized according to Eq. 6.7.11, with the definitions

$$H^* \equiv \frac{H}{(\frac{3}{2}kR^2)^{1/3}} \quad \text{and} \quad p_r^* \equiv \frac{p_r}{W/\pi R^2} \tag{6.7.12}$$

We see that the model does not fit the data very well for values of the reduced pressure p_r^* in the range of 1–2. In this range the disk is barely lifted from the surface, and the pressure drop across the radius of the disk is significant compared to the reservoir pressure. Hence the approximation $p(r) \ll p_r$ is not good in this region, and the model fails.

The data shown in Fig. 6.7.3 correspond to Reynolds numbers below 10, where

$$\text{Re} = \frac{\rho u_z^o H}{\mu} \tag{6.7.13}$$

Hinch and Lemaitre provide a model for the high Reynolds number regime that is quite successful, but we do not review it here.

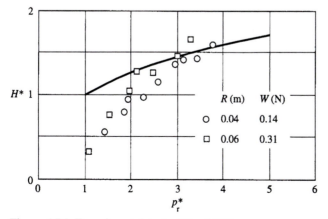

Figure 6.7.3 Experimental test of Eq. 6.7.11.

EXAMPLE 6.7.1 *Design of a Coating Thickness Monitor (Well, Almost)*

The analysis just carried out may be used to design a device for the measurement of coating thickness. Figure 6.7.4 suggests the concept and operation of the device. A flexible film is coated with a thickness T of some solid coating. An example would be the gelatin coating on a polyester photographic film base. At some point the coated film is conveyed continuously across a rigid base. A small circular porous disk is lowered to a position close to and parallel to the coated surface. A carefully controlled flowrate Q of air is maintained through the porous disk, and the pressure p_r^* is measured as a function of time, by means of a transducer capable of measuring very rapid changes in pressure. According to the model developed in this section, the thickness of the air film H can be related to the measured pressure p_r^*. If H_o represents the measured thickness of the air gap in the absence of a coating, then the coating thickness T follows from

$$T = H_o - H \tag{6.7.14}$$

We will need a second relationship, namely, between the pressure and the flowrate Q. This follows from Eqs. 6.7.3 and 6.7.7:

$$Q = 2\pi R \int_0^H u_r(z, r = R)\, dz \tag{6.7.15}$$

and

$$u_r(z, r = R) = -\frac{1}{2\mu} \left| \frac{\partial p}{\partial r} \right|_{r=R} (z^2 - Hz) = \frac{-3kRp_r}{\mu H^3}(z^2 - Hz) \tag{6.7.16}$$

After performing the integration indicated in Eq. 6.7.15 we find

$$Q = \frac{\pi R^2 k p_r}{\mu} \tag{6.7.17}$$

Something is very wrong here! *The dependence of the pressure on the flowrate does not include any dependence on the gap spacing H!* How can this be?

We have just completed a classical example of "throwing the baby out with the bathwater." What was the key assumption that led to the solution (Eq. 6.7.7) for $p(r)$? It was that p_r greatly exceeded $p(r)$. This assumption is equivalent to assuming that the entire pressure drop required to maintain the flow Q takes place across the porous disk, and that there is no significant resistance to flow within the air gap itself. Then it is no wonder that the model we end up with, Eq. 6.7.17, reflects this by showing that the $Q(p_r)$ relationship is independent of the gap thickness H. If we are going to design a film thickness monitor based on the resistance to flow through a narrow air gap, we will have to design the system so that the air gap *does* exert a significant resistance to

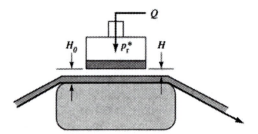

Figure 6.7.4 Conceptual design of a coating thickness monitor.

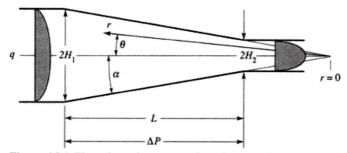

Figure 6.8.1 Flow through a converging planar region.

flow, and we will have to modify the model presented above. In other words, we will have to solve Eq. 6.7.6, and we will need a criterion that guarantees the desired sensitivity of the $Q(p_r)$ relationship to changes in H. The solution to Problem 6.34 yields the required mathematical results.

6.8 FLOW THROUGH A CONVERGING PLANAR REGION

In this next example of an almost parallel flow, we select the case of flow between converging (or diverging) flat plates.[5] Figure 6.8.1 shows the geometry. Intuitively, we expect that the angle α will determine the extent to which the equations for nearly parallel flow lead to a useful model. For later reference, let's look at the solution for the parallel plate case first. We consider steady isothermal Newtonian flow and assume that there is no flow in the direction normal to the page. We have already solved this problem (Example 4.5.2), and when appropriate we will look at some features of the solution. In particular, we will be interested in the velocity profile

$$u_z = \frac{-CH^2}{2\mu}\left[1 - \frac{y^2}{H^2}\right] \tag{6.8.1}$$

and the pressure–flowrate relationship

$$-C = \frac{\Delta P}{L} = \frac{3\mu Q}{2WH^3} \tag{6.8.2}$$

For the parallel case, H is just the uniform (half) spacing between the planes, and y is the coordinate axis normal to the planes, with origin along the line of symmetry; Q/W is the volumetric flowrate per unit width normal to the page.

For the model developed here, we do not really assume that the flow is nearly parallel. We actually account for the fact that the flow field is converging. In the analysis of the converging geometry, it proves convenient to use a plane polar (cylindrical) coordinate system. The origin ($r = 0$) is at the line of convergence of the two bounding planes. For our model to make sense physically, we will have to exclude the origin from the solution. Otherwise we would require an infinitely large flowrate to conserve mass as the flow moves into an area of infinitesimally small cross section. Hence the first approximation of the model lies in our definition of the region over which we study

---

[5] We follow closely the treatment given by Denn in Chapter 10 of *Process Fluid Mechanics* (Prentice-Hall, Englewood Cliffs, 1980).

this flow. We imagine that the converging region connects two plane parallel regions, as shown in Fig. 6.8.1. We will neglect entrance and exit effects as the flow makes a transition from the parallel to the converging region, and out into the downstream parallel region. We can expect, then, that the use of the model we ultimately develop will have at least one restriction: $L \gg H_1$. For this two-dimensional flow, the continuity equation in polar (cylindrical) coordinates is simply (from Eq. 4.1.8b)

$$\frac{1}{r}\frac{\partial(ru_r)}{\partial r} + \frac{1}{r}\frac{\partial u_\theta}{\partial \theta} = 0 \tag{6.8.3}$$

We introduce another assumption at this point. The flow is taken to be strictly radial, in the sense that the velocity component u_θ is assumed zero everywhere. Of course this could not be true near the transition regions to and from the parallel sections. Then Eq. 6.8.3 immediately yields

$$ru_r \neq g(r) \tag{6.8.4}$$

Hence, at most, ru_r is a function only of θ:

$$ru_r = f(\theta) \tag{6.8.5}$$

Equation 6.8.5 is a direct consequence of the assumption that $u_\theta = 0$ everywhere. If this is a poor assumption, we will not be able to satisfy the momentum equations, which we have yet to write. Hence our first task is to see if Eq. 6.8.5 is compatible with conservation of momentum. From the appropriate entries in Table 4.3.1 we have

$$\rho u_r \frac{\partial u_r}{\partial r} = -\frac{\partial p}{\partial r} + \mu \left(\frac{\partial}{\partial r}\left[\frac{1}{r}\frac{\partial}{\partial r}(ru_r) \right] + \frac{1}{r^2}\frac{\partial^2 u_r}{\partial \theta^2} \right) \tag{6.8.6}$$

and

$$0 = -\frac{1}{r}\frac{\partial p}{\partial \theta} + \frac{2\mu}{r^2}\frac{\partial u_r}{\partial \theta} \tag{6.8.7}$$

When Eq. 6.8.5 is substituted for the radial velocity, we obtain two equations for the two unknowns: p and u_r:

$$\frac{\partial p}{\partial r} = \frac{\mu}{r^3}\frac{d^2 f}{d\theta^2} + \frac{\rho f^2}{r^3} \tag{6.8.8}$$

and

$$\frac{\partial p}{\partial \theta} = \frac{2\mu}{r^2}\frac{df}{d\theta} \tag{6.8.9}$$

Note that we are writing $df/d\theta$ with an ordinary, not partial, derivative notation. Since the variable r appears in the equations above, we have no assurance that f is indeed independent of r. Keep in mind that Eq. 6.8.5 is a consequence of an *assumption* (that the flow is strictly radial). *The assumption that this is so, and the use of the appropriate notation, do not make it so.*

We may eliminate the pressure between Eqs. 6.8.8 and 6.8.9 by differentiating Eq. 6.8.8 with respect to θ, and 6.8.9 with respect to r, whereupon subtraction yields the result

$$\frac{d^3 f}{d\theta^3} + \left(4 + \frac{2\rho}{\mu}f \right)\frac{df}{d\theta} = 0 \tag{6.8.10}$$

Now we see that the variable r does not appear in this equation for f. We conclude that the momentum equations *can be satisfied* by a velocity field (Eq. 6.8.5) that is consistent with the assumption that $u_\theta = 0$ everywhere. This does not mean that the resulting solution will be a good model of the flow—only that the solution will satisfy conservation of momentum.

Since Eq. 6.8.10 is third order we need three boundary conditions. Two of them are

$$\text{no slip on the planes } \theta = \pm\alpha: \quad f(\alpha) = f(-\alpha) = 0 \tag{6.8.11}$$

For a third condition we will specify the flowrate (per unit width) q. This boundary condition requires that f satisfy Eq. 6.8.12, where q is *negative* in the converging flow case, since u_r is in the direction opposite the positive r direction.

$$q = \int_{-\alpha}^{\alpha} u_r r\, d\theta = \int_{-\alpha}^{\alpha} f(\theta)\, d\theta \tag{6.8.12}$$

Equation 6.8.10 is nonlinear, and although an analytical solution is possible in this particular case, the solution is in terms of functions (elliptic integrals) that must be evaluated numerically. We will introduce another approximation at this point and find an analytical solution in terms of more simply evaluated functions. First, we nondimensionalize Eq. 6.8.10. The boundary conditions will simplify if we normalize the angular coordinate θ to the half-angle α:

$$\phi = \frac{\theta}{\alpha} \tag{6.8.13}$$

The velocity function f can be normalized using q:

$$F_1 = \frac{f}{q} \tag{6.8.14}$$

but this is not the best choice we can make. While it is true that F_1, so defined, is dimensionless, it is better to define a dimensionless f using a characteristic parameter that is of the same order of magnitude as f. To see this more clearly, let's look at Eq. 6.8.12. From the mean-value theorem of integral calculus, we know that the mean value of f satisfies

$$\int_{-\alpha}^{\alpha} f(\theta)\, d\theta = 2\alpha \bar{f} = q \tag{6.8.15}$$

Hence a characteristic value of f over the region of flow is of the order of q/α. (The factor of 2 is irrelevant to the discussion of order of magnitude.) For this reason, we nondimensionalize f by

$$F = \frac{\alpha f}{q} \tag{6.8.16}$$

With these definitions, Eq. 6.8.10 may be written in the form

$$\frac{d^3 F}{d\phi^3} + (4\alpha^2 + \text{Re}\, F)\frac{dF}{d\phi} = 0 \tag{6.8.17}$$

where a Reynolds number appears, defined as

$$\text{Re} = \frac{2\rho q \alpha}{\mu} \tag{6.8.18}$$

Note that since this Reynolds number is written in terms of q, it does not contain a velocity or length scale, as is usually the case. With these definitions, the boundary conditions become

$$\text{no slip on the planes } \phi = \pm 1: \quad F(1) = F(-1) = 0 \tag{6.8.19}$$

$$\int_{-1}^{1} F(\phi)\, d\phi = 1 \tag{6.8.20}$$

Now we are ready to introduce some approximations that permit us to solve Eq. 6.8.17 in terms of simple functions. One of the advantages of the nondimensional formulation is that it displays the parameters that control, in some sense, the behavior of the solutions. In the case of Eq. 6.8.17 these parameters are the Reynolds number Re and the convergence half-angle α. If the Reynolds number is sufficiently small, it should be possible to drop the term ReF from Eq. 6.8.17. This will yield a *linear* ordinary differential equation, which is quite easily solved.

But what criterion do we impose for neglecting this term? This is where the choice to nondimensionalize f and θ using parameters that are characteristic values of these variables comes in. By this choice, F is of order unity, as are the derivatives $dF/d\phi$ and $d^3F/d\phi^3$. Hence the condition

$$\text{Re } F \frac{dF}{d\phi} \ll \frac{d^3F}{d\phi^3} \tag{6.8.21}$$

will hold if Re $\ll 1$. If we then drop the Reynolds number term from Eq. 6.8.17, we can solve

$$\frac{d^3F}{d\phi^3} + 4\alpha^2 \frac{dF}{d\phi} = 0 \tag{6.8.22}$$

with the boundary conditions given by Eqs. 6.8.19 and 6.8.20. The solution is easily found. Define a new variable

$$Y = \frac{dF}{d\phi} \tag{6.8.23}$$

Then Eq. 6.8.22 becomes the harmonic equation

$$\frac{d^2Y}{d\phi^2} + 4\alpha^2 Y = 0 \tag{6.8.24}$$

and so

$$Y = \frac{dF}{d\phi} = A' \sin 2\alpha\phi + B' \cos 2\alpha\phi \tag{6.8.25}$$

One more integration yields

$$F = A \sin 2\alpha\phi + B \cos 2\alpha\phi + C \tag{6.8.26}$$

(Careful! A and B are not the same as A' and B' of Eq. 6.8.25.) Boundary conditions 6.8.19 require that $A = 0$ and $C = -B \cos 2\alpha$. The integral condition, Eq. 6.8.20, leads to

$$\int_{-1}^{1} F(\phi)\, d\phi = 1 = \int_{-1}^{1} B(\cos 2\alpha\phi - \cos 2\alpha)\, d\phi = \frac{B(\sin 2\alpha - 2\alpha \cos 2\alpha)}{\alpha} \tag{6.8.27}$$

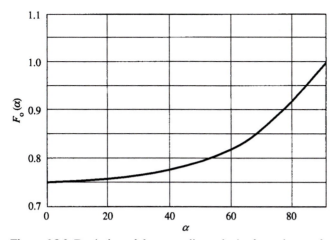

Figure 6.8.2 Deviation of the centerline velocity from the parallel plate value, as a function of α.

Finally, the solution for F is

$$F(\phi) = \frac{\alpha(\cos 2\alpha\phi - \cos 2\alpha)}{\sin 2\alpha - 2\alpha \cos 2\alpha} \tag{6.8.28}$$

It is instructive to compare this solution to that for parallel plates, which is just the case of $\alpha = 0$. Equation 6.8.28 is indeterminate for $\alpha = 0$, but application of L'Hôpital's rule yields the result

$$F(\phi) = \tfrac{3}{4}(1 - \phi^2) \tag{6.8.29}$$

While it is more revealing to plot the velocity profiles $F(\phi)$ (see Problem 6.47), a simple comparison to the $\alpha = 0$ case can be made by looking at the maximum (centerline) velocity at some radial position r, as a function of the angle α. This is the function $F_o(\alpha)$ shown in Fig. 6.8.2. For values of α up to about 20° the centerline velocity (and, it turns out, the whole profile) is essentially independent of α, and given by the $\alpha = 0$ solution. In this sense we could say that the flow is almost parallel up to an angle of 20°.

In keeping with the usual language of fluid dynamics, we refer to the solution to Eq. 6.8.17, for the case of vanishing Re, as the *creeping flow* solution. It is interesting to look back at the definition of this particular converging flow Reynolds number, which we now write as:

$$\mathrm{Re} = \frac{2\rho q}{\mu}\alpha \tag{6.8.30}$$

Normally we think of "small Reynolds number" as corresponding to a combination of low flowrate and high viscosity. We see here that in this flow, and with the particular definition of Re we have taken, we could have a low Reynolds number at a high flowrate, as long as the angle α is small. Does this make sense physically? The answer is Yes! As α gets small (i.e., as we approach the parallel plate case), the inertial terms disappear entirely from the problem because there is no longer any spatial acceleration of the flow: that is, because $\partial u_r/\partial r$ vanishes *exactly* in the parallel flow. Hence, regardless of the flowrate, the left-hand side of the Navier–Stokes equations would vanish in the plane parallel flow case (unless $\rho q/\mu$ is itself so large that the flow makes a transition to turbulence!).

EXAMPLE 6.8.1   *Reynolds Numbers for Flow Toward and Through a Converging Planar Duct*

A viscous liquid is flowing through a duct of uniform rectangular cross section. The width and height of the duct are $W = 0.1$ m and $2H = 0.01$ m, respectively. The viscosity is $\mu = 0.1$ Pa·s, and the density is $\rho = 1000$ kg/m³. The volumetric flowrate is 360 L/h. What is the Reynolds number in the parallel duct?

The parallel duct is joined to a planar converging duct of the same width $W = 0.1$ m. This duct converges, at an angle of $\alpha = 30°$, to a smaller parallel duct of height $2H = 0.002$ m. what is the Reynolds number in this duct?

For plane parallel flow through a very wide duct, the usual definition of the Reynolds number would be

$$Re_l = \frac{2\rho H \overline{U}}{\mu} \tag{6.8.31}$$

where $\overline{U}$ is the average velocity through the duct. The average velocity is found from the flowrate as

$$\overline{U} = \frac{Q}{2WH} = \frac{360/3600 \text{ L/s} \times 10^{-3} \text{ m}^3/\text{L}}{2 \times 0.1 \text{ m} \times 0.005 \text{ m}} = 10^{-1} \text{ m/s} \tag{6.8.32}$$

and the Reynolds number follows as

$$Re_l = \frac{2 \times 1000 \times 0.005 \times 0.1}{0.1} = 10 \tag{6.8.33}$$

For the converging duct we use Eq. 6.8.30 for the Reynolds number. The flowrate through this duct is the same as that of the parallel duct leading into it, since the mass flow must be conserved. Recalling that for *converging* flow, $q < 0$, we write the flowrate per unit width simply as

$$-q = \frac{Q}{W} = \frac{360/3600 \text{ L/s} \times 10^{-3} \text{ m}^3/\text{L}}{0.1 \text{ m}} = 10^{-3} \text{ m}^2/\text{s} \tag{6.8.34}$$

The angle α is 30°, but this must be converted to radians, and so $\alpha = 0.525$ rad. Hence the Reynolds number is of absolute value

$$Re = \frac{2 \times 1000 \times 10^{-3}}{0.1} \times 0.525 = 10.5 \tag{6.8.35}$$

We see that depending on the angle of convergence, the magnitude of the Reynolds number in the converging section could be greater or less than that in the parallel entrance section. There is no physical significance to the relative values, which depend on the specific definitions taken for Re.

Equation 6.8.29 was presented as the parallel flow solution by taking the limit of the general angle solution (Eq. 6.8.28) as the angle became very small. We may generate that solution directly from Eq. 6.8.22 with the assumption that for small angles the second term would be smaller than the first, and therefore an approximate solution, valid for small angles, should follow as the solution to

$$\frac{d^3 F}{d\phi^3} = 0 \tag{6.8.36}$$

Of course this is solved immediately by integration:

$$F(\phi) = A + B\phi + C\phi^2 \tag{6.8.37}$$

Upon imposing the boundary conditions we find the constants A, B, and C and the solution takes the form

$$F(\phi) = \tfrac{3}{4}(1 - \phi^2) \tag{6.8.38}$$

which is identical, as it should be, to Eq. 6.8.29. However, we now regard Eq. 6.8.38 as the small angle solution, rather than as the parallel planes solution, because it comes from an order of magnitude argument on the relative magnitudes of the terms in Eq. 6.8.22. In other words, while Eq. 6.8.29 is the exact solution for parallel planes, Eq. 6.8.38 (the same format) is the approximate solution for converging planes but for small angles of convergence. We belabor this point because it lets us look at Eq. 6.8.17 as the basis for a series of approximate solutions to this converging flow problem. By putting the dynamic equations into this nondimensional format (i.e., by writing Eq. 6.8.17 with our particular choice of nondimensionalization), we can make more rational statements about what we mean by "low Reynolds number," and "nearly parallel flow," and we can lean less heavily on the intuitive arguments that lead to the same simplifications.

In the other models we developed for pressure-driven flows, one of the key features we have sought is the pressure drop–flowrate relationship. An example is Eq. 6.8.2 for the plane parallel duct. With the creeping flow solution for the converging duct available, let's discuss the pressure loss associated with this flow. We can obtain the pressure by integration of Eq. 6.8.8. Since f is independent of r this is immediately integrated to yield

$$p = -\frac{1}{2r^2}\left(\rho f^2 + \mu \frac{d^2 f}{d\theta^2}\right) + C \tag{6.8.39}$$

where C is a constant of integration with respect to r, although C could be a function of θ. However, if we differentiate Eq. 6.8.39 with respect to θ, and then use Eq. 6.8.10 to simplify some of the terms, we find that C is not a function of θ after all.

The main point to note from Eq. 6.8.39 is that the pressure is a function of θ at any position r. We have not encountered this situation before, and its implication is that the pressure drop required to drive this flow will depend on the angular position θ; it is a "local" variable for this nonparallel flow field. But physically we do not impose a *local* pressure at one end of a duct. So what is the relationship of this local pressure to the pressure required to maintain the flow at a given flowrate q?

Our main reason for wanting information about the pressure drop is that it is related

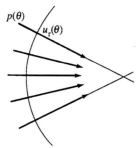

Figure 6.8.3 Detail of flow across a surface at constant r.

to the *power* required to maintain the flow. Look at Fig. 6.8.3, as we discuss the power required to move fluid across a surface normal to the flow. Imagine a small "parcel" of fluid moving at velocity u_r across some surface r = constant. The force (per unit of surface area) "pushing" this parcel across the surface is just the stress in the direction of motion (that's T_{rr}). For a Newtonian fluid

$$T_{rr} = p - \mu \Delta_{rr} = p - 2\mu \frac{\partial u_r}{\partial r} = p + \frac{2\mu f}{r^2} \tag{6.8.40}$$

With Eq. 6.8.39 in hand for p, and with our solution for f, we find

$$T_{rr} = C - \frac{1}{2r^2}\left(\rho f^2 + \mu \frac{df^2}{d\theta^2} - 4\mu f\right) \tag{6.8.41}$$

The power (which is the rate of doing work) required to move this parcel of fluid is the product of the velocity normal to the surface (that's just u_r) times the corresponding force (per unit width) acting to oppose the motion. Over the *entire* surface at any radial position r_1 this is given by the integral

$$P_1 = \int_{-\alpha}^{\alpha} u_r T_{rr} r_1 \, d\theta = \int_{-\alpha}^{\alpha}\left[C - \frac{1}{2r_1^2}\left(\rho f^2 + \mu \frac{df^2}{d\theta^2} - 4\mu f\right)\right] f \, d\theta \tag{6.8.42}$$

At some radial position further downstream, say at $r = r_2$, the power is just

$$P_2 = \int_{-\alpha}^{\alpha} u_r T_{rr} r_2 \, d\theta = \int_{-\alpha}^{\alpha}\left[C - \frac{1}{2r_2^2}\left(\rho f^2 + \mu \frac{df^2}{d\theta^2} - 4\mu f\right)\right] f \, d\theta \tag{6.8.43}$$

The net power required to push fluid through the volume element bounded by the arcs r_1 and r_2 is just the difference of these two quantities:

$$P = \left(\frac{1}{2r_2^2} - \frac{1}{2r_1^2}\right)\int_{-\alpha}^{\alpha}\left(\rho f^2 + \mu \frac{df^2}{d\theta^2} - 4\mu f\right) f \, d\theta \tag{6.8.44}$$

This can be simplified (see Problem 6.49) after integrating Eq. 6.8.10 once, using Eq. 6.8.11 as a boundary condition, and introducing the definition of q. In terms of the dimensionless variables, the result is found to be

$$P = \frac{\mu q^2}{2\alpha^3}\left(\frac{1}{r_2^2} - \frac{1}{r_1^2}\right)\left[8\alpha^2 \int_{-1}^{1} F^2 \, d\phi - F''(1)\right] \tag{6.8.45}$$

where

$$F''(1) = \frac{\alpha}{q}\left(\frac{d^2 f}{d\theta^2}\right)_{\theta=\alpha} \tag{6.8.46}$$

Before proceeding further, let's look at the corresponding result for plane Poiseuille flow. The pressure drop–flowrate relationship is given by Eq. 6.8.2, which we now write as

$$\Delta P = \frac{3\mu L}{2H^3}(-q) \tag{6.8.47}$$

(Recall once again that q is negative for converging flow. We will maintain this sign convention in the limiting case of parallel planes, so that as the angle gets small, the proper sign is found for the limit.) Since the pressure is uniform in the direction normal to the flow, for the parallel plate case, the power is simply $q\Delta P$, or

$$P_{\parallel} = \frac{3\mu L}{2H^3} q^2 \tag{6.8.48}$$

We will now force Eq. 6.8.45 into this form. The length L is given by (see Fig. 6.8.1)

$$L = \frac{H_1 - H_2}{\tan \alpha} \tag{6.8.49}$$

and of course

$$H_1 = r_1 \sin \alpha \tag{6.8.50}$$

with a similar equation for H_2. With these relationships, and a few lines of tedious algebra, we can express the power in the form

$$P = \frac{3\mu L}{2H_1^3} q^2 \frac{2C_R(1 + C_R)}{3} \Phi(\alpha) \tag{6.8.51}$$

where the "contraction ratio" C_R is

$$C_R = H_1/H_2 \tag{6.8.52}$$

and

$$\Phi(\alpha) \equiv \frac{\sin^3 \alpha}{2\alpha^3 \cos \alpha} \left[8\alpha^2 \int_{-1}^{1} F^2 \, d\phi - F''(1) \right] \tag{6.8.53}$$

This format breaks P into a product of three factors: the lead factor is the power requirement for a parallel plate channel, with the upstream separation H_1 as the length scale; the second factor is explicit in terms of the contraction ratio; and the third factor $\Phi(\alpha)$ depends on the convergence angle.

For the creeping flow case, the function $F(\phi)$ is given by Eq. 6.8.28, and the integration indicated in Eq. 6.8.53, as well as the term $F''(1)$, can be worked out with a little effort. The final result is

$$\Phi(\alpha) = \frac{[4\alpha(1 + \cos^2 2\alpha) - 4\cos 2\alpha \sin 2\alpha] \sin^3 \alpha}{\cos \alpha (\sin 2\alpha - 2\alpha \cos 2\alpha)^2} \tag{6.8.54}$$

The function $\Phi(\alpha)$ is plotted in Fig. 6.8.4.

It is not difficult to show that Eq. 6.8.51 behaves as expected (we obtain Eq. 6.8.48) in the limit of $C_R \to 1$ (which of course also implies $\alpha \to 0$). A somewhat more interesting case is that of a fixed contraction ratio, but a small convergence angle. Physically this

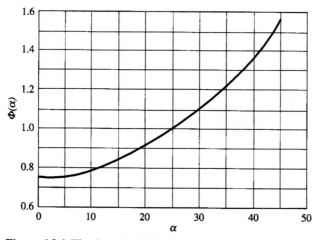

Figure 6.8.4 The function $\Phi(\alpha)$ from Eq. 6.8.54.

implies that the length L grows large. The result is found to be

$$P = \frac{3\mu L}{2H_1^3} q^2 \frac{C_R(1 + C_R)}{2} \tag{6.8.55}$$

In terms of the spacings H_1 and H_2, this may be written in the form

$$P = \frac{3\mu L}{2\langle H \rangle^3} q^2 \tag{6.8.56}$$

where

$$\langle H \rangle^3 = \frac{H_1^2 H_2^2}{\frac{1}{2}(H_1 + H_2)} \tag{6.8.57}$$

may be thought of as (the cube of) a specific average spacing, defined so that Eq. 6.8.57 is otherwise identical to the parallel duct solution. We see that this peculiar average is related to the arithmetic mean spacing, but it is clearly not equal to it.

Recall that we began this discussion with the goal of learning something about the pressure drop across the ends of a converging duct. We have learned, in fact, that since the pressure is a function of angular position across the duct, the concept of "pressure drop" is not clearly defined for this flow. We can, however, calculate the power associated with the converging flow. We suppose that since we have ignored entrance effects as the fluid makes a transition from the upstream parallel section into the duct, and the exit effects as it leaves the duct and enters the downstream parallel section, the pressure drop could be measured across a length that just excludes the boundaries of the converging duct. We might further assume that this pressure drop could then be calculated by ignoring the power required to move the fluid through and across the two transition regions, and thus it would be given by Eq. 6.8.51 with $P = q\Delta P$:

$$\Delta P = \frac{3\mu L}{2H_1^3} (-q) \frac{2C_R(1 + C_R)}{3} \Phi(\alpha) \tag{6.8.58}$$

for the creeping flow case. Clearly the error in such a model would depend on the extent to which the flow in these transition regions deviates from the assumed radial flow. An assessment of that error would require numerical solution of the full Navier–Stokes equations. One such solution [Black and Denn, *J. Non-Newtonian Fluid Mech.*, **1**, 83 (1976)] is shown in Fig. 6.8.5, which indicates that the streamlines of the flow are nearly radial except within a very short distance upstream of the two transition regions.

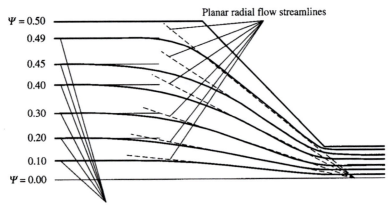

Figure 6.8.5 Streamlines for flow through a planar converging duct.

Pressure Requirement for Flow Through a Nonuniform Planar Duct

A viscous liquid ($\mu = 100$ Pa·s, $\rho = 1000$ kg/m³) is flowing through a planar duct of rectangular cross section. The width W, which is uniform across all sections, is 0.1 m. Figure 6.8.6 gives the detailed geometry. The volumetric flowrate is 360 L/h. Give an estimate of the pressure required to maintain the flow.

In all cases we will neglect entrance/exit effects as the fluid makes transitions from one section to another. The data of this example are identical to those of Example 6.8.1, except that the liquid in this case is higher in viscosity by a factor of 1000. Hence the Reynolds number is well below unity, and we expect that the creeping flow solution is valid. For the plane parallel sections, the pressure drop is given by

$$\Delta P_i = \frac{3\mu L_i}{2H_i^3} q \tag{6.8.59}$$

where q is given by Eq. 6.8.34: $q = 0.001$ m²/s. (We will now use the *magnitude* of q and not bother to regard it as a negative quantity for converging flow.) For the converging section, we use Eq. 6.8.58. If we neglect the transition regions, the pressure drops across the three sections are additive, and we find

$$\Delta P = \frac{3\mu q}{2} \left[\frac{L_1}{H_1^3} + \frac{L_2}{H_2^3} + \frac{L}{H_1^3} \frac{2C_R(1 + C_R)}{3} \Phi(\alpha) \right] \tag{6.8.60}$$

The axial length of the converging section is found as

$$L = \frac{H_1 - H_2}{\tan \alpha} = \frac{0.005 - 0.001}{0.5774} = 0.0069 \text{ m} \tag{6.8.61}$$

The contraction ratio is $C_R = 5$. We calculate (from Eq. 6.8.54) $\Phi(\alpha) = 1.091$ for $\alpha = 30°$. Then, from Eq. 6.8.60, the pressure drop is estimated to be

$$\Delta P = \frac{3(100)(0.001)}{2} \left[\frac{0.05}{0.005^3} + \frac{0.02}{0.001^3} + \frac{0.0069}{0.005^3} \frac{2(5)(1 + 5)}{3} 1.091 \right]$$

$$= 3.24 \times 10^6 \text{ N/m}^2 \approx 470 \text{ psi} \tag{6.8.62}$$

The dominant term in the pressure drop is seen to be from the narrow parallel downstream region. The converging section contributes about 6% additional pressure loss. Hence any error in estimating the loss across the converging section will have a minor effect on the total loss.

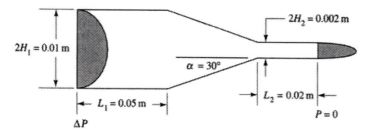

Figure 6.8.6 Planar converging duct for Example 6.8.2 (not to scale).

6.9 LAMINAR FLOW THROUGH A LEAKY TUBE: THE PERTURBATION METHOD OF APPROXIMATION

Another method of obtaining approximate solutions to certain fluid dynamics problems is called the "perturbation method." We can illustrate the technique by returning to the problem modeled in Section 6.2: laminar flow in a leaky tube. The key to the method illustrated in that section was the assertion that Eq. 6.2.6 held. We forced the solution for the axial velocity profile to look like the solution for the nonleaky tube (Poiseuille flow) in the sense that we assumed that the radial dependence of the axial velocity profile was parabolic. This assumption defined an axial function $f(z)$. We found a differential equation, based on conservation of mass and momentum, that $f(z)$ had to satisfy for Eq. 6.2.6 to hold. That equation was Eq. 6.2.12. But the appearance of the radial variable in that equation violated the assumption (Eq. 6.2.6) that the axial profile could be expressed as a product of a parabolic function of r and some unknown function of z. We ignored that fact by observing that if we threw away the r-dependent term in Eq. 6.2.12, we could proceed to a solution. This process defined a parameter β, the magnitude of which would determine the accuracy of the approximation. In this section we solve the same problem, but by a more orderly approach.

The critical starting point in this method is in the nondimensionalization of the continuity and Navier–Stokes equations. We define the following dimensionless variables:

$$s = \frac{r}{R} \qquad \zeta = \frac{z}{L} \qquad v_z = \frac{u_z}{U_o} \tag{6.9.1}$$

We take the average velocity U_o at the tube entrance, $z = 0$, to be specified. (This may not be the case in all problems of this type. See Problem 6.67.) With these definitions Eq. 6.2.3 becomes

$$0 = \frac{1}{s} \frac{\partial}{\partial s} s \left[\frac{u_r}{(R/L)U_o} \right] + \frac{\partial v_z}{\partial \zeta} \tag{6.9.2}$$

Now we introduce the first approximation. It is that we will consider only flows for which the geometry satisfies

$$\varepsilon = \frac{R}{L} \ll 1 \tag{6.9.3}$$

In other words, we consider only "long" tubes. By virtue of the definitions given in Eqs. 6.9.1, we know that s is of order one and that the second term in Eq. 6.9.2 is of order one, since both v_z and ζ are of that order. Hence it must be that

$$\frac{u_r}{\varepsilon U_o} = v_r \tag{6.9.4}$$

is of order one. Otherwise the two terms of Eq. 6.9.2 could not sum to zero. Equation 6.9.4 provides a suitable definition of the dimensionless *radial* velocity. With this definition we find that u_r is of the order of εU_o. Hence, and as expected, the radial velocity will be much smaller than the axial velocity if the leakage is distributed over a long axial length relative to the tube radius (i.e., $\varepsilon \ll 1$).

With these definitions, Eqs. 6.2.1 and 6.2.2 take the forms

$$0 = -\frac{\partial \varphi}{\partial s} + \varepsilon^2 \left[\frac{\partial}{\partial s} \left(\frac{1}{s} \frac{\partial}{\partial s} (s v_r) \right) + \varepsilon^2 \frac{\partial^2 v_r}{\partial \zeta^2} \right] \tag{6.9.5}$$

and

$$0 = -\frac{\partial \varphi}{\partial \zeta} + \frac{1}{s}\frac{\partial}{\partial s}\left(s\frac{\partial v_z}{\partial s}\right) + \varepsilon^2 \frac{\partial^2 v_z}{\partial \zeta^2} \tag{6.9.6}$$

(Remember that we are also assuming creeping flow as soon as we write Eqs. 6.2.1 and 6.2.2 with no inertial terms on the left-hand side.) To remove all other parameters from the equations, a dimensionless pressure has been defined as

$$\varphi = p\frac{\varepsilon R}{\mu U_o} \tag{6.9.7}$$

With Eqs. 6.2.4 and 6.2.5, the boundary condition for the radial velocity at the tube surface becomes

$$v_r = \Lambda \varphi \quad \text{at} \quad s = 1 \tag{6.9.8}$$

where a dimensionless permeability (leakage) coefficient is defined as

$$\Lambda = \frac{K\mu}{\varepsilon^2 R} \tag{6.9.9}$$

The other boundary conditions are

$$v_z = 0 \quad \text{along} \quad s = 1 \tag{6.9.10}$$

and

$$v_z \quad \text{and} \quad v_r \quad \text{are finite along} \quad s = 0 \tag{6.9.11}$$

It is easy to see that if we set $\varepsilon = 0$ in the dimensionless continuity and Navier–Stokes equations (Eqs. 6.9.2, 6.9.5, and 6.9.6) *and if we also set $\Lambda = 0$ in Eq. 6.9.8*[6] we recover the Poiseuille flow solution:

$$v_{rP} = 0 \tag{6.9.12}$$

$$v_{zP} = \frac{1}{4}\left(-\frac{d\varphi}{d\zeta}\right)_P (1 - s^2) \tag{6.9.13}$$

$$\left(\frac{d\varphi}{d\zeta}\right)_P = \text{constant} \tag{6.9.14}$$

For nonzero values of the leakage coefficient Λ, we see that these solutions to Eqs. 6.9.2, 6.9.5, and 6.9.6 depend only on the space coordinates s and ζ, and on the parameters ε and Λ. We can now develop an approximation scheme for the dynamics of this flow, in which we regard Λ as fixed, and not necessarily small, and the parameter ε to be small, but not zero. Obviously we expect that the solutions will be of the forms

$$v_z = f_z(s, \zeta; \Lambda, \varepsilon)$$
$$v_r = f_r(s, \zeta; \Lambda, \varepsilon) \tag{6.9.15}$$
$$\varphi = f_p(s, \zeta; \Lambda, \varepsilon)$$

As we did in Section 6.2, we introduce a specific assumption about the functional form of the solutions. Physically, we expect that the effect of the leakage on the internal

---

[6] This also requires the observation that if there is no leakage, and if $\varepsilon \ll 1$, then the flow is fully developed and the radial velocity is everywhere zero. See Problem 6.59.

flow will depend on the geometry, in the sense that if the leakage flow is distributed over a very long tube, the internal flow will be perturbed only slightly from the Poiseuille solution. We expect also that the solutions will depend on the parameter ε in some "smooth" way. If this is true, we should be able to express the solutions in the forms of power series in ε:

$$v_z = v_{z,0} + \varepsilon v_{z,1} + \varepsilon^2 v_{z,2} + \cdots \tag{6.9.16}$$

$$v_r = v_{r,0} + \varepsilon v_{r,1} + \varepsilon^2 v_{r,2} + \cdots \tag{6.9.17}$$

$$\varphi = \varphi_0 + \varepsilon \varphi_1 + \varepsilon^2 \varphi_2 + \cdots \tag{6.9.18}$$

We refer to the parameter ε as a *perturbation parameter*. Equations 6.9.16–6.9.18 are called "regular perturbation expansions" of each function. The functions $v_{z,i}$, $v_{r,i}$, and φ_i are all unknown, and take their names from the order of the ε coefficient; that is, $v_{r,2}$ is called the second-order expansion function of v_r, and φ_1 is called the first-order expansion function of φ, and $v_{z,0}$ is called the zeroth-order expansion function of v_z.

The perturbation method of solution now proceeds along the following lines. The functions given by Eqs. 6.9.16–6.9.18 are substituted into Eqs. 6.9.2, 6.9.5, and 6.9.6. For example, Eq. 6.9.6 becomes

$$
\begin{aligned}
0 = &-\frac{\partial \varphi_0}{\partial \zeta} + \frac{1}{s}\frac{\partial}{\partial s}\left(s\frac{\partial v_{z0}}{\partial s}\right) + \varepsilon^2 \frac{\partial^2 v_{z0}}{\partial \zeta^2} \\
&- \varepsilon \frac{\partial \varphi_1}{\partial \zeta} + \varepsilon \frac{1}{s}\frac{\partial}{\partial s}\left(s\frac{\partial v_{z1}}{\partial s}\right) + \varepsilon^3 \frac{\partial^2 v_{z1}}{\partial \zeta^2} \\
&- \varepsilon^2 \frac{\partial \varphi_2}{\partial \zeta} + \varepsilon^2 \frac{1}{s}\frac{\partial}{\partial s}\left(s\frac{\partial v_{z2}}{\partial s}\right) + \varepsilon^4 \frac{\partial^2 v_{z2}}{\partial \zeta^2} \cdots
\end{aligned}
\tag{6.9.19}
$$

A similar equation follows from Eq. 6.9.5, but we do not write it here. Neither have we written the expansion of the continuity equation. Since these equations are supposed to be valid for any arbitrary choice of ε, they can be satisfied only if each coefficient of ε^0, ε^1, ε^2, etc. is identically zero. Hence Eq. 6.9.19 generates a set of equations:

$$0 = -\frac{\partial \varphi_0}{\partial \zeta} + \frac{1}{s}\frac{\partial}{\partial s}\left(s\frac{\partial v_{z0}}{\partial s}\right) \tag{6.9.20}$$

$$0 = -\frac{\partial \varphi_1}{\partial \zeta} + \frac{1}{s}\frac{\partial}{\partial s}\left(s\frac{\partial v_{z1}}{\partial s}\right) \tag{6.9.21}$$

$$0 = \frac{\partial^2 v_{z0}}{\partial \zeta^2} - \frac{\partial \varphi_2}{\partial \zeta} + \frac{1}{s}\frac{\partial}{\partial s}\left(s\frac{\partial v_{z2}}{\partial s}\right) \tag{6.9.22}$$

and so on, corresponding to successive powers of ε. A similar set of equations follows from Eq. 6.9.5, but we do not write them here.

We must also write the boundary conditions in terms of these perturbation expansions. From Eq. 6.9.10 we find

$$v_z = 0 = v_{z,0} + \varepsilon v_{z,1} + \varepsilon^2 v_{z,2} + \cdots \text{ along } s = 1 \tag{6.9.23}$$

Hence each expansion function v_{zi} satisfies

$$v_{z,i} = 0 \quad \text{along} \quad s = 1 \tag{6.9.24}$$

Equation 6.9.11 is unchanged and holds for each velocity expansion function. For the radial velocity function, the wall boundary condition becomes

$$v_r = v_{r0} + \varepsilon v_{r,1} + \varepsilon^2 v_{r,2} + \cdots = \Lambda[\varphi_0 + \varepsilon \varphi_1 + \varepsilon^2 \varphi_2 + \cdots] \text{ at } s = 1 \tag{6.9.25}$$

and so for each expansion function φ_i we find

$$v_{ri} = \Lambda\varphi_i \qquad \text{at} \quad s = 1 \tag{6.9.26}$$

Now we can proceed to a solution for the expansion functions. We omit much of the detail. The zeroth-order equations are

$$0 = \frac{1}{s}\frac{\partial}{\partial s}(sv_{ro}) + \frac{\partial v_{zo}}{\partial \zeta} \tag{6.9.27}$$

$$0 = -\frac{\partial \varphi_o}{\partial \zeta} + \frac{1}{s}\frac{\partial}{\partial s}\left(s\frac{\partial v_{zo}}{\partial s}\right) \tag{6.9.28}$$

$$0 = -\frac{\partial \varphi_o}{\partial s} \tag{6.9.29}$$

It is important to keep in mind that there is no assumption here, as there is in arriving at the Poiseuille flow solution, that the flow is fully developed. Equation 6.9.29 does permit us to integrate Eq. 6.9.28 directly, with respect to s, and the result is

$$v_{zo} = \frac{1}{4}\left(\frac{d\varphi_o}{d\zeta}\right)(s^2 - 1) \tag{6.9.30}$$

While Eq. 6.9.30 looks like the Poiseuille flow solution, it is not the same. The fundamental distinction is that the pressure gradient is no longer constant. In this sense, Eq. 6.9.30 is equivalent to Eq. 6.2.6. But again there is a distinction. Equation 6.2.6 is the assumed *complete* solution for the axial velocity profile. Equation 6.9.30 is only the first term in the perturbation expansion of that function. The connection to Eq. 6.2.6 is even clearer if we find the average axial velocity at any position z. This follows from

$$\langle v_{zo}\rangle = \frac{\int_0^1 2\pi s v_{zo}\,ds}{\int_0^1 2\pi s\,ds} = 2\int_0^1 \frac{1}{4}\left(\frac{d\varphi_o}{d\zeta}\right)(s^2 - 1)s\,ds = -\frac{1}{8}\frac{d\varphi_o}{d\zeta} \tag{6.9.31}$$

Then Eq. 6.9.30 may be written in the form

$$v_{zo} = 2\langle v_{zo}\rangle(1 - s^2) \tag{6.9.32}$$

which is identical to Eq. 6.2.6. Again, we have the distinction that Eq. 6.2.6 is the assumed *complete* solution for the axial velocity profile, while Eq. 6.9.32 is only the first term in the perturbation expansion of that function.

After substituting Eq. 6.9.30 for the axial velocity term in the continuity equation, we can solve for the radial velocity by integration. The result is

$$v_{ro} = -\frac{1}{4}\left(\frac{d^2\varphi_o}{d\zeta^2}\right)\left(\frac{s^3}{4} - \frac{s}{2}\right) \tag{6.9.33}$$

If we now introduce Eq. 6.9.33, evaluated at the wall ($s = 1$) into the wall flux boundary condition (Eq. 6.9.26), the result is a differential equation for the dimensionless pressure function:

$$\frac{1}{16}\left(\frac{d^2\varphi_o}{d\zeta^2}\right) - \Lambda\varphi_o(\zeta) = 0 \tag{6.9.34}$$

Solution to this equation provides the pressure profile, in the zeroth-order approximation to the full perturbation series solution. We can solve this equation, subject to the following boundary conditions on pressure at the two ends:

$$\varphi_o(\zeta) = 0 \qquad \text{at} \quad \zeta = 1 \tag{6.9.35}$$

and

$$\varphi_o(\zeta) = \Delta\varphi_o \qquad \text{at} \quad \zeta = 0 \tag{6.9.36}$$

Here we introduce the dimensionless pressure drop driving the flow, which is given by

$$\Delta\varphi_o = \frac{\Delta P \varepsilon R}{\mu U_o} \tag{6.9.37}$$

Either ΔP or U_o is known and specified, but not both. The relationship between them gives us an analog to Poiseuille's law for a leaky tube. (See Problem 6.65.)

The solution to Eq. 6.9.34 (with the boundary conditions given in Eqs. 6.9.35 and 6.9.36) may be written in terms of either hyperbolic or exponential functions. At this stage we write it in the exponential form

$$\frac{\varphi_o}{\Delta\varphi_o} = \frac{e^{\lambda(1-\zeta)} - e^{-\lambda(1-\zeta)}}{e^\lambda - e^{-\lambda}} \tag{6.9.38}$$

where the format is simplified if we define

$$\lambda = 4\Lambda^{1/2} = \frac{4\beta}{\varepsilon} \tag{6.9.39}$$

and β is defined by Eq. 6.2.18.

Equation 6.9.38 is not identical to Eq. 6.2.23, and there is no reason that it should be. Equation 6.2.23 includes the approximation that β is small (Eq. 6.2.13). Equation 6.9.32 does not require that restriction. We can show that Eqs. 6.9.38 and 6.2.23 approach each other in the limit as $\beta \to 0$. (So can you. See Problem 6.63.) Once we have Eq. 6.9.38, we may calculate the velocity components, using Eq. 6.9.33 for the radial velocity and Eq. 6.9.30 for the axial velocity. We will not present those details here, but several problems at the end of the chapter permit us to explore those relationships.

By way of a summary, we make the following points. We have derived the first term of a perturbation expansion as an approximation to the solution for the flow and pressure fields in laminar Newtonian flow through a leaky tube. Additional terms can be evaluated at considerable effort. Through this additional effort, we can estimate the error made in retaining terms through a particular order, such as the zeroth-order solution worked out here. The perturbation expansion is not restricted to small leakage flows, even in the zeroth-order expansion. The only restriction is on the value of the perturbation parameter ε.

6.10 ROLL COATING

Figure 6.10.1 shows the geometry of a very challenging and interesting problem in fluid dynamics. A long cylinder or roll rotates about its axis while partially submerged in a viscous Newtonian liquid. The roll is close to a large planar vertical surface. The closest approach of the two surfaces is H_o, as shown. If the liquid wets the surface, and if the surface speed U is sufficient, a thin film of the liquid will be entrained on the roll and carried around until it reenters the liquid reservoir. The goal of a model of this system is a relationship between the liquid film thickness at the top of the roll, H, and the physical, geometrical, and operating parameters of the system. As we will see, lubrication theory makes possible an analytical approach to this problem.

To begin, we will consider only systems such that H_o is very small in comparison to the roll radius R, so that the submerged region between the surfaces is "almost parallel." We will adopt a Cartesian coordinate system, as shown in Fig. 6.10.1. We will account for the presence of gravity, but will assume that the flow is at such a low Reynolds

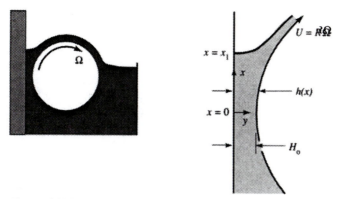

Figure 6.10.1 Sketch for analysis of roll coating.

number (Re $= UH_o/\nu$) that the inertial terms of the Navier–Stokes equations may be dropped. In fact, since the "almost parallel" assumption implies that $u_y \ll u_x$ and $\partial u_x/\partial x \ll \partial u_y/\partial y$, we simply begin with the "lubrication" equation:

$$0 = -\frac{\partial p}{\partial x} + \mu \frac{\partial^2 u_x}{\partial y^2} - \rho g \qquad (6.10.1)$$

A nondimensionalization that removes all parameters from Eq. 6.10.1 follows when we define

$$\xi = \frac{x}{(RH_o)^{1/2}} \qquad \eta = \frac{y}{H_o} \qquad u = \frac{u_x}{U}$$

$$P = \frac{pH_o^{3/2}}{\mu U R^{1/2}} \qquad F = \frac{\rho g H_0^2}{\mu U} \qquad (6.10.2)$$

and the result is

$$\frac{\partial^2 u}{\partial \eta^2} = \frac{\partial P}{\partial \xi} + F \qquad (6.10.3)$$

This is essentially the same equation we arrived at for the planar slider bearing (see Section 6.1) except that the nondimensionalization is different, and we are including gravity in the roll coating analysis. Otherwise, we proceed in a manner very similar to that presented earlier.

For this nearly parallel flow we assume that P is independent of η. This permits us to integrate Eq. 6.10.3 twice with respect to η to give u in the form

$$u = \frac{P'}{2}(\eta^2 - h^*\eta) + \frac{\eta}{h^*} \qquad (6.10.4)$$

where we define

$$P' = \frac{dP}{d\xi} + F \quad \text{and} \quad h^* = \frac{h(x)}{H_o} \qquad (6.10.5)$$

Note that since h^* varies with x (i.e., with ξ), Eq. 6.10.4 gives u as a function of both η and ξ. In writing Eq. 6.10.4 we have used the no-slip boundary conditions:

$$u = 0 \quad \text{at} \quad \eta = 0$$
$$u = 1 \quad \text{at} \quad \eta = h^*(\xi) \qquad (6.10.6)$$

As in the analysis in Section 6.1, we define a dimensionless flowrate as

$$\lambda = \frac{Q/UW}{H_o} = \frac{H}{H_o} = \int_0^{h^*} u(\xi, \eta)\, d\eta \tag{6.10.7}$$

where Q/W is the volume flowrate per axial width W. In writing Eq. 6.10.7 we assume that at the top of the roll the liquid moves as a rigid body with the linear roll speed U and thickness H. Upon performing the integration implied by Eq. 6.10.7, using Eq. 6.10.4 for u, and rearranging, we find

$$P' = \frac{dP}{d\xi} + F = 12\left(\frac{1}{2h^{*2}} - \frac{\lambda}{h^{*3}}\right) \tag{6.10.8}$$

which, except for notation and the addition of the gravitational term, is the same as Eq. 6.1.24.

Integration of Eq. 6.10.8 with respect to ξ yields the following expression for the pressure distribution, $P(\xi)$:

$$P(\xi) - P(-\xi_o) + (\xi - \xi_o)F = 12\int_{-\xi_o}^{\xi}\left(\frac{1}{2h^{*2}} - \frac{\lambda}{h^{*3}}\right) d\xi \tag{6.10.9}$$

For the lower limit on the integral we choose a position $-\xi_o$ so far upstream (i.e., below the line $x = 0$) that the only pressure is hydrostatic. In terms of our dimensionless variables, this implies

$$P(-\xi_o) = -F\xi_o \tag{6.10.10}$$

Consequently, from Eq. 6.10.9, the pressure distribution along the direction of motion is

$$P(\xi) = -F\xi + 12\int_{-\xi_o}^{\xi}\left(\frac{1}{2h^{*2}} - \frac{\lambda}{h^{*3}}\right) d\xi \tag{6.10.11}$$

To proceed further, we make two approximations. One is that $-\xi_o$ is large enough that $1/h^{*2} \ll 1$, and the integral above is then insensitive to the exact value of $-\xi_o$ used. Hence we set the lower limit of the integral to $-\xi_o = -\infty$. The second approximation is with respect to the geometry. It is not difficult to show that

$$h(x) = R + H_o - (R^2 - x^2)^{1/2} \tag{6.10.12}$$

The use of this function for h^* in the integral of Eq. 6.10.11 does not permit a simple analytical integration. However, in the region where $(x/R)^2 \ll 1$, the approximation

$$h^* = 1 + \frac{\xi^2}{2} \tag{6.10.13}$$

is valid and does permit an integration (using integral tables) that leads to

$$P(\xi) = \frac{(3 - 4.5\lambda)\xi}{1 + \frac{1}{2}\xi^2} - \frac{3\lambda\xi}{(1 + \frac{1}{2}\xi^2)^2} + (3 - 4.5\lambda)\sqrt{2}\tan^{-1}\left(\frac{\xi}{\sqrt{2}}\right) + \frac{(3 - 4.5\lambda)\pi}{\sqrt{2}} - F\xi \tag{6.10.14}$$

Keep in mind that λ is still an undetermined constant. Hence we need another equation that involves λ. So far, we have imposed no boundary conditions on pressure. It's time we did so.

At the meniscus where the liquid film separates from the planar surface, the flow field may be quite complex, and the position of the "contact line" ξ_1 is unknown. Surface tension effects, due to the curvature of the meniscus, will give a pressure just within the liquid at the meniscus that is below atmospheric pressure. The simplest approximation that we can make is to ignore surface tension effects at the meniscus and assume that the pressure at the position $\xi = \xi_1$ is atmospheric. Hence one condition on pressure may be written

$$P(\xi_1) = 0 \tag{6.10.15}$$

Then Eq. 6.10.14 becomes

$$0 = \frac{(3 - 4.5\lambda)\xi_1}{1 + \frac{1}{2}\xi_1^2} - \frac{3\lambda\xi_1}{(1 + \frac{1}{2}\xi_1^2)^2} + (3 - 4.5\lambda)\sqrt{2}\tan^{-1}\left(\frac{\xi_1}{\sqrt{2}}\right) + \frac{(3 - 4.5\lambda)\pi}{\sqrt{2}} - F\xi_1 \tag{6.10.16}$$

At this point we have a relationship between the dimensionless flowrate λ (or equivalently, the dimensionless coating thickness H/H_o) and the unknown position ξ_1. We need one more physical statement about the system that involves ξ_1.

We have arrived at a common sticking point in the development of mathematical models of complex physical systems. We don't know enough about the physics, or we cannot describe certain elements of the physics in simple enough terms, to produce a model that is free of unknown parameters. In this case, ξ_1 is unknown. Further progress requires action. One choice is to give up! This fluid dynamics problem may be too complex to yield a simple model, and we may have to resort to numerical methods of solving the Navier–Stokes equations. Alternatively, we can make a bold physical statement that leads to a model that, while suspect, is testable. We decide to illustrate the latter choice here. Then we will test the resulting model against some experimental data.

The approximation we make is the following. We have already assumed that the pressure in the liquid, at the meniscus, is atmospheric. If the pressure were atmospheric *throughout* the liquid, this flow field would just be planar drag flow between nearly parallel surfaces. For planar drag flow, the velocity profile would be linear, and u would satisfy

$$\frac{\partial^2 u}{\partial \eta^2} = 0 \tag{6.10.17}$$

Upon inspection of Eq. 6.10.3 and the definition of P', we see that this condition would correspond to a condition on P' of the form

$$P'(\xi) = 0 \tag{6.10.18}$$

We will adopt this as a boundary condition on P' at the separation line ξ_1, and write

$$P'(\xi_1) = 0 \tag{6.10.19}$$

From Eq. 6.10.8 we find

$$h^* = 2\lambda \quad \text{at} \quad \xi = \xi_1 \tag{6.10.20}$$

or, using this result with Eq. 6.10.13,

$$\xi_1 = [2(2\lambda - 1)]^{1/2} \tag{6.10.21}$$

Now we can substitute this relationship into Eq. 6.10.16, and the result is an *algebraic* equation for the desired quantity, λ, as a function of the parameter F. That equation can be solved numerically, though not analytically, much more easily than we can solve the Navier–Stokes equations for this flow numerically. The result is shown as the curve in Fig. 6.10.2. We find that *according to this model*, λ varies between 0.5 and 0.606—a very limited range—as the gravitational parameter F varies as shown. Some experimental data, obtained with a viscous oil as the liquid, are shown in Fig. 6.10.2. These data,

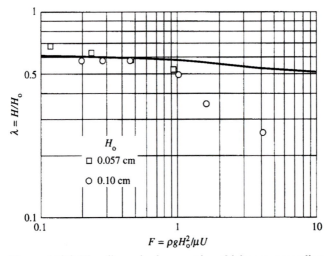

Figure 6.10.2 The dimensionless coating thickness, according to our simple model. The data are from Sullivan and Middleman [*Chem. Eng. Commun.*, **3**, 469 (1979)]. The physical properties of the oil are $\mu = 0.105$ Pa·s, $\rho = 800$ kg/m^3, and $\sigma = 0.035$ N/m. The roll radius was $R = 2.5$ cm.

and additional data shown in the reference cited in the caption, indicate that the model is fairly accurate for small values of the gravitational parameter F. As viscous effects grow smaller ($\mu U/H_o < \rho g H_o$), however, another phenomenon, not accounted for in the model, causes the coating thickness to lie well below that predicted by the model. Surface tension effects seem to be the most likely cause of the failure of the model in this range, and this issue is discussed in the reference cited. Here, we can conclude only that with some effort we can produce a model of *limited* validity, using the basic concepts introduced in this chapter. Nevertheless, the model does work well in a range of parameters of interest.

SUMMARY

Some important fluid dynamics problems involve flow fields that cannot be approximated as one-dimensional but still permit an important simplification, namely, that they are *nearly parallel.* The classic example is the lubrication flow with which we begin this chapter. It is the converging (two-dimensional) nature of that flow that gives rise to the essential character of the flow (the development of a pressure distribution that can support a load). Hence the one-dimensional approximation would "throw out the baby with the bathwater" and lead to a model incapable of describing what is important about this flow. Other examples follow (among them, laminar flow through a leaky tube, spreading of a viscous drop, and converging flow) in which the so-called lubrication approximation leads to models of real engineering utility.

In Section 6.5 we are introduced to a method of analysis called the integral method, and this is applied to yield a model of the development of the flow field and film thickness for a thin liquid film draining down a plane. Section 6.9 presents a method of analysis called the perturbation method. The point of these two brief excursions into mathematical analysis is to show that many methods are available to us for developing approximate solutions to complex flow problems.

PROBLEMS

6.1 Suppose we define a dimensionless pressure using H_o as the length scale (instead of H_1, as in Eq. 6.1.6). Show that

$$\bar{p}' = \frac{pH_o}{\mu U} \qquad \text{(P6.1.1)}$$

has a maximum value given by

$$\bar{p}'_{max} = \frac{3\Lambda'}{2} \frac{\kappa(1-\kappa)}{1+\kappa} \qquad \text{(P6.1.2)}$$

where

$$\Lambda' = \frac{L}{H_o} \qquad \text{(P6.1.3)}$$

Show that there is a geometry, defined by a specific value of κ, that maximizes the maximum pressure under the bearing. Give that value of κ. Regard L and H_o as fixed. What angle θ does this correspond to, for $\Lambda' = 10$, 30, and 100?

6.2 Derive Eqs. 6.1.29 and 6.1.35. *Hint:* The integrations will be simpler if you convert to an integral over h instead of over x, and use Eq. 6.1.20. For reducing the algebra, Eq. 6.1.27 will also be helpful.

6.3 Return to Eq. 6.1.35. For fixed values of L and H_o, find the value of κ at which the load Π is maximized. Is this also the κ at which the maximum pressure is maximized? (See Problem 6.1.)

6.4 Suppose that the slider bearing described in Example 6.1.1 must support a load of 1000 pounds (force) per inch width. Assume that $\theta = 3°$ is maintained and that the bearing position adjusts itself to a value of the gap H_o to support the load. What is the value of H_o for this case? What is the value of H_1 for this case? Do you see any problems with this design? How does your answer change if the load is reduced by a factor of 10?

6.5 A viscous liquid is to be coated onto a moving film as sketched in Fig. P6.5. The desired coating thickness is $H_\infty = 0.01$ cm. Specify values of H_1 and H_o that give this value of H_∞, *while minimizing the pressure force* developed under the blade.

6.6 Design an apparatus for measuring the linear speed of a film that is pulled across a rigid support, as shown in Fig. P6.6. Use the following concepts:

> A rigid solid "blade" W is constrained between a pair of frictionless supporting guides G. These guides permit the blade to move vertically, but maintain the angle θ of the face of the blade, relative to the support.

> A viscous Newtonian liquid is deposited on the film, upstream of the blade. Pressure generated by the lubrication flow between the blade and the film forces the blade to a vertical position that is displaced a distance D from its position when there is no motion.

Specify design parameters for this apparatus that will permit the measurement of velocities U in the range of 100–1000 cm/s. Give the final design equation that relates U to the measured value of D.

6.7 A viscous liquid is being blade-coated onto a moving surface, as shown in Fig. P6.5.
a. Find the final liquid film thickness H_∞ in millimeters.
b. Find the vertical force required to hold the blade in place, in newtons. (The blade has a width W in the z direction.)

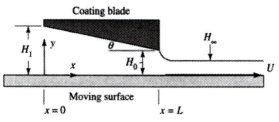

Figure P6.5 A coating blade.

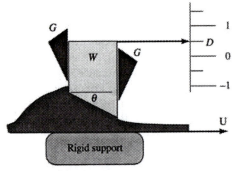

Figure P6.6

Use the following data:

$$U = 1 \text{ m/s} \quad L = 1 \text{ cm} \quad W = 10 \text{ cm}$$
$$H_1 = 1 \text{ mm} \quad H_o = 0.5 \text{ mm} \quad \mu = 10 \text{ Pa} \cdot \text{s}$$
$$\rho = 1000 \text{ kg/m}^3$$

6.8 Calculate the Reynolds number for the flow in Example 6.2.1.

6.9 Derive Eq. 6.2.17. Does Eq. 6.2.5 [with Eq. 6.2.9 and the subsequent solution for $f(z)$] lead to Eq. 6.2.24? Should it?

6.10 Does Eq. 6.2.26 show the expected dependence on tube length L?

6.11 Hollow fiber membranes can become plugged when microscopic particles in the feed water lodge in the pores of the membrane wall. One factor that aids in reducing plugging is maintenance of a high shear rate at the inner wall of the fiber. Calculate the shear rate of the feed flow in Example 6.2.1.

6.12 Kidney dialysis is carried out by passing blood from a patient through a hollow fiber unit similar to the one described in Example 6.2.1. A pressure p_o of 60 mmHg is available to pump the blood through the unit. The hollow fiber bed consists of a number (N_t) of membrane tubes, each of length L and inner radius R. The water permeability is given by $K = 10^{-13}$ (m³/s · m²)/ (N/m²). Design a system that will meet the following criteria:

Total blood flow	0.1 L/min
Maximum shear stress inside the hollow fiber tube	200 Pa
Water loss over 3 hours of dialysis	< 1 L

Take blood to be Newtonian with a viscosity at operating temperature of 0.004 Pa · s.

6.13 A hollow fiber membrane system is de-signed along the lines sketched in Fig. P6.13. Predict the flowrate of permeate liquid as a function of applied pressure P. Ignore osmotic effects. "Bad" water enters a pressure vessel and is maintained at a uniform pressure P, which drives pure water across a set of parallel loops of semipermeable hollow fiber membranes. Figure P6.13 shows one such loop. The pure water passes along the fiber axis to the two outlet ends of each fiber, which are at zero pressure relative to P. The total (end-to-end) permeable length of a single loop of fiber is L, and it has an inner diameter D_i. Assume the fiber wall is thin compared to D_i. Assume that the system consists of N identical loops. Assume P is uniform throughout the pressure vessel.

6.14 A hollow fiber water purifier is set up along the lines suggested in Fig. 6.2.2. The feed rate Q_{feed} and the available pressure p_o are fixed. The length L of the fibers is fixed, but we have available a choice of fibers of different internal radii, R, though all are of the same permeability K. We want to explore how the total permeate flow (the productivity of the module) depends on the choice of fiber radius. Prepare a plot of the permeate output Q_p as a function of fiber radius R. The total *volume* of fibers will be held constant, so the number of fibers in a module, n_f, will have to satisfy $n_f R^2 = $ constant. Use the data of Example 6.2.1 as the "base case," and calculate and plot Q_p versus R for ($25 < R < 100 \ \mu$m). What conclusions can you draw from this result?

6.15 Modify Eqs. 6.3.7 and 6.3.8 for the case of square holes. Make use of Fig. 5.3.2 in solving this problem.

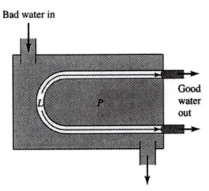

Figure P6.13 A "looped hollow fiber" system.

6.16 How does the factor α change if the holes are square? Combine this result with that of Problem 6.15 and calculate a new value for K and compare it to the value given in Eq. 6.3.10. Trace this change through the analysis of the hydrocephalus shunt, and give a corrected version of Eq. 6.3.27 for the square hole case. What is the percentage of error associated with the use of the circular hole equations, and does this change any conclusions we draw from Fig. 6.3.4?

6.17 Derive Eq. 6.3.18.

6.18 Examine the form taken by Eq. 6.3.24 in the limit of very small values of the parameter λ. Is this the expected result? Explain your answer.

6.19 Predict the coefficient in Eq. 6.3.27 when the perforations are 40 μm squares. Test this against the data given in Fig. 7 of Cho and Back [*J. Biomech.*, **22**, 335 (1989)].

6.20 How sensitive is the coefficient in Eq. 6.3.27 to the inside diameter of the shunt, D? Answer in two ways: What is the algebraic dependence of the coefficient on D? (Is it linear, quadratic, *etc.*?) For the range of D values given by Cho and Back (0.460–0.580 mm), what is the ratio of the coefficients calculated at these extremes?

6.21 Return to our model for "leaky" Hagen–Poiseuille flow (Eqs. 6.2.6 and 6.2.17). Show that

$$\Delta_{rz} \approx \frac{\partial u_z}{\partial r}$$

by finding an expression for

$$\frac{\partial u_z}{\partial r} \bigg/ \frac{\partial u_r}{\partial z} \quad \text{in terms of } \beta$$

6.22 A viscous droplet, of initial volume $V = 20 \times 10^{-9}$ m^3, spreads slowly across a smooth planar surface, which it wets. How long a time is required for the contact radius of the drop to reach $R = 1$ cm? At this time, what is the height of the drop above the surface, at the drop center?

$$\text{Take } \nu = 0.003 \text{ m}^2/\text{s}$$

6.23 Derive the criterion given by Eq. 6.4.15.

6.24 Equation 6.4.16 is based on the approximation of the spreading drop as a disk. This is not necessary, since the shape is given by Eq. 6.4.12. Using Eq. 6.4.12, find the ratio of the height of the drop at the center to the mean height. State clearly your definition of the mean height.

6.25 A viscous liquid flows upward through the inside of a tube and then overflows the top, as shown in Fig. P6.25. If the volumetric flowrate

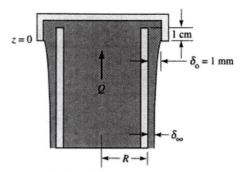

Figure P6.25 (Not to scale)

is $Q = 10^{-6}$ m^3/s and the outside radius of the tube is $R = 12$ mm, what vertical distance downstream from the plane $z = 0$ is required for the film thickness to come within 10% of its fully developed value? What is the fully developed value? The viscosity of the liquid is 0.03 Pa·s, and the density is 980 kg/m^3. Assume $\delta_\infty \ll R$.

6.26 Instead of using Eq. 6.5.19, carry through the analysis of the entrance length required for uniform film thickness for the following velocity profile:

$$u_x(x, y) = 2U_s(x)\left\{\frac{y}{\delta(x)} - \frac{1}{2}\left[\frac{y}{\delta(x)}\right]^2\right\} \quad \textbf{(P6.26)}$$

How sensitive is the predicted entry length to the choice of profile?

6.27 Give the derivations of Eqs. 6.5.22 and 6.5.23.

6.28 Solve Eq. 6.5.23 and put it into the form of Eq. 6.5.26.

6.29 Using the results given in Section 6.5, show that for a given value of Δ_o, the entry length may be expressed in the form

$$\frac{x_{1.1}}{\delta_\infty} = 0.25 X_{1.1} \text{ Re} \quad \textbf{(P6.29)}$$

where Re is the Reynolds number defined in Chapter 4, Eq. 4.4.28.

6.30 The model presented in Section 6.5 is taken from the work of Stücheli and Özisik [*Chem. Eng. Sci.*, **31**, 369 (1976)], who compare this model to data obtained and presented by Whitaker and Cerro (*Chem. Eng. Sci.*, **29**, 963 (1974)]. Figure P6.30 shows a set of Whitaker and Cerro's data for the case $\Delta_o = 0.47$. In this case (i.e., $\Delta_o < 1$), the film is introduced at a *higher* average velocity than that permitted by

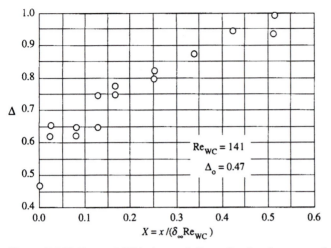

Figure P6.30 Data of Whitaker and Cerro for development of film thickness.

gravity-driven flow, and the film decelerates and gets *thicker* (rather than thinner, as in Fig. 6.5.3).
a. Using Eq. 6.5.26, compare the theoretical model to these data.
b. Note that Whitaker and Cerro define the Reynolds number used in Fig. P6.30 as

$$Re_{WC} \equiv \frac{\rho U_{s\infty} \delta_\infty}{\mu} \qquad (P6.30.1)$$

Show that

$$Re_{WC} \equiv \frac{3}{8} Re \qquad (P6.30.2)$$

where Re is the Reynolds number defined in Chapter 4, Eq. 4.4.28.

6.31 The model of the rotating disk skimmer presented in Section 6.6 is for the case of a disk in contact with only a single liquid phase. Suppose, instead of Fig. 6.6.2, the situation were more accurately represented by Fig. P6.31, where H is the thickness of the oil film floating on the water, but not the thickness of the entrained film on the skimmer. Suggest a method of plotting data on entrainment rate as a function of rotational speed and oil film thickness, using nondimensional variables. What set of dimensionless groups is sufficient to describe the dynamics of this system?

6.32 A rotating disk skimmer will be used to remove oil from a bath, as in Fig. P6.32. Design a system capable of removing oil at a rate of one L/h. Assume that oil is recovered from both sides of the disk by a scraper just above the interface on the plunging quadrant of the disk. The oil properties are $\mu = 0.05$ Pa·s, $\sigma = 0.025$ N/m, and $\rho = 1200$ kg/m^3.

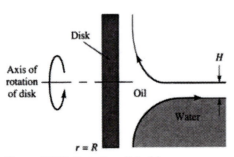

Figure P6.31 Rotating disk skimmer.

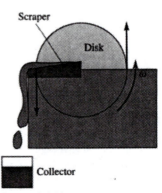

Figure P6.32

6.33 The data in Fig. 6.6.5 are taken from the upper curve of Fig. 6 of Christodoulou et al. [*J. Fluids Eng.,* **112,** 476 (1990)]. The disk radius was $R = 15$ cm and the disk was half-submerged (i.e., the quiescent liquid interface was along the horizontal diameter of the disk). Entrained liquid was scraped from both sides of the disk. The liquid had a density of 874 kg/m³, a viscosity of 0.026 Pa·s, and a surface tension of 0.03 N/m. Rotational speeds varied from 20 to 80 rpm. Give the range of capillary numbers to which these data correspond.

6.34 Show that with the following choices of nondimensional variables,

$$\omega = K\frac{r}{R} \qquad \theta = 1 - \frac{p}{p_\mathrm{r}}$$
$$K^2 = \frac{12kR^2}{H^3} \qquad\qquad \textbf{(P6.34.1)}$$

Eq. 6.7.6 may be transformed to the following format:

$$\frac{1}{\omega}\frac{d}{d\omega}\left(\omega\frac{d\theta}{d\omega}\right) - \theta = 0 \qquad \textbf{(P6.34.2)}$$

with the boundary conditions

$$\theta = 1 \quad \text{at} \quad \omega = K$$
$$\frac{d\theta}{d\omega} = 0 \quad \text{at} \quad \omega = 0 \qquad \textbf{(P6.34.3)}$$

Equation P6.34.2 is a Bessel equation, the solution of which is

$$\theta = \frac{I_0(\omega)}{I_0(K)} \qquad \textbf{(P6.34.4)}$$

where $I_0(\omega)$ is a "modified Bessel function of the first kind, of order zero." In problems that require you to manipulate this Bessel function (e.g., Problem 6.35), it is useful to have the following facts:

$$\frac{dI_0(\omega)}{d\omega} = -I_1(\omega) \qquad \textbf{(P6.34.5)}$$

where $I_1(\omega)$ is a "modified Bessel function of the first kind, of order one." An important integral relationship is

$$\int \omega I_0(\omega)\, d\omega = \omega I_1(\omega) \qquad \textbf{(P6.34.6)}$$

The two functions $I_0(\omega)$ and $I_1(\omega)$ are plotted in Fig. P6.34.

For small values of their arguments, it is useful

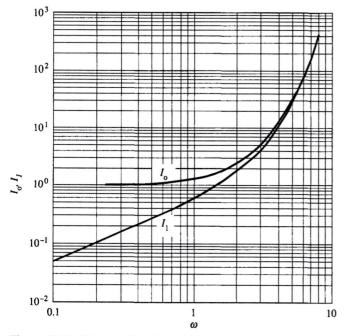

Figure P6.34 The modified Bessel functions of orders zero and one.

to have the following approximations available:

$$I_0(\omega) = 1 + \frac{\omega^2}{4}$$

$$I_1(\omega) = \frac{\omega}{2}$$

(P6.34.7)

6.35 Solve Eq. 6.7.6 for the case of a significant pressure variation under the disk; in particular, the approximation $p(r) \ll p_r$ is not good. (See Problem 6.34 for some of the mathematical necessities.) Cast the solution in the form of a functional relationship between H^* and p_r^*, and compare the solution to the data of Fig. 6.7.3. Show that in the limit of $H^* \ll 1$ the solution reduces to the form

$$H^* = 2^{1/3}(p_r^* - 1)^{2/3} \qquad \text{(P6.35)}$$

6.36 Return to the design problem of Example 6.7.1. Using the mathematical background provided in Problem 6.34, show that the relationship between the reservoir pressure p_r and the flowrate Q may be cast in the form

$$\frac{p_r \pi R^2 k}{Q\mu} = \frac{KI_0(K)}{2I_1(K)} \qquad \text{(P6.36)}$$

Plot the right-hand side of this equation as a function of K. When this function becomes nearly independent of K, it is no longer possible to determine H from the p_r/Q relationship with any precision. Suggest a minimum value for K that yields a good design for a film thickness device. State clearly your criterion for selecting this minimum K.

6.37 Return to the design problem of Example 6.7.1. The volumetric flowrate Q that appears in the analysis is measured at the exit of the disk ($r = R$), hence at atmospheric pressure. How is this related to the volumetric flowrate in the reservoir (i.e., at the reservoir pressure p_r)?

6.38 Return to the design problem of Example 6.7.1. Suppose we wish to measure variations in the coating thickness of the order of 100 μm. The porous disk is prepared by drilling holes of diameter 50 μm through a thin disk. The array of holes is on a square pitch, and the center-to-center distance of the holes is 100 μm. With the results of Problem 6.36 as a basis, specify design and operating conditions for a useful device.

6.39 Return to the design problem of Example

6.7.1. A porous disk is prepared by drilling holes of diameter 50 μm through a disk of 1 mm thickness. The array of holes is on a square pitch, and the center-to-center distance of the holes is 100 μm. What is the k value for this disk, for the flow of air at 25°C? Does the compressibility of the air have to be accounted for? State carefully how Q is defined in your definition of k. At what reservoir pressure p_r does the airflow through the disk become turbulent?

6.40 The arguments by which we transform Eq. 6.1.11 to Eq. 6.1.18, or drop the underlined term in Eq. 6.7.2 may not seem very compelling or intuitive. The point also can be argued through a nondimensionalization different from that illustrated in Section 6.1. Instead of the nondimensionalization suggested in Eqs. 6.1.5, replace the definition of $\tilde{x}$ by $\tilde{x} = x/L$. Leave the definition of $\tilde{y}$ unchanged. Now show that Eq. 6.1.18 follows with the assumptions of low Reynolds number, as before, and the additional assumption that $H_1/L \ll 1$. Nondimensionalize Eq. 6.7.2 in an analogous manner, and state conditions for the neglect of the underlined term.

6.41 Select one of the following topics, and present a bibliography of at least five journal articles that treat the topic problem. Cite each reference in the following format:

Jones, A., B. Smith, and C. Miller, "Title of the Paper," *Journal Name*, **vol. number,** first page (year).

1. Pressure-driven radial flow between stationary parallel disks
2. The effect of surface tension on the spreading of drops on surfaces
3. Coating onto a moving surface (see Fig. P6.5)
4. Flow through a converging tube
5. Squeezing flow between approaching parallel disks

Restrict your search to papers published in a 5-year period centered about the year of your birth plus the last digit of your Social Security number.

Here is a starting list of appropriate journals:

AIChE Journal
Chemical Engineering Science
Journal of Colloid and Interface Science
Journal of Fluids Engineering

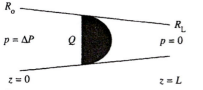

$p = \Delta P \qquad Q \qquad p = 0$

$z = 0 \qquad\qquad z = L$

Figure P6.42 Laminar flow through a tapered tube.

Journal of Fluid Mechanics
Journal of Multiphase Flow

6.42 Develop a simple analytical model for "almost parallel" tube flow through a tapered tube, as shown in Fig. P6.42. Put your final result in the form

$$\frac{\Delta P}{L} = \frac{8\mu Q}{\pi \overline{R}^4} F(\kappa) \qquad \textbf{(P6.42.1)}$$

and give an analytical expression for $F(\kappa)$. In this expression, $\overline{R}$ is the arithmetic average tube radius, and

$$\kappa \equiv \frac{R_L}{R_o} \qquad \textbf{(P6.42.2)}$$

Is the assumption of very low Reynolds number required? Are there any restrictions on the geometry, and if so, what geometrical shape factors determine the accuracy of the "almost parallel" assumption?

6.43 Develop an analytical model for laminar flow through a *leaky* tapered tube that is closed at its smaller end. The geometry is as shown in Fig. P6.42, except that the tube wall $R(z)$ is permeable, according to Eq. 6.2.4, and the end at $z = L$ is closed and impermeable. Put the final result in the form of a $Q(\Delta P)$ relationship.

6.44 Do Problem 6.43, but let the closed end be permeable, with the same K value as the tube wall.

6.45 Do Problem 6.43, but let the end at $z = L$ be open to the atmosphere ($p = 0$).

6.46 Just before Eq. 6.8.21 we assert that F is of order unity, as are the derivatives $dF/d\phi$ and $d^3F/d\phi^3$. Evaluate this statement, using the solution for $F(\phi)$ given by Eq. 6.8.28. Do the same for the inequality given in Eq. 6.8.21. Give an expression for the Reynolds number, in terms of the design and operating parameters, that satisfies

$$\text{Re} \frac{F\, dF/d\phi}{d^3F/d\phi^3} = 0.01 \qquad \textbf{(P6.46)}$$

and provide, thereby, a criterion for "creeping flow."

6.47 Plot the velocity profile (in terms of the function F given as Eq. 6.8.28), for $\alpha = 5°$, $20°$, and $45°$.

6.48 Carry out the proof that C is not a function of θ, as described just after Eq. 6.8.39.

6.49 Derive Eq. 6.8.45.

6.50 Show analytically (i.e., not simply by computation) that the function

$$\Phi(\alpha)$$
$$= \frac{[4\alpha(1 + \cos^2 2\alpha) - 4\cos 2\alpha \sin 2\alpha] \sin^3 \alpha}{\cos \alpha\, (\sin 2\alpha - 2\alpha \cos 2\alpha)^2}$$
$$\textbf{(P6.50)}$$

approaches the asymptotic value $\Phi = \frac{3}{4}$ in the limit as $\alpha \to 0$. How small must α be before Φ is within 10% of its asymptotic value?

6.51 Derive Eq. 6.8.55.

6.52 What percentage error is made in using Eq. 6.8.48 with the arithmetic average H to calculate the power requirement for flow through a converging duct, for the case of a convergence angle $\alpha = 20°$, and contraction ratios C_R in the range of 1.2–5.0?

6.53 Instead of using Eq. 6.8.57, write $\langle H \rangle$ in the form

$$\langle H \rangle = \tfrac{1}{2}(H_1 + H_2)G(C_R) \qquad \textbf{(P6.53)}$$

and give the function $G(C_R)$.

6.54 A highly viscous Newtonian liquid is to be extruded onto a moving surface, as suggested in Fig. P6.54. The extrusion die is designed as sketched in Fig. 6.8.6. The flowrate and liquid properties are as described in Example 6.8.2. Design constraints require that the die outlet be a planar parallel duct of the height shown in Fig.

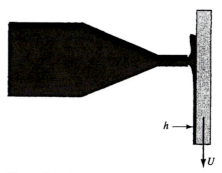

Figure P6.54 Coating die for producing a thin film on a moving surface.

6.8.6, and width $W = 0.1$ m. It is necessary to reduce the pressure drop to a value of 100 psi.
a. What length L_2 is required? What are the most questionable features of the model that you use?
b. The desired coated film thickness is $h = 0.0004$ m. What should the speed U of the substrate be?

6.55 Consider Example 6.8.2, but reverse the flow so that liquid moves from the narrow duct through a *diverging* region into the larger duct. Do you expect the model provided in Section 6.8 to yield a different pressure drop? Does the direction of flow alter your intuitive assessment of the validity of the model's main assumptions? Be specific in your response. List the key assumptions, and describe any reason that the assumptions would depend on the direction of flow.

6.56 Begin with Eq. 6.8.10 (i.e., assume that the radial flow assumption is valid). Prove the following assertion, strictly from inspection of the equation itself, and not from its solution: the flow is "reversible" only in the limit of creeping flow. By a "reversible" flow we mean that the streamlines of the flow, and the velocity profiles at any position, are independent of the direction of the flow (i.e., whether the flow is converging or diverging) except for a change in sign corresponding to the change in direction.

6.57 Prove that $\Phi(\alpha)$ (Eq. 6.8.54) approaches $\Phi(\alpha) = 0.75$ in the limit of vanishing α. Find an approximation formula for $\Phi(\alpha)$ valid for small, but nonzero, α.

6.58 Return to Eq. 6.8.29, which is valid only for small angle α. Using the relationships between Cartesian and polar coordinates, and their approximations valid for the case of small angle α:

$$\frac{y}{r} = \sin\theta \approx \theta \qquad \text{(P6.58a)}$$

$$\frac{H}{r} = \sin\alpha \approx \alpha \qquad \text{(P6.58b)}$$

$$\frac{z}{r} = \cos\theta \approx 1 \qquad \text{(P6.58c)}$$

show that Eq. 6.8.29 is, as expected, identical to Eq. 6.8.1.

6.59 A statement immediately following Eq. 6.9.11 concerns the behavior of the solution for $\varepsilon \to 0$ and $\Lambda = 0$. Suppose K is finite, but ε is

very small. Does the boundary condition Eq. 6.9.8 create any problems? Since $\Lambda \to \infty$ as $\varepsilon \to 0$, does the leakage flow remain bounded in this case?

6.60 Show that Eq. 6.9.30 leads to the following:

$$-\left(\frac{d\varphi_0}{d\zeta}\right)_{\zeta=0} = 8 \qquad \text{(P6.60)}$$

6.61 Derive Eq. 6.9.33.
6.62 Derive Eq. 6.9.38.
6.63 Show that Eqs. 6.9.38 and 6.2.23 approach each other in the limit as $\beta \to 0$.
6.64 Using Eq. 6.9.38, find the local leakage flux, and compare the result to Eq. 6.2.24. Are the two results identical?
6.65 Using Eq. P6.60 from Problem 6.60, find the relationship between ΔP and the average flow velocity into the entrance, U_0, and compare your result to Eq. 6.2.22. Are they the same? What value of what parameter determines how closely Eq. 6.2.22 approximates the zeroth-order solution?
6.66 Rework Example 6.2.1, but use the model developed in Section 6.9 instead. In particular, find the fraction of the feed that is produced as permeate, ϕ_L, and compare it to the result given earlier (Eq. 6.2.28). Then repeat Example 6.2.1, but use a K value 100 times larger. Again, compare the result from using Eq. 6.2.26 to that which you obtain using the analysis from Section 6.9.
6.67 Go back to the beginning of Section 6.9. Set up, and solve, a model for the following situation, illustrated in Fig. P6.67. A "leaky" tube, closed at one end, is embedded within a stagnant liquid that is under a uniform pressure P_0. Liquid permeates the wall of the tube and travels toward

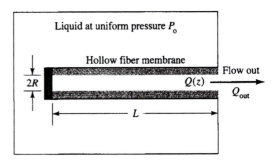

Figure P6.67

the open end. The flow is laminar, isothermal, and Newtonian. Present a model for the volumetric flow that exits the open end of the tube. Use the zeroth-order perturbation approximation to the equations that define this flow.

6.68 In the development given in Section 6.2, we can calculate the leakage flux by two routes: (a) using Eqs. 6.2.5, 6.2.9, and the solution for the function f, Eq. 6.2.17, and (b) Eq. 6.2.4 and the solution (Eq. 6.2.23) for $p(z)$. Problem 6.9 allows us to demonstrate that we do not get the same result by the two methods. Examine the same issue using the zero-order solutions of Section 6.9, namely, Eq. 6.9.38 for the pressure function and Eq. 6.9.33 for the radial velocity function. Do we get the same solutions for the leakage flux?

6.69 Show that the leakage flux that follows from the analysis of Section 6.9 (see Problem 6.68) behaves as expected on the limit of $L/R \to \infty$.

6.70 Derive Eqs. 6.10.12, 6.10.13, and 6.10.14.

6.71 When Eq. 6.10.20 is substituted into Eq. 6.10.16 we find a transcendental equation for $\lambda(F)$. The solution, obtained numerically, yields the line drawn on Fig. 6.10.2. Carry out the numerical solution and verify the line shown. Show *analytically* that λ approaches 0.5 as F grows large.

6.72 At the end of Section 6.10 we comment that surface tension effects, neglected in the analysis, are the probable cause of the failure of the simple model presented. We have a model for coating onto a surface withdrawn from a liquid bath in which surface tension effects are accounted for. The result of that model is Eq. 6.6.7.

a. Show that Eq. 6.6.7 leads to

$$\lambda = \frac{H}{H_o} = 0.946 \, (FCa^{-1/3})^{-1/2} \quad \textbf{(P6.72.1)}$$

where the capillary number is defined as

$$Ca = \frac{\mu U}{\sigma} \quad \textbf{(P6.72.2)}$$

b. The data below are from Sullivan and Middleman [*Chem. Eng. Commun.*, **3**, 469 (1979)]. Test this set against Eq. P6.72.1. The physical properties of the liquid are $\mu = 0.105$ Pa·s, $\rho = 800$ kg/m³, and $\sigma = 0.035$ N/m. The roll radius was $R = 2.5$ cm.

H_o (cm)	Ω (rpm)	H (cm)	H_o (cm)	Ω (rpm)	H (cm)
0.057	10	0.03	0.10	7	0.026
	20	0.033		18	0.036
	40	0.036		28	0.049
	80	0.038		64	0.058
				100	0.058
				145	0.058

H_o (cm)	Ω (rpm)	H (cm)
0.193	48	0.068
	94	0.090
	184	0.102

Does the result make sense physically?

6.73 A continuous tape (similar to photographic film) is conveyed through a liquid bath. At some point the tape is drawn over a supporting cylinder. Contact between the tape and the cylinder is prevented by the pressure generated in the converging/diverging region, as shown in Fig. P6.73. Develop a mathematical model for the relationship of the separation distance h_o to the normal force exerted on the tape **F** by the upper planar block B, in terms of the relevant parameters of the system. Assume that the liquid is Newtonian and that the speed of the tape is so low that the flow is laminar and inertial effects may be neglected. The upper side of the tape is in direct contact with the block B. Assume that the block and the cylinder extend a length W in the direction normal to the page that is large compared to h_o. Let $L \gg h_o$ be the length of the

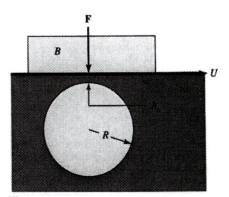

Figure P6.73 Tape drawn across a cylinder, under a plane.

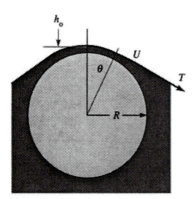

Figure P6.75 Tape drawn across a cylinder, with a wrap angle 2θ.

block in contact with the tape, in the direction of motion of the tape.

6.74 Repeat the calculations of Example 6.1.2, but for the case of a 5° angle. What is the effect of this change on the speed U at which the spacing H_o falls below 50 μm?

6.75 A continuous tape (similar to photographic film) is conveyed through a liquid bath. At some point the tape is drawn over a supporting cylinder to change the direction of motion of the tape. Contact between the tape and the cylinder is prevented by the pressure generated in the converging/diverging region, as shown in Fig. P6.75. Develop a mathematical model for the relationship of the minimum separation distance h_o to the tension exerted on the tape T, in terms of the relevant parameters of the system. Assume that the liquid is Newtonian and that the speed of the tape is so low that the flow is laminar and inertial effects may be neglected. The tension T is defined as the force in the direction of motion of the tape, per unit width of the tape.

6.76 For a slider bearing for which $\kappa = 0.2$ and $\theta = 15°$, plot the velocity profiles at $\check{x} = 0, 1$, and 2.5.

6.77 Repeat Example 6.8.2, but change the angle α to $\alpha = 3°$.

6.78 Show, beginning with Eq. 6.1.19, that the velocity profiles at the entrance to the bearing region exhibit the behavior sketched in Fig. P6.78.

6.79 Show that Eq. 6.1.44 has the behavior

$$\frac{(5\kappa - 18\lambda)\tan^2\theta}{\dfrac{3\kappa(\kappa - 1)}{\kappa + 1}} \rightarrow 4.33\tan^2\theta \qquad \text{for } \kappa \rightarrow 0$$

$$(P6.79)$$

For what value of θ does $\mathfrak{R}$ exceed 0.1?

Fixed bearing

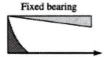

Moving surface
$\kappa > 1/2$ $\qquad\qquad$ $\kappa = 1/2$ $\qquad\qquad$ $\kappa < 1/2$

Figure P6.78

Chapter 7

Unsteady Flows

In most of the problems considered to this point we have made the assumption of a "steady state" flow. While in many important fluid dynamics problems the flow is in fact steady in time, there are *unsteady-state* problems that are of great interest. In addition, many steady state flows of technological interest begin from rest, and it is important to know how much time is required for the transient part of the dynamics (the "start-up" period) to decay to the point that the flow may be treated as if it were steady for all time. We begin with a treatment of a transient problem in a simple geometry. This is followed by consideration of a stability problem, which by definition is unsteady. A good portion of this chapter is then devoted to quasi-steady analyses, in which we learn how it is possible to treat unsteady flows by simpler steady state models.

7.1 A TRANSIENT PRESSURE FLOW

Imagine that a long tube is attached to the bottom of a reservoir. Both the tube and the reservoir are filled with liquid. The cross section of the tube is circular. At times $t < 0$ the liquid is everywhere at rest—there is no flow. Suddenly, at times $t > 0$, a constant pressure difference ΔP is imposed across the ends of the tube, and the liquid begins to move from the reservoir into the tube, and thence down the tube axis. We would like to be able to estimate the time required for the flow within the tube to reach its steady state.

We will assume laminar flow at low Reynolds numbers and consider a tube so long that the entrance region has little effect on the character of the flow. We assume that the only velocity component is $u_z(r, t)$. The equations of conservation of mass and momentum reduce to the following forms for isothermal incompressible Newtonian flow:

$$\rho \frac{\partial u_z}{\partial t} = -\frac{\partial \mathscr{P}}{\partial z} + \mu \frac{1}{r} \frac{\partial}{\partial r}\left(r \frac{\partial u_z}{\partial r}\right) \tag{7.1.1}$$

and

$$0 = \frac{\partial u_z}{\partial z} \tag{7.1.2}$$

321

(Keep in mind that $\mathscr{P}$ is a modified pressure that includes any hydrostatic contribution. See Eq. 3.2.35.)

It is useful, in solving transient problems, to have the steady state solution available at the beginning of the solution procedure. The steady velocity in Hagen–Poiseuille flow (Chapter 3: Section 3.2) may be written in the form

$$u_z^s = u_z^{max} \left[1 - \left(\frac{r}{R} \right)^2 \right] \qquad (7.1.3)$$

and

$$-\frac{\partial \mathscr{P}}{\partial z} = \frac{\Delta \mathscr{P}}{L} = \text{constant} = \frac{4 u_z^{max} \mu}{R^2} \qquad (7.1.4)$$

We denote the centerline or maximum velocity by u_z^{max}. Alternatively, as in earlier chapters, we can write expressions relating the pressure gradient to the flow using the average velocity, which is just half of u_z^{max}, or the volume flowrate Q.

We will look for a solution to Eq. 7.1.1 which is the sum of the steady state solution and an unknown transient function denoted by $U(r, t)$:

$$u_z = u_z^s + U(r, t) \qquad (7.1.5)$$

We are going to assume that the transient *pressure* profile is identical to the corresponding steady state profile, so that even for the unsteady flow

$$-\frac{\partial \mathscr{P}}{\partial z} = \frac{\Delta \mathscr{P}}{L} = \frac{\Delta P - \rho g_z L}{L} = \text{constant} \qquad (7.1.6)$$

If these assumptions (Eqs. 7.1.5 and 7.1.6) do not make sense, we will find out quickly enough, since we will be unable to get a solution to Eq. 7.1.1 that satisfies them. Hence we substitute these assumed constraints on the solutions into Eq. 7.1.1 and seek a solution that can satisfy physically meaningful boundary conditions. Upon substitution of Eq. 7.1.5 with Eqs. 7.1.3, and Eq. 7.1.6, into Eq. 7.1.1, we find that the function $U(r, t)$ must be a solution to

$$\rho \frac{\partial U}{\partial t} = \mu \frac{1}{r} \frac{\partial}{\partial r} \left(r \frac{\partial U}{\partial r} \right) \qquad (7.1.7)$$

We can nondimensionalize this equation by defining the following variables:

$$s = \frac{r}{R} \qquad \Phi = \frac{U}{u_z^{max}} \qquad \tau = \frac{\mu t}{\rho R^2} \qquad (7.1.8)$$

Equation 7.1.7 then takes the form

$$\frac{\partial \Phi}{\partial \tau} = \frac{1}{s} \frac{\partial}{\partial s} \left(s \frac{\partial \Phi}{\partial s} \right) \qquad (7.1.9)$$

The choice of nondimensional variables given in Eqs. 7.1.8 was made to yield an equation (Eq. 7.1.9) free of parameters. Whenever possible, we nondimensionalize equations to remove individual parameters from them, leaving behind the minimum number of *dimensionless* parameters—none if possible.

We need two boundary conditions and an initial condition on Φ. The no-slip condition requires that

$$u_z = 0 \qquad \text{on} \quad r = R \qquad (7.1.10)$$

Using Eqs. 7.1.5 and 7.1.3 we see that this requires that

$$\Phi = 0 \qquad \text{on} \quad s = 1 \qquad\qquad \textbf{(7.1.11)}$$

The velocity profile must be symmetric about $r = 0$. Mathematically we require that

$$\frac{\partial \Phi}{\partial s} = 0 \qquad \text{along} \quad s = 0 \qquad\qquad \textbf{(7.1.12)}$$

The initial condition on the velocity is that the liquid be at rest initially. If we set $u_z = 0$ in Eq. 7.1.5, it follows that

$$\Phi = -(1 - s^2) \qquad \text{at} \quad \tau = 0 \qquad\qquad \textbf{(7.1.13)}$$

Equation 7.1.9 along with the boundary and initial conditions given here is a special class of partial differential equation known as a *Sturm–Liouville equation*. It is usually solved by the method of separation of variables. We will not present the details of the solution procedure here. The solution involves Bessel functions (we met these particular functions in Section 2.4 of Chapter 2), and the solution is expressed as an infinite series that takes the form

$$\Phi = -8 \sum_{n=1}^{\infty} \frac{J_o(\lambda_n s)}{\lambda_n^3 J_1(\lambda_n)} \exp(-\lambda_n^2 \tau) \qquad\qquad \textbf{(7.1.14)}$$

The coefficients λ_n are the so-called *eigenvalues* of the differential equation. In this case they are the zeros of the Bessel function J_o. That is, the λ_n are the infinite set of roots of

$$J_o(\lambda_n) = 0 \qquad\qquad \textbf{(7.1.15)}$$

These roots or "zeros" are tabulated in most math tables, where we find

$$\lambda_1 = 2.41, \qquad \lambda_2 = 5.52, \qquad \lambda_3 = 8.65, \qquad \lambda_{n+1} = \lambda_n + \pi \qquad \text{for} \quad n \geq 3 \qquad \textbf{(7.1.16)}$$

The function Φ gives the transient part of the velocity profile. The rate at which the velocity along the tube axis ($s = 0$) builds to its maximum value gives a measure of how rapidly the steady state profile is achieved. In view of the definition of Φ, we examine the behavior of $\Phi(0, \tau)$, and observe how quickly this function becomes *small* compared to unity.

Setting $s = 0$ in Eq. 7.1.14, we find

$$\Phi(0, \tau) = -8 \left[\frac{1}{\lambda_1^3 J_1(\lambda_1)} \exp(-\lambda_1^2 \tau) + \frac{1}{\lambda_2^3 J_1(\lambda_2)} \exp(-\lambda_2^2 \tau) + \cdots \right] \qquad \textbf{(7.1.17)}$$

Fortunately this series converges very rapidly, except for very small values of the time τ, and we need consider only the first term of the series for values of τ larger than 0.2. Figure 7.1.1 shows the decay of $\Phi(0, \tau)$, and we conclude that the steady profile is achieved for $\tau \approx 1$, since $\Phi(0, \tau)$ falls well below 0.01 by that time. The conclusion we draw is that the transient disappears in a *dimensionless* time of the order of $\tau = 1$. In terms of real time, the transient disappears, or steady flow is achieved, after a time given by

$$t_{\infty} = \frac{\rho R^2}{\mu} \qquad\qquad \textbf{(7.1.18)}$$

Similar results are found for other *viscous-dominated* transient flows, such as axial annular flow, and tangential annular flow. As a general rule of thumb, a time of the

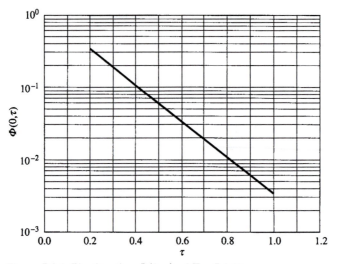

Figure 7.1.1 The function $\Phi(0, \tau)$ of Eq. 7.1.17.

order of that given by Eq. 7.1.18 is required to establish steady flow from rest, when R is taken as the "equivalent radius" of the specific geometry. Since we are looking at an order-of-magnitude estimate for t_∞, we may simply define the equivalent radius R_{eq} as twice the ratio of the area normal to flow to the wetted perimeter. For example, for flow through the annular region between concentric cylinders of radii κR and R, we find

$$R_{eq} = 2\left[\frac{\pi(R^2 - \kappa^2 R^2)}{2\pi(R + \kappa R)}\right] = (1 - \kappa)\,R \tag{7.1.19}$$

The factor of 2 outside the brackets is chosen to ensure that we recover the expected result $R_{eq} = R$ for a circular cross section.

EXAMPLE 7.1.1   *Start-Up of a Capillary Viscometer*

A capillary viscometer with a capillary radius $R = 0.2$ mm is available to measure the viscosity of a series of viscous liquids. The test liquid is available in small quantities and is very expensive. Your technician reports that it takes about 3 minutes for the effluent from the viscometer to achieve a steady flowrate. He insists that this is due to the high viscosity of the liquids, which "slows down the rate at which steady flow is reached." Comment on this conclusion. The technician's preliminary data indicate that the viscosities are of the order of 10 Pa·s.

We can use Eq. 7.1.18 to estimate the time required to achieve steady flow. We have no information on the liquid density, so we will use a nominal value of 1000 kg/m³, suitable to most liquids. Then we find

$$t_\infty = \frac{\rho R^2}{\mu} = \frac{1000(2 \times 10^{-4})^2}{10} = 4 \times 10^{-6}\,\text{s!} \tag{7.1.20}$$

The steady flow is achieved instantaneously, according to our model. Furthermore, high viscosity does not retard the achievement of steady flow—it promotes it. The observed transient must be due to another factor, probably associated with the mechanism by which pressure is imposed on the system, or the mechanism for its measurement.

Figure 7.2.1 Breakup of a laminar jet into droplets.

7.2   STABILITY OF A LAMINAR LIQUID JET

When a liquid in laminar flow issues from a horizontal capillary, a jet is formed which, at sufficiently high velocities, remains nearly horizontal for some distance from the exit. A common observation is that while the jet remains axisymmetric, its surface does not remain cylindrical (i.e., of constant radius). Instead, bulges begin to grow until they cause the jet to be disrupted into a series of droplets. Figure 7.2.1 depicts the phenomenon. This is an example of a *stability* problem, a point we will clarify shortly.

The mechanism of this instability lies in the presence of a finite surface tension. We discussed surface tension earlier in the context of a force per unit length, and we derived the Young–Laplace equation from a force balance on an interface with curvature. However, we may also regard surface tension as the *energy per unit of surface area* associated with an interface. We need to adopt this perspective because, in general, stability problems are related to *energy* considerations. When a system is perturbed in such a way that its energy is *increased,* it will usually attempt to return to the original, *lower,* energy state. We would call that a stable system. However, if the perturbed system can move toward a state of *lower* energy than that of the initial state, the original system is unstable. Hence we should consider the question of whether a cylinder has more, or less, area than a sphere. This is not a clearly stated question, as we will see in a moment.[1]

A simple example can be constructed, with reference to Fig. 7.2.2. We imagine an infinitely long cylinder of radius R_c, and we consider a finite *section* of axial length Λ of that cylinder. We ask the question: Does that section have a larger, or smaller, area than a sphere of the same volume?

When we equate the volume of the cylindrical section,

$$V_o = \pi R_c^2 \Lambda \tag{7.2.1}$$

to the volume of a sphere

$$V_o = \tfrac{4}{3}\pi R_s^3 \tag{7.2.2}$$

we find that the "volume equivalent" sphere has a radius given by

$$R_s = (\tfrac{3}{4} R_c^2 \Lambda)^{1/3} \tag{7.2.3}$$

Now we examine the areas of the cylindrical section and the "volume equivalent" sphere:

$$A_s = 4\pi R_s^2 = 4\pi (\tfrac{3}{4} R_c^2 \Lambda)^{2/3} \tag{7.2.4}$$

and

$$A_c = 2\pi R_c \Lambda \tag{7.2.5}$$

---

[1] Some questions that appear to be well defined, in fact are not. This is an important point, and we will see other examples periodically throughout this text.

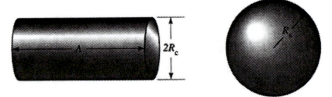

Figure 7.2.2 Transformation of a finite section of a long cylinder into a sphere of equal volume.

The ratio of the area of the sphere to that of the cylindrical section is

$$\frac{A_s}{A_c} = 4\pi \frac{(\frac{3}{4} R_c^2 \Lambda)^{2/3}}{2\pi R_c \Lambda} = \left(\frac{9}{2} \frac{R_c}{\Lambda}\right)^{1/3} \tag{7.2.6}$$

It follows then that the "equivalent volume" sphere has a smaller surface area, hence a lower surface energy, only if Λ is greater than $4.5R_c$. In the context of our definition of stability, a section of the cylinder shorter than $4.5R_c$ is stable This is a somewhat oversimplified view of stability, but it will be helpful in the discussion to follow.

We now return to the problem of the liquid jet moving horizontally at a uniform velocity U. We ignore any influence of the ambient medium. This is reasonable for a low speed jet, for which the shear stress exerted by the surrounding air is negligible. We adopt a cylindrical coordinate system fixed in the jet (i.e., moving with the velocity U). In that coordinate system the undisturbed jet is stationary. Hence the velocity vector has zero axial and radial components, and the pressure is everywhere uniform, and given by

$$p^{\circ} = \frac{\sigma}{R} \tag{7.2.7}$$

We now carry out an analysis of the stability of the jet to small disturbances. In a stability analysis we assume that the velocity components, the pressure, and the jet radius are somehow perturbed from their current values. We then solve the Navier–Stokes equations for the ensuing velocity, pressure, and radius that follow upon that perturbation. If these variables decay with time back to their initial values, we say that the system was stable to these disturbances. If these quantities grow without bound, we say that the system was unstable.

If we assume that the perturbed jet remains axisymmetric, we may write the dynamic equations in the form

$$\rho \frac{\partial u_z}{\partial t} = -\frac{\partial p}{\partial z} + \mu \left[\frac{1}{r} \frac{\partial}{\partial r} r \left(\frac{\partial u_z}{\partial r} \right) + \frac{\partial^2 u_z}{\partial z^2} \right] \tag{7.2.8}$$

and

$$\rho \frac{\partial u_r}{\partial t} = -\frac{\partial p}{\partial r} + \mu \left[\frac{\partial}{\partial r} \left(\frac{1}{r} \frac{\partial}{\partial r} r u_r \right) + \frac{\partial^2 u_r}{\partial z^2} \right] \tag{7.2.9}$$

Note that we have dropped the *nonlinear* terms from the left-hand side of the Navier–Stokes equations. We are assuming that inertial effects are not significant. This may be reasonable for viscous jets, but it is not clear that it is a good assumption in general. Note, as well, that we are keeping the acceleration terms, $\partial/\partial t$. Since a stability problem is by its very nature a transient problem, we cannot formulate it with the steady state equations. Ultimately, the test of the assumption to linearize the equations lies in examination of reality. Hence we will make the assumption now, in the interest of simplifying the mathematics, and examine its validity later in Section 7.2.2, where we compare the nonlinear inertial terms to the acceleration terms.

Of course we must add the continuity equation

$$\frac{1}{r}\frac{\partial}{\partial r}ru_r + \frac{\partial u_z}{\partial z} = 0 \tag{7.2.10}$$

At this point we see that the mathematical formulation yields three equations in the three dynamic variables. The Navier–Stokes equations are *second-order* partial differential equations. Because they are linear, we will be able to get analytical solutions. Unfortunately the solutions will have a very complicated-looking format resembling an explosion in a Bessel function factory. For this reason, and others, we will solve this problem along lines we have not illustrated up to this point.

We ask the question: Can we neglect viscous effects, and if so, why have we not done that in any earlier examples? We are moved to suggest such neglect because the viscous terms are *second order* (not nonlinear—just second order—make sure you understand the distinction). Hence *if it makes sense* to throw these terms out, we can simplify the mathematics by means of reducing the order of the differential equation.

When would it *not* make sense to throw away the viscous terms? If we review the various fluid dynamics problems illustrated up to this point, we will find that practically every one involved viscous effects as the *only* resistive force to the motion. For example, to find the pressure drop–flowrate relationship in a long tube of uniform cross section (we have looked at several such problems), we must realize that if there were no viscous effects, there would be no pressure drop. Hence the viscous terms were the only significant terms in the formulation of the problem; had we dropped the viscous terms, we would have obtained trivial solutions. There are exceptions to be found in earlier examples.

In Section 4.4 of Chapter 4 we looked at bubble growth in a viscous liquid. We maintained the inertial terms down to Eq. 4.4.44, and then dropped them to consider growth in a very viscous liquid. However, we *could* drop the viscous terms in that equation, but retain the inertial terms and still have a physically sensible problem. In fact, Eq. 4.4.44 with no viscous terms provides a very useful model for growth and collapse of bubbles in water, where inertial effects dominate viscous effects.

In most of the earlier problems in which we dropped the inertial terms, we had relied on the assumption that, for example, the flow was fully developed, in which case the inertial terms vanished exactly. Another example to reconsider is lubrication flow, and other "almost parallel" flows, where the inertial terms exist by virtue of the geometry of the flow. We may choose to neglect them and live with the consequences. Thus we conclude that the relative roles of inertial and viscous effects have something to do with viscosity, and something to do with geometry. We will not deal with this point further here, but we do point out that flow theory for fluids of vanishingly small viscosity is very important. It forms the foundation of aerodynamics, for example.

We turn, then, to an analysis of the stability of a liquid jet with surface tension, but in the absence of viscous effects. We call this an "inviscid flow analysis."

7.2.1 Stability of an Inviscid Jet

The continuity equation is unaffected by any assumption about viscosity. The dynamic equations, upon setting the viscous terms to zero, take the forms

$$\frac{\partial u_r^o}{\partial t} = -\frac{1}{\rho}\frac{\partial p^o}{\partial r} \tag{7.2.11}$$

and

$$\frac{\partial u_z^\circ}{\partial t} = -\frac{1}{\rho}\frac{\partial p^\circ}{\partial z} \qquad (7.2.12)$$

We use the superscript $^\circ$ to denote the inviscid approximation. We begin with the *assumption* that there is a function ϕ such that

$$u_r^\circ = \frac{\partial \phi}{\partial r} \qquad \text{and} \qquad u_z^\circ = \frac{\partial \phi}{\partial z} \qquad (7.2.13)$$

Can we find a function ϕ that yields velocities that satisfy the dynamic equations? Not under all conditions. The flow we are examining, however, happens to be one for which such a function ϕ exists. Rather than discussing the point in general here, we will find a function ϕ and confirm that the dynamic equations are satisfied.

When Eqs. 7.2.13 are substituted into both dynamic equations, we obtain a pair of equations that are satisfied by a pressure function of the form

$$p^\circ = -\rho\frac{\partial \phi}{\partial t} + \frac{\sigma}{R} \qquad (7.2.14)$$

The constant of integration becomes σ/R so that the pressure is strictly due to surface tension in the stationary case ($\phi = 0$).

When Eqs. 7.2.13 are substituted into the continuity equation, we find that ϕ must be a solution to

$$\frac{1}{r}\frac{\partial}{\partial r}\left(r\frac{\partial \phi}{\partial r}\right) + \frac{\partial^2 \phi}{\partial z^2} = 0 \qquad (7.2.15)$$

We expect that ϕ is a function of r, z, and t. Based on our observations, we know that the disturbance to the radius is periodic along the z axis and that it grows monotonically in time. For this reason we seek a solution that has the form

$$\phi = \Phi(r)\exp(ikz + \alpha t) \qquad (7.2.16)$$

We choose to write the periodic z dependence in the more compact format $\exp(ikz)$ instead of in terms of sine or cosine functions. The time dependence grows or dampens exponentially according to the magnitude and sign of α, which is a growth rate parameter. We call k the *wavenumber,* and α the *growth rate* of the disturbance. We have hypothesized that ϕ has the form given in Eq. 7.2.16. In particular, we have assumed that the surface has been disturbed or perturbed in a spatially periodic manner along the z axis. When we substitute this into the continuity equation (now in the form of Eq. 7.2.15), we find that the assumed forms of the z and t functions do satisfy the equation and that Φ must be the solution to

$$\frac{1}{r}\frac{d}{dr}\left(r\frac{d\Phi}{dr}\right) - k^2\Phi = 0 \qquad (7.2.17)$$

There is a form of Bessel equation. (See the discussion in Chapter 6, Section 6.7, on the same equation, and Problem 6.34.) If we study this class of ordinary differential equation, we find that the solutions are the "hyperbolic" or "modified" Bessel functions I_0 and K_0. The K_0 function is unbounded at $r = 0$, so a solution involving K_0 cannot satisfy a boundary condition that requires that the velocity be bounded along $r = 0$. Hence we write the solution in the form

$$\Phi = AI_0(kr) \qquad (7.2.18)$$

where A is a constant of integration. When this relationship is introduced into Eq. 7.2.16, we may write the dynamic variables from Eqs. 7.2.13 and 7.2.14 as follows:

$$p^\circ = A\rho\alpha I_0(kr)\exp(ikz + \alpha t) + \frac{\sigma}{R} \tag{7.2.19}$$

$$u_r^\circ = -AkI_1(kr)\exp(ikz + \alpha t) \tag{7.2.20}$$

$$u_z^\circ = AikI_0(kr)\exp(ikz + \alpha t) \tag{7.2.21}$$

(We have used the fact that for this kind of Bessel function, $dI_0(kr)/dr = -kI_1(kr)$, where I_1 is another kind of Bessel function.) We will write the disturbance to the jet radius as

$$R^\circ = R + \zeta^\circ(z, t) \tag{7.2.22}$$

where ζ° is connected to the radial velocity of the surface by[2]

$$\frac{\partial\zeta^\circ}{\partial t} = u_r^\circ \quad \text{at} \quad r = R \tag{7.2.23}$$

This permits us to write the surface disturbance as

$$\zeta^\circ = -A\frac{k}{\alpha}I_1(kR)\exp(ikz + \alpha t) \tag{7.2.24}$$

We now impose a boundary condition on the pressure function, namely, the Young–Laplace equation

$$p^\circ = \sigma\left(\frac{1}{R_1} + \frac{1}{R_2}\right) \quad \text{at} \quad r = R \tag{7.2.25}$$

The radii of curvature are given by the following expressions. Normal to the axis the cross section is circular, with radius R° given by Eq. 7.2.22 above. Hence

$$\frac{1}{R_1} = \frac{1}{R + \zeta^\circ} = \frac{1}{R(1 + \zeta^\circ/R)} \approx \frac{1 - \zeta^\circ/R}{R} \tag{7.2.26}$$

In the plane of the axis the radius of curvature is given approximately by (see Problem 7.21)

$$\frac{1}{R_2} = -\frac{\partial^2\zeta^\circ}{\partial z^2} \tag{7.2.27}$$

Then the pressure is given by

$$p^\circ = \frac{\sigma}{R} - \frac{\sigma}{R^2}\left(\zeta^\circ + R^2\frac{\partial^2\zeta^\circ}{\partial z^2}\right) \tag{7.2.28}$$

This expression is valid only for small-amplitude disturbances such that the approximation that $\zeta^\circ/R \ll 1$ is valid. When this expression for p° is equated to that given in Eq. 7.2.19, and Eq. 7.2.24 is used for ζ°, we find the following:

$$\frac{\sigma}{R^2}(1 - k^2R^2)\frac{k}{\alpha}I_1(kR) = \rho\alpha I_0(kR) \tag{7.2.29}$$

---

[2] The boundary conditions given by Eqs. 7.2.23 and 7.2.25, should really be evaluated at $r = R + \zeta^\circ$. This complicates application of the boundary condition, since ζ is unknown. We make the approximation that for ζ°/R small enough, the simpler boundary condition will not give rise to a large error.

Note that the integration constant A has disappeared. What we are left with is a relationship between the growth rate α and the wave number of a disturbance to the jet surface, k. Solving for α we find

$$\alpha^2 = \frac{\sigma k}{\rho R^2}(1 - k^2 R^2)\frac{I_1(kR)}{I_0(kR)} \tag{7.2.30}$$

Recall that we stated that if α is positive, the disturbances grow exponentially. We find that the ratio of Bessel functions (I_1/I_0) is positive for all arguments. Hence α is positive so long as $kR < 1$. (If $kR > 1$ then α^2 is negative and α is imaginary; hence the time dependence is periodic.) We conclude that any periodic disturbance to the jet having a wavenumber that satisfies the condition $0 < kR < 1$ will lead to instability of the jet. For a given value of kR in that range, the disturbance will grow at a rate given by Eq. 7.2.30.

It is interesting to compare the result of this analysis with the simple energy argument presented earlier, indicating that a portion of a jet shorter than $4.5R$ would be stable. What is the relationship of this result to the restriction on kR? First we must note (from the form of Eq. 7.2.16) that a disturbance of wavenumber k corresponds to a disturbance of *wavelength* $2\pi/k$. Hence the restriction on kR corresponds to the restriction that the jet is stable to disturbances such that (see Fig. 7.2.2 for the definition of Λ)

$$\frac{2\pi R}{\Lambda} = kR > 1 \tag{7.2.31}$$

or

$$\frac{\Lambda}{R} \leq 2\pi \tag{7.2.32}$$

The factor 2π is different from the factor 4.5 but is of the same magnitude, and so is roughly consistent. It is more significant that qualitatively we find that a detailed fluid dynamic analysis predicts that if a disturbance corresponds to a wavelength that is too small, the jet is stable, in agreement with the energy analysis given earlier. There is no reason for the two analyses to agree. Neither is exact. The energy analysis assumes that the jet breaks up into a set of equal-sized drops. In fact, it is commonly observed that large drops, separated by small drops, are formed. (See Fig. 7.2.1) The fluid dynamic analysis is based on the assumption that the disturbance is very small. While this might hold in the early stages of growth, eventually the disturbance becomes comparable to the jet radius, and the theory breaks down.

Figure 7.2.3 shows the functional dependence of α on kR, where α has been nondimensionalized. The most significant feature of this plot is the occurrence of a maximum value of α, at a value of $kR = 0.697$. Its value is

$$\alpha^{max} = 0.34\left(\frac{\sigma}{\rho R^3}\right)^{1/2} \tag{7.2.33}$$

What is the physical meaning of this maximum?

A jet of liquid issuing from a capillary is subject to environmental noise that arises from ever-present vibrations in building structures. These vibrations may correspond to a spectrum of frequencies at the exit of the capillary, and this spectrum in turn creates disturbances in the jet flow and in the radius that correspond to a spectrum of wavelengths. According to our theory, some disturbances will damp out and others will grow. Among those that grow, there is one wavelength that grows most rapidly. To connect the theory to observations, we make the assumption that the jet breaks down

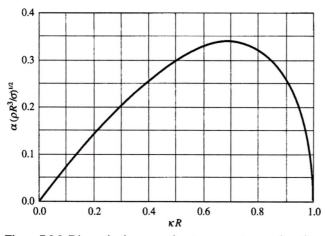

Figure 7.2.3 Dimensionless growth rate parameter as a function of the dimensionless wavenumber of a disturbance.

from the most rapidly growing disturbance. Hence we should observe uniformly spaced drops, and the spacing L should satisfy the following constraint:

$$\frac{L}{2\pi R} = \frac{1}{0.69} \tag{7.2.34}$$

or

$$L = 9.1R = 4.55D \tag{7.2.35}$$

Figure 7.2.1 shows a drawing made from a high-speed flash photograph of jet breakup. It is difficult to make precise measurements, but roughly we see that the distance between successive large drops is 1.1 cm and the jet diameter is approximately 0.25 cm. These estimates yield $L = 4.4D$, in excellent agreement with the theory.

If we assume that the observed disturbance corresponds to α^{max}, we predict that the radius of the disturbance grows as

$$\zeta^\circ = \zeta_i \exp\left(\alpha^{max}t\right) \tag{7.2.36}$$

where ζ_i is the magnitude of the disturbance corresponding to α^{max}. We do not know the magnitude of this initial disturbance. (It is presumably microscopic, and too small to see on a photograph of a jet.) From this expression we find the time for the disturbance to break up the jet, based on the assumption that this corresponds approximately to the time at which $\zeta^\circ = R$. The result is

$$t^* = \frac{L^*}{U} = \frac{1}{\alpha^{max}} \ln \frac{R}{\zeta_i} = \frac{C}{\alpha^{max}} \tag{7.2.37}$$

Note that if the jet is traveling at the constant speed U, the distance from the capillary exit at which the first drops appear is $L^* = Ut^*$. The constant C replaces the unknown ratio of the initial disturbance to the radius of the jet. If we now nondimensionalize the breakup length L^* with the jet diameter, we find

$$\frac{L^*}{D} = 1.04\, C \left(\frac{\rho U^2 D}{\sigma}\right)^{1/2} = 1.04\, C\, We^{1/2} \tag{7.2.38}$$

The Weber number arises naturally in the analysis. At this stage we have a testable mathematical model. Let's do it.

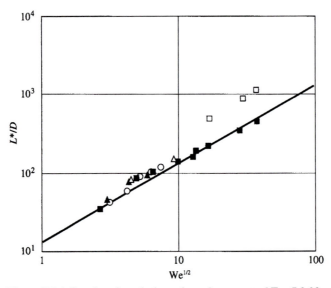

Figure 7.2.4 Breakup length data plotted as a test of Eq. 7.2.38. The line is for $C = 13$. Data from R. P. Grant, Ph.D. thesis, University of Rochester, 1965.

Figure 7.2.4 shows data for breakup length as a function of jet velocity, plotted according to the dimensionless groups suggested by Eq. 7.2.38. Except for one set of data, the model correlates all the data when a value of $C = 13$ is selected. The "exceptional" set of data is for a liquid with a viscosity of 0.16 Pa·s. The other data are for low viscosity liquids (0.001–0.026 Pa·s). It seems reasonable to assume that a high viscosity will retard the rate of growth of the disturbance and thus lead to a larger breakup length. We conclude that the theory for an *inviscid* liquid does a very good job of predicting the observed parametric behavior of the breakup length, even up to viscosities an order of magnitude greater than that of water. The theory has been extended to viscous liquids, but we will not present it here.

Let's look now at an application of this stability analysis to a practical design problem.

EXAMPLE 7.2.1   *Design of a Liquid Electrical Contact*

Suppose it is necessary to create and control an electrical contact to a moving surface without imposing any frictional wear on the surface. A proposed approach to this problem is to use a jet of conducting liquid, such as a salt solution, as the conductor. Then it is essential that this liquid "wire" not break its connection to the surface. Figure 7.2.5 shows the concept. Given our foregoing stability discussion, it should be clear that the length L must be less than the breakup length of the jet. Suppose that the following design constraints have been imposed:

L can be no more than 0.1 m. The liquid flowrate must be less than 10^{-5} m³/s. The jet diameter must be in the range $1 < D < 3$ mm.

A liquid has been selected for which $\rho = 1000$ kg/m³, $\sigma = 0.060$ N/m, and $\mu = 0.0013$ Pa·s.

This is a typical engineering design problem. Within a given set of constraints we have some freedom to select the design and operating variables for the system. The

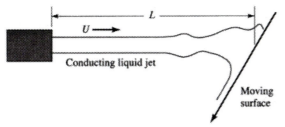

Figure 7.2.5 A liquid jet is used to provide a conducting path to a moving surface.

constraints specified are often determined by mechanical, materials, and economic considerations. The people who impose those constraints may not be aware of any fluid dynamic considerations that make their constraints difficult, or even impossible, to achieve. We have to use our knowledge of fluid dymamics to evaluate other potential problems in this design problem.

For example, we learned very early (see the discussion following Fig. 2.1.10) that unless the Weber number is sufficiently large, a coherent jet cannot be obtained at all—liquid will simply drip out of the capillary. Hence we must operate above a minimum Weber number. If we try to design our system based on the analysis of this section, the jet must be initially smooth and steady as it exits the capillary. This means that we want to use a laminar jet. Hence we must operate below a maximum Reynolds number.

In this design problem we have several free design parameters. They are the separation distance L, the capillary diameter D, and the jet flowrate Q. With reference to Fig. 7.2.6, we will set up a graph representing a "design space" in the coordinates D and L, with Q as a parameter. Anywhere on this graph are pairs of (D, L) coordinates, for a given Q, that either satisfy or do not satisfy the constraints on the operation of the system. We want to find on this graph an "operating window," or "design window," that meets the constraints.

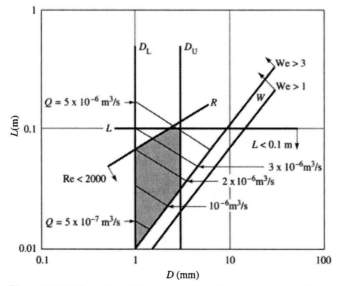

Figure 7.2.6 Lines that define the "operating window" in Example 7.2.1.

One "edge" of the window is the horizontal line L, corresponding to the constraint $L < 0.1$ m. Two "side" boundaries of the window are the vertical lines D_L and D_U that represent the demand that D lie between upper and lower boundaries (3 and 1 mm). These three boundaries arise from the geometric constraints. Now we must look at the fluid dynamics.

The relevant fluid dynamics are contained in Eq. 7.2.38, which relates the breakup length L^* to the speed and diameter of the jet. Using the empirical coefficient ($C = 13$) that permits us to fit data in Fig. 7.2.4, we may write the following expression for $L^*(D, U)$:

$$L^* = 13 \left(\frac{\rho}{\sigma}\right)^{1/2} UD^{3/2} \qquad (7.2.39)$$

or, in terms of Q,

$$L^* = \frac{52}{\pi} \left(\frac{\rho}{\sigma}\right)^{1/2} QD^{-1/2} \qquad (7.2.40)$$

We do not want to design a system that is right on the edge of the instability of the jet. We will choose L so that $L = 0.5 L^*$. Using this criterion, and the physical properties listed earlier, we find the $L(D, U)$ relationship as

$$L^* = 1068QD^{-1/2} \qquad \text{in SI units} \qquad (7.2.41)$$

For a fixed value of Q, this describes a straight line for L versus D, on log–log coordinates, with a slope of -0.5. Such lines are drawn on Fig. 7.2.6 for several choices of Q. If we pick one of these lines, say the one for $Q = 5 \times 10^{-6}$ m³/s, and take the upper left end, which corresponds to a small capillary diameter D, we would find a correspondence to large values of both Re and We. This expectation follows from the inverse dependence on D, *at fixed Q*, of both Re and We:

$$\text{Re} \equiv \frac{DU\rho}{\mu} = \frac{4Q\rho}{\pi D\mu} \qquad (7.2.42a)$$

$$\text{We} \equiv \frac{\rho U^2 D}{\sigma} = \frac{16\rho Q^2}{\pi^2 \sigma D^3} \qquad (7.2.42b)$$

Hence as we "travel down" a line of constant Q from left to right (increasing D), the Reynolds number will eventually fall below a critical value, about 2000, and the jet flow will be laminar. But if we continue too far along that line, the Weber number will fall below a critical value, of order unity, such that the flow drips but does not jet. Along any line of constant Q there are two points that correspond to Re = 2000 and We = 1. The locus of these points along several Q lines defines the lines marked R and W on Fig. 7.2.6. Note that the R line reduces the operating window in its upper left corner, while the W line does the same to the lower right corner. Because the Weber number criterion on the dripping-jetting transition is less clearly established for horizontal jets, we show two lines, one at We > 1 and one at We > 3, just to indicate the sensitivity of the window to this degree of uncertainty.

Based on the criteria examined so far, we conclude that any choice of D, L, and Q within the operating window shown shaded in Fig. 7.2.6 would yield an acceptable design. In view of some of the uncertainties mentioned, reasonable choices would be

$$D = 1 \text{ mm} \qquad \text{and} \qquad Q = 10^{-6} \text{ m}^3/\text{s}$$

which are well within the Re and We constraints.

7.2.2   Assessment of Some Assumptions of the Stability Analysis

To obtain an analytical solution to the jet stability problem, we neglected the inertial terms in the Navier–Stokes equations, compared to the acceleration terms. Hence we now want to assess the validity of the assumption that ratios such as $\mathfrak{R}$ defined in Eq. 7.2.43 below are small; that is,

$$\mathfrak{R} \equiv \frac{u_z^o \dfrac{\partial u_z^o}{\partial z}}{\dfrac{\partial u_z^o}{\partial t}} \ll 1 \tag{7.2.43}$$

We will use Eq. 7.2.21, from which we find

$$\frac{\partial u_z^o}{\partial t} = \alpha u_z^o \tag{7.2.44}$$

and

$$u_z^o \frac{\partial u_z^o}{\partial z} = ik(u_z^o)^2 \tag{7.2.45}$$

Then

$$\mathfrak{R} = \frac{iku_z^o}{\alpha} = \frac{ikRu_z^o}{\alpha R} \tag{7.2.46}$$

From Eq. 7.2.33 we know that the fastest growing disturbance corresponds to

$$\alpha R = 0.34 \left(\frac{\sigma}{\rho R}\right)^{1/2} \tag{7.2.47}$$

and this occurs at $kR = 0.69$. Hence the magnitude of $\mathfrak{R}$ is

$$|\mathfrak{R}| = \frac{0.69 \, |u_z^o|}{0.34 \left(\dfrac{\sigma}{\rho R}\right)^{1/2}} \approx 2 \left[\frac{\rho |u_z^o|^2 R}{\sigma}\right]^{1/2} = 2 \, \mathrm{We}^{1/2} \tag{7.2.48}$$

In writing this final result we recognize that the bracketed term is a Weber number for the jet.

Now we ask the question: Is $\mathfrak{R}$ small? That is, is the Weber number for the jet small? We learned in Chapter 2 that unless the Weber number exceeds a value of about 3, we do not even get a jet—instead, the liquid simply drips out of the tube. But the Weber number relevant to that issue is based on the average velocity of the fluid leaving the tube. The velocity that appears in Eq. 7.2.48 is the *disturbance velocity* that is superimposed on the motion of the jet. How do we estimate that velocity?

The instability considered here implies that the liquid in the jet rearranges itself (i.e., it flows) in a way that leads to the formation of drops. The drops are spaced a distance $\Lambda = 2\pi/k$. We can form a rough estimate of the disturbance velocity by considering the liquid to move a distance $\Lambda/2$ in the time t^* given by Eq. 7.2.37. Hence

$$|u_z^o| \approx \frac{\frac{1}{2}\Lambda}{t^*} = \frac{\frac{1}{2}(2\pi/k)}{t^*} = \frac{\pi R/0.69}{C/\alpha^{\max}} = \frac{\pi}{0.69C}\alpha^{\max}R$$

$$= \frac{\pi}{0.69C} 0.34 \left(\frac{\sigma}{\rho R}\right)^{1/2} = \frac{0.34\pi}{0.69 \times 13} \left(\frac{\sigma}{\rho R}\right)^{1/2} = 0.12 \left(\frac{\sigma}{\rho R}\right)^{1/2} \tag{7.2.49}$$

It follows, then, from Eq. 7.2.48, that

$$|\Re| = 2|u_z^o| \left(\frac{\rho R}{\sigma}\right)^{1/2} \approx 0.24 \qquad (7.2.50)$$

While this number is less than unity, it is not small enough to allow us to conclude that the neglect of the inertial terms is justified. But there is one more point to be noted, namely, that the growth of the disturbance on the jet, according to Eq. 7.2.21, is *exponential* in time. Hence over most of the history of the growth of the disturbance, the velocity u_z^o is considerably less than the value estimated from Eq. 7.2.49. This suggests that the neglect of the inertial terms is a good approximation in the early stages of the breakdown of the jet. The good agreement of experimental data with the theory is consistent with this conclusion.

The assessment of the validity of approximations that are made in the development of a mathematical model is an essential stage of the modeling process. Often, as in the case described here, it is difficult to draw a clear message from the analysis. Regardless of the difficulty, it is essential in an assessment that we have a simple concept of the physics of the flow. This leads to *some estimate, which is better than no estimate,* of the relative magnitudes of the terms we are comparing. In the absence of experimental data on jet stability, we would have been left with the task of arguing whether $|\Re| \approx 0.24$ is small enough to yield a useful model of the physics of interest.

7.3   QUASI-STEADY FLOWS

Some unsteady flows have a character that permits us to obtain a good approximation to their dynamics by using a so-called quasi-steady analysis. We have already introduced the quasi-steady approach in some earlier examples (can you find them?), but with no justification. We can illustrate the concept best by looking immediately at an example of development of a model of a transient flow.

EXAMPLE 7.3.1   *Model of Draining of a Tank Through a Capillary*

Figure 7.3.1 shows the physics of the problem. A tank holding some volume V of a viscous liquid is draining through a capillary or other outlet. The flow is being driven by the hydrostatic head $H(t)$, and the head decreases as the volume of liquid in the tank decreases. Hence the outlet flow slows as draining proceeds. This is clearly an unsteady flow.

The dynamic equations for the flow through the capillary are

$$\rho \frac{\partial u_z}{\partial t} = -\frac{\partial p}{\partial z} + \mu \frac{1}{r} \frac{\partial}{\partial r}\left(r \frac{\partial u_z}{\partial r}\right) \qquad (7.3.1)$$

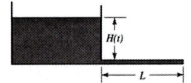

Figure 7.3.1 A tank drains through a capillary.

and

$$0 = \frac{\partial p}{\partial r} \tag{7.3.2}$$

and the continuity equation, even in an unsteady flow, still gives

$$0 = \frac{\partial u_z}{\partial z} \tag{7.3.3}$$

In writing these equations we are neglecting entrance effects, just as we did in the corresponding steady state analyses. Hence we are assuming that the flow is laminar, and u_z is the only nonzero component of velocity. We acknowledge that u_z is a function of time, in writing Eq. 7.3.1. The pressure along the capillary axis, $p(z, t)$ is also a function of time, since the only pressure source, the hydrostatic head, changes as the tank drains:

$$p = \rho g H(t) \text{ at } z = 0 \tag{7.3.4}$$

where $z = 0$ is the entrance to the capillary, whose length is L.

The same arguments applied earlier (Section 3.2) to the steady state flow in a long capillary apply here and lead to the conclusion that

$$-\frac{\partial p}{\partial z} = C(t) = \frac{\rho g H(t)}{L} \tag{7.3.5}$$

Note that now C is a function of time, but not of z. Hence the pressure still falls off linearly down the capillary axis, but the slope (the pressure gradient) varies with time.

If this result is put into Eq. 7.3.1 we have

$$\rho \frac{\partial u_z}{\partial t} = \frac{\rho g H(t)}{L} + \mu \frac{1}{r} \frac{\partial}{\partial r} \left(r \frac{\partial u_z}{\partial r} \right) \tag{7.3.6}$$

which has *two* unknowns. The second required equation comes from an overall statement of conservation of mass: the volume change of liquid in the tank, hence the change in height H of the liquid level in the tank, is related to the velocity u_z by

$$Q = -\frac{dV}{dt} = -A_T \frac{dH}{dt} = \int_0^R 2\pi r u_z \, dr \tag{7.3.7}$$

(We are assuming that the cross-sectional area of the tank normal to the flow, A_T, is uniform.)

Equations 7.3.6 and 7.3.7 are a pair of "coupled integrodifferential equations." "Coupled" means that both variables $H(t)$ and $u_z(r, t)$ appear in *each* equation. "Integro" refers to the appearance in the dynamic equation for H of an integral of u_z. Because of the mathematical complexity, we seek an approximate solution to this pair of equations through the introduction of an assumption that has some physical plausibility. Specifically, we assume that if the flow is slow enough, in some sense that must be quantified later, the Hagen–Poiseuille velocity field holds, but with the constant pressure replaced by the time-varying pressure. This is what we mean by a "quasi-steady flow." We use a steady state model for one particular feature (the velocity profile–pressure gradient relationship, in this case) of an unsteady but slowly varying flow. Hence we write (cf. Eq. 3.2.33)

$$u_z = -\frac{C(t)R^2}{4\mu} \left[1 - \left(\frac{r}{R} \right)^2 \right] \tag{7.3.8}$$

and, using Eq. 7.3.7

$$Q = \frac{\pi R^4}{8\mu} \frac{\rho g H(t)}{L} = - A_\text{T} \frac{dH}{dt} \qquad (7.3.9)$$

Equation 7.3.8 is the steady state relationship of the analysis. It is the "quasi-steady" part of the model. Equation 7.3.9 contains the time dependence, through the time derivative dH/dt. We may now solve this differential equation for $H(t)$ and find

$$\frac{H}{H_\text{o}} = e^{-\tau} \qquad (7.3.10)$$

where a dimensionless time is defined by

$$\tau \equiv \frac{\pi \rho g R^4}{8\mu L A_\text{T}} t \qquad (7.3.11)$$

and we used the initial condition that $H = H_\text{o}$ at $t = 0$.

We now have a predictive model for the time required to drain the tank. If we define that time by setting $H = 0$, we find, of course, that an infinite time is required. This "answer" is typical of problems in which the system is driven to some equilibrium by a "force" that continually decreases as equilibirum is approached. In this case, the tank drains under the action of gravity, and the hydrostatic head decreases as the tank drains. Such an "asymptotic" process requires an infinite time to reach equilibrium. While the "answer" is formally correct, it is not useful. What we need then is an estimate of the time for the tank to be "nearly" drained. If we were satisfied with 90% drainage, we would find that

$$\tau = 2.3 \text{ when } H/H_\text{o} = 0.10 \qquad (7.3.12)$$

For 99% drainage we would find

$$\tau = 4.6 \text{ when } H/H_\text{o} = 0.01 \qquad (7.3.13)$$

There is a factor of 2 in the ratio of these two times, and each is correct; but the two are based on different criteria of "empty." We will adopt an arbitrary choice here, but one that is useful for engineering design purposes and is easy to remember. We will agree that the tank is *practically* empty when $\tau = 3$, which corresponds to $H/H_\text{o} = 0.05$, or 95% empty.

Thus we define a time to empty the tank as (inverting Eq. 7.3.11 for t)

$$t_\infty \equiv \frac{24\mu L A_\text{T}}{\pi \rho g R^4} \qquad (7.3.14)$$

It is interesting to note that this time is independent of the initial head in the tank. How is it possible that two tanks filled to different heights, but otherwise identical with respect to their outlets, can drain in the same time? Think about it!

To what degree is the foregoing "quasi-steady" model valid and useful? One way to examine this question is to recognize that the quasi-steady model is obtained by setting the time derivative to zero in Eq. 7.3.1. Physically, we are asserting that the acceleration term is not significant compared to the viscous term. Hence the ratio

$$\Phi = \frac{\rho \dfrac{\partial u_z}{\partial t}}{\mu \dfrac{1}{r} \dfrac{\partial}{\partial r}\left(r \dfrac{\partial u_z}{\partial r}\right)} \qquad (7.3.15)$$

should be small compared to one, and this should provide us with a criterion based on some value of a dimensionless group. Using Eqs. 7.3.5 and 7.3.8, we find

$$\Phi = \left| \frac{\rho R^2}{4\mu} \frac{1}{C} \frac{dC}{dt} \right| = \left| \frac{\rho R^2}{4\mu} \frac{1}{H} \frac{dH}{dt} \right| \tag{7.3.16}$$

When Eq. 7.3.9 is used for dH/dt, we may write this as

$$\Phi = \frac{\pi R^6 g}{32 A_T L \nu^2} \ll 1 \tag{7.3.17}$$

This becomes a criterion for validity of the quasi-steady approximation *for this specific flow*. Note that this criterion is written in terms of parameters that are known prior to draining the tank. Hence we may use Eq. 7.3.14 for a first estimate of the drainage time t_∞, calculate Φ from Eq. 7.3.17, and decide whether the estimate of t_∞ is likely to be accurate based on whether Φ is much smaller than unity.

EXAMPLE 7.3.2   *A Quasi-Steady Model of a Viscous Squeezing Flow*

In the case of a quasi-steady analysis (Example 7.3.1), the time dependence comes from the time-dependent pressure that drives the flow. Now we look at transient behavior that arises from a time-dependent change in the geometry. Figure 7.3.2 shows the flow of interest to us.

Two parallel circular disks of radius R are driven toward each other at a speed given by $-dH/dt$. The initial disk separation is $2H_o$, and the disks remain parallel as H changes. Note that the relative speed of one disk toward the other is $2\dot{H}$. We want to find the pressure that resists this transient squeezing motion.

We begin with the assumption of laminar axisymmetric flow. This is clearly a two-dimensional flow with velocity components u_r and u_z. Many of the systems in which this kind of squeezing flow occurs involve highly viscous liquids. We will neglect inertial effects for this reason; a parameter that permits us to evaluate the importance of the inertial terms will be found later.

The dynamic and continuity equations take the forms

$$\rho \frac{\partial u_r}{\partial t} = -\frac{\partial p}{\partial r} + \mu \left[\frac{\partial}{\partial r} \left(\frac{1}{r} \frac{\partial}{\partial r} (r u_r) \right) + \frac{\partial^2 u_r}{\partial z^2} \right] \tag{7.3.18}$$

$$\rho \frac{\partial u_z}{\partial t} = -\frac{\partial p}{\partial z} + \mu \left[\frac{1}{r} \frac{\partial}{\partial r} \left(r \frac{\partial u_z}{\partial r} \right) + \frac{\partial^2 u_z}{\partial z^2} \right] \tag{7.3.19}$$

$$\frac{1}{r} \frac{\partial}{\partial r} (r u_r) + \frac{\partial u_z}{\partial z} = 0 \tag{7.3.20}$$

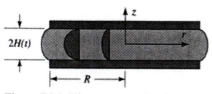

Figure 7.3.2 Viscous squeezing flow between two parallel circular disks. The disks move toward each other at the same speed $\dot{H}$ relative to the fixed plane $z = 0$.

We need to think for a moment about the relative magnitudes and roles of the axial and radial velocities. The axial velocity anywhere within the fluid is certainly no greater than the axial velocity ($\dot{H} = -dH/dt$) of the upper disk. This motion displaces liquid at a rate

$$Q = 2\pi R^2 \dot{H} \tag{7.3.21}$$

The average radial velocity U_R at the periphery of the disks (i.e., at $r = R$) must be such as to balance this flowrate, so

$$Q = 2\pi R \times 2H \times U_R \tag{7.3.22}$$

When we equate these two expressions, we find

$$\frac{U_R}{\dot{H}} = \frac{R}{2H} \tag{7.3.23}$$

If we restrict our analysis to geometries in which the plates are very close together, relative to the disk radius, or

$$\frac{R}{H} \gg 1 \tag{7.3.24}$$

we can regard this flow as one that is "almost parallel" in the sense that the radial velocity is much greater than the axial velocity. Equation 7.3.3 provides an *overall* criterion for applying this approach. In certain regions of the flow field, regardless of the value of R/H, the radial velocity is *smaller* than the axial velocity. One such region lies along the disk surfaces, where the no-slip condition requires that the radial velocity vanish, while the axial velocity is $\pm\dot{H}$. Since this approximation is not uniformly valid over the whole flow, it is not clear what effect this might have on the validity of a model based on the approximation. As usual, we can and will find out by obtaining such a model and testing it against observations.

The consequence of the approximation that u_z is small can be seen on inspection of Eq. 7.3.19, which leads to the statement that

$$\frac{\partial p}{\partial z} = 0 \tag{7.3.25}$$

or simply

$$p = p(r, t) \tag{7.3.26}$$

If we now turn our attention to Eq. 7.3.18 we see that the pressure distribution arises from two physical sources: one due to viscous effects and the other due to the unsteady nature of this flow. The liquid is always being accelerated, and a portion of the pressure field arises in response to this acceleration. This leads us to introduce a second approximation. We will assume that viscous effects dominate the inertial (accelerative) effects, and we will drop the time derivative from Eq. 7.3.18. Thus we are going to solve for the velocity and pressure from the continuity equation (Eq. 7.3.20) and the *quasi-steady* form of Eq. 7.3.18:

$$0 = -\frac{dp}{dr} + \mu \left[\frac{\partial}{\partial r} \left(\frac{1}{r} \frac{\partial}{\partial r} (r u_r) \right) + \frac{\partial^2 u_r}{\partial z^2} \right] \tag{7.3.27}$$

Now let's think about the axial velocity u_z, which is assumed to be smaller than u_r. Look at the continuity equation (Eq. 7.3.20). It may seem tempting to just throw out the u_z term. We can't do this, since although we may claim that u_z is small, it does not follow from this that its derivative in the z direction is small. Instead, we make the following argument.

We are asserting that u_z is small throughout the flow field, although we have admitted that it is the only nonzero velocity along the disk surfaces. Since the disks are rigid bodies, the velocity u_z along those surfaces is independent of radial position. We now make the assumption that u_z is *everywhere independent of r*. This lets us write the continuity equation in the form

$$\frac{1}{r}\frac{\partial}{\partial r}(ru_r) = -\frac{du_z}{dz} = f(z,t) \tag{7.3.28}$$

Now we can integrate this equation with respect to r, since the function f is not a function of r if u_z is not a function of r, and the result is

$$u_r = \tfrac{1}{2}rf(z,t) \tag{7.3.29}$$

When this result is substituted into Eq. 7.3.27, we find

$$\frac{dp}{dr} = \mu\frac{\partial^2 u_r}{\partial z^2} = \frac{\mu r}{2}\frac{\partial^2 f}{\partial z^2} \tag{7.3.30}$$

We may now integrate the left-hand "half" of this pair of equations twice with respect to z. We find, using boundary conditions that $f = 0$ on $z = \pm H$,

$$u_r = \frac{rf}{2} = \frac{H^2}{2\mu}\left(-\frac{dp}{dr}\right)\left[1 - \left(\frac{z}{H}\right)^2\right] \tag{7.3.31}$$

Over any region $[0, r]$, the rate of displacement of fluid must satisfy Eq. 7.3.32.[3]

$$\int_{-H}^{H} 2\pi r u_r\, dz = 2\pi R^2 \dot{H} \tag{7.3.32}$$

(If we use $r = R$ we recover Eq. 7.3.21.) Now we use Eq. 7.3.31 for u_r, perform the indicated integration over z, and find

$$-\frac{1}{\mu r}\frac{dp}{dr} = \frac{3\dot{H}}{2H^3} \tag{7.3.33}$$

When this result is substituted into Eq. 7.3.31, we find $u_r (r, z, t)$ in terms of $H(t)$. Of greater immediate interest is the pressure profile $p(r)$, which follows by integrating Eq. 7.3.33:

$$p = \frac{3\dot{H}\mu R^2}{4H^3}\left[1 - \left(\frac{r}{R}\right)^2\right] \tag{7.3.34}$$

(We have taken the pressure at the edge of the disk to be the ambient pressure, and we arbitrarily set the ambient pressure to zero.) This pressure resists the movement of the disks toward each other, and so an external force is required to drive them together. This force is obtained by integrating the normal stress T_{zz} across the radius of the disks:

$$F = 2\pi\int_0^R T_{zz|z=H}\, r\, dr \tag{7.3.35}$$

---

[3] Remember! The rate at which the two disks approach each other is $2\dot{H}$, not $\dot{H}$ itself, because each disk has the speed $\dot{H}$ toward the other.

For a Newtonian fluid, we must calculate the total normal stress from

$$T_{zz} = p - 2\mu \frac{\partial u_z}{\partial z}$$

(7.3.36)

From the solution for the velocity field it follows that

$$2\mu \frac{\partial u_z}{\partial z} = -2\mu \frac{1}{r} \frac{\partial}{\partial r}(r u_r) = \frac{3\mu \dot{H}}{H}\left[1 - \left(\frac{z}{H}\right)^2\right]$$

(7.3.37)

Note how we used the continuity equation to replace the left-hand side of Eq. 7.3.37 with a term involving u_r, which we already have solved for in Eq. 7.3.31. This makes it unnecessary to find u_z. However, u_z follows directly from integration of the continuity equation.

But in the force balance of Eq. 7.3.35, it is the value of the stress at the disk surface (i.e., along $z = \pm H$) that appears. Hence the viscous part of the total normal stress (Eq. 7.3.37) vanishes on the solid surface, and we simply calculate the force from

$$F = 2\pi \int_0^R p(r)\, r\, dr = \frac{3\pi R^4 \mu \dot{H}}{8H^3}$$

(7.3.38)

This result, known as the Stefan equation, gives the relationship between the motion of the disks and the force opposing that motion. Figure 7.3.3 shows a test of the Stefan equation for two Newtonian liquids of different viscosities. In all cases the value of R was 15.2 cm, and the initial *half*-separation H_o ranged from approximately 3 mm to 10 mm. Disk speeds were 0.017–0.17 mm/s. The data are in good agreement with the theoretical prediction based on the Stefan equation.

If the disks are driven toward each other under *constant force* conditions, Equation 7.3.38 may be solved for the separation as a function of time. We write Eq. 7.3.38 in the form

$$\dot{H} = -\frac{dH}{dt} = \frac{8F_o H^3}{3\pi R^4 \mu}$$

(7.3.39)

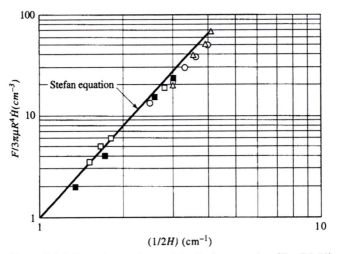

Figure 7.3.3 Experimental test of the Stefan equation (Eq. 7.3.38); viscosities were 18 and 33 Pa · s (solid and open symbols), respectively. Shirodkar and Middleman, *J. Rheol.*, **26**, 1 (1982).

(Now we use F_o to represent the magnitude of the constant force that is driving the two surfaces together.) The solution is easily found as

$$\frac{1}{H^2} - \frac{1}{H_o^2} = \frac{16F_o t}{3\pi R^4 \mu} \tag{7.3.40}$$

where H_o is the initial (half) spacing of the disks.

In our derivation of the Stefan equation, we have neglected the inertial terms and the acceleration term of the Navier–Stokes equations. We expect that this assumption is good for *high viscosity, slow squeezing*. A criterion for this can be established by calculating the neglected terms, using Eq. 7.3.31 for the radial velocity, and comparing the neglected terms to the viscous terms. The result should be an inequality stated in terms of a certain dimensionless grouping of the parameters of the system. We leave this as a homework exercise (Problem 7.12). Instead, we examine inertial effects by looking at the other extreme of behavior: squeezing of a liquid with no viscous resistance.

7.3.1 Inertial Effects in the Absence of Viscosity

We suppose now that the fluid is inviscid: $\mu = 0$. The radial component of the Navier–Stokes equation, in the absence of viscous stresses, takes the form

$$\rho \left[\frac{\partial u_r}{\partial t} + u_r \frac{\partial u_r}{\partial r} + u_z \frac{\partial u_r}{\partial z} \right] = -\frac{\partial p}{\partial r} \tag{7.3.41}$$

The continuity equation is unchanged:

$$\frac{1}{r} \frac{\partial}{\partial r}(r u_r) = -\frac{\partial u_z}{\partial z} = C(t) \tag{7.3.42}$$

If the liquid is without viscosity there is no mechanism to force it to satisfy a no-slip condition along the disk surface. Hence it is squeezed without shear, and the radial velocity is independent of the axial position z. The velocity field is as sketched in Fig. 7.3.4.

Since the radial velocity does not vary with z, it follows from integrating the right-hand half of Eq. 7.3.42 that the axial velocity is simply (using the boundary condition $u_z = -\dot{H}$ at $z = H$)

$$u_z = -\frac{\dot{H}}{H} z \tag{7.3.43}$$

and the radial velocity then follows from the other half of Eq. 7.3.42 as

$$u_r = \frac{\dot{H}}{2H} r \tag{7.3.44}$$

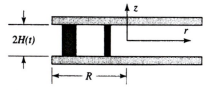

Figure 7.3.4 Radial velocity for shear-free flow.

When these functions are substituted into the radial momentum equation (Eq. 7.3.41), we find that the pressure gradient is

$$-\frac{\partial p}{\partial r} = \frac{\rho r}{H}\left(\frac{\ddot{H}}{2} - \frac{\dot{H}^2}{4H}\right) \tag{7.3.45}$$

For the case of constant speed squeezing, where $\dot{H}$ = constant, the pressure is found to be

$$p = \frac{\rho r^2 \dot{H}^2}{8H^2} \tag{7.3.46}$$

and the inertial force is

$$F_I = \int_0^R 2\pi r\, p(r)\, dr = \frac{\rho\pi R^4 \dot{H}^2}{16H^2} \tag{7.3.47}$$

for the case of constant speed squeezing. This force arises strictly from the unsteady nature of the flow. In a real, viscous, flow, there would be such a force, but it would not have this exact form, since the assumed flat velocity profile (Eq. 7.3.44) would not be valid. A viscous liquid would have to satisfy the no-slip condition. Hence Eq. 7.3.47 does not give the correct magnitude of the neglected forces in our viscous quasi-steady analysis. Nevertheless it indicates how the parameters of the problem combine to yield the inviscid inertial contribution to the force. Hence a rough estimate of the relative importance of the unsteady terms in the quasi-steady squeezing flow analysis can be obtained from the ratio (see Eq. 7.3.38 for the viscous force $F = F_V$)

$$\Theta \equiv \frac{F_I}{F_V} = \frac{\rho H \dot{H}}{6\mu} \tag{7.3.48}$$

If this ratio is small compared to unity, our quasi-steady analysis should be accurate.

Design of a Viscous Damper

The position of an electronic measuring device is maintained by an electromechanical control system in such a way that the device and support structure are isolated from contact with any solid surface. Figure 7.3.5 shows the geometry.

In the event of an electrical failure, the device and its support structure would fall toward the lower surface, damaging the device. A viscous liquid is held in a reservoir for the purpose of damping (slowing) any fall of the package that might occur. Other considerations indicate that as long as the velocity of fall does not exceed 1 mm/s at any time, the device will not be damaged. Select a liquid viscosity, and a damper radius, to ensure that this criterion will be met. The total mass of the device and support is 5

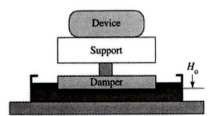

Figure 7.3.5 Schematic of a viscous damper.

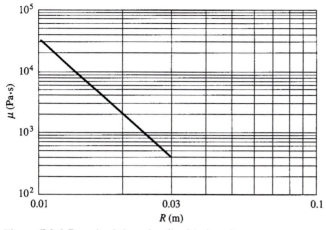

Figure 7.3.6 Required damping liquid viscosity.

kg. The floating height of the damper is controlled at $H_o = 0.2$ cm. The damper radius cannot exceed 3 cm. Note that our design criterion is based on the velocity of fall, not on the velocity at which the package hits the bottom surface.

Equation 7.3.39 provides the relationship among the relevant dependent and independent variables. We see that the maximum velocity occurs at the start of the fall, so we set $H = H_o$ in the equation. The constant force driving the surfaces together is the weight Mg. Then the liquid viscosity and the damper radius must satisfy the following equation:

$$R^4 \mu = \frac{8 \, Mg \, H_o^3}{3\pi \dot{H}} = \frac{8(5)(9.8)(0.002)^3}{3\pi \, (0.001)} = 3.3 \times 10^{-4} \, \mathrm{m^4 \cdot Pa \cdot s} \qquad (7.3.49)$$

From this result we may prepare a simple plot (Fig. 7.3.6) of the required viscosity of the damping liquid as a function of the damper radius.

We see that the required viscosities are extremely high, and that they depend very strongly on the radius of the damper. We should use the largest value of R that corresponds to an available liquid of this level of viscosity.

EXAMPLE 7.3.4   *Squeeze-Film Lubrication in the Knee*

The human knee and hip are examples of "diarthrodial joints." Under periodic heavy loads these joints provide nearly frictionless relative motion between bones of the joint, with no significant wear of the joint surfaces. Figure 7.3.7 is a simplified sketch of the knee joint.

Synovial fluid, which lubricates the joints, is a solution of proteins and polysaccharides, of which hyaluronic acid is thought to be the major functional component. The lubricating capacity of the joint is determined largely by the dynamics of the squeezing flow of the synovial fluid. The articular cartilage also plays an important role because it is not a passive "no-slip" surface. Articular cartilage is porous, and fluid-filled, and the high pressure generated within the synovial fluid causes fluid to move across the boundary of the cartilage. This motion contributes to the dynamics of the joint. An analysis of this problem is given in a paper by Hou *et al.* [*J. Biomech.*, **25**, 247 (1992)].

We will examine a simple application of Eq. 7.3.40 to the dynamics of the knee joint. We select the following geometrical and physical properties as representative of the

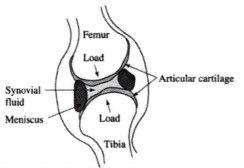

Figure 7.3.7 The knee joint.

knee joint:

$$H_o = 100 \ \mu\text{m} \qquad R = 25 \ \text{mm} \qquad F = 500 \ \text{N} \qquad \mu = 1 \ \text{Pa} \cdot \text{s}$$

We want to know how long it will take for the lubricating fluid film to be reduced from its initial value of $H_o = 100 \ \mu\text{m}$ to a value of $H = 20 \ \mu\text{m}$. From Eq. 7.3.40 we find the time as

$$t = \frac{9\pi R^4 \mu}{2FH_o^2} = \frac{9\pi(0.025)^4 \ (1)}{2(500)(0.0001)^2} = 1.1 \ \text{s} \qquad (7.3.50)$$

Note the very strong dependence of the time on the radius R of the load support.

While the geometry and mechanics of the knee are much more complex than a simple squeeze flow between parallel disks, we find that the biomechanics literature takes the simple squeeze film as a starting point for subsequent analyses. It is commonly the case that the simple models used in this introductory text provide the background necessary for the study of more complex problems. Another example of this, drawn from the physiology–biomechanics area, follows in Section 7.4. Before turning to that section, we look at one more example of a quasi-steady analysis that is based on some steady flows we have already considered.

EXAMPLE 7.3.5   *Draining of a Film of Liquid from a Vertical Plate*

Figure 7.3.8 defines the problem of interest. By some means, a uniform film of liquid of initial thickness H suddenly begins to drain from the vertical surface, under the force of gravity. The goal of the model we wish to produce is the estimation of the time required for the liquid to drain from the plate.

This problem has a number of common elements with the film flow considered earlier in Section 4.4.2, which was a *steady state* flow, with a film of uniform thickness created by the steady supply of fresh liquid at the top of the plate. In the case now being considered, there is no resupply of liquid. The initial amount of liquid ultimately drains completely off the plate.

If we assume that the film thickness $h(x, t)$ varies *gradually* in the x direction, we might hope for an almost parallel flow field. If we ignore surface tension effects on the pressure field and agree to restrict our attention to viscous thin films for which the

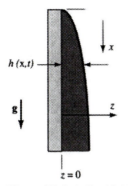

Figure 7.3.8 A liquid film drains from a vertical plate.

Reynolds number is small, then the Navier–Stokes equations reduce to

$$\frac{\partial u}{\partial t} = \nu \frac{\partial^2 u}{\partial z^2} + g \qquad (7.3.51)$$

where u is the x component of velocity. Note the implication that the viscous stress in the direction of the flow is smaller than that normal to the flow, which permits the neglect of the term $\partial^2 u/\partial x^2$ in writing the right-hand side of Eq. 7.3.51. This is similar to the "lubrication approximation" discussed in Chapter 6. It is part of the assumed "almost parallel" character of the flow.

Now we introduce the "quasi-steady" assumption: the flow field at any position x along the plate, for any film thickness $h(x, t)$, is the same as the steady flow for uniform film thickness:

$$u = \frac{g}{\nu}\left(hz - \frac{z^2}{2}\right) \qquad (7.3.52)$$

We have obviously imposed a no slip condition at $z = 0$, and a no shear stress condition at $z = h(x, t)$. Except for notation, this velocity profile is the same as that of Eq. 4.4.26, as we would expect.

Now we examine conservation of mass in this flow. Look at Fig. 7.3.9. The flowrate (per unit width in the y direction, normal to the page) across any plane at x is found from

$$q = \int_0^{h(x,t)} u(z,t)\, dz = \frac{gh^3}{3\nu} \qquad (7.3.53)$$

If we examine any two planes x and $x + dx$, we will realize that any difference between the flow in and out of this volumetric region of film must appear as a change in film thickness[4]:

$$q_x - q_{x+dx} = \frac{\partial h}{\partial t}\, dx \qquad (7.3.54)$$

or

$$\frac{\partial h}{\partial t} = -\frac{\partial q}{\partial x} = -\frac{\partial}{\partial x}\frac{gh^3}{3\nu} = -\frac{gh^2}{\nu}\frac{\partial h}{\partial x} \qquad (7.3.55)$$

---

[4] Actually, Eq. 7.3.54 is an approximation, subject to the assumption that the slope of the free surface is small (i.e., $\partial h/\partial x \ll 1$). See Problem 7.15.

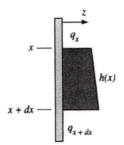

Figure 7.3.9 Schematic for a mass balance on a draining film.

This is a nonlinear partial differential equation. If we try a solution of the form

$$h = Ax^\alpha t^\beta \tag{7.3.56}$$

and substitute this in Eq. 7.3.55, we find that α and β must satisfy

$$\beta Ax^\alpha t^{\beta-1} = -\frac{g}{\nu}(A^2 x^{2\alpha} t^{2\beta})\alpha Ax^{\alpha-1}t^\beta \tag{7.3.57}$$

or, upon equating the exponents of the x and t terms:

$$\begin{aligned} x: \quad & \alpha = 2\alpha + \alpha - 1 \quad && \text{or} \quad && \alpha = \tfrac{1}{2} \\ t: \quad & \beta - 1 = 2\beta + \beta \quad && \text{or} \quad && \beta = -\tfrac{1}{2} \end{aligned} \tag{7.3.58}$$

Equating the *coefficients* (not the exponents) on both sides of Eq. 7.3.57, we find A as

$$A = \left(\frac{\nu}{g}\right)^{1/2} \tag{7.3.59}$$

It is easy to verify that

$$h = \left(\frac{\nu x}{gt}\right)^{1/2} \tag{7.3.60}$$

satisfies Eq. 7.3.55. But what about boundary conditions on Eq. 7.3.55? We can see that this "solution" has the following properties:

At time $t = 0$ the film thickness is infinite.
At the top of the plate, $x = 0$, the film thickness is zero for any time $t > 0$.

Neither of these conditions seems to be consistent with our physical expectations. What is wrong, then, with this model?

In earlier discussions of the philosophy of modeling, we indicated that if we introduce "poor" assumptions in defining a model that we "build," we often obtain as a result a model that does not behave in accordance with our observations or expectations. In effect, that is what has happened here. The quasi-steady assumption and the assumption of a small slope to the free surface are not uniformly valid over the whole length of the plate, or over the whole "history" in time of the drainage. Often, in a situation such as this, we must "dump" the inadequate model and attempt a solution to a more complete (less restrictive) formulation of the continuity equation and the equations of motion. Before doing so, let's look at Fig. 7.3.10, which presents some experimental data of the type we wish this model could describe. I don't think we are prepared to dump this model!

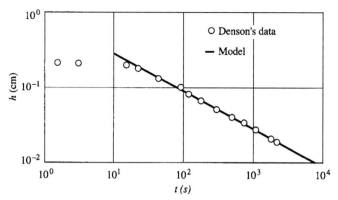

Figure 7.3.10 Comparison of the simple drainage model (Eq. 7.3.60) to the data of Denson [*Ind. Eng. Chem. Fundam.*, **9**, 443 (1970)]. The film thickness was measured at a position 2.9 cm below the top edge of a plate 5.8 cm long. The liquid had a kinematic viscosity of $\nu = 261$ cm²/s.

What's going on here? First we say that the model is not consistent with our expectations, and then we find that the model does an excellent job of fitting the data, *at least after an initial period of time in the drainage history*. This observation suggests that the model is good at "long times," at least at a position far from the top edge of the plate. But we might have expected this, since our day-to-day experience with the drainage of viscous liquids suggests that the small slope approximation is a good idea. Certainly for a liquid high enough in viscosity, the neglect of inertial effects should be valid.

On the basis of this observation, we conclude that the simple drainage model might be quite useful if our primary interest is in the behavior at long times, when the film is more than, for example, 50% drained from the plate, and as long as we are not concerned with the film thickness very near the top of the plate. Problem 7.16 presents some additional data that reinforce our confidence in this model.

7.4   TRANSIENT FLOW IN THE MICROCIRCULATION

The network of fine capillaries that carry blood through skeletal muscle, and through organs, is referred to as the microcirculation. Because of their small diameters (of the order of 10–100 μm), these vessels experience flow that is laminar, and indeed at very low Reynolds numbers. Since the arterial flow that feeds the capillaries is pulsatile, the microcirculation reflects this unsteady-state behavior. A further complication, however, arises from the nature of the capillaries: they are elastic—not rigid tubes. Hence the radius of the vessel is a function of time, and since the pressure varies along the axis of the vessel, the radius varies along the axis as well. As a result, blood flow in the microcirculation can be a complex function of time. We can, however, use our knowledge of fluid dynamics to build a model for transient flow in the microcirculation. We follow to some degree a more detailed paper by Schmid-Schönbein *et al.* [*Biorheology*, **26**, 215 (1989)] and also a study by these authors reported elsewhere [*J. Biomech. Eng.*, **112**, 437 (1990)]. Our goal is to learn how the outflow from a capillary region responds in time to a transient arterial pressure. This is a matter of interest to physiologists, hence to bioengineers.

We begin with the assumption that flow in the capillaries is quasi-steady and almost parallel. We would expect the quasi-steady assumption to be reasonable if the frequency of the pulsating pressure is small with respect to a time scale based on viscous parameters. Thus we expect that the parameter

$$\text{Wo} \equiv \frac{\omega R_o^2}{\nu} \ll 1 \tag{7.4.1}$$

will determine the degree to which the quasi-steady approximation holds. Wo is called the Womersley number, and since the mean radius R_o of a capillary is so small, this condition is always met in the microcirculation. We regard blood as a Newtonian fluid in this model, even though non-Newtonian phenomena occur in the microcirculation because the diameter of red cells is of the same order of magnitude as the diameter of the vessel.

For an elastic tube whose radius can vary because of axial variations in internal pressure relative to the external pressure (which we will take to be constant in space and time in this simple model), changes in cross-sectional area with time reflect an axial variation of volume flowrate along the axis:

$$\frac{\partial A}{\partial t} = -\frac{\partial Q}{\partial z} \tag{7.4.2}$$

(This follows from a simple mass balance on a variable volume element.)

We will take the vessel wall to be a *linearly* elastic material with respect to a specific form of the strain measure E, which we define as

$$E(t) \equiv \frac{1}{2}\left[\left(\frac{R}{R_o}\right)^2 - 1\right] = \frac{1}{\alpha}p(t) \tag{7.4.3}$$

This equation set first defines E, then states that the change in cross-sectional area of the capillary, at any axial position, relative to an equilibrium state defined by the area πR_o^2, is just proportional to the internal pressure in the capillary, at any axial position, relative to the external (extravascular) pressure. This is an *elastic constitutive equation*, and it defines the elastic coefficient α that is characteristic of the capillary wall. In other words, α is a material property, but its definition is based on this choice of E as a measure of the deformation of the capillary.

We should note here that the radius denoted R_o is the radius of the capillary at the reference pressure $p = 0$. This is an *arbitrary* reference state for pressure, but this choice does not affect the results we obtain from the model.

With the quasi-steady assumption, we take Poiseuille's law to hold, so we write

$$Q(t) = -\frac{\pi R(t)^4}{8\mu}\frac{\partial p(t)}{\partial z} \tag{7.4.4}$$

With only the *definition* of E, and no other assumptions, we may write Eqs. 7.4.2 and 7.4.4 as

$$Q = -\frac{A_o^2(2E + 1)^2}{8\pi\mu}\frac{\partial p}{\partial z} \tag{7.4.5}$$

and

$$2A_o\frac{\partial E}{\partial t} = -\frac{\partial Q}{\partial z} \tag{7.4.6}$$

where we have dropped the notation that reminds us that Q, E, and p are functions of time. These two equations, along with the constitutive equation in the form

$$p = \alpha E \tag{7.4.7}$$

yield a set of three nonlinear partial differential equations for the dynamics of this system. In particular, we would like to find the outlet flowrate (Q at $z = L$) for a specified transient inlet pressure [$p = p(t)$ at $z = 0$].

It is instructive to look first at the transient behavior for a *rigid* capillary. In that case we must solve the transient Navier–Stokes equations for Poiseuille flow subject to a step change in upstream pressure. This is exactly the problem solved at the beginning of this chapter, and we will refer back to that model as needed.

If we substitute Eq. 7.4.3 into Eq. 7.4.6 we find

$$\frac{2A_\text{o}}{\alpha}\frac{\partial p}{\partial t} = -\frac{\partial Q}{\partial z}$$

(7.4.8)

When we use Eq. 7.4.5 to eliminate Q we obtain an equation for pressure:

$$\frac{\partial P}{\partial t} = \frac{\alpha A_\text{o}}{16\pi\mu}\frac{\partial}{\partial z}\left[P^2\frac{\partial P}{\partial z}\right]$$

(7.4.9)

where we have defined a dimensionless pressure as

$$P \equiv 1 + \frac{2p}{\alpha}$$

(7.4.10)

We may write this equation in the form

$$\frac{\partial P}{\partial t} = \frac{\alpha A_\text{o}}{48\pi\mu}\frac{\partial^2 P^3}{\partial z^2}$$

(7.4.11)

At this point we can introduce a dimensionless axial variable

$$\xi \equiv \frac{z}{L}$$

(7.4.12)

where L is the length of a capillary. Having done this, we find that a good choice for a dimensionless time variable is

$$\tau \equiv \frac{\alpha A_\text{o}}{16\pi\mu L^2}t$$

(7.4.13)

because it lets us reduce Eq. 7.4.11 to a form free of parameters:

$$\frac{\partial P}{\partial \tau} = \frac{1}{3}\frac{\partial^2 P^3}{\partial \xi^2}$$

(7.4.14)

This is a nonlinear partial differential equation to which no analytical solution is available. We can gain some insight into the physics of this system if we make the assumption that the *variation* in capillary radius is small in some sense. Specifically, we will assume that

$$\frac{2p}{\alpha} \ll 1$$

(7.4.15)

which guarantees (note Eq. 7.4.3) that $R(t)$ will not vary greatly from R_o.

Then we will take the P^3 term in the second derivative of Eq. 7.4.14 to be

$$P^3 \equiv \left(1 + \frac{2p}{\alpha}\right)^3 \approx 1 + \frac{6p}{\alpha} = 3P - 2$$

(7.4.16)

and, after substituting this for P^3 in Eq. 7.4.14, solve

$$\frac{\partial P}{\partial \tau} = \frac{\partial^2 P}{\partial \xi^2}$$

(7.4.17)

We can solve this linear equation easily once we have defined a set of initial and boundary conditions.

We choose to look at the following transient problem:

The pressure within the vessel is initially uniform, such that the value of the dimensionless P is

$$P = P_v \quad at \quad t = 0 \, (\tau = 0) \tag{7.4.18}$$

The pressure at the downstream (venous) end of the capillary is maintained constant such that

$$P = P_v \quad at \quad z = L \, (\xi = 1) \tag{7.4.19}$$

At the arterial end the dimensionless pressure is suddenly *changed from* P_v *to a higher value* P_a:

$$P = P_a \quad at \quad z = 0 \, (\xi = 0) \quad for \, all \, \tau > 0 \tag{7.4.20}$$

The solution to Eq. 7.4.17, with these boundary conditions, is found by the method of separation of variables and may be written in the form of an infinite series:

$$\frac{P_a - P(\xi)}{P_a - P_v} = \frac{p_a - p(\xi)}{p_a - p_v} = \xi + \frac{2}{\pi} \sum_{k=1}^{\infty} \frac{1}{k} \exp(-k^2 \pi^2 \tau) \sin \pi k \xi \tag{7.4.21}$$

Using Eqs. 7.4.3 and 7.4.4 we may find the volumetric flowrate at the two ends of the capillary, as a function of time. The results, in terms of a dimensionless flowrate $\tilde{Q}$ are

$$\tilde{Q}_o \equiv \frac{8\mu L Q(0)}{\pi R_o^4 (p_a - p_v)} = \left(1 + \frac{p_a}{2\alpha}\right)^2 \left[1 + 2 \sum_{k=1}^{\infty} \exp(-k^2 \pi^2 \tau)\right] \tag{7.4.22}$$

at the arterial end, and

$$\tilde{Q}_L \equiv \frac{8\mu L Q(L)}{\pi R_o^4 (p_a - p_v)} = \left(1 + \frac{p_v}{2\alpha}\right)^2 \left[1 + 2 \sum_{k=1}^{\infty} (-1)^k \exp(-k^2 \pi^2 \tau)\right] \tag{7.4.23}$$

at the venous end.

These results are plotted in Fig. 7.4.1 for the case of $p_v = 0$, and $p_a/2\alpha = 0.1$. At the venous end, the flowrate gradually rises to a new equilibrium value. Because of the

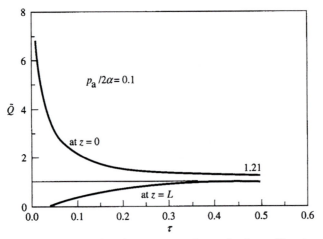

Figure 7.4.1 Transient flow response in an elastic capillary to a sudden step increase in arterial pressure.

choice of nondimensionalization used here, $\bar{Q}$ approaches unity at the venous end. At the arterial end the flowrate undergoes a sudden jump because of the step change of pressure that is applied. From a purely mathematical point of view, we have imposed—for an instant of time—a discontinuity in the pressure at the entrance to the capillary. This corresponds to an infinite pressure gradient at the entrance, initially. Since the flowrate is proportional to the pressure gradient, we find that the flowrate is infinite at time $t = 0$ and then falls toward its equilibrium value. Of course, this is a mathematical artifact and does not occur physically: there would be a sudden jump in flowrate, but not to an infinite value. This mathematical point is resolved by realizing that it is physically impossible to increase the pressure at the arterial end infinitely fast. In other words, if we use a physically impossible boundary condition of an *instantaneous* step change in pressure at $z = 0$, we should expect to find a flowrate based on the pressure gradient that reflects that artifact.

We see from Fig. 7.4.1 that the rise and fall of the two flowrates toward their respective equilibrium values are complete by about $\tau = 0.4$. From the definition of τ we can evaluate a time scale for this system that serves as an indication of how rapidly flowrate variations follow pressure variations. Using $\tau = 0.4$ as the criterion, we find

$$t_\infty = \frac{16\pi\mu L^2}{\alpha A_o}\tau = \frac{20\,\mu L^2}{\alpha A_o} \tag{7.4.24}$$

For a typical capillary we may use the following parameters:

$$L = 500\,\mu\text{m} \qquad R_o = 5\,\mu\text{m}$$
$$\alpha = 7 \times 10^4\,\text{Pa} \qquad \mu = 0.003\,\text{Pa} \cdot \text{s}$$

This yields a value of t_∞ of the order of 10^{-2} s. Since this is a very short time in comparison to the period of the pulsating arterial flow, the outlet flow would essentially be proportional to the inlet flow with no significant time lag. In an organ, however, the region between an arteriole that supplies the blood and a venule that collects it may be thought of as a series of connected capillaries, or an equivalent capillary whose length may be as much as 100 times that of the value for L used above. Since t_∞ varies as the square of L, we might expect transients of the order of 1–100 seconds in duration. This estimate is consistent with data presented by Schmid-Schönbein *et al.* in the references cited earlier.

In Section 7.1 we considered transient Poiseuille flow and showed that a time scale for the rate at which the velocity field (and so, presumably, the volume flowrate) reaches equilibrium following a step change in pressure is given by Eq. 7.1.18. We note that this time decreases as the viscosity increases: viscous fluids reach equilibrium, in response to changes in external pressure, faster than less viscous fluids. Just the opposite is the case for flow in an *elastic* tube, as Eq. 7.4.24 indicates. Why is this so?

The primary reason lies in the physical origin of the transients in the two problems. In the case of the elastic tube, the transient arises from the change in radius of the elastic wall of the vessel. Fluid must move if it is to continue to occupy the volume within the tube, but the viscosity of the fluid retards this motion. Hence the more viscous the fluid, the more slowly equilibrium is reached. Another way to think about this is to note that the constitutive equation for the vessel wall (Eq. 7.4.7) has no time dependence. Any change in pressure, locally, is accompanied by an instantaneous change in vessel radius. The pressure change itself, however, varies axially as well as in time in accordance with Eq. 7.4.9. Flow then occurs in accordance with Eq. 7.4.5.

In the case of the *rigid* tube it is not necessary for fluid to move to occupy a changing volume, since the volume is rigidly fixed. The transient is determined strictly by the rate of transfer of momentum through the fluid, across the radius. The flow is fully developed along the axis, as required by the continuity equation. The more viscous the fluid, the more rapidly momentum is transported. Hence the more viscous fluid reaches equilibrium faster in the rigid tube.

Indeed, we should take note of Eq. 7.4.4. It really represents a *quasi-steady* approximation for the volume flowrate, since it states that the flowrate $Q(t)$ is proportional to the pressure gradient $\partial p/\partial z$. It appears that we have thrown away the momentum transfer part of the transient behavior in our model. This makes sense only if the time scale for momentum transfer is much smaller than that for the response of the flow to a changing radius. In Section 7.1 we found that the time scale for momentum transfer in Poiseuille flow is given approximately by

$$t'_\infty = \frac{\rho R^2}{\mu} \tag{7.4.25}$$

With the data given above for a capillary, we find (using $\rho = 1000 \text{ kg/m}^3$) that t'_∞ is of the order of only 10^{-5} s, and is independent of capillary length. Hence it *is* reasonable to use Eq. 7.4.4. The transient behavior of such small capillaries is indeed dominated by the motion of the fluid in response to the change in radius of the vessel wall.

7.5 THE LEVELING OF A SURFACE DISTURBANCE ON A THIN FILM

Figure 7.5.1 is a schematic diagram of the interface between the recording head and the disk in a magnetic recording system. The recording head is attached to a carrier called a "slider," which "flies" on an air bearing created by the flow of air dragged between the disk surface and the slider by the motion of the disk. The velocity of the slider relative to the disk is of the order of 10 m/s, and the separation of the slider from the disk is only a few hundred *nanometers*. In an attempt to achieve higher recording densities, this flying height will be reduced even further in future disk drives.

To prevent disk/slider contacts that could destroy data, the disk surface is lubricated with a very thin film (30–50 Å) of a viscous oil. A number of interesting and important fluid dynamics problems arise in systems of this type. Because the lubricating film thickness is of the order of molecular sizes, certain phenomena are not accounted for by the Navier–Stokes equations, as we have written them. We will not deal with those issues here.[5] Instead, we are going to look at one aspect of lubricant behavior, and model that behavior on the assumption that classical fluid dynamics serves to account for the phenomena of interest.

We imagine the following situation. Contact of the slider with the lubricant (but not the disk) has occurred, and as a result there is a "furrow" in the lubricant layer that forms in the wake of the slider. Figure 7.5.2 shows this furrow in cross section. We want to develop a model that indicates how quickly the lubricant flows back into the furrow to restore the protection of that region from any subsequent interaction with the slider.

To develop a simple analytical model, we will take a very simple model of the initial disturbance to the lubricant film. It turns out that the mathematics simplifies if we

---

[5] Force terms must be added to the Navier–Stokes equations that account for the van der Waals interaction between the molecules of the liquid and the solid substrate. In a "thick" film these forces play no role. When the film is of molecular dimensions, however, the van der Waals force cannot be ignored.

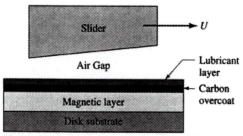

Figure 7.5.1 Schematic of the slider/disk interface.

assume that the disturbance to the film thickness is sinusoidal. Hence, with reference to Fig. 7.5.3, we write the film thickness in the form

$$H(x) = \overline{H} + h \sin kx \tag{7.5.1}$$

where $\overline{H}$ is the mean film thickness and k is the wavenumber of the disturbance. The wavenumber is the inverse of the wavlength λ, or more exactly,

$$k \equiv \frac{2\pi}{\lambda} \tag{7.5.2}$$

We will assume that the amplitude of the disturbance, h, is small compared to the mean film thickness, so that

$$h \ll \overline{H} \tag{7.5.3}$$

We proceed, utilizing two now familiar methods of approximate analysis: lubrication theory (i.e., the assumption of almost parallel flow) coupled with the assumption of a quasi-steady flow. Hence we begin with the steady state form of the dynamic equation in the x direction, neglecting inertial and unsteady terms, and assuming $\partial^2 u_x/\partial x^2 \ll \partial^2 u_x/\partial y^2$:

$$0 = \mu \frac{\partial^2 u_x}{\partial y^2} - \frac{\partial p}{\partial x} \tag{7.5.4}$$

In the y direction, for this nearly parallel flow, we have simply

$$\frac{\partial p}{\partial y} = 0 \tag{7.5.5}$$

Because the lubricant film is so thin, we neglect any effect of gravity. We assume that surface tension provides the dominant force for restoration of the uniform film. Then we take the pressure to be

$$p(x, t) = \frac{\sigma}{R(x,t)} \tag{7.5.6}$$

Figure 7.5.2 A "furrow" created by contact of the slider with the lubricant film.

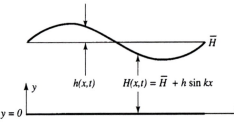

Figure 7.5.3 Idealized sinusoidal disturbance to the lubricant film.

where the radius of curvature for this two-dimensional surface may be calculated from

$$\frac{1}{R} = -\frac{d^2H}{dx^2} \tag{7.5.7}$$

when Eq. 7.5.3 holds.

To find the velocity field we integrate Eq. 7.5.4 twice with respect to y. The result is

$$u_x = \frac{p'}{2\mu} y^2 + \frac{a}{\mu} y + b \tag{7.5.8}$$

where $p' = \partial p / \partial x$ and the integration constants a and b are found from the no-slip condition,

$$u_x = 0 \quad \text{at} \quad y = 0 \tag{7.5.9}$$

and the assumption that there is no significant shear stress on the free surface:

$$\mu \frac{\partial u_x}{\partial y} = 0 \quad \text{at} \quad y = H(x,t) \approx \overline{H} \tag{7.5.10}$$

Note how we simplify this boundary condition with the approximation that we apply it at $y = \overline{H}$, rather than at the *unknown* time-varying surface $y = H(x, t)$. We easily find a and b, with the result that the velocity field is given by

$$u_x = \frac{p'}{2\mu} y^2 - \frac{p'\overline{H}}{\mu} y \tag{7.5.11}$$

Equation 7.5.11 is a quasi-steady solution. The time dependence is implicit, through the time dependence of H, which will change the curvature R and thereby the pressure $p(x, t)$. Now we have to connect the changing amplitude H to the velocity field. This follows through writing a form of conservation of mass, and we have already carried out this analysis in Section 7.3.

From the first half of Eq. 7.3.55, but using the current notation, we find

$$-\frac{\partial H}{\partial t} = \frac{dQ}{dx} \tag{7.5.12}$$

The volumetric flowrate Q is related to the velocity through Eq. 7.3.53. Again, changing to the current notation this takes the form

$$Q = \int_0^{H(x,t)} u_x \, dy \approx \int_0^{\overline{H}} u_x \, dy = -\frac{p'\overline{H}^3}{3\mu} \tag{7.5.13}$$

Hence we find

$$\frac{dQ}{dx} = -\frac{p''\overline{H}^3}{3\mu} = -\frac{\partial H}{\partial t} \tag{7.5.14}$$

From Eqs. 7.5.6 and 7.5.7 we find

$$p = -\sigma H'' = \sigma h k^2 \sin kx \tag{7.5.15}$$

and so

$$p'' = -\sigma h k^4 \sin kx \tag{7.5.16}$$

We may now write Eq. 7.5.14 in the form

$$\frac{\partial H}{\partial t} = \frac{\partial h}{\partial t} \sin kx = -\frac{dQ}{dx} = \frac{p''\overline{H}^3}{3\mu} = -\frac{\sigma h k^4 \sin kx \, \overline{H}^3}{3\mu} \tag{7.5.17}$$

or, finally,

$$\frac{1}{h}\frac{\partial h}{\partial t} = -\frac{\sigma k^4 \overline{H}^3}{3\mu} \equiv -\beta \tag{7.5.18}$$

Thus we find that the disturbance decays exponentially, according to

$$h = h_o \exp(-\beta t) \tag{7.5.19}$$

The parameter β is a decay rate for the disturbance. Our analysis shows that the decay rate gets small inversely with the magnitude of the viscosity, and is proportional to the surface tension. The decay rate is rapid for large wavenumbers (small wavelengths), and the decay rate is a very strong function of the mean film thickness. Hence we expect that a disturbance to an extremely thin film, of the type of interest in this example, will decay very slowly. Let's look at some numbers.

EXAMPLE 7.5.1   *Decay Rate of a Small Surface Disturbance*

Estimate the time required for the disappearance of a small disturbance to the surface of a thin film of lubricant. The surface tension and viscosity are 0.020 N/m and 0.2 Pa·s, respectively. The disturbance is approximately sinusoidal, with a wavelength of $\lambda = 100\ \mu$m. Compare the results for two initial film thicknesses: 1 and 0.1 μm.

First of all, what do we mean by "disappearance"? The decay of the disturbance is exponential and asymptotic. As Eq. 7.5.19 shows, an infinite time is required for "complete" decay. Hence we take as a working definition an arbitrary extent of decay, such that $h/h_o < 0.05$ for $t > t^*$. From Eq. 7.5.19, this requires that

$$\beta t^* = 3 \tag{7.5.20}$$

or

$$t^* = \frac{3}{\beta} = \frac{9\mu}{\sigma k^4 \overline{H}^3} \tag{7.5.21}$$

A wavelength of $\lambda = 100\ \mu$m corresponds to a wavenumber of

$$k = \frac{2\pi}{\lambda} = \frac{2\pi}{100 \times 10^{-6}\,\mathrm{m}} = 6.3 \times 10^4\ \mathrm{m}^{-1} \tag{7.5.22}$$

Then we find

$$t^* = \frac{9 \times 0.2}{0.02(6.3 \times 10^4)^4(10^{-6})^3} = 6.1 \text{ seconds} \tag{7.5.23}$$

For the second case, of a film 10 times smaller, the required time is 1000 times larger, or 6100 seconds.

In the case of lubrication of a magnetic recording head/disk interface, even the "short" time is too long. For a typical hard disk drive, the period of revolution is of the order of 10 ms. Hence, unless the film defect "heals" almost instantaneously, the slider will pass over a partially starved lubricant film, with a resulting higher probability of damage if contact occurs. Our model suggests that one should use a lubricant of lower viscosity so that "defects" will heal faster. Unfortunately this is not a good choice. First, lower viscosity lubricants will have a smaller load-bearing capacity, and it will be easier for the head to "crash" through the lubricant. In the second place, since the high centrifugal force associated with the rotation of the disk tends to produce a radial flow of lubricant off the disk, it will be difficult to keep the lubricant on the disk for its intended service life. One "solution" to the second problem is to develop lubricants that bond chemically to the topmost solid layer of the disk.

EXAMPLE 7.5.2 *Retention of a Lubricant Film on a Spinning Disk*

While we are considering the problem of lubrication, we should keep in mind that the application of interest involves the behavior of a thin liquid film on a rapidly rotating disk. What if centrifugal force could induce the lubricant to flow radially outward, eventually depleting the disk of sufficient lubricant to maintain effective protection? In this example, we look at the unsteady behavior of the film thickness in the presence of a centrifugally induced radial flow. Our goal is to develop a model for the film thickness as a function of time.

We will consider an axisymmetric flow of a viscous liquid. Because of the relatively high viscosity of a lubricant, and in view of the very thin film thickness, it seems quite reasonable to assume that the flow within the thinning film is laminar. Furthermore, because of the application, we expect that the only cases of interest are those in which the film thickness decreases very slowly with time. Hence the quasi-steady concept seems natural in this example. At this point we could examine the continuity and dynamic equations for laminar, axisymmetric, low Reynolds number flow and derive a model for flow in the film. Actually, we have already done most of the necessary work. Go back to Chapter 4, Example 4.5.3. Up to the derivation of Eq. 4.5.66, all the assumptions used in that analysis would hold in the problem we are considering here. The difference between Example 4.5.3 and the lubricant retention problem arises first at the point that we make any statement about whether we resupply liquid to the disk. In Example 4.5.3 we do, and Eq. 4.5.67 provides a balance between the resupply rate Q and the loss due to radial flow. Because of the resupply, the problem treated in Example 4.5.3 is a steady state problem. In the current example we have an unsteady-state problem, and the volume flowrate of the lubricant film off the disk causes the depletion of the film thickness. The mathematical formulation of that statement is simply

$$Q = 2\pi R \int_0^{H_R} u_r(z, R) \, dz = -\pi R^2 \frac{d\overline{H}}{dt} \tag{7.5.24}$$

which equates the volumetric flow across the outer edge of the disk ($r = R$) to the rate of change in volume of lubricant. Note that on the right-hand side of this balance we use the mean value of H. Since Eq. 4.5.66 holds for the case under consideration here, as long as the quasi-steady approximation is valid, we can write down the integral above by inspection of Eq. 4.5.67, after replacing r by R in that expression. The result is

$$-\pi R^2 \frac{d\overline{H}}{dt} = \frac{2\pi\rho\omega^2 R^2}{3\mu} H_R^3 \tag{7.5.25}$$

where H_R is the lubricant film thickness of the edge of the disk. A suitable initial condition for this first-order differential equation is

$$\overline{H} = H_R = H_o \quad \text{at} \quad t = 0 \tag{7.5.26}$$

This is essentially the assumption that the initial film thickness is uniform across the disk.

Without further assumptions we cannot solve Eq. 7.5.25 because it contains two unknowns: the average film thickness across the entire disk and the film thickness at the edge. We can, however, assume that the film thickness, initially uniform according to the choice of initial condition (Eq. 7.5.26), *remains uniform* throughout the history of depletion. In this case, Eq. 7.5.25 may be written in the form

$$-\pi R^2 \frac{dH}{dt} = \frac{2\pi\rho\omega^2 R^2}{3\mu} H^3 \tag{7.5.27}$$

Now we have a single unknown, $H(t)$, and we simply integrate this equation to find

$$\frac{1}{H^2} - \frac{1}{H_o^2} = \frac{4\rho\omega^2}{3\mu} t \tag{7.5.28}$$

or

$$\frac{H(t)}{H_o} = \left(1 + \frac{4\rho\omega^2 H_o^2}{3\mu} t\right)^{-1/2} \tag{7.5.29}$$

For long times such that $H \ll H_o$, Eq. 7.5.28 simplifies to

$$H(t) = \frac{1}{\omega}\left(\frac{3\mu}{4\rho}\right)^{1/2} t^{-1/2} \tag{7.5.30}$$

This model is easily tested, and an example of experimental data is shown in Fig. 7.5.4. We observe that under conditions of interest to this application (i.e., retention of a

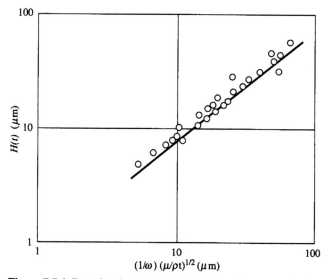

Figure 7.5.4 Data for depletion of a lubricant film on a spinning disk. (Goodwin, unpublished data.)

viscous lubricant as a thin film on a rotating disk), the model mimics the observations quite well.

<hr>

EXAMPLE 7.5.3   *Rate of Loss of Lubricant*

A disk of radius 8.5 cm spins at a rotational speed of 400 rpm. It is initially lubricated with a uniform thin film of oil, of thickness $H_o = 10 \ \mu$m. The oil has a density of $\rho = 900$ kg/m^3 and a viscosity at the temperature of service of $\mu = 0.02$ Pa$\cdot$s. How long can the disk spin at this speed before 10% of the lubricant is lost?

For a uniform film, a loss of 10% of the lubricant corresponds to a reduction in film thickness to $H/H_o = 0.9$. We simply solve Eq. 7.5.29 for t:

$$t = \frac{3\mu}{4\rho\omega^2 H_o^2}\left[\left(\frac{H(t)}{H_o}\right)^{-2} - 1\right] = \frac{3 \times 0.02}{4 \times 900 \left(\dfrac{400}{2\pi \times 60}\right)^2 (10^{-5})^2}[(0.9)^{-2} - 1] = 3.1 \times 10^4 \text{ s}$$

$$\text{(7.5.31)}$$

This is just under 10 hours. The impact of reducing the initial film thickness, or operating at higher rotational speeds, can be explored easily with this model.

SUMMARY

Many flows of engineering importance are unsteady in time. In some cases, as in Sections 7.1 and 7.4, our primary interest is in estimating the length of time required for a flow to achieve its steady state character. In other cases—for example, the stability analysis given in Section 7.2—there is no steady state, and our interest is in the dynamics of the time variation itself. Many flows are intrinsically unsteady, but their character permits the application of a so-called quasi-steady approximation. Examples include the draining and squeezing flows modeled in Section 7.3. In Section 7.5 we combine several of the ideas presented in earlier chapters and find that the transient leveling of a surface disturbance to a very thin film can be modeled by means of the lubrication analysis coupled with a consideration of the role of surface tension in controlling the pressure that drives the flow within the film. In all these cases the essential feature of the models illustrated is the application of simplifying assumptions that retain the main physical character of the flow while leading us to equations that can be solved analytically to yield testable predictions.

PROBLEMS

7.1 A fluid initially at rest and bounded by two rigid parallel planes is suddenly set in motion by the movement of the lower surface ($y = 0$) at the constant speed U in its own plane. At steady state this flow has the velocity profile

$$u(y, t) = U\left(1 - \frac{y}{d}\right) \quad \text{(P7.1.1)}$$

where d is the space between the plates.

a. Formulate the dynamic equations and boundary and initial conditions for the transient part of this flow.

b. The solution of these equations is known to be

$$u(y, t) = U\left(1 - \frac{y}{d}\right)$$
$$- \frac{2U}{\pi}\sum_{n=1}^{\infty}\frac{1}{n}\exp\left(-n^2\pi^2\frac{vt}{d^2}\right)\sin\frac{n\pi y}{d}$$

$$\text{(P7.1.2)}$$

Give the value of the dimensionless time $\tau = \nu t/d^2$ at which the velocity along $y = d/2$ is within 1% of its steady state value. Is the result consistent with the statement following Eq. 7.1.18?

7.2 When viscosity is accounted for in the theory of the stability of a cylindrical jet the analog to Eq. 7.2.38 may be written in the form

$$\frac{L^*}{D} = C\,\text{We}^{1/2}\left(1 + 3\frac{\text{We}^{1/2}}{\text{Re}}\right) \quad \textbf{(P7.2.1)}$$

The dimensionless group

$$Z = \frac{\text{We}^{1/2}}{\text{Re}} \quad \textbf{(P7.2.2)}$$

is called the Ohnesorge number.

How much more stable is a jet of liquid of viscosity $\mu = 0.162$ Pa · s, compared to a water jet, all other parameters being equal? Assume $D = 2$ mm.

7.3 A vertical jet of liquid ($\rho = 1200$ kg/m³, $\sigma = 0.06$ N/m, $\mu = 0.01$ Pa · s) issues from a capillary of diameter $D = 2$ mm at an average velocity of $U = 2.5$ m/s. We want to photograph the jet at the point where the first drop breaks away from the coherent jet. About how far downstream from the capillary exit does this occur? How much does this distance change if the surface tension is reduced by a factor of 2? If the viscosity is increased by a factor of 2?

7.4 A cylindrical tank of diameter 1 m and height 1 m is filled to the top with a liquid of viscosity 0.01 Pa · s and density 1200 kg/m³. The tank drains through a horizontal capillary, as in Fig. 7.3.1. If the capillary is 0.1 m long and has a diameter of 6 mm, how much time is required to drain 95% of the initial liquid contents? Is the flow laminar? Does the liquid drip from the end of the capillary, or does it leave the capillary as a coherent jet? If you wanted to halve the time to drain, what diameter capillary would you use?

7.5 Repeat Problem 7.4, but suppose that the capillary through which the tank drains is *vertical*.

7.6 Address the question that follows Eq. 7.3.14.

7.7 A criterion for application of the quasi-steady model of draining is given in terms of a parameter Φ defined in Eq. 7.3.17. Give the value of Φ for the conditions of Problem 7.4. A second condition is that the specific flow model we used in deriving Eq. 7.3.14 is Hagen–Poiseuille flow.

What criteria determine the validity of this assumption, and are they met in the conditions of Problem 7.4?

7.8 Using Eqs. 7.3.31 and 7.3.34, and the continuity equation, find an expression for the axial velocity profile $u_z(z, t)$. Derive an expression for the ratio of the axial to radial velocity at any point in the flow.

7.9 Derive the form of the radial velocity component as an explicit function of r and z, from Eqs. 7.3.31 and 7.3.33. Plot the radial velocity, normalized to $\dot{H}$, at $r = R/3$ and $r = R$. Do two cases: $R/H = 10$ and 20.

7.10 Prove that consistent with the assumptions made in developing our model of laminar Newtonian squeezing flow, the force opposing the squeezing may be written in terms of the *shear stress* at the disk surface, in the form

$$F = \frac{\pi}{H}\int_o^R r^2 T_{rz}\big|_{z=H}\,dr \quad \textbf{(P7.10)}$$

where $T_{rz} = -\mu(\partial u_r/\partial z)$.

Hint: Begin with the integral in Eq. 7.3.38, do an integration by parts to introduce an integral of dp/dr, and then use Eq. 7.3.31 to introduce the shear stress at $z = H$.

7.11 Two disks, of radii 10 cm, are initially separated by a gap of 2 mm, within which is a liquid with a viscosity of 0.1 Pa · s. The disks are suddenly subjected to a constant load of 100 kg. For how long can that load be supported if the disks are not permitted to get any closer together than 0.2 mm?

7.12 Equation 7.3.48 gives a ratio of inertial to viscous forces in the case of constant speed squeezing. Another such ratio may be obtained by substituting the solutions we have obtained for u_r and u_z into the left-hand side of the radial momentum equation (Eq. 7.3.41), and evaluating the resulting terms relative to the viscous terms (the right-hand side of Eq. 7.3.27). If you manage to keep your wits about you, you will find a criterion for neglecting inertial and unsteady terms. Do so, and compare your result to Eq. 7.3.48.

7.13 Begin with Eq. 7.3.39 and develop an expression for the deceleration of a disk driven by constant force toward a second parallel disk. In Example 7.3.3, suppose the design constraint was in terms of deceleration, instead of velocity. Give

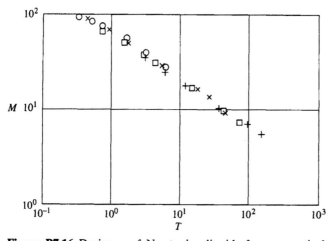

Figure P7.16 Drainage of Newtonian liquids from a vertical plate. M is the percentage of mass remaining after the dimensionless time T.

the analog to Fig. 7.3.6 for the case that the deceleration must not exceed $g = 9.8$ m/s^2.

7.14 Someone suggests that we add the force due to fluid inertia (Eq. 7.3.47) to the force due to friction (Eq. 7.3.38) to produce a model that accounts for both effects. Why is this a poor idea?

7.15 Equations 7.3.53 and 7.3.54 are approximations, valid in the limit of small slope $\partial h/\partial x$. The more exact result is

$$\frac{\partial q}{\partial x} = -\frac{\partial h}{\partial t} + u_x(h, t)\frac{\partial h}{\partial x} \qquad \textbf{(P7.15)}$$

Derive this by starting with the continuity equation for this flow. Integrate the continuity equation across the film and solve for $u_z(h, t)$. (You will need the Leibniz rule for differentiating an integral with a variable upper limit; see Eq. 6.5.13.) Make the connection that $u_z(h, t) = \partial h/\partial t$.

7.16 Keeley et al. [*J. Non-Newtonian Fluid Mech.*, **28**, 213 (1988)] presented the data shown in Fig. P7.16. They measured the rate of drainage of films of four different viscous liquids from a square vertical plate (150 mm on a side). The liquids had viscosities in the range 0.13–5.21 Pa · s. They plotted the percentage of mass (labeled M in the figure) left on a plate after a time t, in terms of a dimensionless time which they defined as

$$T = \frac{gH^2}{\nu L}t \qquad \textbf{(P7.16)}$$

The initial film dimensions in all cases were $H = 1$ mm and $L = 150$ mm. Does the simple drainage model (Eq. 7.3.60) lead to a model for the $M(T)$ behavior that describes the long time portion of these data?

7.17 Find the analog to the Hagen–Poiseuille law for flow of a Newtonian fluid through a linearly elastic capillary. To do so, solve Eq. 7.4.11 under the assumption that the flow is at steady state. Put your result in the form of Poiseuille's law with a multiplicative factor that accounts for the role of vessel elasticity:

$$Q = \frac{\pi R_o^4}{8\mu}\frac{P_a - P_v}{L}(1 + \psi) \qquad \textbf{(P7.17)}$$

Determine the factor ψ.

7.18 A solid cylinder falls freely, under the influence of gravity, coaxially through a concentric tube filled with a viscous liquid (Fig. P7.18). Develop models for $H(t)$, the separation of the bottom of the cylinder from the bottom of the container. Do two cases:

a. $\kappa \ll 1$ and $H/\kappa R \ll 1$

b. $1 - \kappa \ll 1$ and $H/\kappa R \ll 1$

7.19 Consider a slider flying over a magnetic recording disk at a relative speed of 10 m/s, with a minimum clearance of 300 nm, and an angle of 5°. If the drive is in air at 25°C, what is the Reynolds number for this flow? Use Re = $(Uh_{\min}\rho/\mu)$ as the definition of the Reynolds number.

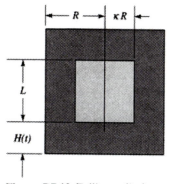

Figure P7.18 Falling cylinder near the bottom of its container.

7.20 Compare the pressures arising from gravity, and surface tension, in a lubricant film whose surface is perturbed as in Fig. 7.5.3, for the case that the disturbance amplitude is $h = 30$ Å and the wavelength is $\lambda = 100$ μm.

7.21 We used the approximation

$$\frac{1}{R} = -\frac{d^2H}{dx^2} \qquad \text{(P7.21.1)}$$

in considering the curvature of a two-dimensional thin film with a disturbance of small slope. Show that the exact result, for a two-dimensional disturbance of arbitrary slope, is

$$\frac{1}{R} = \frac{-dH^2/dx^2}{[1 + (dH/dx)^2]^{3/2}} \qquad \text{(P7.21.2)}$$

Begin by isolating a differential arc of a curve $y = H(x)$. Represent that arc as the equivalent section of a circular arc of radius R, as suggested in Fig. P7.21. Equation P7.21.2 follows from a simple trigonometric development.

7.22 Derive Eq. 7.2.14.

7.23 A lubricant film is placed on a disk that rotates at 1000 rpm. The initial film thickness is 0.05 μm. The lubricant density is 850 kg/m^3. We want 50% of the lubricant to remain on the disk after 2 years of constant service. What is the required viscosity of the lubricant?

7.24 Data for pulsatile flow of a Newtonian fluid through a smooth straight capillary are presented by Back [*J. Biomech.*, **27**, 169 (1994)] as part of his study of flow through catheterized coronary arteries. He used a sugar/water solution with a kinematic viscosity of $\nu = 0.035$ cm^2/s; the inside

diameter of the capillary was 1.93 mm, and the pulsation frequency was 1.5 Hz. Calculate the Womersley number (Eq. 7.4.1) and compare the data for mean pressure gradient as a function of mean flowrate, given below, to a time-averaged form of Poiseuille's law (based on Eq. 7.4.4).

Q (mL/min)	36	50	72.6	87
$\Delta P/L$ (mmHg/cm)	0.5	0.78	1.1	1.33

7.25 Brindley *et al.* [*J. Non-Newtonian Fluid Mech.*, **1**, 19 (1976)] present the data shown in Fig. P7.25 for squeezing between two disks under a constant force. Compare these data to an appropriate model of this transient flow.

7.26 In Chapter 4 we modeled steady radial flow between two circular disks. Do a quasi-steady model of the *transient* filling of the space between two circular disks for the geometry described in Example 4.5.1. The goal of the model is to yield an expression for the radius of the filled region $R^*(t)$ and the pressure $P(t)$ required to maintain a constant volumetric flowrate. As an initial condition, take the radius to be R_i.

a. Plot $P(t)$ and $R^*(t)$ in some dimensionless format.

b. Suppose a process requires that the region between two disks of radius 9 cm and separation 0.316 cm be filled with a Newtonian liquid of viscosity 10^5 poise in a time of 1 second. The radius of the central feed tube is 0.5 cm. What is the pressure at the entrance to the disks when fill occurs?

c. What force acts to separate the disks at the instant of fill?

7.27 Develop a model for the unsteady dripping from a system such as shown in Fig. 2.2.8 of

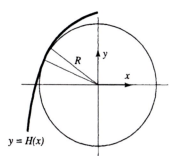

Figure P7.21 Representation of a section of a curve by the arc of a circle.

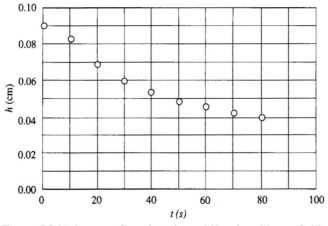

Figure P7.25 Squeeze flow data for a 140-poise silicone fluid, under a steady force $F = 423{,}800$ dyn; disk radius $= 3.75$ cm.

Chapter 2. Allow for the possibility that the back pressure due to surface tension is significant in comparison to the hydrostatic head. Assume that the drop shape is always a sector of a sphere. It will be useful to apply the following goemetrical formula to Fig. P7.27.

$$V = \frac{\pi h^2(3r - h)}{3} = \frac{\pi h(h^2 + 3a^2)}{6} \quad \text{(P7.27)}$$

7.28 Equation 7.2.34 gives us a model for the spacing between drops formed from an unstable, low viscosity jet of liquid. Use this result to show that the resulting drop radius is approximately twice that of the capillary from which the jet issues. How does this compare to the size of drops formed by permitting a liquid to drip slowly from a capillary? Plot the ratio of drop radius to capillary radius for water drops either jetting, or dropping, from capillaries with radii in the range 20–1000 μm.

7.29 Derive Eq. 7.4.2.

7.30 In Section 7.5 we used the "almost parallel flow" assumption to reduce the Navier–Stokes equations to a simple form that leads to the solution for u_x given by Eq. 7.5.11. This approach implies two inequalities:

$$\Re_1 \equiv \frac{u_y}{u_x} \ll 1 \quad \text{(P7.30.1)}$$

and

$$\Re_2 \equiv \frac{\partial^2 u_x/\partial x^2}{\partial^2 u_x/\partial y^2} \ll 1 \quad \text{(P7.30.2)}$$

Show that

$$\Re_1 = O\left(\frac{\overline{H}}{\lambda}\right) \quad \text{(P7.30.3)}$$

and

$$\Re_2 = O\left(\frac{\overline{H}}{\lambda}\right)^2 \quad \text{(P7.30.4)}$$

where O means "is of the order of"

7.31 Go back to Eq. 7.3.42. The term $C(t)$ was written as a function only of time, but we offered no physical argument to support that assumption. Present an argument based on the geometry of this flow

We find the pressure distribution from Eq. 7.3.45, with Eq. 7.3.46 as the result. But there is

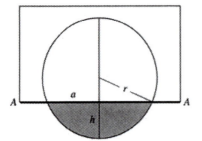

Figure P7.27 V is the volume of the spherical segment below the line AA. The sphere has a radius r, a is the radius of the base of the sphere on the line AA, and h is the height below the line AA.

a contribution to pressure that varies in the z direction. Show this by beginning with the fact that

$$dp = \frac{\partial p}{\partial r} dr + \frac{\partial p}{\partial z} dz \qquad \text{(P7.31.1)}$$

Then write the axial (z) component of the Navier–Stokes equation (neglecting gravity) and

show that Eqs. 7.3.43 and 7.3.44 imply that

$$\frac{\partial p}{\partial z} = \frac{\rho z}{H}\left(\ddot{H} - \frac{\dot{H}^2}{H}\right) \qquad \text{(P7.31.2)}$$

Use this result to find the missing contribution to the pressure distribution, and then demonstrate that except in the region $[0 < r < H]$, the axial variation of pressure is negligible.

Chapter 8

The Stream Function

In this chapter we examine flows that are two-dimensional by virtue of symmetries in the confining boundaries, and we define something called the "stream function," which contains all the information required to describe the kinematics of such flows. Stream function "portraits" are an aid in interpreting the dynamics of these two-dimensional flows.

8.1 DEFINITION OF THE STREAM FUNCTION

The Navier–Stokes and continuity equations are the basic equations of fluid dynamics for incompressible isothermal Newtonian flow. The dependent variables are the three components of velocity, and the pressure. All other features of the dynamics, such as local stresses and forces acting on the boundaries, follow from a knowledge of these so-called *primitive variables*.

There is a wide class of flows that are either *one-* or *two-dimensional* by virtue of geometrical symmetry associated with the boundaries of the flow. We have already seen several examples, such as Stokes flow around a solid sphere (a two-dimensional flow), and laminar fully developed flow in a long capillary (a one-dimensional flow). In the case of steady incompressible one- or two-dimensional flows there is an alternative approach to solving for the dynamics of the flow. It involves the definition of a scalar function of the kinematics called the *stream function*.

If we consider the dynamic equations for steady, incompressible, two-dimensional flow in Cartesian coordinates, we begin with the following set of nonlinear second-order partial differential equations (when we ignore gravitational effects):

$$\frac{\partial u_x}{\partial x} + \frac{\partial u_y}{\partial y} = 0 \tag{8.1.1}$$

$$\rho \left(u_x \frac{\partial u_x}{\partial x} + u_y \frac{\partial u_x}{\partial y} \right) = -\frac{\partial p}{\partial x} + \mu \left(\frac{\partial^2 u_x}{\partial x^2} + \frac{\partial^2 u_x}{\partial y^2} \right) \tag{8.1.2}$$

$$\rho \left(u_x \frac{\partial u_y}{\partial x} + u_y \frac{\partial u_y}{\partial y} \right) = -\frac{\partial p}{\partial y} + \mu \left(\frac{\partial^2 u_y}{\partial x^2} + \frac{\partial^2 u_y}{\partial y^2} \right) \tag{8.1.3}$$

We now define a function ψ such that

$$u_x = \frac{\partial \psi}{\partial y} \quad \text{and} \quad u_y = -\frac{\partial \psi}{\partial x} \tag{8.1.4}$$

When these expressions for the velocity components are substituted into the continuity equation and the indicated differentiations of the velocities are performed, we find

$$\frac{\partial^2 \psi}{\partial x \, \partial y} - \frac{\partial^2 \psi}{\partial y \, \partial x} \equiv 0 \tag{8.1.5}$$

This equation is satisfied for *any* continuous function $\psi(x, y)$. Hence the continuity equation is of no further use to us. We have lost an equation, but we have also reduced the number of dependent variables from three to two (ψ and p). We now substitute the stream function expressions for the velocity components into the two dynamic equations. We will not show the algebra here, but it should be clear that since Eqs. 8.1.2 and 8.1.3 are second-order in the velocities, they will be third-order in the stream function. Each equation will contain ψ and p. If we differentiate Eq. 8.1.2 (which has a term $\partial p/\partial x$) with respect to y, and differentiate Eq. 8.1.3 (which has a term $\partial p/\partial y$) with respect to x, each equation will then have fourth-order derivatives of ψ, and each will have a term $\partial^2 p/\partial y \, \partial x$. Then we may subtract the two equations and eliminate the pressure terms. This results in a single fourth-order nonlinear differential equation in the single variable ψ. The form of the equation is found to be[1]

$$\frac{\partial \psi}{\partial x} \frac{\partial (\nabla^2 \psi)}{\partial y} - \frac{\partial \psi}{\partial y} \frac{\partial (\nabla^2 \psi)}{\partial x} = \nu \nabla^4 \psi \tag{8.1.6}$$

where

$$\nabla^2 \psi \equiv \frac{\partial^2 \psi}{\partial x^2} + \frac{\partial^2 \psi}{\partial y^2}$$

$$\nabla^4 \psi \equiv \frac{\partial^4 \psi}{\partial x^4} + 2 \frac{\partial^4 \psi}{\partial x^2 \partial y^2} + \frac{\partial^4 \psi}{\partial y^4} \tag{8.1.7}$$

and the kinematic viscosity is defined by

$$\nu = \frac{\mu}{\rho} \tag{8.1.8}$$

The good news is that we have reduced three equations to one equation for a single unknown, ψ. The bad news is that the one equation—Eq. 8.1.6—is formidable. The main complication arises from the terms on the left-hand side, the nonlinear inertia terms. For low Reynolds number flows we could drop the nonlinear terms and have a fourth-order *linear* partial differential equation to solve.

We have already dealt with a linear fourth-order differential equation in our derivation of Stokes' law for flow about a solid sphere (Eq. 4.4.130). In fact, we actually used a stream function approach in that problem, without defining the stream function as such.

Is the trade-off (a single equation but higher order) worth it? The stream function has a geometrical interpretation that makes it an important and useful characteristic of the flow field. In Cartesian coordinates we may write

$$d\psi = \frac{\partial \psi}{\partial x} dx + \frac{\partial \psi}{\partial y} dy = -u_y \, dx + u_x \, dy \tag{8.1.9}$$

---

[1] This is another example in which you have a choice. You can carry out the mathematical steps outlined in this paragraph, and satisfy yourself that we do indeed obtain Eq. 8.1.6. There is nothing difficult about doing this. It is tedious and boring, however. Or you can accept Eq. 8.1.6 as correct. Choose.

Along a line of constant ψ, for which $d\psi = 0$, we find from Eq. 8.1.9 that

$$\left[\frac{dy}{dx}\right]_{\psi_1} = \frac{u_y}{u_x} \tag{8.1.10}$$

With reference to Fig. 8.1.1, we see that a line of constant ψ is a *streamline*—a line whose tangent at any point along the streamline is collinear with the velocity vector **u** at that point. Hence each streamline shows the path of a "tracer particle" in the flow. The streamlines of a flow often provide more immediately useful information about the flow field than the actual velocity profile itself. An example of such streamlines is shown in Fig. 1.1.7 of Chapter 1.

Another property of streamlines is quite useful to us. The rate of mass flow (per unit width in the z direction) between two streamlines is given by

$$\dot{m} = \int_{\psi_1}^{\psi_2} \rho u \, dn \tag{8.1.11}$$

where u is the magnitude of **u** and dn is a line normal to any streamline ψ. The vector **dn** has components dx and dy. It is not difficult to see (from Fig. 8.1.1) that

$$\frac{dx}{dn} = -\frac{u_y}{u} \quad \text{and} \quad \frac{dy}{dn} = \frac{u_x}{u} \tag{8.1.12}$$

while from Eq. 8.1.9

$$\frac{d\psi}{dn} = -u_y \frac{dx}{dn} + u_x \frac{dy}{dn} \tag{8.1.13}$$

Since

$$u^2 = u_x^2 + u_y^2 \tag{8.1.14}$$

it follows that

$$\frac{d\psi}{dn} = u \tag{8.1.15}$$

This permits us to write Eq. 8.1.11 as

$$\dot{m} = \int_{\psi_1}^{\psi_2} \rho \, d\psi = \rho(\psi_2 - \psi_1) \tag{8.1.16}$$

We have proven that the mass flow between a pair of streamlines is proportional to the difference in the values of the stream functions of those two streamlines. Hence, if we plot only streamlines whose values differ by a constant increment, the velocity is

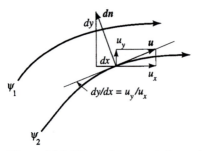

Figure 8.1.1 Lines of constant ψ are streamlines.

highest in the regions where the streamlines are closest together. Thus one can inspect a streamline plot (but only when *equal increments* in ψ are chosen) and easily see regions of very high or very low velocity.

While we have used Cartesian coordinates in the examples and developments presented here, all the ideas hold for other coordinate systems. We show this with an example of an axisymmetric flow.

EXAMPLE 8.1.1 *The Stream Function Analysis for Poiseuille Flow in a Tube*

For this flow, the continuity equation has the form given in Table 4.1.1 by Eq. 4.1.8b (without the θ-derivative term). We now make the observation that if we define a function ψ by

$$u_z = \frac{1}{r} \frac{\partial \psi}{\partial r} \quad \text{and} \quad u_r = -\frac{1}{r} \frac{\partial \psi}{\partial z} \tag{8.1.17}$$

then the continuity equation is satisfied automatically. Hence Eq. 8.1.17 provides the definition of the stream function for axisymmetric flow in cylindrical coordinates.

For small Reynolds numbers and steady state flow, the Navier–Stokes equations in cylindrical coordinates reduce to

$$0 = -\frac{\partial p}{\partial r} + \mu \left\{ \frac{\partial}{\partial r} \left[\frac{1}{r} \frac{\partial}{\partial r} (r u_r) \right] + \frac{\partial^2 u_r}{\partial z^2} \right\} \tag{8.1.18}$$

and

$$0 = -\frac{\partial p}{\partial z} + \mu \left\{ \frac{1}{r} \frac{\partial}{\partial r} \left(r \frac{\partial u_z}{\partial r} \right) + \frac{\partial^2 u_z}{\partial z^2} \right\} \tag{8.1.19}$$

If we now introduce the function ψ defined by Eqs. 8.1.17, we eliminate the velocity components from the Navier–Stokes equations. At this point we have two equations with two unknowns, ψ and p. However, p can be eliminated simply by differentiating Eq. 8.1.18 with respect to z, differentiating Eq. 8.1.19 with respect to r, and then subtracting one equation from the other. The result is a fourth-order partial differential equation for $\psi(r, z)$, which takes the form

$$0 = \left(\frac{\partial^2}{\partial r^2} - \frac{1}{r} \frac{\partial}{\partial r} + \frac{\partial^2}{\partial z^2} \right) \left(\frac{\partial^2 \psi}{\partial r^2} - \frac{1}{r} \frac{\partial \psi}{\partial r} + \frac{\partial^2 \psi}{\partial z^2} \right) \tag{8.1.20}$$

If we assume fully developed flow, we expect ψ to be a function only of the radial coordinate r. Hence ψ is a solution to

$$0 = \left(\frac{d^2}{dr^2} - \frac{1}{r} \frac{d}{dr} \right) \left(\frac{d^2 \psi}{dr^2} - \frac{1}{r} \frac{d\psi}{dr} \right) \tag{8.1.21}$$

This is an ordinary differential equation, and we would solve it by trying a function $\psi = r^n$. It is not difficult to see that three values of n "work": $n = 0, 2,$ and 4. Hence

$$\psi = a + br^2 + cr^4 \tag{8.1.22}$$

satisfies Eq. 8.1.21, where a, b, c are constants to be determined by boundary conditions. One boundary condition should express no-slip on $r = R$. From

$$u_z = 0 \quad \text{on} \quad r = R \tag{8.1.23}$$

we infer

$$-\frac{1}{r}\frac{d\psi}{dr} = 0 \qquad \text{on} \quad r = R \tag{8.1.24}$$

This yields

$$c = -\frac{1}{2}\frac{b}{R^2} \tag{8.1.25}$$

A second boundary condition, symmetry about the axis, takes the form

$$\frac{du_z}{dr} = 0 \qquad \text{on} \quad r = 0 \tag{8.1.26}$$

or

$$\frac{d}{dr}\left(\frac{1}{r}\frac{d\psi}{dr}\right) = 0 \tag{8.1.27}$$

This is automatically satisfied by Eq. 8.1.22 and does not give an additional relationship among the constants a, b, c. At this point, then, we have

$$\psi = a + b\left(r^2 - \frac{1}{2}\frac{r^4}{R^2}\right) \tag{8.1.28}$$

A further property of the stream function can be deduced by considering the volumetric flowrate between two arbitrary radii, given by

$$q_{12} = \int_{r_1}^{r_2} 2\pi r u_z(r)\, dr = \int_{r_1}^{r_2} 2\pi \frac{d\psi}{dr}\, dr = 2\pi[\psi(r_2) - \psi(r_1)] \tag{8.1.29}$$

This states that the difference between the values of ψ at two radii is proportional to the flowrate through the annulus bounded by the two radii. Another way of looking at this result is to observe that if we draw a set of streamlines whose values differ by a fixed constant, the volumetric flowrate through the annular region bounded by each adjacent pair of streamlines will be identical. This is the analog, in cylindrical coordinates, of the result given earlier in Eq. 8.1.16.

If we consider the streamline along the tube axis ($r = 0$), and the streamline along the tube wall ($r = R$), Eq. 8.1.29 gives us

$$\psi(R) - \psi(0) = \frac{Q}{2\pi} = \frac{UR^2}{2} \tag{8.1.30}$$

where Q is the total volumetric flowrate through the tube, and $U = Q/\pi R^2$ is the average velocity. From Eq. 8.1.28, we may now find

$$b = U \tag{8.1.31}$$

This permits us to express ψ as

$$\psi = a + UR^2\left[\left(\frac{r}{R}\right)^2 - \frac{1}{2}\left(\frac{r}{R}\right)^4\right] \tag{8.1.32}$$

The constant a cannot be fixed by the physics of this problem. This follows from the observation (see Eq. 8.1.17) that if ψ is a stream function, then $\psi +$ any constant is still a stream function. Hence the value of a is arbitrary, and we are free to choose a convenient value; the simplest choice is

$$a = 0 \tag{8.1.33}$$

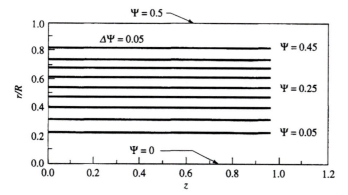

Figure 8.1.2 Streamlines for Poiseuille flow in a tube.

in consequence of which we may write

$$\psi = UR^2 \left[\left(\frac{r}{R} \right)^2 - \frac{1}{2} \left(\frac{r}{R} \right)^4 \right] = \frac{Q}{\pi} \left[\left(\frac{r}{R} \right)^2 - \frac{1}{2} \left(\frac{r}{R} \right)^4 \right] \qquad (8.1.34)$$

We may now nondimensionalize the stream function by defining

$$\frac{\pi \psi}{Q} \equiv \Psi = \left(\frac{r}{R} \right)^2 - \frac{1}{2} \left(\frac{r}{R} \right)^4 \qquad (8.1.35)$$

This stream function has values in the range $[0 < \Psi < 0.5]$. If we choose equal increments for Ψ, say $\Delta\Psi = 0.05$, we may plot the streamlines for this flow. These are shown in Fig. 8.1.2.

Streamline plots for such a simple flow—fully developed one-dimensional flow—are not very interesting. The axis of the tube coincides with the streamline $\Psi = 0$, and the tube wall is the position of the streamline $\Psi = 0.5$. Streamlines are drawn for equal increments in Ψ. All the volume flowrates through the annuli bounded by any adjacent pair of streamlines are equal. Because of the cylindrical geometry, the areas between pairs of streamlines vary as r^2. The velocity is largest near the axis of the tube, but the streamlines $\Psi = 0$ and $\Psi = 0.05$ ($\Delta\Psi = 0.05$) are spaced far apart because there is relatively little area in the central portion of the tube. Near the tube wall ($r/R = 1$) the velocity is getting very small, and so the streamlines $\Psi = 0.45$ and $\Psi = 0.5$ ($\Delta\Psi = 0.05$, again) are spaced far apart so that the flow between them can match that between other pairs of lines.

8.2 STREAMLINES FOR OTHER AXISYMMETRIC FLOWS

8.2.1 Flow Around a Sphere

In Chapter 4 we derived a solution for the velocity field that arises when a solid sphere is approached by a uniform steady Newtonian flow, at vanishingly small Reynolds numbers. In spherical coordinates the stream function is defined so that

$$u_\theta = -\frac{1}{r \sin \theta} \frac{\partial \psi}{\partial r} \qquad \text{and} \qquad u_r = \frac{1}{r^2 \sin \theta} \frac{\partial \psi}{\partial \theta} \qquad (8.2.1)$$

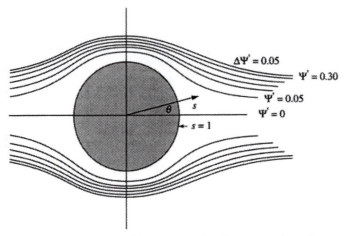

Figure 8.2.1 Streamlines (Ψ') for creeping flow around a sphere.

It is not difficult to show that the solution obtained in Chapter 4, Eqs. 4.4.123, corresponds to a stream function that we may write in the form

$$\frac{2\psi}{UR^2} = s^2 \sin^2\theta\, F(s) \tag{8.2.2}$$

The stream function vanishes on the sphere surface, $s = 1$, since $F(s) = 0$ at $s = 1$. Far from the sphere, as s grows large compared to unity, the stream function becomes

$$\Psi' = \frac{2\psi}{UR^2} = (s \sin\theta)^2 \tag{8.2.3}$$

From Fig. 4.4.6 we see that $s \sin\theta = $ constant is a line parallel to the z axis. Hence the streamlines become straight lines parallel to the external flow, far from the sphere, as expected. We can plot curves of constant Ψ', which are the streamlines, and these are shown in Fig. 8.2.1, for selected values of Ψ'.

8.2.2   Flow in a CVD Reactor

In the microelectronics industry a number of commercial reactors are used for the growth of thin solid films of silicon from gaseous precursors. The process is known as chemical vapor deposition, or CVD. Among the characteristics of these CVD reactors are the complexity of the detailed gas flow through them and the importance of the flow field on the quality of the solid film produced. Figure 8.2.2 shows a common configuration for a CVD reactor of the "bell jar" type. Gas is fed through a central opening in the lower (heated) surface, circulates through the volume of the reactor, and exits from an annular manifold along the perimeter of the lower surface. The flow field is too complex to model by analytical methods, but for a given (assumed) set of boundary conditions, it is possible to solve the continuity and Navier–Stokes equations numerically.

Specifically, we can solve the continuity and Navier–Stokes equations numerically under the assumptions of isothermal incompressible Newtonian flow; Fig. 8.2.3 shows streamline plots for two different Reynolds numbers. (The stream function is defined in Eq. 8.1.17 for axisymmetric flow.) Note the appearance of regions of closed streamlines, which represent regions wherein the fluid is circulating but not passing out of the

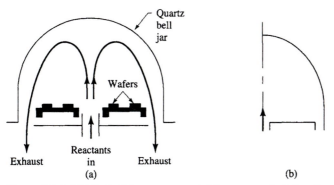

Figure 8.2.2 (a) The bell jar reactor and (b) a geometrical simplification of the boundaries for numerical simulation of the flow field.

reactor subsequently. The direction of flow of the reactants across the wafers plays a very important role in the uniformity of the thickness of the solid film that grows on the wafers. The flows are in opposite directions in the two cases shown in Fig. 8.2.3.

8.2.3   Flow in an Occluded Blood Vessel

The flow of blood through vessels that are partly occluded provides another example of an axisymmetric flow for which the streamlines aid in interpretation of the fluid dynamics. Such an occlusion (called a "stenosis" in the medical literature) may arise from deposition of material on the interior of the vessel surface. An occlusion may also develop from pathological growth of cellular material from the vessel surface into the bloodstream. Physiologically, these occlusions are rarely axisymmetric, and as a consequence it is difficult to produce mathematical models of the flow field. In what follows, we assume the presence of an *axisymmetric stenosis* and present numerical solutions for the main features of the flow. We assume that the flow is laminar, isothermal, steady, and Newtonian.

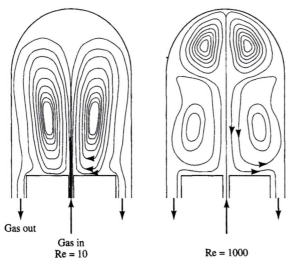

Figure 8.2.3 Streamlines for gas flow through an isothermal bell jar reactor.

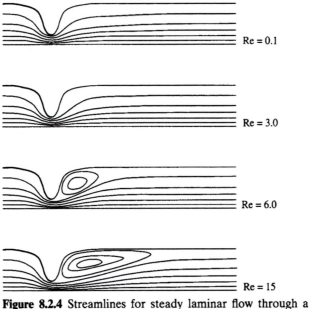

Re = 0.1

Re = 3.0

Re = 6.0

Re = 15

Figure 8.2.4 Streamlines for steady laminar flow through a tube with a stenosis. Middleman and English, unpublished work.

Streamline plots were obtained from a numerical solution of the continuity and Navier–Stokes equations, subject to a given (assumed) set of boundary conditions. In the case shown (Fig. 8.2.4), we simply assumed axisymmetry and no-slip at the tube surface. The velocity profile at the upstream inlet of the tube ("upstream" is to the left in the figure) was taken to be parabolic, corresponding to a specified flowrate. We observe that as the Reynolds number grows large compared to unity, the flow field exhibits a recirculation region just downstream of the narrowing of the tube. The size of this recirculating region is found to increase as the Reynolds number increases. The presence of the recirculating region is a consequence of the inertial terms in the Navier–Stokes equations.

Flow through a stenosis gives rise to an excess pressure drop relative to that for the same flowrate through a straight tube. The excess pressure loss associated with the stenosis is plotted in Fig. 8.2.5, in terms of a friction loss coefficient defined as

$$C_f = \frac{\Delta P}{\rho U^2} \tag{8.2.4}$$

In the limit of "creeping flow" (i.e., vanishingly small Reynolds numbers), the loss coefficient so defined is an inverse function of Reynolds number and depends only on the geometry of the stenosis. Hence $C_f \text{Re}$ is a constant for a given geometry. As Reynolds number increases, inertial effects cause the $C_f \text{Re}$ product to increase, as shown in Fig. 8.2.5.

Often the availability of the streamline plots aids in the interpretation of friction loss plots such as Fig. 8.2.5. For example, we may find that the Reynolds number at which the $C_f \text{Re}$ product begins to rise above its creeping flow limit corresponds to the transition of the flow field from smooth, nearly parallel streamlines to one in which recirculation regions appear.

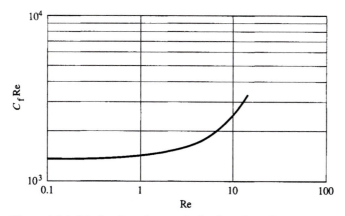

Figure 8.2.5 Friction loss for steady laminar flow through a tube with a stenosis. Middleman and English, unpublished work.

8.2.4   Radial Flow Between Parallel Disks

In Section 4.5 we considered steady radial flow between parallel disks to be unidirectional on the basis of the assumption that the entrance region near the axis, where the flow turns the corner, could be ignored. If we were to drop this assumption, we would find that the two-dimensional Navier–Stokes equations would not yield an analytical solution except for vanishingly small Reynolds numbers, and even for that case the analytical solution would be so computationally complex that we could not easily calculate any desired features of the flow field. Such a flow would yield to numerical methods, however, from which the streamlines are easily obtained. Figure 8.2.6 shows an example of streamlines for this flow at several Reynolds numbers. From such streamline "portraits" we can easily conclude that the assumption of unidirectional flow would be quite reasonable, for this geometry (i.e., for specific choices of R/H and R/R_i), at Reynolds

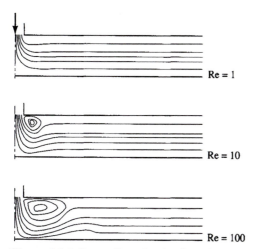

Figure 8.2.6 Streamlines for laminar flow between parallel disks: $Re = 2Q\rho/\pi R_i \mu$. Middleman and English, unpublished work.

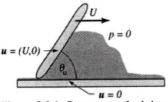

Figure 8.3.1 Geometry of wiping.

numbers of 1 or less. At a Reynolds number of 10 there is a significant region of recirculating flow, but over most of the area of this system the flow is still unidirectional. When the Reynolds number exceeds a value of a few hundred, the inertia of the fluid as it enters the space between the disks exerts a strong influence on the flow far downstream. This example shows us that the streamline formulation can be a substantial aid in the interpretation of the flow field and can serve as a guide to developing a "strategy" of modeling similar problems.

8.3 STREAMLINE ANALYSIS FOR A WIPING FLOW

Let's develop a streamline solution for a two-dimensional flow of some technological interest. Figure 8.3.1 shows the geometry and boundary conditions for a flow field that arises when a planar blade pushes a viscous liquid across a planar surface. Such a flow field is a simplified model for the more complex flow that occurs when viscous liquids are wiped off a surface. We will assume that the wiper blade makes a frictionless contact with the lower plane and that there is no leakage of fluid under the blade. We are interested in deriving a streamline "portrait" of this flow, and ultimately we would like to be able to predict the pressure distribution acting on the blade. In a real system, as opposed to the idealization considered here, this pressure distribution could deflect the blade and permit leakage, reducing the efficiency of wiping.

To begin, we note that this is an unsteady flow. The velocity field at some point in the fluid varies with time as the blade approaches that point. However, a simple transformation to a moving coordinate system yields a steady state problem. We simply choose to fix a coordinate system in the moving plane, as shown in Fig. 8.3.2. In this coordinate system, the wiper blade is fixed, and the lower plane moves with the velocity $-U$ in its own plane. With this transformation, the boundary conditions on velocity are as follows, in a plane polar coordinate system:

$$u_r = -U \quad \text{and} \quad u_\theta = 0 \quad \text{on} \quad \theta = 0 \tag{8.3.1}$$

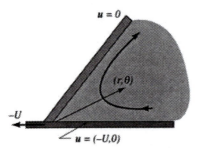

Figure 8.3.2 Transformation to a moving coordinate system.

$$u_r = 0 \quad \text{and} \quad u_\theta = 0 \quad \text{on} \quad \theta = \theta_o \tag{8.3.2}$$

To develop the stream function formulation, we begin with the definition of the stream function appropriate to planar polar coordinates. The choice of a polar coordinate system may not seem logical here, but we find that the resulting analysis is simpler in polar coordinates than in Cartesian. From the continuity equation, Eq. 4.1.8b, for $\mathbf{u} = (u_r, u_\theta)$, we see that if ψ is defined so that[2]

$$-\frac{\partial \psi}{\partial r} = u_\theta \quad \text{and} \quad \frac{1}{r}\frac{\partial \psi}{\partial \theta} = u_r \tag{8.3.3}$$

then the continuity equation is satisfied. Next we have to obtain the stream function formulation of the Navier–Stokes equations appropriate to this flow. We use the same methodology illustrated in Example 8.1.1, except that in this case we consider the r and θ components of the dynamic equations, instead of the r and z components. The result is a partial differential equation for the stream function, which we may write (with the assumption of "creeping flow") in the form

$$\nabla^2(\nabla^2\psi) = 0 \tag{8.3.4}$$

where the ∇^2 operator is defined by

$$\nabla^2 \equiv \frac{\partial^2}{\partial r^2} + \frac{1}{r}\frac{\partial}{\partial r} + \frac{1}{r^2}\frac{\partial^2}{\partial \theta^2} \tag{8.3.5}$$

Equation 8.3.4 requires specification of four boundary conditions. These follow from the boundary conditions on velocity, given in Eqs. 8.3.1 and 8.3.2, and they take the forms

$$\frac{1}{r}\frac{\partial \psi}{\partial \theta} = -U \quad \text{and} \quad \frac{\partial \psi}{\partial r} = 0 \quad \text{on } \theta = 0 \tag{8.3.6}$$

and

$$\frac{1}{r}\frac{\partial \psi}{\partial \theta} = 0 \quad \text{and} \quad \frac{\partial \psi}{\partial r} = 0 \quad \text{on } \theta = \theta_o \tag{8.3.7}$$

Equation 8.3.4 can be solved analytically by introducing a function $f(\theta)$ defined in such a way that

$$\psi = rf(\theta) \tag{8.3.8}$$

It is not difficult (just tedious) to show that the function f satisfies a fourth-order ordinary differential equation of the form

$$f + 2f'' + f^{\text{iv}} = 0 \tag{8.3.9}$$

The general solution of this equation is of the form

$$f = A \sin \theta + B \cos \theta + C\theta \sin \theta + D\theta \cos \theta \tag{8.3.10}$$

After imposing the boundary conditions given in Eqs. 8.3.6 and 8.3.7, we find the solution to be

$$-(\theta_o^2 - \sin^2\theta_o)\frac{f(\theta)}{U} = \theta_o^2 \sin \theta - (\theta_o - \sin \theta_o \cos \theta_o)\,\theta \sin \theta - (\sin^2\theta_o)\,\theta \cos \theta \tag{8.3.11}$$

---

[2] Note the difference between this formulation, in planar polar coordinates, and the formulation given in Eqs. 8.1.17 for axisymmetric cylindrical coordinates.

This completes the mathematical analysis of this flow. The important task is now the physical interpretation of the solution. We can explore the following questions:

What do the streamlines look like?
How do we get the pressure distribution on the wiper blade?

The first question is conceptually simple. The streamlines are lines of constant ψ, so we write the stream function, using Eqs. 8.3.8 and 8.3.11, and find the (r, θ) coordinates for various values of ψ. First we will nondimensionalize the stream function, and the radial coordinate r. A suitable definition of a dimensionless radial coordinate is

$$s \equiv \frac{U}{\nu} r \qquad \text{(8.3.12)}$$

and a dimensionless stream function is

$$\Psi \equiv \frac{-\psi}{\nu} (\theta_o^2 - \sin^2 \theta_o) = s[\theta_o^2 \sin \theta - (\theta_o - \sin \theta_o \cos \theta_o)\theta \sin \theta - (\sin^2 \theta_o)\theta \cos \theta]$$

$$\text{(8.3.13)}$$

For a fixed value of the blade angle θ_o we solve for the coordinates (s, θ) that correspond to particular values of Ψ. For example, for $\theta_o = \pi/2$ we find

$$-\Psi = s \left[\frac{\pi^2}{4} \sin \theta - \frac{\pi}{2} \theta \sin \theta - \theta \cos \theta \right] \qquad \text{(8.3.14)}$$

In Fig. 8.3.3 we show streamlines corresponding to the values $\Psi = 0, -1, -2, -3$.

When the stream function is available, we can find the velocity components. For example, the radial velocity follows from

$$u_r = \frac{1}{r} \frac{\partial \psi}{\partial \theta} = \frac{U}{\pi^2/4 - 1} \left[\left(\frac{\pi^2}{4} - 1 \right) \cos \theta - \frac{\pi}{2} \theta \cos \theta - \left(\frac{\pi}{2} - \theta \right) \sin \theta \right] \qquad \text{(8.3.15)}$$

for the case $\theta_o = \pi/2$.

To find the pressure distribution along the blade surface, we can go back to the radial component of the Navier–Stokes equations for this flow and write

$$\frac{\partial p}{\partial r} = \mu \left[\frac{\partial}{\partial r} \frac{1}{r} \frac{\partial}{\partial r} (r u_r) + \frac{1}{r^2} \frac{\partial^2 u_r}{\partial \theta^2} - \frac{2}{r^2} \frac{\partial u_\theta}{\partial \theta} \right] \qquad \text{(8.3.16)}$$

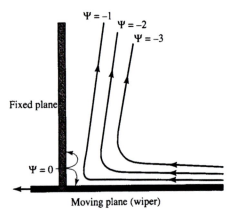

Figure 8.3.3 Streamlines near a vertical blade wiping a horizontal planar surface. Coordinate system fixed with the blade.

Instead of proceeding further with the mathematics, note that $\partial p/\partial r$ will vary with r as $1/r^2$. Hence the pressure becomes infinite as the corner ($r = 0$) is approached. Furthermore, the force obtained by integrating the pressure with respect to r along the blade surface $\theta = \pi/2$ will vary as $1/r$. Hence, according to this model, the force acting on the blade will be infinite. This apparently nonphysical result of the analysis follows directly from an earlier nonphysical assertion that there is no leakage at the corner.[3] If a leak were to be allowed for, the infinite pressure problem would disappear. This type of analysis is well beyond the scope of an introductory presentation of fluid dynamics.

SUMMARY

Sometimes the most important thing we need to know about a flow field is obtainable from a "streamline portrait" of the flow, a picture of the paths that would be followed by tiny tracer particles in the flow. An analytical (or numerical) solution for the velocity and pressure distributions would yield this information; in flows of certain kinds, however, we have a more direct route to obtaining these "portraits." We find that we can define something called the "stream function" for two-dimensional flows and that we can replace the continuity and Navier–Stokes equations with a single equation for the stream function. This approach has the advantage of producing the desired information directly; it is not necessary to derive the streamline portrait from the solution for the flow field. The trade-off is that we go from a set of second-order partial differential equations (Navier–Stokes, along with the continuity equation) to a single fourth-order partial differential equation. We present just a few examples of stream function analyses in this brief chapter.

PROBLEMS

8.1 Give the stream function formulation (Eq. 8.1.6) and find the stream function for the following flows:

a. Plane Couette flow
Fluid is confined between infinite parallel planes, one of which moves at constant speed in its own plane. There is no pressure gradient. Plot streamlines of equal increments.

b. Plane Poiseuille flow
Fluid is confined between infinite parallel planes that are stationary. Flow occurs due to imposition of a steady state pressure gradient. Plot streamlines of equal increments.

8.2 Prove the assertion following Eq. 8.1.21: that is, derive Eq. 8.1.22.

8.3 Give the stream function formulation in cylindrical coordinates (but not the solution) of the problem illustrated in Chapter 3, Section 3.3.2. Include the ends of the system. What boundary conditions must be satisfied?

8.4 Stream functions are often defined so that they will take on the value $\psi = 0$ on solid boundaries. What choice of the constant a in Eq. 8.1.32 would give us such a stream function?

8.5 Liquid issues in laminar fully developed flow from a tube of circular cross section (radius R_o) and the flow impinges on a flat rigid surface, as shown in Fig. P8.5. The entire region $0 < z < B$ is filled with the liquid that issues from the tube. This is called a "submerged jet." Write

---

[3] Be careful about drawing generalizations from this aspect of this flow. In our earlier example of axial annular flow in a closed container (Section 3.3.2), leakage of fluid through the region where the inner cylinder exits the container was not permitted, and no such problem arose.

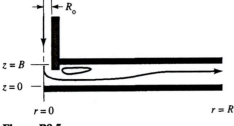

$z = B$

$z = 0$

$r = 0$ $r = R$

Figure P8.5

out the stream function formulation (but not the solution) of this problem. What boundary conditions would you impose at $r = R$, assuming that $R \gg B$? Do the steady state "creeping flow" case.

8.6 Liquid enters, fills, and leaves a cylindrical tube as shown in Fig. P8.6. Write out the stream function formulation (but not the solution) of this problem. Do the steady state "creeping flow" case. What boundary condition would you impose at $z = 0$? Can you impose a boundary condition at $z = L$? Discuss this point.

8.7 Show that Eq. 8.2.1, the definition of the stream function in spherical coordinates, is consistent with the continuity equation.

8.8 Verify Eq. 8.2.2.

8.9 One sometimes finds the following expression given as the stream function for steady two-dimensional creeping flow about a cylinder of radius a:

$$\psi = Uy \left[1 - \frac{a^2}{x^2 + y^2} \right] \qquad \textbf{(P8.9)}$$

where x and y are Cartesian coordinates centered on the axis of the cylinder. How is the stream function defined for this flow? What is the nature of this flow? Specifically, what boundary conditions are satisfied by u_θ and u_r on the cylinder surface, and far from the surface?

8.10 Verify Eq. 8.3.4.
8.11 Derive Eq. 8.3.9, and verify that Eq. 8.3.10 is the general solution.
8.12 Draw the streamlines for $\Psi = -1/2$, and $-3/2$, on Fig. 8.3.3.
8.13 For the wiping flow described in Section 8.3, the flowrate normal to the circular arc $r = a$ is found from

$$q = \int_0^{\pi/2} u_r(r, \theta) a \, d\theta \qquad \textbf{(8P13.1)}$$

Find q and interpret the result physically.
8.14 Prove the assertion following Eq. 8.3.16. Find an expression for the pressure distribution on the vertical surface.
8.15 Take our stream function solution, Eq. 8.3.14, and transform the stream function to Cartesian coordinates. From the resulting function $\Psi(x, y)$, show that the y component of velocity is

$$u_y(x, y) = \frac{\pi}{2} \frac{1}{1 + (x/y)^2} - \tan^{-1}\left(\frac{y}{x}\right) + \frac{y/x}{1 + (y/x)^2}$$

$$\textbf{(P8.15.1)}$$

Plot $u_y(x)$ for $y = 1, 2$, and 4. Find the value of the integral

$$q = \int_0^\infty u_y(x, y) \, dx \qquad \textbf{(P8.15.2)}$$

What does the result mean physically?
8.16 Return to the analysis of the flow through a slider bearing (Section 6.1) and show that the stream function is given by

$$\Psi = \bar{y} - \frac{2\bar{y}^2}{\eta} + \frac{3\lambda\bar{y}^2}{\eta^2} + \frac{\bar{y}^3}{\eta^2}\left(1 - \frac{2\lambda}{\eta}\right) \qquad \textbf{(P8.16)}$$

For the case $\kappa = 0.2$, $\theta = 15°$, give the value of the stream function along the upper stationary surface. Do this without using Eq. P8.16. Plot

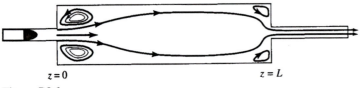

$z = 0$ $z = L$

Figure P8.6

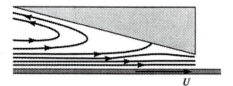

Figure P8.16 Streamlines for flow under a slider bearing.

streamlines for four values of ψ: 0.085, 0.170, 1/6, and 0.2. The results should appear as suggested in the sketch in Fig. P8.16. What is the physical significance of the $\psi = 1/6$ streamline?

8.17 On a copy of Fig. 8.2.1, draw the streamline for $\Psi' = 1.0$, from the center of the sphere to a position 10 radii downstream from the center.

8.18 Review the analysis of planar converging flow in Chapter 6, Section 6.8. Find the stream functions for the planar Poiseuille flow region and for the planar radial flow region. Draw the $\Psi = 0.4$ streamline for each of these regions, and compare your results to the corresponding line on Fig. 6.8.5. What nondimensionalization of Ψ is used to yield $\Psi = 0.5$ on the solid boundary? Is a *positive* value of Ψ consistent with the definition of ψ given in Eq. 8.3.3?

Chapter **9**

Turbulent Flow and the Laminar Boundary Layer

Thus far we have considered laminar flows, and we have discovered a wide variety of laminar flows simple enough to permit the development of mathematical models that are useful for the purposes of engineering analysis and design. But many flows of technological significance are *turbulent*. In this chapter we will describe some of the characteristics of turbulent flows, and the difficulties in developing simple but realistic mathematical models of turbulence that have predictive capability.

9.1 TURBULENT FLOW

We begin with a "thought experiment." Imagine that water is caused to flow through a long straight pipe of uniform circular cross section. We are able to control the rate Q at which the water is pumped into the entrance of the pipe. Midway down the axis of the pipe ($z = L/2$) we have a set of measuring devices so small that they do not disturb the flow in any way. (Remember, this is a *thought* experiment.) One device measures the pressure at the axis of the pipe, one measures the velocity components u_r and u_z on the axis, and one measures the shear stress T_{rz} acting on the inside wall of the pipe (i.e., at $r = R$). Initially, in this thought experiment, the flow-rate is so low that laminar fully developed flow prevails, and we could predict the measured quantities by applying the Hagen–Poiseuille law. Thus, if the volume flowrate were $Q = $ constant, and the pressure were $p = 0$ at the downstream end, we would find that (see Eqs. 3.2.33 and 3.2.34)

$$p\big|_{z=L/2} = \frac{4\mu L Q}{\pi R^4} \tag{9.1.1}$$

$$u_z\big|_{r=0} = \frac{2Q}{\pi R^2} \tag{9.1.2}$$

$$u_r\big|_{r=0} = 0 \tag{9.1.3}$$

and

$$T_{rz}\big|_{r=R} = \frac{4\mu Q}{\pi R^3} \tag{9.1.4}$$

Since Q is controlled to be constant with respect to time, these measured quantities would also be constant in time.

In this thought experiment, we now raise the flowrate Q to the point that the Reynolds number Re $= 4Q\rho/\pi D\mu$ becomes quite large, say Re $= 5000$. We are now well into the turbulent flow regime, and our measuring instruments no longer produce steady signals. Instead we would observe very "noisy" data, as sketched in Fig. 9.1.1. The axial velocity would fluctuate about some mean value. Conservation of mass would require that this mean value be

$$\bar{u}_z = \frac{Q}{\pi R^2} \tag{9.1.5}$$

We also observe that even though the flow is *axial* in the mean, with no *mean* radial velocity, there would now be a nonzero fluctuating *radial* velocity, u_r, as Fig. 9.1.1 suggests. The instantaneous pressure would fluctuate as well, and its mean value would be observed to exceed that given by Eq. 9.1.1. Although we do not include it in Fig. 9.1.1, the shear stress at the pipe wall would behave in a manner similar to the pressure fluctuation, and its mean value would also exceed the laminar flow prediction.

The primary characteristic of turbulence is that the kinematic and dynamic variables are *random fluctuating* functions of time. These fluctuations, as in the case of any time-varying signal, have a steady state mean value, if the flowrate Q is steady, when they are averaged over a time scale that is sufficiently long in comparison to some characteristic time scale of the fluctuations in the flow. This averaging would be performed, in our thought experiment, by integrating the instantaneous signal with respect to time over some time *interval* τ:

$$\bar{u}_z \equiv \frac{1}{\tau} \int_t^{t+\tau} u_z \, dt \tag{9.1.6}$$

In addition, these fluctuations would be characterized by their "amplitude" or "intensity" about the mean value, which we would normally define for a random signal as the *root-mean-square value*

$$u_{z,\text{rms}} = \left[\frac{1}{\tau} \int_t^{t+\tau} (u_z - \bar{u}_z)^2 \, dt \right]^{1/2} \tag{9.1.7}$$

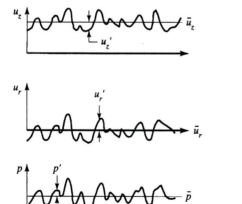

Figure 9.1.1 Instantaneous values of measured characteristics of a turbulent pipe flow.

Because the observed signals show a *random* time variation, we cannot define a single frequency characteristic of the fluctuations. It is possible to do a "spectral analysis" of the signal, much as we take white light and decompose it into a spectrum that shows distribution of light intensity with respect to wavelength. (Using the value of the speed of light, we can convert wavelength to frequency.) The frequency spectrum would provide another, independent characterization of the fluctuating variable. Hence any variable, such as a particular component of velocity, would be characterized by its mean, its intensity about the mean, and its frequency spectrum.

These ideas are much too complex to discuss in any detail in an introductory text. For most applications we are interested primarily in the *average* values of velocities, pressure, and stresses in the fluid, and how these averages depend on operating conditions such as the externally imposed flowrate and pressure drop in a system. While we have introduced these concepts through the thought experiment of turbulent flow through a pipe, the same ideas hold generally for turbulent flows whether they are bounded (pipe flow, flow over an automobile or a baseball) or largely unbounded (e.g., atmospheric and oceanographic turbulence).

From a practical point of view, much progress has been made in modeling turbulent flows by establishing a very large observational base, which is then used to support the development of empirical models that relate various turbulent averages to macroscopic parameters. Many of these models introduce coefficients that cannot be obtained *independently* outside the turbulent observation itself.

9.1.1   Time-Averaging and the Reynolds Stresses

Although we cannot solve the Navier–Stokes equations for a turbulent flow, we can learn something by "time-averaging" the equations. We begin by defining the instantaneous values of velocity components and pressure as the sums of their means plus a fluctuating quantity. (Note Fig. 9.1.1.) We will work in a Cartesian coordinate system. (Similar results would follow if another choice of coordinate system were made.) Hence we define

$$u_z = \bar{u}_z + u_z'$$
$$u_x = \bar{u}_x + u_x'$$
$$u_y = \bar{u}_y + u_y'$$
$$p = \bar{p} + p' \tag{9.1.8}$$

We will take each of the Navier–Stokes equations, substitute the expressions above for the velocities and pressure, and then *time-average* the entire equation. In doing the time-averaging we must first take note of several properties of the averaging process. To begin, we define the time-average of any scalar function S as (cf. Eq. 9.1.6.)

$$\bar{S} \equiv \frac{1}{\tau} \int_t^{t+\tau} S \, dt \tag{9.1.9}$$

If we write *any* function as the sum of the mean, plus a fluctuating quantity, the average becomes

$$\bar{S} = \frac{1}{\tau} \int_t^{t+\tau} (\bar{S} + S') \, dt = \bar{\bar{S}} + \bar{S'} \tag{9.1.10}$$

But, by definition, $\bar{S}$ is independent of time, so the average $(\bar{\bar{S}})$ of $\bar{S}$ is just $\bar{S}$ itself. Hence

$$\bar{S'} \equiv 0 \tag{9.1.11}$$

and so the fluctuations must sum to zero.

The average of a *product* of two fluctuating quantities, call them T and S, is not related uniquely to the averages of the individual quantities. To see this, consider the average of the product of TS:

$$\overline{TS} \equiv \frac{1}{\tau} \int_t^{t+\tau} (\overline{T} + T')(\overline{S} + S')\, dt = \overline{\overline{T}\,\overline{S}} + \overline{T'\,\overline{S}} + \overline{\overline{T}S'} + \overline{T'S'}$$
$$= \overline{T}\,\overline{S} + \overline{T'S'}$$

(9.1.12)

Note in particular that

$$\overline{TS} \neq \overline{T}\,\overline{S}$$

(9.1.13)

When these averaging rules are applied to the z component of the Navier–Stokes equation for an *incompressible* fluid, we find

$$\frac{\partial}{\partial t}\rho\overline{u}_z + \frac{\partial}{\partial x}\rho\overline{u}_z\overline{u}_x + \frac{\partial}{\partial y}\rho\overline{u}_z\overline{u}_y + \frac{\partial}{\partial z}\rho\overline{u}_z\overline{u}_z = -\frac{\partial\overline{p}}{\partial z} + \mu\,\nabla^2\overline{u}_z + \rho g_z$$
$$- \frac{\partial}{\partial x}\rho\,\overline{u_z'u_x'} - \frac{\partial}{\partial y}\rho\,\overline{u_z'u_y'} - \frac{\partial}{\partial z}\rho\,\overline{u_z'u_z'}$$

(9.1.14)

We find similar expressions for the x and y components of the time-averaged Navier–Stokes equations. We can also express the time-averaged equations in cylindrical and spherical coordinate systems.

If we time-average the continuity equation for an incompressible fluid, we find (in Cartesian coordinates)

$$\frac{\partial\overline{u}_x}{\partial x} + \frac{\partial\overline{u}_y}{\partial y} + \frac{\partial\overline{u}_z}{\partial z} = 0$$

(9.1.15)

We can now make the following conclusions. When we time-average the Navier–Stokes and continuity equations, we recover equations that are identical to the original set, except that the instantaneous variables are replaced by their *average* values, *and an additional set of terms arises* in the Navier–Stokes equations. This additional set, the last line of Eq. 9.1.14, represents stresses associated with the fluctuating part of the velocity field. Hence turbulent fluctuations *alter* the time-averaged stresses in the fluid. These additional stresses are called *Reynolds stresses*. We may also write them individually as turbulent momentum fluxes, defined as

$$\overline{T}_{zz} = \rho\,\overline{u_z'u_z'}, \ \ \overline{T}_{zx} = \rho\,\overline{u_z'u_x'}, \ldots$$

(9.1.16)

This format is a good reminder that although we refer to these terms as "stresses," they correspond physically to the convection of momentum by the velocity fluctuations. A major task of turbulence modeling is to relate these Reynolds stresses to the time-averaged velocity and the time-averaged velocity gradients. This has been achieved largely through empiricisms that introduce parameters that must be determined from the data. We will describe some simple examples of this kind of turbulence modeling. More realistic and successful models are highly complex.

9.1.2   Reynolds Stress Models

In considering Reynolds stress models, we have a situation very much like that which we faced after the derivation of the momentum equations in Chapter 4. Equation 4.3.18 required additional information regarding the relationship of the stresses **T** to the velocity gradient tensor **Δ**. This information was supplied through the assumption of a specific *constitutive equation*, such as that of the Newtonian fluid—Eq. 4.3.19. In fact, one of the first proposals of a *turbulent* constitutive equation was a simple analogy to Newton's law of viscosity, in which the *turbulent part* of the shear stress, for example, was taken to be proportional to the time-averaged velocity gradient:

$$\overline{T}_{zx} = -\mu_t \frac{\partial \overline{u}_z}{\partial x} \tag{9.1.17}$$

for the case of a flow that is one-dimensional in the mean. This model introduces a turbulent viscosity parameter μ_t that is *not a fluid property*, in contrast to the case of Newton's law of viscosity. We know, for example, that the turbulent fluctuations vary as a function of distance from a solid surface, since the presence of the solid tends to suppress these fluctuations. Hence the turbulent viscosity would have to be a function of position. It might also depend on which turbulent stress we were considering. There could be a different μ_t for each component of **T**. Thus the model is not very useful, except conceptually in focusing our attention on the observation that in some loose sense a turbulent flow is more viscous than a laminar flow because there are additional stresses that accompany the deformation of the fluid.

One of the simplest models worth examining is expressible in the form

$$\overline{T}_{zx} = -\rho l^2 \left| \frac{\partial \overline{u}_z}{\partial x} \right| \left(\frac{\partial \overline{u}_z}{\partial x} \right) \tag{9.1.18}$$

where l is an empirical parameter with the units of length. Equation 9.1.18 is called "Prandtl's mixing length theory," and l is the "mixing length." The mixing length is usually taken to be a function of position in the flow, and generally as a linear function of the distance from a solid boundary.

With Eq. 9.1.18 we can now "solve" a turbulent flow problem.

EXAMPLE 9.1.1 *The Time-Averaged Velocity Profile in a Pipe*

We will use Prandtl's mixing length hypothesis to yield a dynamic equation that can be solved for the mean velocity profile for a fluid in turbulent flow in a pipe or tube. We make the following assumptions:

The mean flow is fully developed and steady state, so

$$\begin{aligned} \overline{u}_z &= \overline{u}_z(r) \\ \overline{u}_r &= \overline{u}_\theta = 0 \end{aligned} \tag{9.1.19}$$

The mixing length is a linear function of distance from the tube wall:

$$l = \kappa s \qquad \text{where} \quad s \equiv R - r \tag{9.1.20}$$

and κ is a parameter of the model. Then Eq. 9.1.18, in cylindrical coordinates, gives the *turbulent* portion of the shear stress as

$$\overline{T}_{rz} = \rho \kappa^2 s^2 \left(\frac{d\overline{u}_z}{ds} \right)^2 \tag{9.1.21}$$

The axial component of the time-averaged Navier–Stokes equations (in cylindrical coordinates) takes the form

$$0 = \frac{\Delta \overline{\mathscr{P}}}{L} + \mu \frac{1}{r} \frac{d}{dr} \left(r \frac{d\overline{u}_z}{dr} \right) - \frac{1}{r} \frac{d}{dr} (r \overline{T}_{rz}) \tag{9.1.22}$$

We can simplify this model if we first apply it to the region far from the tube wall, where we *assume* that the turbulence dominates the viscous shear stress and therefore drop the viscous term from Eq. 9.1.22. Then we can integrate this equation directly for the turbulent shear stress distribution and find

$$\overline{T}_{rz} = \frac{\Delta\overline{\mathscr{P}}R}{2L}\frac{r}{R} = T_{R}\frac{r}{R} \tag{9.1.23}$$

(We have used the boundary condition of zero mean shear stress on the axis of symmetry, $r = 0$.) T_{R} is the time-averaged shear stress at the tube wall. (See Problem 9.8.)

Using Eq. 9.1.21 we may write Eq. 9.1.23 as

$$\rho\kappa^2 s^2 \left(\frac{d\bar{u}_z}{ds}\right)^2 = T_{R}\left(1 - \frac{s}{R}\right) \tag{9.1.24}$$

Thus we have a differential equation for the average velocity:

$$\frac{d\bar{u}_z}{ds} = \left(\frac{T_{R}}{\rho\kappa^2}\right)^{1/2}\frac{(1 - s/R)^{1/2}}{s} \tag{9.1.25}$$

With a table of integrals we could write out an analytical solution for the average axial velocity. We leave this as a homework exercise (Problem 9.1). Instead, we illustrate the "classical" approach to this problem, which begins by dropping the s/R term in the right-hand side. This makes no sense, physically, since we neglected the viscous stress with the argument that we would apply the resulting model to the region *far* from the wall, where s is *nearly R*.

This is a good point to pause and take a look at the flow field from the perspective of the assumed physics of turbulent momentum transport. In Fig. 9.1.2 we see that the flow field has been divided into three sections. Near the wall we assume that viscous effects dominate—that the flow is nearly laminar. Far from the wall, in the central core of the tube flow we are considering, turbulence dominates the transport of momentum, and viscous effects are negligible. These two regions are separated by a zone (called the "buffer zone") where both viscous and turbulent stresses contribute to the momentum transfer. Unfortunately, we have no *a priori* basis for defining *where* these regions are. We *assume that they exist,* and we will estimate their positions from an examination of experimental data for the time-averaged velocity field in a turbulent flow.

Returning, then, to the analysis, we drop the s/R term in Eq. 9.1.25 and solve for the average velocity by integrating this differential equation from the outer edge of the buffer zone, $s = s_1$, to some arbitrary position $s > s_1$. This introduces a constant of integration that corresponds to the velocity at s_1. Keep in mind that we do not know

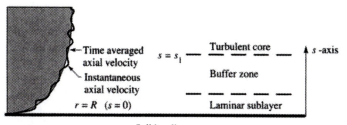

Figure 9.1.2 Regions in a turbulent flow near a surface.

s_1 or the velocity at s_1. The result may be written in a very compact nondimensional format as

$$u^+ = u_1^+ + \frac{1}{\kappa} \ln \frac{s^+}{s_1^+} \tag{9.1.26}$$

where we define

$$u^+ \equiv \frac{\overline{u}_z}{u_*} \quad \text{and} \quad s^+ \equiv \frac{s u_* \rho}{\mu} \tag{9.1.27}$$

and

$$u_* \equiv \left(\frac{T_R}{\rho} \right)^{1/2} \tag{9.1.28}$$

is called the "friction velocity."

This model contains three empirical constants: u_1^+, s_1^+, and κ. We may write the model as

$$u^+ = \left(u_1^+ - \frac{1}{\kappa} \ln s_1^+ \right) + \frac{1}{\kappa} \ln s^+$$
$$= A + B \ln s^+ \tag{9.1.29}$$

Before we evaluate the model represented by Eq. 9.1.29, which is limited to the turbulent core, we should develop a model valid near the wall—within the laminar sublayer. We go back to Eq. 9.1.22 and make the approximation that the flow near the wall is dominated by viscous effects. Then the wall shear stress is simply

$$\overline{T}_R = \frac{\Delta \overline{\mathscr{P}} R}{2L} = \mu \frac{d \overline{u}_z}{ds} \bigg|_{s=0} \tag{9.1.30}$$

It is not difficult to confirm that this corresponds to the nondimensional form

$$1 = \frac{du^+}{ds^+} \bigg|_{s=0} \tag{9.1.31}$$

A velocity profile of the form

$$u^+ = s^+ \tag{9.1.32}$$

satisfies this equation. In effect, all we have done is show that near the wall, a good approximation to the velocity profile is a linear function, the slope of which corresponds to the wall shear stress.

Equation 9.1.32 has no undetermined constants. If we ignore the presence of the buffer region, we can require that these two models for velocity—Eqs. 9.1.29 and 9.1.32—match at a point defined by $s^+ = s_1^+$. This matching condition, applied to experimental data, permits us to determine the model coefficients.

Extensive experimental studies of the mean velocity across the radius of a pipe through which turbulent flow is occurring show that Eq. 9.1.29 fits the data quite well for large values of s^+. Figure 9.1.3 compares these data with the two models for the velocity profile being matched arbitrarily at the point $s^+ = 10$.

Much more extensive comparisons of experimental data to this model are available, as well as similar models for the velocity profile. Usually the simple two-region model illustrated here, in which we ignore the presence of the buffer zone, fits the data less

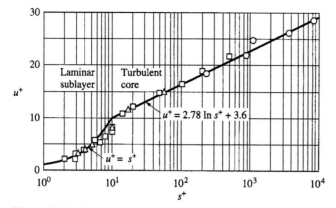

Figure 9.1.3 Test of the simple model of the time-averaged axial velocity in turbulent pipe flow.

well in the neighborhood of $s^+ = 10$. This failure is alleviated by providing a model for the buffer region itself. We will not review that approach. It should be apparent that data over a wide range of the distance parameter s^+ are accounted for reasonably well by the analysis outlined here.

Note that with the choice of coefficients illustrated in Fig. 9.1.3 we may write Eq. 9.1.29 as

$$u^+ = 2.78 \ln s^+ + 3.6 \tag{9.1.33}$$

Note also that we use the natural logarithm (ln) in this equation, but the data are plotted, as usual, on base 10 (log) coordinates on the horizontal axis.

While we have consistently argued in favor of nondimensionalization as a tool for simplifying the procedure of modeling, we have been looking at an example in which the nondimensionalization tends to obscure the physical meaning of the variables. In particular, the position variable s was not made dimensionless with some characteristic length scale for the flow, such as the radius of the pipe. As a result it is not clear what a very large value of s^+ corresponds to, and s^+ does not take on a simple value at the axis of the pipe. In fact, from its definition (Eq. 9.1.27) we see that at the pipe axis

$$s_R^+ \equiv \frac{R u_* \rho}{\mu} \tag{9.1.34}$$

which is in the form of a Reynolds number, except that it is based on the friction velocity u_*, instead of the average velocity U. Thus the value of s^+ at the pipe axis depends on the pressure drop per unit axial length for the flow. We will come back to this point shortly, but it would be useful to have a relationship between the average velocity and the pressure drop driving the flow, in the turbulent regime where the Hagen–Poiseuille law does not hold.

We show now that we can predict the pressure drop–flowrate relationship for turbulent flow from Eq. 9.1.33. We will do this in terms of the *friction factor f* defined earlier in Chapter 5 (see Eq. 5.3.29).

EXAMPLE 9.1.2   *The Turbulent Friction Factor*

The friction factor is defined as

$$f \equiv \frac{T_R}{\frac{1}{2}\rho U^2} = \frac{\Delta \overline{\mathscr{P}} R}{\rho U^2 L} \tag{9.1.35}$$

(Except for notation, this is Eq. 5.3.29.)

The friction *velocity* may be written as

$$u_*^2 \equiv \frac{T_R}{\rho} = \frac{f U^2}{2} \tag{9.1.36}$$

or

$$\frac{u_*^2}{U^2} = \frac{f}{2} \tag{9.1.37}$$

(Hence we see that the dimensionless friction velocity, u_*/U, is related in a very simple way to the friction factor.) The average velocity U may be related to the mean velocity profile through

$$\pi R^2 U = \int_0^R 2\pi r \bar{u}_z \, dr \tag{9.1.38}$$

We will introduce Eq. 9.1.33 into this equation for the velocity profile, replace r by $R - s$, and make use of Eq. 9.1.37. The integral becomes, after a few lines of algebra,

$$\frac{1}{\sqrt{f}} = \sqrt{2} \int_0^1 (1 - \sigma)(2.78 \ln \sigma + 3.6 + 2.78 \ln \mathrm{Re} \sqrt{f/8}) \, d\sigma \tag{9.1.39}$$

where

$$\sigma = \frac{s}{R} \quad \text{and} \quad \mathrm{Re} = \frac{2UR\rho}{\mu} \tag{9.1.40}$$

After the integration, we find

$$\frac{1}{\sqrt{f}} = 1.97 \ln (\mathrm{Re} \sqrt{f}) - 2.45 = 4.52 \log (\mathrm{Re} \sqrt{f}) - 2.45 \tag{9.1.41}$$

This expression does not do a particularly good job of fitting data for $f(\mathrm{Re})$ for turbulent pipe flow. However, an empirical adjustment of the coefficients *does* permit an expression of this form to fit data very well. The usual form is given as

$$\frac{1}{\sqrt{f}} = 4 \log (\mathrm{Re} \sqrt{f}) - 0.4 \tag{9.1.42}$$

and this is sometimes referred to as the Prandtl resistance law.

One reason for the failure of Eq. 9.1.41 lies in our complete neglect of the buffer region. A second arises from our decision not to introduce the laminar sublayer expression at all when performing the integration shown in Eq. 9.1.38. Instead, we integrated the turbulent core expression all the way out to the pipe wall. We have already noted that, at the other end, the turbulent core expression does not give the right gradient at $r = 0$. What is remarkable, then, is that this simple theory ("model" is really a better word) gives a result that has a *functional* relationship of f to Re that is at all realistic.

9.1.3 The Friction Factor for Pipe Flow: Laminar and Turbulent

Since flow through long pipes and tubes occurs so commonly in engineering systems, it is very useful to provide a friction factor model that covers flows ranging from laminar through turbulent. Equation 9.1.35 (from Eq. 5.3.29) is our basic definition of the friction factor for flow through a pipe or tube of circular cross section. In the laminar flow regime we have the Hagen–Poiseuille law (assuming that the entrance effects are minor, i.e., for "long tubes"):

$$\frac{\Delta P}{L} = \frac{8\mu Q}{\pi R^4} = \frac{8\mu U}{R^2} \qquad \text{for laminar flow} \qquad \textbf{(9.1.43)}$$

From the definition of f, this leads to

$$f_{\text{lam}} = \frac{\Delta P R}{\rho U^2 L} = \frac{8\mu}{\rho U R} = \frac{16}{\text{Re}} \qquad \text{for laminar flow} \qquad \textbf{(9.1.44)}$$

Note that the Reynolds number for pipe flow is usually defined with the *diameter,* not the radius, as the length scale. This is an arbitrary convention, but it is the one that we will follow, consistent with most other textbooks.

With Eqs. 9.1.44 and 9.1.42, we may now plot a friction factor–Reynolds number curve that should be useful over a wide range of flowrates. This curve is shown in Fig. 9.1.4. Because Eq. 9.1.42 is algebraically complex (it is not explicit in f), we use a simpler but accurate curve-fit as shown in the figure. The most common simple expression for f in the turbulent regime is given by

$$f = 0.079 \, \text{Re}^{-0.25} \qquad \text{for turbulent flow (Re} > 3500) \qquad \textbf{(9.1.45)}$$

Note that in Fig. 9.1.4 we have marked the region corresponding roughly to $2000 < \text{Re} < 3500$ as a "transition region." We find that in this range of Reynolds numbers the friction factor can depend strongly on the geometry of the inlet region of the pipe or tube. An experimental study of the friction factor in the transition region was carried out using three inlet geometries, and we summarize the results briefly here. The original paper [Ghajar and Madon, *Exp. Thermal Fluid Sci.,* **5,** 129 (1992)] should be consulted for further details. Figure 9.1.5 shows the inlet geometries studied. In all cases the test

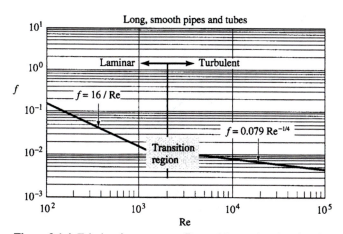

Figure 9.1.4 Friction factor versus Reynolds number for circular pipes and tubes.

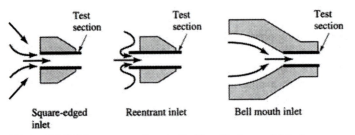

Figure 9.1.5 Inlet geometries studied by Ghajar and Madon.

section was a stainless steel seamless pipe with an inside diameter of 1.58 cm and a length of 6.1 m.

The friction factor data are summarized in Fig. 9.1.6. As we might expect, the friction factor is smallest (by as much as 20%) for the smoothest inlet geometry. We also see that the data for all three geometries agree very well with the laminar and turbulent expressions we have already presented (Eqs. 9.1.44 for laminar flow and 9.1.45 for turbulent flow.) In the absence of other information, we may use Fig. 9.1.6 for the estimation of the friction factor if the Reynolds number falls in the transition region ($2000 < \text{Re} < 3500$). Note that at a Reynolds number of about 3000 the friction factors predicted by the laminar and turbulent correlations differ by nearly a factor of 2. Hence this is a region in which *a priori* design estimates could be in considerable error.

We will leave the topic of turbulence *modeling* here. Much more complex views of turbulence are available which lead to useful predictions of a variety of turbulent transport phenomena. This material, however, is largely beyond the scope of an introductory course in fluid dynamics.

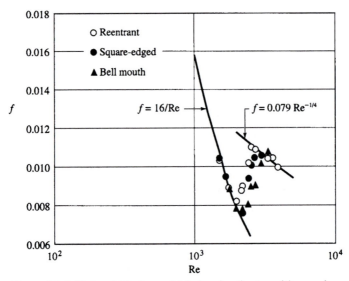

Figure 9.1.6 Data of Ghajar and Madon for the transition region.

EXAMPLE 9.1.3   *Predicting the Flowrate for a Turbulent Pipe Flow*

Water at 25°C flows through a horizontal pipe of length $L = 300$ m and inside diameter $D = 3$ cm. Pressure is supplied by an open reservoir with a constant head of $H = 2$ m above the inlet of the pipe. The downstream end of the pipe is open to the atmosphere. Find the mass flowrate and the Reynolds number achieved in this situation.

In view of the large pipe diameter and the low viscosity of water, we anticipate that the flow is turbulent. We could solve this problem using a friction factor chart (Fig. 9.1.4) or an analytical expression for the f/Re relationship (Eq. 9.1.45). The latter method is straightforward, and by combining Eqs. 9.1.45 and 9.1.35 we can solve for the average velocity from

$$U = \left(\frac{gHD^{1.25}\rho^{0.25}}{0.158\, L\, \mu^{0.25}}\right)^{1/1.75} \tag{9.1.46}$$

For water we take nominal values of $\rho = 1000$ kg/m^3 and $\mu = 1$ mPa · s. The results are

$$U = 0.35 \text{ m/s} \quad \text{and} \quad \mathrm{Re} = \frac{UD\rho}{\mu} = 1.07 \times 10^4 \tag{9.1.47}$$

A graphical method based on Fig. 9.1.4 is less straightforward, because the unknown velocity U appears in both the ordinate and the abscissa. Hence trial-and-error would be required, and this is more difficult than the analytical approach. However, we can

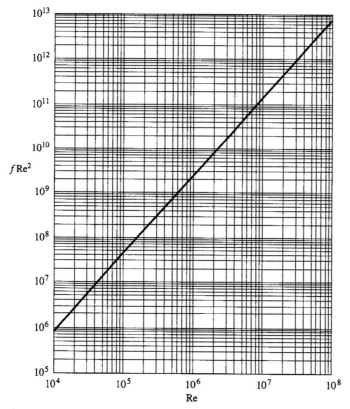

Figure 9.1.7 Plot based on Eq. 9.1.49.

modify Fig. 9.1.4 in a way that avoids the need to use the trial-and-error approach and provides a quick method of estimating the desired flowrate.

We begin by noting that the unknown velocity can be eliminated in the combination

$$f\,\mathrm{Re}^2 = \frac{\rho D^3\,\Delta\mathscr{P}}{2\mu^2 L} \qquad (9.1.48)$$

Now, using Eq. 9.1.45, we may plot $f\,\mathrm{Re}^2$ versus Re from

$$f\,\mathrm{Re}^2 = 0.079\,\mathrm{Re}^{1.75} \qquad (9.1.49)$$

This is shown in Fig. 9.1.7. With the data given in this example, the right-hand side of Eq. 9.1.48 is known:

$$f\,\mathrm{Re}^2 = \frac{1000(0.03)^3\,(1000\times 9.8\times 2)}{2(0.001)^2\times 300} = 9\times 10^5 \qquad (9.1.50)$$

From Fig. 9.1.7 we find

$$\mathrm{Re} = 1.1\times 10^4 \qquad (9.1.51)$$

in good agreement with the result of the analytical method.

We turn, now, to a topic that shares some common ground with the material just completed. The distinctions will become clear as we proceed.

9.2   THE DEVELOPING LAMINAR BOUNDARY LAYER

We begin, as we did in Section 9.1, with a thought experiment. Suppose, as suggested in Fig. 9.2.1, that an otherwise unbounded turbulent flow encounters a solid surface. We know that a real fluid must satisfy a no-slip boundary condition at the solid surface. We might suppose that viscous effects dominate the flow sufficiently close to the surface and that in some region near the surface, the flow is laminar.

> *The distinction between this flow and the one of Section 9.1 is that in the preceding case we considered a fully developed flow, unchanging in the direction of the mean flow. In the present case we consider the* development *of the flow field near the surface, along a coordinate in the direction of the mean flow.*

We assume that the flow within the viscous region is, indeed, laminar. We refer to this region as the *laminar boundary layer*. In the laminar boundary layer the velocity components are not fluctuating. The velocity close to the wall is so slow that flow within this region is truly laminar, no matter how turbulent the flow is far from that surface.

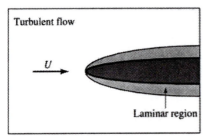

Figure 9.2.1 A turbulent flow encounters a solid obstacle.

By distinction, we assumed in Section 9.1 that there was a laminar *sublayer,* but we continued to use the time-averaged equations for that region. In other words, the laminar sublayer was not treated as a laminar flow. Some authors prefer to use the term *viscous sublayer* to avoid this confusion.

Our goal here is to develop a model of the flow in the laminar boundary layer. This model must give us the boundary layer thickness as a function of position along the solid surface, as well as the shear stress distribution along that surface.

We will consider a simple case here: that of a two-dimensional flow in which the external mean velocity U is parallel to a solid surface that occupies the plane at $y = 0$. We ignore gravity. Figure 9.2.2 shows the geometry. The y and x components of the Navier–Stokes and continuity equations are

$$\rho\left(u_x \frac{\partial u_y}{\partial x} + u_y \frac{\partial u_y}{\partial y}\right) = -\frac{\partial p}{\partial y} + \mu\left(\frac{\partial^2 u_y}{\partial x^2} + \frac{\partial^2 u_y}{\partial y^2}\right) \tag{9.2.1}$$

$$\rho\left(u_x \frac{\partial u_x}{\partial x} + u_y \frac{\partial u_x}{\partial y}\right) = -\frac{\partial p}{\partial x} + \mu\left(\frac{\partial^2 u_x}{\partial x^2} + \frac{\partial^2 u_x}{\partial y^2}\right) \tag{9.2.2}$$

and

$$\frac{\partial u_x}{\partial x} + \frac{\partial u_y}{\partial y} = 0 \tag{9.2.3}$$

Our task now is to solve these equations. The flow in the boundary layer is two-dimensional and steady. We will not neglect the inertial terms. While the flow in the boundary layer is assumed to be laminar, we have no basis for assuming creeping flow, and indeed we know that we want to apply this model to cases in which the creeping flow assumption would be a poor one. However, we will regard this flow as "nearly parallel," much as we did in the analysis of the slider bearing in Chapter 6, except that we treat the pressure in a different manner.

For an almost parallel flow we may assume that the y component of velocity in Eq. 9.2.1 will not contribute to a significant transport of y-directed momentum, from either viscous or inertial terms.

We are essentially setting $u_y = 0$ in Eq. 9.2.1 (but retaining it in Eqs. 9.2.2 and 9.2.3). As a consequence, Eq. 9.2.1 reduces to

$$\frac{\partial p}{\partial y} = 0 \quad \text{within the boundary layer} \tag{9.2.4}$$

For the flow *outside* the boundary layer (which we are assuming is turbulent), since the mean flow is uniform and parallel to the surface at $y = 0$, and viscous effects are negligible (because the outer flow is turbulent), there will be no pressure variation anywhere in the xy plane. In particular,

$$\frac{\partial p}{\partial x} = 0 \quad \text{outside the boundary layer} \tag{9.2.5}$$

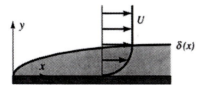

Figure 9.2.2 The laminar boundary layer on a flat plate.

But Eq. 9.2.4 and the statement that there is no pressure variation anywhere in the xy plane outside the boundary layer imply that p does not vary with y anywhere, so if p is independent of x outside the boundary layer, it is independent of x *within* the boundary layer. Hence we drop the pressure term from the x-directed dynamic equation. This reduces the number of dependent variables by one, because pressure is no longer an unknown, and therefore we must also drop one equation. Otherwise the problem would be overspecified from a mathematical point of view. But we have already done so, since Eq. 9.2.1 now takes the form of $0 = 0$.

The boundary layer concept, and Fig. 9.2.2, suggest that the velocity field changes from the no-slip condition at $y = 0$ to the free stream velocity U over a thin region $\delta(x)$ in the neighborhood of the solid surface. This is the origin of the term "boundary layer" for this type of flow. With this in mind, we introduce one further approximation, based on the notion that since the flow is *nearly parallel,* the x-component of velocity varies more gradually in the x direction than it does in the y direction. If this is so, we may neglect the second derivative of u_x with respect to x, in comparison to the second derivative of u_x with respect to y, in the viscous stress terms of Eq. 9.2.2. Note, however, that we retain the convective terms. First of all, while we are assuming laminar flow in the boundary layer, we are not assuming that the Reynolds number is vanishingly small. Furthermore, we have no basis for arguing the relative magnitudes of the two convective terms. Hence we retain both.

Putting these ideas and approximations together, we may now write the dynamic equations for this flow in the forms

$$\rho\left(u_x\frac{\partial u_x}{\partial x} + u_y\frac{\partial u_x}{\partial y}\right) = \mu\frac{\partial^2 u_x}{\partial y^2} \tag{9.2.2'}$$

$$\frac{\partial u_x}{\partial x} + \frac{\partial u_y}{\partial y} = 0 \tag{9.2.3'}$$

If we now introduce the stream function for Cartesian coordinates (Eqs. 8.1.4 of chapter 8), we eliminate the continuity equation (Eq. 9.2.3′) and Eq. 9.2.2′ transforms to the following equation:

$$\frac{\partial\psi}{\partial y}\frac{\partial^2\psi}{\partial x\,\partial y} - \frac{\partial\psi}{\partial x}\frac{\partial^2\psi}{\partial y^2} = \nu\frac{\partial^3\psi}{\partial y^3} \tag{9.2.6}$$

where $\nu = \mu/\rho$. Let's look at boundary conditions before we go any further.

Boundary conditions on velocity are as follows:

no slip: $\qquad\qquad u_x = u_y = 0 \qquad$ on $\quad y = 0$, for $x > 0$ $\qquad$ (9.2.7)
no upstream disturbance
$\quad$ to the external flow: $\qquad u_x = U \qquad$ for $\quad x < 0$ $\qquad\qquad\qquad$ (9.2.8)
effect of the laminar layer
$\quad$ vanishes far from the surface: $\quad u_x \to U \qquad$ for $\quad y \to \infty$ $\qquad\qquad$ (9.2.9)

In terms of the stream function these boundary conditions become

$$\frac{\partial\psi}{\partial y} = -\frac{\partial\psi}{\partial x} = 0 \qquad \text{on} \quad y = 0, \text{for } x > 0 \tag{9.2.10}$$

$$\frac{\partial\psi}{\partial y} \to U \qquad \text{for} \quad x < 0, \text{and } y \to \infty \tag{9.2.11}$$

A number of methods exist for solving Eq. 9.2.6. No analytical solution is possible, because of the nonlinearity of the equation. An "exact" solution is possible through numerical methods. The word "exact" in this case simply means that no further approximation to Eq. 9.2.6 is introduced. Several approximate methods have been developed, as well. The "classical" solution to this problem was carried out by Blasius in 1908. He discovered, by trial and error and intuition, a transformation of variables that reduced Eq. 9.2.6 to an ordinary differential equation. The transformation is with respect to both the dependent and independent variables. It is called a *similarity transformation.* It seems quite mysterious at first glance, although later investigators presented numerous arguments based on geometrical considerations, or on inspectional analysis, that lead to the same transformation.

What Blasius discovered was that if he defined a new function (be careful to distinguish cap Ψ from lowercase ψ)

$$\Psi \equiv \frac{\psi}{(U\nu x)^{1/2}} \qquad (9.2.12)$$

and a combination of the space variables y and x in the form

$$\eta \equiv y \left(\frac{U}{\nu x} \right)^{1/2} \qquad (9.2.13)$$

then Eq. 9.2.6 transformed to the following ordinary differential equation:

$$2\,\Psi''' + \Psi\Psi'' = 0 \qquad (9.2.14)$$

where the primes denote differentiation with respect to η, and Ψ is a function only of η. (You have a choice of accepting this result—Eq. 9.2.14—or investing some time in application of the chain rule of differential calculus. There is no physics involved in going from Eq. 9.2.6 to Eq. 9.2.14. It is simply tedious mathematics. There is no approximation involved in going from Eq. 9.2.6 to Eq. 9.2.14.)

The boundary conditions on Ψ (Eqs. 9.2.10 and 9.2.11) must be transformed, using Eqs. 9.2.12 and 9.2.13, also, and the results are as follows:

$$\begin{aligned} \Psi = \Psi' = 0 \qquad &\text{at} \quad \eta = 0 \\ \Psi' \to 1 \qquad &\text{as} \quad \eta \to \infty \end{aligned} \qquad (9.2.15)$$

Note that we require and have three boundary conditions for this third-order equation. Once a solution for Ψ has been found, the actual velocity component u_x is found from

$$u_x = \frac{\partial \psi}{\partial y} = U \frac{d\Psi}{d\eta} \qquad (9.2.16)$$

Figure 9.2.3 shows the solution $\Psi(\eta)$ and the velocity profile u_x/U as a function of η. (The solution is obtained numerically—an analytical solution is impossible.) Keep in mind that the function Ψ is not the stream function ψ. The two psis do not differ simply by a constant factor. We must use Eq. 9.2.12 to recover the stream function ψ.

Upon inspection of Fig. 9.2.3 we quickly note that the outer edge of the boundary layer, defined as the line $y = \delta(x)$ along which u_x is nearly equal to U, is given approximately by $\eta = 5$. Along the outer edge of the boundary layer we replace y by $\delta(x)$ in the definition of η. Thus we find that the "boundary layer thickness" $\delta(x)$ is such that

$$\delta \left(\frac{U}{\nu x} \right)^{1/2} = 5 \qquad (9.2.17)$$

Another important feature of this flow field that comes from the boundary layer solution is the shear stress that acts along the solid surface. We define a local drag coefficient as

$$C_D \equiv \frac{-\mu(\partial u_x/\partial y)_{y=0}}{\frac{1}{2}\rho U^2} \qquad (9.2.18)$$

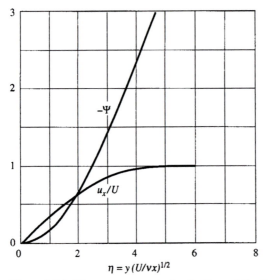

Figure 9.2.3 The x component of velocity and the function Ψ for the laminar boundary layer.

After introducing the similarity transform (Eq. 9.2.13) and using the chain rule a bit, we find

$$C_D \equiv 2\left(\frac{\nu}{xU}\right)^{1/2}\Psi''(0) \qquad (9.2.19)$$

From the numerical solution for the function Ψ, we find that its second derivative, at $\eta = 0$, is $\Psi''(0) = 0.332$. Hence the local drag coefficient is

$$C_D \equiv 0.664\left(\frac{\nu}{xU}\right)^{1/2} = \frac{0.664}{(\mathrm{Re}_x)^{1/2}} \qquad (9.2.20)$$

where we have defined a "local" Reynolds number as

$$\mathrm{Re}_x \equiv \frac{xU}{\nu} \qquad (9.2.21)$$

Note that this Reynolds number *varies* along the length of the solid plane. That is why we call it a "local" Re.

For a planar surface of finite length L in the x direction, we may find the total shear *force* by integration of the local shear stress along the surface:

$$F_s = W\int_0^L -\mu\left(\frac{\partial u_x}{\partial y}\right)_{y=0} dx \qquad (9.2.22)$$

where W is the width of the plane in the z direction. Using Eqs. 9.2.18 and 9.2.19 we may obtain the following result:

$$\frac{F_s/LW}{\tfrac{1}{2}\rho U^2} \equiv \overline{C}_D = \frac{1.328}{(UL/\nu)^{1/2}} = \frac{1.328}{(\mathrm{Re}_L)^{1/2}} \qquad (9.2.23)$$

This equation defines a mean drag coefficient (or friction factor) and is written in terms of a Reynolds number based on the length of the plane.

This completes the analysis of the laminar boundary layer on a plane parallel to an external steady (in the mean) turbulent flow. Note that nothing about turbulence seems to have entered the model. Is the model, therefore, really limited to turbulent flow? Could we not apply it to find the shear stress and boundary layer thickness on a plane aligned parallel to a *laminar* flow? The answers are *Yes* to the first question and *No* to the second.

The planar boundary layer concept implies that *all* the viscous effects are confined to a thin region near the surface and that the external flow is not disturbed outside of that layer. This makes sense only if the external flow is turbulent. If a plane were immersed in a laminar flow that was moving parallel to the plane, the viscous effects would eventually propagate across the entire flow region. There is no such thing as a *steady state* boundary layer in a purely viscous flow.

We have assumed that the flow within the boundary layer is laminar. Such a situation is possible, but there are limitations to the existence of the laminar boundary layer. We observe that somewhere along the x coordinate (the direction of the mean flow), the flow will eventually become unstable to disturbances and will undergo a transition to turbulence. Beyond that point we still have a boundary layer flow, but it involves a *turbulent* boundary layer. The transition to the turbulent boundary layer depends strongly on such factors as the roughness of the surface and the presence of disturbances in the turbulent flow adjacent to the boundary layer. The usual rule of thumb is to consider the boundary layer to be turbulent beyond a downstream distance that corresponds to

$$\text{Re}_x^{\text{trans}} = 5 \times 10^5 \qquad \text{(transition to turbulence)} \qquad \textbf{(9.2.24)}$$

Models for the *turbulent* boundary layer have been developed and tested, but we will not review them here.

So far we have not given any indication of the thickness of a laminar boundary layer. We can see this in the following example.

EXAMPLE 9.2.1 *Thickness of the Laminar Boundary Layer*

In two separate experiments we have water, and then air, flowing in turbulent flow at a speed of $U = 3$ m/s. A large flat plate is aligned parallel to the flow. Over what length of the plate is the boundary layer laminar? What is the boundary layer thickness just before the position where the boundary layer becomes turbulent?

Water   kinematic viscosity at 25°C is 10^{-6} m^2/s

Use the criterion for transition given in Eq. 9.2.24. With

$$\frac{Ux^{\text{trans}}}{\nu} = 5 \times 10^5 \qquad \textbf{(9.2.25)}$$

we find

$$x^{\text{trans}} = \frac{5 \times 10^5 \, \nu}{U} = \frac{(5 \times 10^5)10^{-6}}{3} = 0.17 \text{ m} \qquad \textbf{(9.2.26)}$$

From Eq. 9.2.17 we find

$$\delta = \frac{5}{(U/\nu x)^{1/2}} = \frac{5}{[3/(10^{-6} \times 0.17)]^{1/2}} = 0.0012 \text{ m} \qquad \textbf{(9.2.27)}$$

The boundary layer is about a millimeter thick at the transition point, which occurs 17 cm from the leading edge of the plate.

Air kinematic viscosity at 25°C is 1.5×10^{-5} m²/s

Equation 9.2.26 gives us

$$x^{\text{trans}} = 2.5 \text{ m} \tag{9.2.28}$$

and, from Eq. 9.2.17 we find

$$\delta = 0.018 \text{ m} \tag{9.2.29}$$

The boundary layer is much thicker, in air, than in water at the same fluid speed. This is because the *kinematic* viscosity of air is 10 times greater than that of water. The *kinematic* viscosity is the "diffusivity" for momentum. Momentum diffuses farther into the main flow when the "diffusivity" is larger.

EXAMPLE 9.2.2 *The Mechanics of Water Beasts*

A beast shaped like a flat plate (Fig. 9.2.4) swims through the ocean at a speed of 0.4 m/s. At what rate must it expend energy to maintain this motion? Take the shape to be equivalent to a thin flat plate of width $W = 1$ m and length $L = 1.5$ m. For a simple flat plate moving steadily through the water, the boundary layer would be laminar for a length determined by Eq. 9.2.25:

$$L^{\text{trans}} = \frac{(5 \times 10^5)\nu}{U} = \frac{(5 \times 10^5)10^{-6}}{0.4} = 1.25 \text{ m} \tag{9.2.30}$$

This is less than the equivalent length of the beast, but not by much, so we will use the laminar boundary layer model.

The total force for *both* sides is found from Eq. 9.2.23 (with a factor of 2 for the two sides):

$$F_s = 2 \times \frac{0.664}{(\text{Re}_L)^{1/2}} \rho U^2 LW = 2 \times \frac{0.664}{(6 \times 10^5)^{1/2}} 10^3 (0.4)^2 \times 1.5 \times 1 = 0.41 \text{ N} \tag{9.2.31}$$

(This force is equivalent to a weight of 42 g on Earth.) In calculating the Reynolds number we have used a nominal value of the kinematic viscosity of water of 10^{-6} m²/s.

The power expended in maintaining this motion is the product of the force times the speed:

$$\text{power} = 0.41 \times 0.4 = 0.16 \text{ W} = 2 \times 10^{-4} \text{ hp} \tag{9.2.32}$$

Figure 9.2.4 Drag on a water beast.

This is a very small power. It is an underestimate of the required power because the boundary layer is turbulent over a portion of the surface, and the turbulent boundary layer has a drag coefficient several times that for the laminar boundary layer. In addition, since the beast is not infinitely thin, there will be some *form drag* that must be overcome as well. Probably more important, swimming is an *impulsive motion* in which the beast "pushes" against the water with its fin or winglike surfaces, accelerates to some speed, and then glides until friction reduces its speed to the point that another "stroke" is necessary. See Problem 9.12 for an estimate of the power required to propel a dolphin.

9.3   THE INTEGRAL BOUNDARY LAYER ANALYSIS

There is another method available for solving the equations that describe flow within the laminar boundary layer. In some ways the so-called *integral boundary layer method* is less "dense" than Blasius' "exact" solution, although it is more tedious to read. Instead of solving the Navier–Stokes equations, along with the continuity equation, by a numerical method, we can *assume* an algebraic form for the velocity profile, force this form to satisfy an integral or averaged form of the mass and momentum conservation relationships, and obtain an analytical model for the shear stress (momentum transfer) and the boundary layer thickness. (This is essentially the same method used earlier in Section 6.5 of Chapter 6.)

We begin by rewriting the dynamic equations for the laminar boundary layer along a flat planar surface as

$$\rho \left(u_x \frac{\partial u_x}{\partial x} + u_y \frac{\partial u_x}{\partial y} \right) = \mu \frac{\partial^2 u_x}{\partial y^2} \tag{9.3.1}$$

and

$$\frac{\partial u_x}{\partial x} + \frac{\partial u_y}{\partial y} = 0 \tag{9.3.2}$$

This formulation includes the approximations (discussed in Section 9.2) that delete the pressure gradient $\partial p/\partial x$ and the term $\partial^2 u_x/\partial x^2$ from the right-hand side of Eq. 9.2.2.

Now we are going to do a little bit of mathematical manipulation. First, we multiply the continuity equation by ρu_x and add the whole equation to the dynamic equation—Eq. 9.3.1. This gives us

$$\rho \left(u_x \frac{\partial u_x}{\partial x} + u_y \frac{\partial u_x}{\partial y} \right) + \rho \left(u_x \frac{\partial u_x}{\partial x} + u_x \frac{\partial u_y}{\partial y} \right) = \rho \left(\frac{\partial u_x^2}{\partial x} + \frac{\partial u_x u_y}{\partial y} \right) = \mu \frac{\partial^2 u_x}{\partial y^2} \tag{9.3.3}$$

The "integral boundary layer" analysis takes its name from the integration of Eq. 9.3.3 (and Eq. 9.3.2, as well) across the boundary layer: that is, normal to the plate. This is equivalent to averaging the momentum and mass conservation equations across the boundary layer. To begin, we multiply each term in the dynamic equation above by dy, and then integrate term by term from the surface $y = 0$ to the edge of the boundary layer, which we define now as the line $y = \delta(x)$ where $u_x = U$. The intermediate result is

$$\int_0^{\delta(x)} \frac{\partial u_x^2}{\partial x} \, dy + [u_x u_y]_{y=0}^{y=\delta} = \nu \left[\frac{\partial u_x}{\partial y} \right]_{y=0}^{y=\delta} \tag{9.3.4}$$

We impose the boundary conditions:

$$u_x = 0 \qquad \text{for} \quad y = 0 \tag{9.3.5}$$

and

$$u_x = U \text{ for } y = \delta \tag{9.3.6}$$

and find

$$\int_0^{\delta(x)} \frac{\partial u_x^2}{\partial x} \, dy + U u_{y(y=\delta)} = \nu \left[\frac{\partial u_x}{\partial y} \right]_{y=0}^{y=\delta} \tag{9.3.7}$$

If we also average the continuity equation across the boundary layer, we find

$$\int_0^{\delta(x)} \left(\frac{\partial u_x}{\partial x} + \frac{\partial u_y}{\partial y} \right) dy = \int_0^{\delta(x)} \frac{\partial u_x}{\partial x} \, dy + u_{y(y=\delta)} = 0 \tag{9.3.8}$$

where we have used the boundary condition

$$u_y = 0 \quad \text{for} \quad y = 0 \tag{9.3.9}$$

Now we have to manipulate the x derivatives inside the integrals of Eqs. 9.3.7 and 9.3.8. It is tempting to bring the derivative outside of the integral, but

$$\int_0^{\delta(x)} \frac{\partial u_x}{\partial x} \, dy \neq \frac{d}{dx} \int_0^{\delta(x)} u_x \, dy \tag{9.3.10}$$

The reason for this inequality is that the upper limit of this integral is a function of x. When this is the case, we must use the Leibniz rule for differentiating integrals. (Refer to Chapter 6, Eq. 6.5.13.) This states that

$$\frac{d}{dx} \int_{a(x)}^{b(x)} f(x,y) \, dy = \int_{a(x)}^{b(x)} \frac{\partial f(x,y)}{\partial x} \, dy + f[x,b(x)] \frac{db}{dx} - f[x,a(x)] \frac{da}{dx} \tag{9.3.11}$$

Now we can write the integrals in Eqs. 9.3.7 and 9.3.8 as

$$\int_0^{\delta(x)} \frac{\partial u_x^2}{\partial x} \, dy = \frac{d}{dx} \int_0^{\delta(x)} u_x^2 \, dy - u_{x(y=\delta)}^2 \frac{d\delta}{dx} = \frac{d}{dx} \int_0^{\delta(x)} u_x^2 \, dy - U^2 \frac{d\delta}{dx} \tag{9.3.12}$$

and

$$\int_0^{\delta(x)} \frac{\partial u_x}{\partial x} \, dy = \frac{d}{dx} \int_0^{\delta(x)} u_x \, dy - U \frac{d\delta}{dx} \tag{9.3.13}$$

where we have used Eq. 9.3.6 as a boundary condition.

We now introduce the shear stress at the solid surface:

$$\tau_0 = \mu \left[\frac{\partial u_x}{\partial y} \right]_{y=0} \tag{9.3.14}$$

and use it to write the lower limit of the bracketed term on the right-hand side of Eq. 9.3.7.

At the upper limit of that same term we set the velocity gradient $\partial u_x / \partial y = 0$. We are asserting here, as a boundary condition on u_x, that at $y = \delta(x)$ we must satisfy Eq. 9.3.6 as well as a "smoothness" condition on the velocity. If the x-directed velocity is uniformly $u_x = U$ as we approach $\delta(x)$ from the free stream side of $\delta(x)$, the velocity gradients (or, equivalently, the shear stresses) should match as $\delta(x)$ is approached from either side.

We can finally put all this algebra and calculus together and write Eq. 9.3.7 as

$$\frac{d}{dx} \int_0^{\delta(x)} u_x^2 \, dy - U \frac{d}{dx} \int_0^{\delta(x)} u_x \, dy = -\frac{\tau_0}{\rho} \tag{9.3.15}$$

Equation 9.3.15, which is an averaged version of the momentum conservation equation, was originally developed by von Karman, whose name is associated with several major contributions to the field of fluid dynamics. It has three unknowns: $\delta(x)$, $u_x(x, y)$, and τ_0. Equation 9.3.14 provides a second, independent equation among these unknowns. We are short one equation. Where does it come from? We cannot attempt further imposition of the constraints on mass and momentum conservation because we have already used those equations.

We solve these equations by the following procedure. First, we make an assumption about the nature of the velocity profile. We assume that the profile satisfies what is called a condition of "similarity." The assumption is that at any position x along the plane $y = 0$, the velocity profile has the same "shape" when the y coordinate is scaled by $\delta(x)$:

$$\frac{u_x}{U} = F\left[\frac{y}{\delta(x)}\right] = F(\eta) \tag{9.3.16}$$

where

$$\eta \equiv y/\delta(x)$$

In words, we are asserting[1] that along a line $y(x)$ that represents a constant fraction η of the distance from the solid surface into the boundary layer, the x component of the velocity is a constant fraction of the free stream velocity U. (The two fractions are, however, not the same.)

Next we write Eq. 9.3.15 in the form

$$\frac{d}{dx}\int_0^{\delta(x)} (u_x^2 - Uu_x)\, dy = -\frac{\tau_0}{\rho} \tag{9.3.17}$$

and then introduce Eq. 9.3.16 to find

$$\frac{d}{dx}\rho U^2 \delta \int_0^1 \left(\frac{u_x^2 - Uu_x}{U^2}\right) d\eta = -\tau_0 = \frac{d}{dx}\rho U^2 \delta \int_0^1 (F^2 - F)\, d\eta \tag{9.3.18}$$

In the course of doing this we can move δ across the integral sign because it is not a function of y. Since F is a function only of η, the definite integral above is just some constant. We write this as

$$\int_0^1 (F - F^2)\, d\eta = \alpha \tag{9.3.19}$$

and Eq. 9.3.18 becomes

$$\rho U^2 \alpha \frac{d\delta}{dx} = \tau_0 \tag{9.3.20}$$

Equation 9.3.14 may be written as

$$\tau_0 = \mu\left[\frac{\partial u_x}{\partial y}\right]_{y=0} = \frac{\mu U}{\delta}\left(\frac{dF}{d\eta}\right)_{\eta=0} = \frac{\mu U}{\delta}\beta \tag{9.3.21}$$

where we define

$$\beta \equiv \left(\frac{dF}{d\eta}\right)_{\eta=0} \tag{9.3.22}$$

---

[1] This is consistent with the result of the analysis that leads to Eq. 9.2.17, which when combined with Eq. 9.2.13 yields Eq. 9.3.16.

Now Eq. 9.3.18 may be written in the form

$$\delta \frac{d\delta}{dx} = \frac{\mu\beta}{\rho U\alpha} \tag{9.3.23}$$

This simple ordinary differential equation has the solution

$$\delta = \left(\frac{2\mu\beta}{\rho U\alpha} x\right)^{1/2} \tag{9.3.24}$$

and the local shear stress along the solid surface is found from Eq. 9.3.21 in the form

$$\tau_0 = \frac{\mu U}{\delta}\beta = \left(\frac{\alpha\beta}{2}\frac{\rho\mu U^3}{x}\right)^{1/2} \tag{9.3.25}$$

At this stage we have for δ and τ_0 models that are explicit in the parameters that define the flow, such as U, and the properties μ and ρ. The next issue is: How do we estimate the constants α and β? The simplest approach is to assume an algebraic form for the function $F(\eta)$, the velocity profile. Keep in mind that as long as the boundary conditions we have imposed on F hold, the results for δ and τ_0 (Eqs. 9.3.24 and 9.3.25) will be independent of the choice of function F. Two possible choices of functions F are

$$F_2 = 2\eta - \eta^2 \tag{9.3.26a}$$

$$F_3 = \frac{3\eta}{2} - \frac{\eta^3}{2} \tag{9.3.26b}$$

both of which satisfy Eqs. 9.3.5 and 9.3.6 as well as the "smoothness" condition that

$$\frac{\partial u_x}{\partial y} = 0 \quad \text{at} \quad \eta = 1 \tag{9.3.27}$$

For the cubic function F_3, for example, we can evaluate α and β and show that

$$\frac{\delta}{x} = 4.64 \left(\frac{\nu}{Ux}\right)^{1/2} = 4.64\,\mathrm{Re}_x^{-0.5} \tag{9.3.28}$$

(compare this to Eq. 9.2.15—the "exact" solution) and (note Eqs. 9.2.18 and 9.2.20)

$$C_D = \frac{\mu(\partial u_x/\partial y)_{y=0}}{\frac{1}{2}\rho U^2} = 0.646\,\mathrm{Re}_x^{-0.5} \tag{9.3.29}$$

Recall that the local Reynolds number is defined as in Eq. 9.2.19.

Notice what we have achieved to this point. We have a simple *analytical* model of the dependence of the boundary layer thickness on distance downstream, x. The price we have paid is an uncertain degree of approximation in the assumed velocity field. However, comparison of this result with the more exact analysis of Section 9.2 shows that this simple integral method yields a very accurate representation of the velocity field. This observation serves to give us some confidence in integral methods of analysis, which are applied to other fluid dynamics problems.

The methodology of this integral method is very interesting. It involves what I like to call a *transformation of ignorance*. We traded out lack of knowledge of the velocity function $u_x(x,y)$ for a different unknown function, $\delta(x)$. The point of this method is that it is easier to solve for $\delta(x)$ than it is to solve the Navier–Stokes equations for $u_x(x,y)$. We imposed an approximate velocity field, and the accuracy of the results is the issue. The test of the solution procedure, *and indeed of any modeling exercise*, comes when the predictions of the model are compared to reality. The results given here, as Eqs. 9.3.28 and 9.3.29, are in very good agreement with the "exact" numerical solution to the boundary layer equations, and both agree very well with experimental studies.

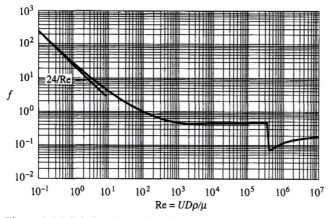

Figure 9.4.1 Friction factor for flow relative to a solid sphere.

9.4 TURBULENT DRAG FOR FLOWS RELATIVE TO BLUNT BODIES

In Chapter 4 we considered laminar flow relative to a rigid sphere. We found expressions for the shear force and pressure force along the line of motion, and combined them into a total dynamic force for laminar flow, F_{SL}, given by Eq. 4.4.130:

$$F_{SL} = 6\pi\mu RU \tag{9.4.1}$$

As the Reynolds number, defined for this flow as

$$\mathrm{Re} = \frac{UD\rho}{\mu} \tag{9.4.2}$$

approaches and exceeds Re = 1, the total force increases more strongly than linearly with the relative velocity, corresponding to an increasingly complex flow field about the sphere. Drag on a sphere is usually presented in terms of a friction factor f or drag coefficient C_D, defined as

$$f = C_D = \frac{\text{total force}}{\frac{1}{2}\rho U^2 \pi R^2} \tag{9.4.3}$$

Keep in mind that the density ρ used here is that of the surrounding fluid. Figure 9.4.1 shows the observed behavior of the friction factor as a function of the Reynolds number. Figure 9.4.2 shows streamlines associated with increasing Reynolds numbers. Changes in the slope of the f/Re function are associated with corresponding changes in the nature of the flow field behind the sphere, in the so-called wake region. These changes are associated with transitions in the character of the boundary layer along the spherical surface, and their complexity is beyond the scope of an introductory fluid dynamics course.[2] Of special note is the observation that at a Reynolds number of approximately 3.4×10^5, the boundary layer becomes turbulent and the wake behind the sphere becomes much smaller. This reduction in the area of the wake region gives rise to a significant reduction in the total drag on the sphere over a small Re range.

---

[2] When a gas bubble rises in a liquid, the situation is even more complex. Because the bubble is so light, the inertia of the surrounding liquid can "push the bubble around," giving rise to a spiraling flow. Documentation of this phenomenon can be found in Karamanev et al. [*AIChE J.*, **42**, 1789 (1996)]. See also Problem 9.21.

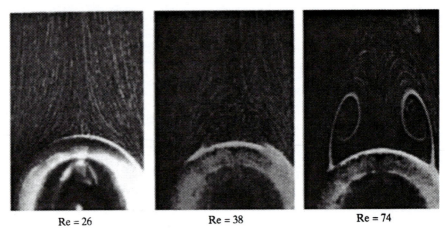

Re = 26   Re = 38   Re = 74

Figure 9.4.2 Streamlines in the wake of a sphere at various Reynolds numbers. Taneda, *J. Phys. Soc. Japan*, **11**, 1104 (1956). (With permission from The Physical Society of Japan.)

We do not need to understand all the details of this flow to see, from Fig. 9.4.1, that we may make use of these observations in practical situations. For example, the sudden drop in the friction factor at a Reynolds number of about 3.4×10^5 suggests that if it were possible, one might design a system to operate in this range with the goal of reducing the force on the sphere. With that idea in mind, we should look at the behavior of high speed flow past a sphere in terms of force versus relative velocity.

EXAMPLE 9.4.1   *Wind Force on a Sphere*

Plot the drag force on a sphere of diameter $D = 4.29$ cm for air speeds in the range 10 to 60 m/s. The sphere has a mass of 45.4 g.

We will take the kinematic viscosity of air to be $\nu = 1.6 \times 10^{-5}$ m²/s. At the lowest speed the Reynolds number is

$$\text{Re} = \frac{UD}{\nu} = \frac{10 \times 0.0429}{1.6 \times 10^{-5}} = 2.7 \times 10^4 \tag{9.4.4}$$

and at the highest speed Re = 1.6×10^5. From Fig. 9.4.1 we see that the drag coefficient is nearly constant over this range of speeds, and has the value

$$C_D = 0.5 \tag{9.4.5}$$

From Eq. 9.4.3 we find

$$\text{force} = \frac{\rho U^2 \pi R^2 C_D}{2} \tag{9.4.6}$$

Figure 9.4.3 shows the dependence of the drag force on the speed. (We use an air density of 1.23 kg/m³.) In the range of interest (≤ 60 m/s) the force increases quadratically with speed. At about 100 m/s, the boundary layer transition occurs, there is a sudden drop in the drag coefficient, and the force drops. There is a narrow range of speeds over which the force remains below the value at 100 m/s. The force then continues to climb beyond that point.

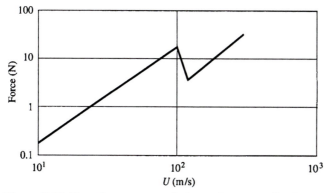

Figure 9.4.3 Drag force versus wind speed on a small sphere.

SUMMARY

Turbulent flows are characterized by velocity and pressure functions that depend on time in a random, chaotic manner. While the Navier–Stokes and continuity equations still are valid for such flows, we are incapable of solving them. If we average them over time, we obtain a set of dynamic equations that look like the Navier–Stokes and continuity equations, but an additional set of terms appears in the Navier–Stokes equations. These terms arise from the interactions of the fluctuating velocity components, and we call them the Reynolds stresses. If the geometry and boundary conditions are simple enough (e.g., turbulent flow through a pipe), solutions to these *time-averaged equations* are possible once some assumptions about the relationship of the Reynolds stresses to the mean flow field have been introduced. The classic example is the Prandtl mixing length theory (Eq. 9.1.18) illustrated in this chapter, which leads to a (mean) velocity profile and a prediction of the dependence of the friction factor on the Reynolds number.

Another approach to modeling turbulent flows is possible in certain cases through *boundary layer theory,* when the flow can be decomposed into two regions—one just adjacent to a solid surface wherein laminar (viscous) flow exists, and the other external to this so-called boundary layer wherein a simple and known mean flow is assumed to exist. The laminar boundary layer theory is developed in this chapter for the simplest case, where the external flow is steady (in the mean) and parallel to a long planar surface. The primary results of the boundary layer theory are predictions for the thickness of the boundary layer as a function of distance along the surface, and the drag exerted on the surface by the flow.

Two methods of solving the boundary layer equations are illustrated. One, due to Blasius, reduces the continuity and Navier–Stokes equations for the boundary layer flow to a single third-order ordinary differential equation in terms of a so-called stream function Ψ. Because the boundary layer equations are intrinsically nonlinear, no analytical solution is possible. Blasius produced a numerical solution from which the desired characteristics of the flow in the boundary layer follow. An alternative approach is the integral method (used earlier in Chapter 6, Section 6.5), which yields results very similar to those that come out of Blasius' method.

For turbulent flows relative to bodies of complex shape, we usually abandon any hope of obtaining a mathematical model. Instead, we are satisfied with dimensionless correlations of data relating the force exerted on the body (usually through a friction factor or drag coefficient) and the Reynolds number characteristic of the external (turbulent) flow.

PROBLEMS

9.1 Carry out the integration required in Eq. 9.1.25 and compare the resulting velocity profile to that given in Eq. 9.1.26. How does *this* profile behave at the pipe centerline?

9.2 Air, and water, in two separate experiments, are flowing at a mean speed of 3 m/s through a pipe of diameter 0.1 m and length 10 m. Use the Prandtl resistance law (Eq. 9.1.42) to find the pressure drop in each case. Find the laminar sublayer thickness in each case. Assume that the temperature is 20°C.

9.3 The Prandtl resistance law (Eq. 9.1.42) is not explicit in *f*. Prepare a log–log plot of *f*(Re) from that expression, in the Re range $[3000, 10^5]$. Show that a good approximation is

$$f = 0.079 \ \mathrm{Re}^{-0.25} \qquad \textbf{(P9-3)}$$

9.4 Water must be pumped through a pipe of length 100 m at a volume flowrate of 10^{-3} m³/s. The water temperature is 25°C. The annualized capital, installation, and maintenance cost of the type of piping to be used is proportional to the pipe length, and to the pipe diameter to a power slightly above unity. Assume that we can represent these costs by

$$\text{pipe cost} = 100 \ D^{1.4}L \ (\$/\text{yr}) \qquad \textbf{(P9.4)}$$

when *D* and *L* are in meters. The power (pumping) cost is proportional to the power expended, which is pressure drop times flowrate. Assume that the pump operates full time over a year at 100% efficiency, and the power cost is $0.15/kW·h.

Prepare a plot of pressure drop versus pipe diameter *D*, and total (pipe plus power) annual cost versus pipe diameter (on the same graph). Look at the range $0.01 < D < 0.1$ m. What conclusions do you draw from this exercise? What flowrate can be produced, and what is the required pressure drop, under economically optimum conditions? What is the cost per gallon pumped?

9.5 A large cylindrical object is to be recovered from the ocean floor. It may be regarded as a thin-walled pipe, open at both ends, and of inner diameter 1 m and outer diameter 1.05 m. Its axial length is 10 m. It has a density of 7000 kg/m³. It will be raised by cable, and pulled so that its axis is vertical.

What is the force on the cable if the pipe is raised at a speed of 0.1 m/s? Remember to include the weight and buoyancy of the pipe. Take the drag coefficient to be

$$\overline{C}_D = \frac{0.04}{\mathrm{Re}_L^{0.2}} \qquad \text{for} \quad 5 \times 10^5 < \mathrm{Re}_L < 10^7 \qquad \textbf{(P9.5)}$$

What is the effect on the cable force if the speed is doubled?

9.6 Repeat Problem 9.2, but use Eqs. 9.2.18 and P9.5 to calculate the pressure drop. How much do the results differ?

9.7 The water beast of Fig. 9.2.4 accelerates to a speed of 0.4 m/s and then glides while resting with no expenditure of energy. Assume that the beast has a uniform thickness of 0.03 m, but neglect any form drag. Assume also that the beast is of the same density as the surrounding water. How long can it glide before its speed falls to 0.1 m/s?

9.8 Equation 9.1.23 is derived by ignoring the viscous stresses in Eq. 9.1.22. It is argued that this might make sense far from the tube wall, where we assume that turbulence dominates viscous shear. Then can we expect that Eq. 9.1.23 is accurate as *r* approaches *R*? In particular, Eq. 9.1.23 states that the average shear stress at the tube wall is

$$T_R = \frac{\Delta \overline{\mathscr{P}} R}{2L} \qquad \textbf{(P9.8)}$$

Show that a simple force balance gives the same result. Does this imply that Eq. 9.1.23 is valid all the way across the tube?

9.9 Give the results that correspond to Eqs. 9.3.28 and 9.3.29, using the quadratic polynomial of Eq. 9.3.26.

9.10 Give a derivation of Eq. 9.1.22. State clearly the definition of T_{rz}.

9.11 In the "integral momentum analysis" of boundary layer flow over a flat plate (parallel to the far-field flow) we used a velocity profile

$$F_3(\eta) = \frac{3\eta}{2} - \frac{\eta^3}{2} \qquad \textbf{(P9.11.1)}$$

We selected this because it has approximately the correct shape. Why would

$$F_1(\eta) = \eta \qquad \textbf{(P9.11.2)}$$

Figure P9.12 Dolphin (*Tursiops gilli*).

be a poor choice for $F(\eta)$? Your answer should state the physical, hence mathematical, demands we would like to place on $F(\eta)$, and then contrast how F_1 and F_3 meet these demands.

9.12 Equation 9.4.3 defines a drag coefficient for a sphere. A more general definition of a drag coefficient for a body moving relative to a fluid is

$$f = C_D = \frac{\text{total force}}{\frac{1}{2}\rho V^2 A_n} \qquad \text{(P9.12)}$$

where A_n is the area of the body in a view normal to the flow. If the drag coefficient is known, it is possible to determine the force required to move a body at a given speed V.

For the dolphin sketched in Fig. P9.12, find the horsepower required to maintain a speed of 12 knots (610 cm/s). A nominal value of C_D for a well-streamlined body is $C_D = 0.06$.

9.13 With regard to Fig. 9.4.1, how fast must a baseball be pitched to cause any slight reduction in its speed to lead to a sudden increase in the friction factor? What would be the implications of such an increase? Is there any evidence that this occurs in practice?

9.14 A baseball is pitched at a speed of 100 mph. To what extent is the speed reduced by the time the ball reaches the batter? The mass and diameter of a baseball are 145 g and 74 mm, respectively.

9.15 A soccer ball is kicked at an initial speed of 40 mph toward the goal, 20 yards away. What is the speed of the ball at the goal?

9.16 The dragonfly pictured in Fig. P9.16 can sustain speeds of 10 m/s for short times. It has a body weight of 2 g. Estimate the power expended to overcome drag on its body (but not the wings). If a 100 kg human expended the same power per body weight as the dragonfly, how many horsepower would be expended? The axial length (from nose to tip of tail) is 0.1 m.

9.17 An experiment is designed to test the laminar boundary layer model (Sections 9.2 and 9.3) for a flat plate. A thin plate is suspended from a spring, as in Fig. P9.17. Under static conditions, the extension of the spring (ΔE) permits a measurement of the weight of the plate. A vertical upward turbulent airflow is then imposed, and the resulting drag on the plate reduces the extension of the spring.

Figure P9.16 Dragonfly.

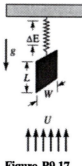

Figure P9.17

Find the extension of the spring under the following conditions:

Mass of plate	$M = 0.001$ kg
Plate length	$L = 0.2$ m
Plate width	$W = 0.1$ m
Plate thickness	$t = 1$ mm
Air speed	$U = 10$ m/s
Air temperature	$T = 288$ K
Air density	$\rho = 1.2$ kg/m^3

Spring extension under the weight of the plate, but with no airflow $\Delta E = 5$ cm

Assume that the spring is Hookeian (linear).

9.18 Give the details that lead from Eq. 9.3.3 to Eq. 9.3.4.

9.19 Find the terminal velocity of an oil drop of diameter 97 μm, and density 1.2 g/cm^3, in air at 25°C and 1 atm pressure.

9.20 If we wish to find the terminal velocity of a solid sphere falling through a fluid under the action of gravity, we cannot use Fig. 9.4.1 directly, since the Re is unknown. Show that a graphical procedure is possible in which the "solution" to the problem comes from the intersection of the $f/$Re curve and the line

$$f = \frac{K}{\text{Re}^2} \qquad \text{(P9.20.1)}$$

where

$$K = \frac{32}{3} \frac{\rho g \, \Delta\rho \, R^3}{\mu^2} \qquad \text{(P9.20.2)}$$

9.21 Karamanev et al. [*AIChE J.*, **42**, 1789 (1996)] present the data tabulated below for the terminal rise velocity of low density spheres of diameter 7 mm through water at 25°C. Take their data, and plot a curve for f versus Reynolds number for comparison to Fig. 9.4.1. Predict the terminal rise velocity of these data, using the so-called standard drag curve for spheres of comparable or greater density than the surrounding fluid:

$$C_D = f = \frac{24}{\text{Re}} (1 + 0.173 \, \text{Re}^{0.657})$$

$$+ \frac{0.413}{1 + 16{,}300 \, \text{Re}^{-1.09}} \quad \text{for} \quad \text{Re} \le 2 \times 10^5$$

$$\text{(P9.21)}$$

Compare the predictions to the data below.

ρ_p (g/cm^3)	0.028	0.183	0.290	0.450	0.590	0.630	0.845	0.916
U (cm/s)	31.9	28.6	25.2	22.6	23.1	23.1	17.0	13.6

9.22 Using Eq. 9.3.16 for $u_x(x,y)$, and choosing F_2 (Eq. 9.3.26a with Eq. 9.3.28), calculate the terms $u_x \partial u_x / \partial x$ and $u_y \partial u_x / \partial y$. Is one negligible compared to the other? Compare $\rho u_x \partial u_x / \partial x$ to $\mu \partial^2 u_x / \partial y^2$, as well. Can you make any argument regarding the relative magnitudes of these two terms?

Chapter **10**

Flow Through Porous Media

We have seen that geometrical simplification plays a key role in permitting us to build mathematical models of interesting flows. Sometimes the geometry of the flow field appears to defy any rational simplification. Indeed great technological importance is attached to the cases of a fluid that must pass through a *particulate or fibrous medium* in which the particles or fibers are nonuniform or have random orientations and nonuniform spacings. Such flows arise in filtration through fibrous filters or dense screens, in underground movement of oil or water through sand or porous rock, and in the flow of gases through tubes packed with catalyst particles. In this chapter we take a look at very simple, but very successful, approaches to modeling the pressure drop–flowrate relationship in such systems.

10.1 FLOW THROUGH A PACKED BED OF SPHERES

Spherical particles are often used as packing in catalyst beds. Sand beds used for filtration contain particles that are not exactly spherical but are rounded to some degree and may be considered spherical, with a mean particle size to characterize them. Flow through such a bed is quite complex, as Fig. 10.1.1 suggests. In examining such a complex flow field, one of the first questions we raise is whether this flow is similar to flow *around* an odd-shaped porous body, or more nearly similar to flow *within* a duct of very nonuniform boundaries. There is no simple answer to this question, and it is not especially productive to spend much time thinking about it. Obviously, the way we choose to view the flow will determine how we choose to model it.

10.1.1 The Equivalent Capillary Model of a Porous Medium

The most common approach to modeling flow through a porous medium is to regard the flow as if it takes place through a set of parallel channels or tubes of very nonuniform geometry. Then a relationship is postulated between some features of the packing geometry and an "equivalent" diameter of the imaginary tube through which each

411

Figure 10.1.1 An element of fluid follows a convoluted path as it moves through a packed bed.

element of fluid passes. One way to think of this is suggested in Fig. 10.1.2, which shows the hypothetical "tube" through which the fluid element of Fig. 10.1.1 passes.

We will imagine that the pressure drop–flowrate relationship for flow through a packed bed is similar to that for fully developed flow through a straight tube. The equivalent diameter of the tube will be based on the "hydraulic radius" concept introduced in Chapter 5, and defined in Eq. 5.3.32. The hydraulic radius is the ratio of the cross-sectional area of the tube to the wetted perimeter. How will we calculate this for a packed bed of spheres?

We assume that the bed is composed of single-sized spheres of diameter D_p. The packing is random, rather than ordered. In a unit volume of the bed (this would be a volume large compared to D_p^3, so that we can define some average characteristics) there is a certain *wetted area* per unit of volume, a_w, and a fraction of empty volume ε, called the *void fraction*.

In addition, any solid particle has a certain surface area per *particle* volume, a_v. For a spherical particle this is simply

$$a_v = \frac{\pi D_p^2}{\dfrac{\pi D_p^3}{6}} = \frac{6}{D_p} \tag{10.1.1}$$

We now make the assumption that characteristics of the bed, such as the wetted perimeter and the fraction of the cross section of the bed that is open to flow, are uniform along the length of the bed in a statistical sense. It follows from this that we may write the hydraulic radius as

$$r_h \equiv \frac{\text{cross section open to flow}}{\text{wetted perimeter}} = \frac{\text{void volume/bed length}}{\text{wetted surface/bed length}}$$
$$= \frac{\text{void volume/bed volume}}{\text{wetted surface/bed volume}} = \frac{\varepsilon}{a_w} \tag{10.1.2}$$

Figure 10.1.2 A hypothetical tube through which an element of fluid passes as it moves through a packed bed.

From its definition we may write a_w as

$$a_w \equiv \frac{\text{wetted surface}}{\text{bed volume}} = \frac{\text{wetted surface}}{\text{particle volume}} \times \frac{\text{particle volume}}{\text{bed volume}} = a_v(1 - \varepsilon) \quad \textbf{(10.1.3)}$$

Then the hydraulic radius has the form

$$r_h = \frac{\varepsilon}{a_v(1 - \varepsilon)} = \frac{\varepsilon D_p}{6(1 - \varepsilon)} \quad \textbf{(10.1.4)}$$

This completes the first task: we have related an important hydrodynamic feature of this complex geometry to easily measured parameters of the bed. Now the question is: How do we use the hydraulic radius?

We originally introduced the hydraulic radius in Chapter 5 by stating that for a duct of odd-shaped cross section, we define the friction factor with the hydraulic radius as the length scale. Equation 5.3.31 gives this definition:

$$f = \frac{2(\Delta P/L)r_h}{\rho U^2} \quad \textbf{(10.1.5)}$$

We will maintain this definition of f, using Eq. 10.1.4 to replace r_h.

Note that f is written in terms of the average velocity U. This may be written in terms of the volume flowrate as

$$Q = UA_\varepsilon = U\varepsilon A_c \quad \textbf{(10.1.6)}$$

where, as in Eq. 10.1.2, we have assumed that at any cross section, the ratio of open area A_ε to the bed cross-sectional area A_c is the same as the ratio of the open volume to the bed volume, ε. We can measure Q, but we do not normally put a probe into the interior of the packing to measure the average velocity U. We define something called the "superficial velocity" U_o, as

$$U_o \equiv U\varepsilon \quad \textbf{(10.1.7)}$$

From Eq. 10.1.6 we see that the superficial velocity is simply $U_o = Q/A_c$. It is a smaller velocity than the actual average velocity within the packing of the bed. Because U_o is easily measured or related to the volume flowrate and the open or unpacked cross-sectional area of the bed, we write the friction factor in terms of U_o. Along with Eq. 10.1.4 this lets us write f as

$$f = \frac{(\Delta P/L)\varepsilon^3 D_p}{3\rho U_o^2(1 - \varepsilon)} \quad \textbf{(10.1.8)}$$

10.1.2 The Laminar Flow Regime

Equation 10.1.8 is simply a definition of f in terms of bed characteristics. So far we have made no attempt to relate pressure drop to flowrate. We do that now through a bold assertion. We assume that in laminar flow the friction factor for flow through a packed bed will be the same as that given by the Hagen–Poiseuille law, when we replace the pipe diameter by $4r_h$ in the Reynolds number. The Hagen–Poiseuille law may be written as

$$f = \frac{16}{\text{Re}} \quad \textbf{(10.1.9)}$$

For the packed bed, we may write the modified Reynolds number as

$$\text{Re}' = \frac{U(4r_h)\rho}{\mu} = \frac{2U_oD_p\rho}{3\mu(1-\varepsilon)} \tag{10.1.10}$$

When we combine Eqs. 10.1.8, 10.1.9, and 10.1.10, we find the following model:

$$\frac{\Delta P D_p \varepsilon^3}{L\rho U_o^2(1-\varepsilon)} = \frac{72(1-\varepsilon)}{U_oD_p\rho/\mu} = \frac{48}{\text{Re}'} \tag{10.1.11}$$

Equation 10.1.11 does not fit available data for slow (laminar) flow of viscous fluids through packed beds. What is remarkable is that with a simple empirical correction *to the numerical coefficient only,* not to the functional forms in which ε, D_p, and so on appear, the model fits data extremely well. We increase the numerical coefficient 72 to 150 and write

$$f \equiv \frac{\Delta P D_p \varepsilon^3}{L\rho U_o^2(1-\varepsilon)} = \frac{150(1-\varepsilon)}{U_oD_p\rho/\mu} \tag{10.1.12}$$

This is called the Blake–Kozeny equation. We use the notation f instead of f to define a modified friction factor, since Eqs. 10.1.8 and 10.1.12 differ by a factor of 3. (One often finds an equation identical to Eq. 10.1.12, but with a coefficient of 180, rather than 150, as the recommended empirical coefficient. This is sometimes called the Carman–Kozeny equation. Still others interchange the designations of the Blake–Kozeny and Carman–Kozeny equations.)

It is not surprising that the numerical coefficient in Eq. 10.1.11 is not consistent with observations. After all, the concept that the fluid is confined within a definable tube with rigid uniform boundaries seems very far removed from reality. It is interesting that the equivalent tube concept leads to the right dependence of pressure drop on the void fraction ε. *Such a result would not come out of dimensional analysis.* Since ε is

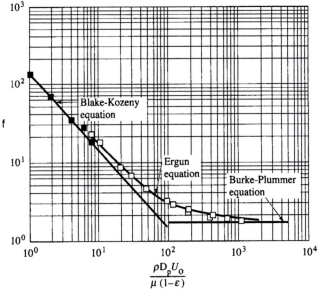

Figure 10.1.3 Friction factor versus Reynolds number for flow through a packed bed of spheres. (The Ergun and Burke–Plummer equations are mentioned in Section 10.1.3.)

dimensionless, *any* function of ε would be permissible according to dimensional arguments. The dependence on ε^3, for example, which is close to what is observed, is predicted through this "equivalent tube" model.

Data for flow through packed beds of spheres are shown in Fig. 10.1.3. Note that the Reynolds number usually used for this type of plot is given in Eq. 10.1.10, but without the factor of 2/3:

$$\text{Re} = \frac{U_o D_p \rho}{\mu(1 - \varepsilon)} \qquad (10.1.13)$$

This is an arbitrary choice, but one that is most commonly followed. The data agree with the Blake–Kozeny equation very well, at low Reynolds numbers, and up to Reynolds numbers as large as 10.

10.1.3   The Turbulent Flow Regime

Next we look at turbulent flow. By analogy to the result given in Chapter 9 (see Fig. 9.1.4) we might expect the friction factor for flow through a packed bed, in the turbulent regime to follow the Prandtl resistance law (Eq. 9.1.42). However, the Prandtl resistance law is valid, and agrees well with observations, only for *smooth* tubes and pipes. A rough inner surface (due, e.g., to corrosion, buildup of mineral deposits, or simply the use of a pipe material that is rough) causes the friction factor to be higher than that shown on Fig. 9.1.4, and to be essentially independent of Reynolds number in the turbulent regime. (We introduce this later in Chapter 11, Section 11.3.) Hence we are led to predict that for turbulent flow

$$\frac{\Delta P D_p \varepsilon^3}{L \rho U_o^2 (1 - \varepsilon)} = \text{constant} \qquad (10.1.14)$$

A value for the constant of 1.75 permits this expression to fit data from a variety of sources. With this coefficient, Eq. 10.1.14 is called the Burke–Plummer equation. It fits data quite well for Reynolds numbers (as defined in Eq. 10.1.13) greater than 1000.

The simplest model we can adopt for fitting the data between Reynolds numbers of 10 and 1000 is obtained by adding the friction factors of the Blake–Kozeny and Burke–Plummer equations to predict

$$f \equiv \frac{\Delta P D_p \varepsilon^3}{L \rho U_o^2 (1 - \varepsilon)} = \frac{150(1 - \varepsilon)}{U_o D_p \rho / \mu} + 1.75 \qquad (10.1.15)$$

This is the line labeled "Ergun eq." in Fig. 10.1.3, to correspond to the usual terminology.

Figure 10.1.3 provides an empirical basis for predicting the performance of packed bed systems with respect to flow resistance. The analytical forms taken by the high and low Reynolds number extremes are predictable, within a multiplicative constant, from simple fluid dynamic ideas. The figure is valid only for single-sized (i.e., uniform diameter) spherical packing, and methods exist for defining "equivalent" particle sizes, permitting the use of this graph for other particle shapes. While the Ergun equation may provide a reasonable starting point for estimating the order of magnitude of the pressure drops that accompany flows through various porous media, it is best to carry out an experimental program, guided by and made more efficient through use of a model such as the Ergun equation.

These cautionary remarks notwithstanding, the model developed here is quite commonly applied to porous media composed of nonspherical particles.

10.1.4 Pressure Drop Across a Fibrous Filter

In the semiconductor industry it is essential that the air in the fabrication facility be extremely clean with respect to particulates. Submicrometer particles are filtered from the air using fibrous filters composed of extremely fine fibers. It is important to be able to predict the pressure drop–flowrate relationship for such filters. Two sets of experimental data are shown in Fig. 10.1.4. The x axis is the "solidity" of the filter, α, which is just $(1 - \varepsilon)$. The y axis is a dimensionless pressure drop, defined as

$$\Phi \equiv \frac{\Delta P}{L} \frac{R^2}{4\alpha\mu U_\mathrm{o}} \tag{10.1.16}$$

where R is the radius of the fibers that make up the filter bed.

Filters for this particular class of application typically use fibers with diameters of the order of 10 μm. As a consequence, even though air is a low viscosity fluid, a Reynolds number based on fiber diameter is very small and the flow is in the laminar regime. We will examine the application of the Blake–Kozeny equation to these data.

To use the Blake–Kozeny equation for a bed of cylindrical fibers, it is necessary to identify an appropriate particle size. For nonspherical particles we usually invert Eq. 10.1.1 and define an equivalent particle diameter based on the area/particle volume factor, a_v:

$$D_\mathrm{p} \equiv \frac{6}{a_\mathrm{v}} \tag{10.1.17}$$

For a long cylinder of radius R, the area/particle volume factor is

$$a_\mathrm{v} = \frac{2}{R} \tag{10.1.18}$$

In terms of the function Φ defined above, the Blake–Kozeny equation now takes the form

$$\Phi = \frac{25\alpha}{6(1 - \alpha)^3} \tag{10.1.19}$$

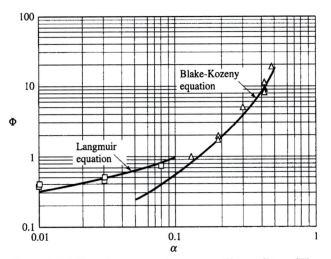

Figure 10.1.4 Data for pressure drop across fibrous filters. (The Langmuir model is discussed in connection with Eq. 10.1.20.) Schaefer, MS thesis, University of Minnesota, 1976.

We see from Fig. 10.1.4 that the Blake–Kozeny equation fits the data reasonably well, except at very low solids fraction. This should not be surprising, since the Hagen–Poiseuille law, which is at the foundation of the Blake–Kozeny equation, cannot be expected to give an accurate estimate for the pressure drop associated with flow past an array of cylinders, especially when the solids fraction becomes very small. In fact, the surprise is that the Blake–Kozeny equation works as well as it does at *any* solids density!

Models for pressure drop across a bed of cylindrical fibers have been based on the flow field about a single cylinder, and about a cylinder surrounded by a regular pattern of like cylinders. The derivation of these models is beyond the scope of our discussion here. One of the more useful models is due to Langmuir, who derived an expression that takes the form

$$\Phi = \left(-\ln \alpha + 2\alpha - \frac{\alpha^2}{2} - 1.5 \right)^{-1} \tag{10.1.20}$$

We can expect this model to be appropriate for small values of α, where the cylinders are relatively isolated from one another. Figure 10.1.4 shows that the model does fit these data quite well.

EXAMPLE 10.1.1   *Finding the Solidity of a Filter*

We wish to estimate the fractional solidity of a commercial filter about which we have the following information. The filter composition is 10 μm Dacron fibers. The filter thickness is 10 cm. The filter assembly has an area normal to the air flow of 0.5 m². The pressure drop across the filter is 2 atm when the flowrate of air is 100 standard cubic feet per minute (scfm). The downstream face of the filter is at atmospheric pressure.

We will begin with the assumption that the solidity α is high enough to allow us to use the Blake–Kozeny equation. We rearrange the equation to the form

$$\frac{25\alpha^2}{6(1 - \alpha)^3} = \frac{\Delta P}{L} \frac{R^2}{4\mu U_o} \tag{10.1.21}$$

The flowrate designation "100 scfm" means that the volumetric flowrate was 100 standard cubic feet per minute, which means 100 ft³/min at standard temperature and pressure (STP: 273 K and 1 atm). Since the pressure drop is given as 2 atm, the absolute pressure on the upstream face of the filter was 3 atm. Hence the volume flowrate at the upstream face was 100/3 = 33.3 ft³/min. This gives us a superficial velocity U_o of

$$U_o = \frac{Q}{A} = \frac{33.3 \text{ ft}^3/\text{min} \times (0.305 \text{ m/ft})^3 \times \frac{1}{60} \text{ s/m}}{0.5 \text{ m}^2} = 0.031 \text{ m/s} \tag{10.1.22}$$

at the upstream face. Since the downstream face is at one atmosphere, the superficial velocity is 0.093 m/s downstream. What U_o should we use in Eq. 10.1.21 for a compressible flow? We noted in Chapter 3 (Eq. 3.4.20) that the form of Poiseuille's law, which is the basis of the Blake–Kozeny equation, holds if we evaluate the density at the *mean absolute pressure*. Hence we will use $U_o = (0.031 + 0.093)/2 = 0.062$ m/s, and solve for α from Eq. 10.1.21:

$$\frac{25\alpha^2}{6(1 - \alpha)^3} = \frac{2 \times 10^5 \text{ Pa}}{0.1 \text{ m}} \frac{(5 \times 10^{-6} \text{ m})^2}{4(1.7 \times 10^{-5} \text{ Pa} \cdot \text{s})(0.062 \text{ m/s})} = 11.9 \tag{10.1.23}$$

The result is $\alpha = 0.54$.

10.2 FLOW THROUGH POROUS (CONSOLIDATED) MATERIALS

Many porous materials have a rigid structure that *cannot* be regarded as particulate; hence there is no simple way to identify and quantify a particle size for use in a model such as the Ergun equation. Materials of this class are called "consolidated" materials. If the density of the solid phase is known, it may be possible to estimate the void fraction of the medium. To the degree that such a porous medium is a "black box" of unknown internal structure, it is not possible to model the flow through such a medium from basic principles.

Instead, a common empirical approach is to note that for laminar flow through porous media, the superficial velocity obeys a simple linear relationship known as Darcy's law:

$$U_o = K_D \frac{\Delta P}{\mu L} \tag{10.2.1}$$

The coefficient K_D is called the Darcy's law permeability (or simply, "permeability"). The permeability is a function of the geometrical structure of the porous medium—the porosity ε, some measure of the size distribution of the pores, and their "connectivity" and "tortuosity." Connectivity refers to the degree to which pores or channels are interconnected, while tortuosity is a measure of the degree to which the fluid is forced to depart from a straight path as it flows through the bed.

If the geometry is well defined, we may calculate the permeability from a simple hydrodynamic model. For example, consider a porous medium that consists of a set of parallel, circular cross section capillaries of two diameters, as sketched in Fig. 10.2.1. In the laminar flow regime we would apply the Hagen–Poiseuille law (Eq. 3.2.34) independently to each capillary. If, in each area A of the cross section, there are n_1 capillaries of diameter D_1 and n_2 of diameter D_2, the volume flowrate conveyed across that area would be

$$\frac{Q}{A} = U_o = \frac{\pi[(n_1/A)D_1^4 + (n_2/A)D_2^4]}{128} \frac{\Delta P}{\mu L} = K_D \frac{\Delta P}{\mu L} \tag{10.2.2}$$

We see that the permeability depends on the number of capillaries per unit of cross-sectional area of each size (n_1/A and n_2/A) and on the sizes of the two capillaries (D_1 and D_2). We also see that K_D has units of length squared.

Most porous media do not have such an evidently simple geometry, and we therefore must use K_D as an empirical parameter to be determined from data on Q versus ΔP. We then use Darcy's law as a model for the flow, in combination with the conservation of mass principle, to solve specific problems. We treat such a problem here, selecting one from the area of physiology.

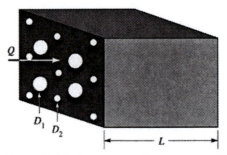

Figure 10.2.1 A porous medium penetrated by straight capillaries of two different diameters.

10.2.1   Fluid Transport in a Tumor

Some tumors may be thought of as porous masses of cells interpenetrated by blood vessels. The walls of these vessels are permeable to plasma and some solutes. Hence fluid from the vessels continually "leaks" into the tumor. (This occurs in normal tissue, as well.) One method of chemotherapy involves the injection into the blood of specific chemicals that travel throughout the vascular system, including the vasculature of the tumor. When these chemicals "leak" into the tissue region of the tumor, they act on the proliferating cells and kill them. One problem with this form of chemotherapy is that the resulting outflow of fluid from the tumor (see Fig. 10.2.2) reduces the ability of the therapeutic chemical to sustain a high and effective concentration inside the tumor.

With this background in mind, we want to examine the fluid dynamics of leakage into a tumor and passage of fluid out of the tumor. We look strictly at the hydrodynamic problem—the *chemistry* of chemotherapy is not considered. Figure 10.2.2 shows the various flows in the interior of the tumor. We will introduce two key assumptions that are not especially realistic from an anatomical point of view but make it possible for us to develop a simple first approximation to the flow field. One is that the blood vessels, the capillaries, are *uniformly* distributed throughout the tumor. The other is that the capillaries are *continuously* distributed throughout the tumor. These ideas are distinct. *Uniformity* simply expresses the condition that any unit volume of the tumor contains the same number of capillaries as any other. These vessels are *discretely* distributed in the sense that the tumor may be thought of as being composed of two materials—tissue and blood vessels. When we assume that the capillaries are *continuously* distributed, we are replacing the real, discrete anatomy with a volume that is homogeneous in the sense that at every point in the tumor there is a source of leakage. It is as if the tumor cells and the capillaries are so intimately "mixed" that the smallest volume element contains both tumor cells and blood vessels.

With these idealizations, our goal is to develop a model of the fluid dynamics inside the tumor. In particular we want to know the tumor's internal pressure distribution, and the velocity profile. We assume laminar, steady, Newtonian incompressible flow. Instead of working through the Navier–Stokes equations, we view the tumor as a porous medium in which Darcy's law holds locally, that is,

$$u_r = u = -\frac{K_D}{\mu}\frac{\partial p}{\partial r} \tag{10.2.3}$$

The mass conservation principle does not take the form of the continuity equation in this model, because we are distributing the leakage uniformly through the volume. This has the effect of introducing a *source* of mass at every point within the medium.

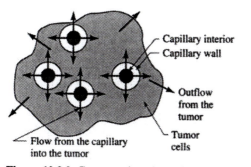

Figure 10.2.2 Cross section through a region of a tumor.

(In a sense, mass is "created" within the volume element, and so it is not conserved.) As a result, the continuity equation takes the form (in spherical coordinates)

$$\nabla \cdot \mathbf{u} = \frac{1}{r^2}\frac{\partial(r^2 u)}{\partial r} = \phi_v \tag{10.2.4}$$

The source term ϕ_v is the rate at which fluid enters an element of volume, by leakage from a capillary, in units of volume/time per unit of volume. We need a model for this leakage function.

Physiologists usually adopt a capillary leakage model in which the efflux across the wall of the capillary is a linear function of the pressure difference between the interior of the capillary and the surrounding tissue. If we ignore osmotic effects (they are easily accounted for, but their appearance will detract from our main goal), this model takes the form

$$\phi_v = L_p a_v [p_i - p(r)] \tag{10.2.5}$$

where a_v is the capillary surface area per unit of tissue volume. The pressure within the capillaries is taken to be p_i, independent of position in the tumor. The pressure inside the tumor is $p(r)$. The coefficient L_p is called the "hydraulic conductivity" of the capillary wall.

When we combine Eqs. 10.2.3, 10.2.4, and 10.2.5, we find

$$\frac{K_D}{\mu}\frac{1}{r^2}\frac{d}{dr}\left(r^2\frac{dp}{dr}\right) + L_p a_v[p_i - p(r)] = 0 \tag{10.2.6}$$

If we define the following dimensionless variables and parameters

$$\frac{R^2 \mu L_p a_v}{K_D} \equiv \beta^2 \qquad \frac{r}{R} \equiv s \qquad \frac{p - p_i}{p_o - p_i} \equiv P \tag{10.2.7}$$

the differential equation for $P(s)$ becomes

$$\frac{1}{s^2}\frac{d}{ds}\left(s^2\frac{dP}{ds}\right) - \beta^2 P = 0 \tag{10.2.8}$$

We have defined R as the tumor radius and p_o as the pressure in the region exterior to the tumor.

We solve Eq. 10.2.8 subject to the following boundary conditions:

$$\frac{dP}{ds} = 0 \quad \text{at} \quad s = 0 \quad \text{(symmetry at the center of the sphere)} \tag{10.2.9}$$

$$P = 1 \quad \text{at} \quad s = 1 \quad (p = p_o \text{ at } r = R) \tag{10.2.10}$$

The solution of this differential equation (subject to these boundary conditions) is found to be (see Problem 10.23)

$$P(s) = \frac{1}{s}\frac{\sinh \beta s}{\sinh \beta} \tag{10.2.11}$$

and the velocity profile may be written in the form (after applying Eq. 10.2.3)

$$\frac{\mu R}{K_D(p_i - p_o)} u(s) = \frac{1}{s^2}\left[\frac{\beta s \cosh \beta s - \sinh \beta s}{\sinh \beta}\right] \tag{10.2.12}$$

From Eq. 10.2.11, and the behavior of the hyperbolic sine function in the limit of small argument, it follows that the pressure at the center of the tumor is

$$P(0) = \frac{\beta}{\sinh \beta} \tag{10.2.13}$$

With this model in hand, we may evaluate the magnitudes of the pressure and flow inside a tumor. Fortunately, physiological data are available to support our use of the model.

EXAMPLE 10.2.1 *Pressure and Flow in a Tumor*

The following parameter values are presented by Baxter and Jain [*Microvasc. Res.*, **37**, 77 (1989)]:

$$L_p = 2.8 \times 10^{-7} \text{ (cm/mmHg} \cdot \text{s)} \qquad \frac{K_D}{\mu} = 4.1 \times 10^{-8} \text{ (cm}^2/\text{mmHg} \cdot \text{s)}$$

$$p_i - p_o = 12 \text{ mmHg (corrected for osmotic effects)} \quad R = 1 \text{ cm} \quad a_v = 200 \text{ cm}^{-1}$$

Note that a value is given for K_D/μ, not for K_D itself. We begin by calculating a value of the parameter β:

$$\beta = \left[\frac{(1^2)(2.8 \times 10^{-7})(200)}{4.1 \times 10^{-8}} \right]^{1/2} = 37 \qquad \text{(10.2.14)}$$

Note that the units of the parameters given above are mixed, since we use millimeters of mercury for pressure, instead of pascals or dynes per centimeter squared, while the other units are cgs units for length. However, the use of mmHg for pressure is consistent; and since β is dimensionless, it is not necessary to convert units to calculate β, since the mmHg units drop out. Had there been a parameter in this model involving force, with dynes used for the force unit, it would have been necessary to convert all units to a consistent set (e.g., cgs) before calculating β.

The velocity at the periphery of the tumor is (setting $s = 1$ in Eq. 10.2.12)

$$u(1) = \frac{4.1 \times 10^{-8}(12)}{1} [36] = 18 \times 10^{-6} \text{ cm/s} = 0.18 \ \mu\text{m/s} \qquad \text{(10.2.15)}$$

The pressure at the center of the tumor follows from Eq. 10.2.13. For large values of β this may be written, using an approximation for sinh β, as

$$P(0) = \frac{\beta}{\sinh \beta} \approx \frac{2\beta}{e^\beta} \ll 1 \qquad \text{for} \quad \beta \gg 1 \qquad \text{(10.2.16)}$$

Hence $p(r = 0) = p_i$, which is 12 mmHg above the exterior pressure p_o.

A consequence of the very large value of β is that P is very small over almost the entire volume of the tumor. This means that there is a very small transcapillary driving pressure, $[p_i - p(r)]$, and as a result there is very little leakage from the capillaries. If that leakage will bring a therapeutic agent into the tumor, the transport of the agent into the tumor will be minimal.

The model outlined here, which follows that of Baxter and Jain, is greatly oversimplified, but it does provide some qualitative insight into the fluid dynamics inside a vascularized tumor, with significant implications for chemotherapy. Baxter and Jain discuss the limitations and implications of their model, and we leave further study of this topic to the interested reader. Of importance here is the practice derived in manipulating the tools with which complex fluid dynamical systems are modeled.

10.2.2   Leakage from a Pressurized Porous Pipe

A long porous-walled pipe, closed at both ends with impermeable (nonporous) covers, is initially charged with air at some elevated pressure. The exterior of the pipe is at an ambient pressure below that inside the pipe. The system is isothermal. The pipe wall has a uniform permeability. Gas leaks from the high pressure side of the pipe to the ambient, with the obvious result that the interior pressure falls with time.

Our first task is to derive a model for the decay of internal pressure with time, in terms of the characteristic properties and geometry of the container. We begin with Eq. 10.2.3 for the radial velocity:

$$u_r = u = -\frac{K_D}{\mu}\frac{\partial p}{\partial r} \tag{10.2.17}$$

Notice that we keep the partial derivative notation here because the pressure will be a function of time.

For a general model, we will write the inside and outside radii as κR and R. The volume flowrate Q' (per unit of pipe length) of gas across the inside surface[1] will be

$$Q' = 2\pi\kappa Ru = -\frac{2\pi\kappa R K_D}{\mu}\left(\frac{\partial p}{\partial r}\right)_{r=\kappa R} \tag{10.2.18}$$

The velocity field must satisfy the continuity equation. Since the fluid is a gas, and the absolute pressure varies from the inside to the outside of the pipe, the compressibility of the gas must be accounted for. Then we write the continuity equation in the form of Eq. 4.1.6:

$$\frac{\partial\rho}{\partial t} = -\nabla\cdot\rho\mathbf{u} \tag{10.2.19}$$

We now introduce two assumptions to simplify Eq. 10.2.19. One is that for a very long pipe, or even for a short one that has closed ends, as in this case, the outflow will be strictly radial, with no variation of the outflow velocity in the axial direction. The second assumption is that of quasi-steady behavior. If the time rate of change of pressure is slow enough, the $\partial\rho/\partial t$ term should not be significant in the continuity equation. Then we write

$$\nabla\cdot\rho\mathbf{u} = \frac{1}{r}\frac{\partial}{\partial r}(r\rho u) = 0 \tag{10.2.20}$$

or

$$r\rho u = A(t) \tag{10.2.21}$$

For a compressible fluid we need a connection between the density and the pressure. This is supplied here with the assumption that air is an ideal gas. If we assume further that the leakage process is isothermal we have

$$\rho = \frac{pM_w}{R_G T} \tag{10.2.22}$$

where M_w is the molecular weight of air ($M_w = 29$), and R_G is the gas constant. When this result is substituted into Eq. 10.2.21 for ρ, and Eq. 10.2.17 is used to eliminate the

---

[1] Since the fluid is compressible, the *volume* flowrate will vary with pressure, hence will be different between the inside and outside surfaces of the pipe. Of course the *mass* flowrate will be constant.

velocity u, we obtain a differential equation for the radial pressure profile through the wall of the pipe:

$$p \frac{\partial p}{\partial r} = -\frac{\mu R_G T A}{K_D M_w} \frac{1}{r} = \frac{1}{2} \frac{\partial p^2}{\partial r} \qquad (10.2.23)$$

Equation 10.2.23 may be integrated easily, and when we use the boundary condition that the external pressure is atmospheric

$$p = p_a \quad \text{at} \quad r = R \qquad (10.2.24)$$

we may solve for and eliminate the coefficient A. Upon completing the algebra, we find

$$\frac{p^2 - p_i^2}{p_a^2 - p_i^2} = \frac{\ln(r/\kappa R)}{\ln(1/\kappa)} \qquad (10.2.25)$$

We use the notation p_i to represent the unknown gas pressure everywhere inside the pipe, which is also $p(\kappa R)$.

Once we have the pressure profile, we can evaluate $\partial p/\partial r$ at the inner surface and find the efflux of gas through the use of Eq. 10.2.17. The pressure gradient at the inside surface of the pipe is found from Eq. 10.2.25 to be

$$\left(\frac{\partial p}{\partial r}\right)_{\kappa R} = \frac{p_a^2 - p_i^2}{2 p_i \kappa R \ln(1/\kappa)} \qquad (10.2.26)$$

and the volume flowrate across the inner surface (per unit length of pipe) follows from Eq. 10.2.18 as

$$Q' = \frac{\pi K_D (p_a^2 - p_i^2)}{\mu (\ln \kappa) p_i} \qquad (10.2.27)$$

So far, we have done this analysis as a steady state (quasi-steady) flow. Now we must account for the correspondence of the loss of gas to a loss in pressure in the pipe. This is done by means of the ideal gas law again. If the gas remains at constant temperature then the number of moles of gas (per unit axial length) n' is given by

$$p_i V'_{\text{pipe}} = n' R_G T \qquad (10.2.28)$$

where

$$V'_{\text{pipe}} = \pi (\kappa R)^2 \qquad (10.2.29)$$

is the volume of gas per unit length. The *molar* efflux is given by

$$Q'C = -\frac{dn'}{dt} = -\frac{V'_{\text{pipe}}}{R_G T} \frac{dp_i}{dt} \qquad (10.2.30)$$

and C is the molar density of the gas:

$$C = \frac{p_i}{R_G T} \qquad (10.2.31)$$

When Eq. 10.2.27 is used in Eq. 10.2.30 we obtain a differential equation for the pressure inside the pipe as a function of time:

$$\frac{dp_i}{p_i^2 - p_a^2} = \beta \, dt \qquad (10.2.32)$$

where the coefficient β is

$$\beta \equiv \frac{K_D}{\mu(\kappa R)^2 \ln \kappa} \tag{10.2.33}$$

Equation 10.2.32 may be integrated (just look it up in a table of integrals) from time $t = 0$, where the interior pressure is labeled $p_i(0)$, to some arbitrary time t, where the interior pressure is $p_i(t)$. The result is found to be

$$2p_a\beta t = \ln\left[\frac{p_i(t) - p_a}{p_i(t) + p_a}\frac{p_i(0) + p_a}{p_i(0) - p_a}\right] \tag{10.2.34}$$

This completes the development of a general model for the rate of loss of pressure due to leakage.

EXAMPLE 10.2.2   *Determination of Permeability from Transient Leakage Data*

A long porous-walled pipe, closed at both ends with impermeable covers, is initially charged with air to an absolute pressure of 10 atm. The exterior of the pipe is at atmospheric pressure. The system is maintained at 25°C. The permeability of the pipe wall is uniform but of an unknown value. The pressure is observed to fall to 5 atm after a time interval of 1 hour. If the outside pipe diameter is 10 cm, and the wall thickness is 5 mm, what is the value of the permeability?

In this example we specify the pressures at two times, and we wish to find K_D, which is in the coefficient β. The right-hand side of Eq. 10.2.34 can be found from the data for this example:

$$p_i(0) = 10 \text{ atm} \quad p_a = 1 \text{ atm} \quad p_i(1 \text{ h}) = 5 \text{ atm}$$

Since only ratios of pressures appear, it is not necessary to convert any units here. Thus we find

$$\frac{p_i(t) - p_a}{p_i(t) + p_a} = \frac{5-1}{5+1} = 0.667$$
$$\frac{p_i(0) + p_a}{p_i(0) - p_a} = \frac{10+1}{10-1} = 1.22 \tag{10.2.35}$$

and so

$$\beta = \frac{\ln(0.667 \times 1.22)}{2(1 \times 1.013 \times 10^5)3600} = -2.8 \times 10^{-10} \text{ (m·s/kg = 1/Pa·s)} \tag{10.2.36}$$

Note that we write the atmospheric pressure in SI units here, and so we will use SI units in the terms that appear in β (Eq. 10.2.33).

The viscosity of air at 25°C is found from Fig. 3.4.3 to be $\mu = 1.8 \times 10^{-5}$ Pa·s, R is given as 0.05 m, and the wall thickness of 5 mm leads to $\kappa = 0.9$. Then

$$K_D = \beta\mu(\kappa R)^2 \ln \kappa = -(2.8 \times 10^{-10})(1.8 \times 10^{-5})(0.045)^2 \times \ln 0.9 = 1.1 \times 10^{-18} \text{ m}^2 \tag{10.2.37}$$

Since permeability is a material property not often cited for common materials, we have no way to assess this result. While the absolute value found is a small number, it is not a *dimensionless* number, and so it is not really meaningful to speak of a dimensioned number as "small." Can we somehow evaluate this permeability result? One approach is to put our model of flow through a packed bed of spheres in the form of Darcy's law. If we do so, using Eq. 10.1.12, we find

$$K_D = \frac{D_p^2 \varepsilon^3}{150(1 - \varepsilon)^2} \qquad \text{for a bed of packed spheres} \qquad (10.2.38)$$

Now let's imagine that we make a very dense but porous ceramic sheet by sintering (melting) a fine powdered glass with a particle size (diameter) of 10 μm in such a way that the bed is consolidated (rigid) but still open. The result is as if the particles had melted and solidified at their contact points while otherwise maintaining their spherical shape and original size. If the void fraction is found in an independent experiment to be $\varepsilon = 0.15$, we have a permeability of

$$K_D = \frac{(10 \times 10^{-6})^2 (0.15)^3}{150(0.85)^2} = 3.1 \times 10^{-15} \, \text{m}^2 \qquad (10.2.39)$$

Intuitively, we would expect that a barrier made by sintering such a fine powder would correspond to a material with a very low permeability. We see that the material of our porous pipe example has a permeability (Eq. 10-2.37) lower than that of this sintered ceramic sheet by more than three orders of magnitude. We conclude that in the usual meaning of the term, this pipe is made of a very impermeable material. Since, however, it obviously leaks air, it is clearly permeable *with respect to air.*

Given the permeability found in Eq. 10.2.37, it would be interesting to discover how permeable this pipe is to a liquid, say water. Given the incompressibility of water, however, we realize that the model for pressure loss with time—Eq. 10.2.34—is not valid. Hence our first task would be the derivation of a model for $p(t)$ appropriate to water, followed by calculation of the leakage rate from the pipe if it were pressurized to 10 atm with water inside. We leave this as a homework exercise (Problem 10.11). It is, however, easy to modify our analysis so that it applies to a steady gas leakage. This is accomplished in the next example.

EXAMPLE 10.2.3   *Determination of Permeability from Steady Leakage Data*

A tube fabricated from sintered metal particles has an axial length of $L = 14$ cm, an outer diameter of 7 cm, and a wall thickness of 0.5 cm. The porosity (void fraction) of the wall is 0.34. One end of the tube is connected to a high pressure air source. The other end is impermeable. Air leaks steadily across the tube wall to the ambient, which is at atmospheric pressure. Assume that the temperature is 300 K. Data are available [Ramachandran et al., *Exp. Fluids,* **18,** 119 (1994)] for the mass flowrate of air across the wall of the tube as a function of the pressure difference across the tube wall. The results are:

Δp (psi)	5	10	19
w (lb$_\text{m}$/s)	1.1	2.2	4.3

Find the Darcy's law permeability K_D of the wall material, in units of meters squared.

We begin by converting Eq. 10.2.27 to a mass flowrate. Since the volumetric flow that appears in that equation is at the inner surface of the pipe, we must use the mass density ρ_i evaluated at the inside pressure p_i. (Remember! We are dealing here with a *compressible* gas.) The result may be written in the form

$$w = -\frac{2\pi K_D L \rho_i}{\mu \ln \kappa} \frac{(p_i - p_a)(p_i + p_a)}{2p_i} = -\frac{2\pi K_D L}{\mu \ln \kappa} \frac{\rho_i}{p_i} \Delta p \bar{p} \qquad (10.2.40)$$

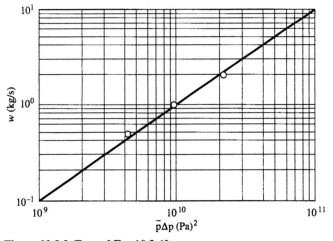

Figure 10.2.3 Test of Eq. 10.2.40.

where Δp is the pressure difference across the tube wall, and $\bar{p}$, the average of the internal and ambient pressures, is an *absolute* pressure—not a gage pressure. For an ideal gas, we may write

$$\frac{\rho_i}{p_i} = \frac{M_w}{R_G T} \tag{10.2.41}$$

According to this model, then, if Darcy's law holds, we should find that when plotted as w versus $\Delta p \bar{p}$ the data yield a linear function, and the slope will be proportional to K_D. Figure 10.2.3 shows the data above, converted to SI units, plotted as a test of Eq. 10.2.40: the line has the form

$$w \text{ kg/s} = 10^{-10} \bar{p} \, \Delta p \tag{10.2.42}$$

with the pressures in pascals, from which the permeability is found as

$$K_D = -10^{-10} \left(\frac{\mu \ln \kappa}{2\pi L} \frac{p_i}{\rho_i} \right) (\text{m}^2) \tag{10.2.43}$$

For an ideal gas at constant temperature, the pressure-to-density ratio is constant. At ambient pressure, assuming $T = 300$ K and one atmosphere pressure, this ratio is

$$\frac{p_a}{\rho_a} = \frac{p_i}{\rho_i} = \frac{1.013 \times 10^{-5} \, \text{Pa}}{1.2 \, \text{kg/m}^3} = 8.44 \times 10^4 \, \text{m}^2/\text{s}^2 \tag{10.2.44}$$

Using $\mu = 1.85 \times 10^{-5}$ Pa·s, $L = 0.14$ m, and $\kappa = 3/3.5 = 0.86$, we find

$$K_D = -10^{-10} \left[\frac{(1.85 \times 10^{-5}) \ln 0.86}{2\pi \times 0.14} 8.44 \times 10^4 \right] = 2.7 \times 10^{-11} \, \text{m}^2 \tag{10.2.45}$$

10.3 THE DYNAMICS OF PRINTING ON A POROUS SURFACE

A key event common to many printing processes is the transfer or "wicking" of ink into the porous substrate—paper in most cases. Image sharpness depends on the rate of wicking of the ink into the substrate, relative to the rates of subsequent phenomena such as "setting" of the ink by drying or radiation curing, printing of the next image on the other side of the paper (as in the case of paper currency), and contact of the printed surface with downstream surfaces in the process line. In this section we will develop a mathematical model of wicking.

Wicking is a complex phenomenon that depends on the *structure* of the porous medium through which fluid moves. By "structure" we mean factors such as porosity, the size and orientation of the fibers or particles that make up the consolidated medium, and possibly the surface chemistry of the structural material itself. The simplest model of wicking that mimics some of the features of a complex medium is based on flow into a long straight capillary, as suggested in Fig. 10.3.1. We can develop a simple model of wicking into a porous medium based on a single capillary model.

The driving force for wicking is the capillary pressure difference p_c which, for a small-diameter capillary, is well approximated (using the Young–Laplace equation—Eq. 2.2.7) by

$$p_c = \frac{2\sigma}{r_c}\cos\theta_d \tag{10.3.1}$$

where σ is the surface tension of the liquid. In this equation we allow for a *dynamic* contact angle θ_d different from the equilibrium contact angle θ_c. It is known that the dynamic contact angle depends on the rate of rise of liquid into the capillary.

Because the flow is opposed by gravity, the net pressure driving force is given by

$$\Delta p = \frac{2\sigma}{r_c}\cos\theta_d - \rho g x \tag{10.3.2}$$

where x is the height of rise measured from the surface of the reservoir.

If we assume that the liquid is Newtonian and that the flow is very slow (i.e., at low Reynolds number), we may write the average velocity of liquid rising in the capillary from Poiseuille's law (see Section 3.2) as

$$v = \frac{dx}{dt} = \frac{r_c^2 \Delta p}{8\mu x} \tag{10.3.3}$$

Since the velocity v is just dx/dt, we are led to a differential equation in the form

$$\frac{dx}{dt} = \frac{\sigma r_c \cos\theta_d}{4\mu x} - \frac{\rho g r_c^2}{8\mu} \tag{10.3.4}$$

We now introduce two simplifications. First, we replace the dynamic contact angle by the *equilibrium* contact angle and set that angle equal to a constant. Next, we assume

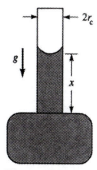

Figure 10.3.1 Wicking into a single capillary.

that the liquid completely wets the inside of the capillary. This is equivalent to the assumption that the equilibrium contact angle is $\theta = 0$. With these approximations, we may solve the dynamic equation for capillary rise (Eq. 10.3.4) *analytically* and find the solution in the form

$$T = -H^2 \left[\frac{X}{H} + \ln\left(1 - \frac{X}{H}\right) \right] \tag{10.3.5}$$

where the following parameters and variables have been defined:

$$T = \frac{t}{\tau} \qquad \tau = \frac{\mu r_c}{4\sigma}$$

$$X = \frac{4x}{r_c} \qquad H = \frac{4h}{r_c} \tag{10.3.6}$$

$$h = \frac{2\sigma}{\rho g r_c}$$

We note that T, X, and H are dimensionless. τ is a characteristic time scale for wicking, and h is the equilibrium capillary rise. H is just the inverse of a Bond number (Eq. 2.1.10) except for a factor of 2. Figure 10.3.2 shows the $X(T)$ behavior according to Eq. 10.3.5. Note that Eq. 10.3.5 implies that X/H is a function of T/H^2.

If gravity is negligible, as is often the case, a simpler analytical model can be derived from Eq. 10.3.4 by dropping the term with g, with the result that

$$\frac{x}{r_c} = \left(\frac{\sigma}{2\mu r_c} t\right)^{1/2} \tag{10.3.7}$$

or, in dimensionless terms,

$$X = (2T)^{1/2} \tag{10.3.8}$$

Equation 10.3.8 has the property that X increases *continuously* with time. If the capillary is in contact with a *finite* volume of fluid, such as a drop in contact with the end of the capillary, X stops increasing because the liquid supply is eventually depleted. Hence Eq. 10.3.8 holds until the drop has disappeared into the capillary.

Equation 10.3.5 has been tested against data on the rate of capillary rise into single capillaries. The data reveal a number of discrepancies relative to this simple model. One problem is that the equilibrium contact angle is often different from zero, and the

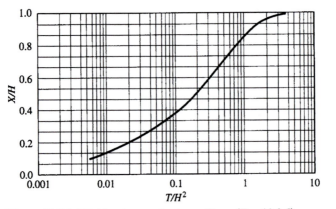

Figure 10.3.2 Wicking into a single capillary (Eq. 10.3.5).

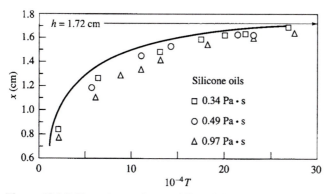

Figure 10.3.3 Experimental test of Eq. 10.3.5: capillary radius $R = 0.025$ cm; $\sigma = 20.5$ dyn/cm. Joos et al., *J. Colloid Interface Sci.*, **136**, 189 (1990).

dynamic contact angle is an unknown function of the rate of rise itself. Another is that the results depend on whether the capillary has been prewet with the liquid or is dry prior to the intrusion of the liquid into the capillary.

Figure 10.3.3 shows the experimental data of Joos *et al.*, who used silicone oils of varying viscosities to test the prediction of Eq. 10.3.5. The data plotted lie consistently below the curve, largely because the dynamic contact angle chosen by these investigators is much larger than the assumed value (we assumed $\theta_d = 0$), which lowered the capillary pressure driving the flow. Nevertheless, Eq. 10.3.5 provides a model that is useful for quick estimation of the time required to achieve wicking into a capillary. Now we can return to the development of a model of printing.

10.3.1   Wicking of a Drop into a Porous Surface

We can now apply our capillary model to the problem of determining the rate of wicking of a drop, initially sitting on a porous surface, into the porous medium below the surface (Fig. 10.3.4). Our goal is to find the volume flowrate of liquid across the surface, and in particular we want to calculate the time required for the drop to disappear across that surface. We assume that if the drop is small enough, gravity will not affect the wicking process. As a first approximation, we regard the contact area to be independent of time, to have radius R_o, and to be open to a bundle of capillaries of radius r_c. The surface is characterized by a fractional open area given by

$$\varepsilon = \frac{N_c \pi r_c^2}{\pi R_o^2} \tag{10.3.9}$$

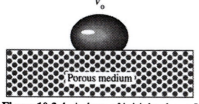

Figure 10.3.4 A drop of initial volume V_o sits on a porous substrate.

where N_c is the number of capillaries that open into the "footprint" of the drop. For each individual capillary, the velocity of flow is (from Eq. 10.3.7)

$$v = \frac{dx}{dt} = \left(\frac{\sigma r_c}{8\mu t}\right)^{1/2} \tag{10.3.10}$$

which corresponds to a volume flowrate Q of

$$\pi R_o^2 v \varepsilon = Q = -\frac{dV}{dt} \tag{10.3.11}$$

where $V(t)$ is the volume of the drop, above the porous surface, at any time after the start of wicking. We obtain a differential equation for $V(t)$ by combining Eqs. 10.3.10 and 10.3.11. The solution is easily found in the form

$$V(t) = V_o - \pi R_o^2 \varepsilon \left(\frac{\sigma r_c}{2\mu} t\right)^{1/2} \tag{10.3.12}$$

where V_o is the initial volume of the drop.

The time for the wicking process to be complete follows upon setting $V(t) = 0$ in Eq. 10.3.12. This assumes that the entire drop is wicked into the porous medium. We find, after some algebraic rearrangement, that a nondimensional wicking time may be written in the form

$$\frac{\sigma t_w}{\mu L} = \left(\frac{V_o}{\pi R_o^2 \varepsilon}\right)^2 \frac{2}{r_c L} \tag{10.3.13}$$

where t_w is the actual "wicking time." A length scale L appears in this nondimensionalization, but we do not define L at this point. We do not need to define it because although L appears in Eq. 10.3.13, it divides out of both sides of the equation.

According to this model, the wicking time should depend on the initial drop size, through the terms V_o and R_o. The wick properties enter through the porous medium parameters r_c and ε, neither of which is known *a priori*. The model predicts that when the wicking time is nondimensionalized as

$$T_w = \frac{\sigma t_w}{\mu L} \tag{10.3.14}$$

we should find T_w to vary as V_o^2/R_o^4 for a given porous medium. Since V_o and R_o are likely to be related to the same length scale L, we see, from Eq. 10.3.13, that T_w varies as L, when L is defined as some characteristic length scale such as $V_o^{1/3}$.

Data on the wicking times of drops varying in initial size, viscosity, and surface tension are available to provide a test of Eq. 10.3.13. Since the parameters that characterize the wicking material are not known, we can test the proposition only to a limited extent. In what follows, L is defined specifically as the "equivalent" drop radius, calculated from the drop volume V_o as

$$L = \left(\frac{3V_o}{4\pi}\right)^{1/3} \tag{10.3.15}$$

The initial drop footprint radius R_o depends on the drop volume V_o or, equivalently, on L. In a separate set of experiments, the relationship was observed to be approximately linear, and within 10% was given by

$$R_o = L \tag{10.3.16}$$

If we adopt this approximation, we may write Eq. 10.3.13 in the form

$$\frac{\sigma t_w}{\mu V_o^{1/3}} = \left(\frac{1.37}{r_c \varepsilon^2}\right) V_o^{1/3} \tag{10.3.17}$$

Note that the term in parentheses contains all the unknown information about the structure of the porous medium, represented as a simple bed of parallel capillaries having a single uniform radius.

Figure 10.3.5 shows unpublished data of Chien and Middleman indicating that the model is in good agreement with the observations. The data on t_w are plotted against V_o, in accordance with Eq. 10.3.17. These data were obtained using a single medium, a porous ultra–high molecular weight polyethylene filter available in 1/8 inch thick sheets, as the model for the wick or paper. Most of the data are for drops of silicone oils of various initial volumes. Silicone oils are available in a range of viscosities, with other physical properties such as surface tension and contact angle essentially independent of viscosity. A substantial set of data for another liquid, castor oil, is also shown. For each liquid the viscosity is fixed, and the only variable is the drop size.

The dependence of the wicking time on surface tension, viscosity, and drop size is approximately accounted for by the theory when the data are plotted in this format. When, as in this case, a single porous medium is studied (so that ε and r_c are fixed), the simple model outlined here predicts that all the data for a particular liquid should fall on a single curve. Considering the range of viscosities (8–34 poise) and surface tensions (19–58 dyn/cm), each set of data may be regarded as clustering about a single line. The predicted slope (from Eq. 10.3.17) of $1:3$ is not observed exactly, but the observed values are close to that expected from our simple model. The observation that the data for the two types of liquid separate into two "lines" suggests the failure of the assumption that the wicking liquid completely wets the porous material (i.e., that the contact angle $\theta_d = 0$). Contact angles were not measured for these liquids on this material.

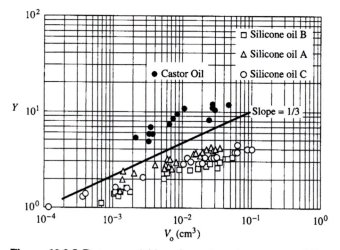

Figure 10.3.5 Data on wicking time plotted as a test of Eq. 10.3.17: Y stands for the left-hand side of Eq. 10.3.17.

10.4 AIRFLOW THROUGH SOIL: A PROBLEM IN ENVIRONMENTAL DECONTAMINATION

In a number of sites throughout the world, inadvertent chemical spills have led to contamination of the soil. A variety of techniques have been proposed, and are in use, for remediation of a contaminated site. In this section we consider some of the issues associated with the development of a model of the rate at which a site may be returned to an acceptable level of cleanliness. The discussion will lean on several concepts introduced in this and earlier chapters.

We will take a case of a specific method of soil remediation called "vacuum venting." Figure 10.4.1 defines some of the characteristics of this technique. A hole is drilled into the soil and filled with a pipe, which is perforated over some axial length at one end, as shown. This creates a well. A pump above ground pulls a vacuum and sucks air through the soil, into the pipe, and thence out of the ground. If a volatile organic compound (VOC) enters the soil, the contaminant is removed with the air. A process unit downstream of the pump would then further treat the removed VOC so that it is not simply vented into the environment. This downstream unit (not shown) could be a gas absorber specific to the contaminant, an incinerator, or any of several other devices.

The fluid dynamics problem of interest to us here is the following:

What factors determine the spatial extent of the region, surrounding the well, from and through which the contaminated air is removed? If this region does not extend very far from the well, it would be necessary to have many such wells in the ground. If the opposite is true, a smaller number of wells would suffice.

By solving a well-defined fluid dynamics problem, we can provide some useful information to the design engineers responsible for site remediation. In this section we consider only the fluid dynamics issues. In the second volume of this text, *Introduction to Mass and Heat Transfer,* we consider a related problem and examine more carefully the factors that control the removal of the VOC from the gas phase of the soil bed, as well as from the solid material upon and within which the VOC might be absorbed.

As usual, we will present a set of assumptions that define the problem we will solve. Because our goal, at this stage, is to introduce a conceptual approach, we will simplify the geometry and the physics considerably. Once the general approach has been outlined, the obvious complications (i.e., the realities of the physics) can be discussed and dealt with.

Our first assumption is that the soil may be regarded as a porous medium whose properties are spatially and directionally uniform. We would refer to such a soil as homogeneous and isotropic.[2] We take the airflow through the soil to obey Darcy's law, Eq. 10.2.1, but written in a vector format:

$$\mathbf{U}_o = -\frac{K_D}{\mu}\nabla P \tag{10.4.1}$$

The continuity equation in the porous medium takes the form

$$\nabla \cdot \mathbf{U}_o = 0 \tag{10.4.2}$$

if we assume that the fluid (air) is incompressible[3] (see Problem 10.14). It follows then

---

[2] And nonexistent! Real soils are far from homogeneous, since they usually include rocks, layers of clay, and other geological features.

[3] This will be a good approximation for cases characterized by a pressure drop through the soil that is small relative to atmospheric pressure. Problem 10.15 deals with the compressible fluid case.

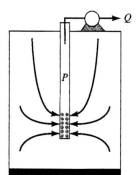

Figure 10.4.1 A vacuum venting well.

that the pressure field must satisfy

$$\nabla^2 P = 0 \tag{10.4.3}$$

To introduce some physical concepts that will be useful to us, we simplify the geometry of the problem to the point of allowing an analytical solution to the pressure field. Then the flow field follows from Eq. 10.4.1. Our goals are to be able to describe certain features of the flow field that are important and to explain how these features depend on the parameters of the problem.

We consider the following case. A long cylindrical pipe, of radius R, is buried in a porous material whose permeability K_D is uniform spatially. The pipe surface is perforated along its entire length, and a reduced pressure, relative to atmospheric pressure, is maintained along the perforated boundary. This suction pressure draws air through the porous medium into the pipe. Figure 10.4.2 defines the geometry.

If we assume symmetry about the pipe axis, the pressure field is a function only of the radial and axial coordinates. Hence we must find the solution to

$$\nabla^2 P = 0 = \frac{1}{r}\frac{\partial}{\partial r}\left(r\frac{\partial P}{\partial r}\right) + \frac{\partial^2 P}{\partial z^2} \tag{10.4.4}$$

One boundary condition on pressure reflects the assumption that there is a uniform vacuum pressure within the pipe, so that along the surface of the pipe we require

$$P = -P_v \quad \text{at} \quad r = R, 0 < z < H_o \tag{10.4.5}$$

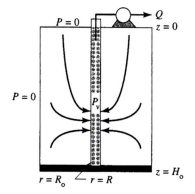

Figure 10.4.2 Simplified model of a vacuum well.

We expect that at some large distance $r = R_o$ from the pipe surface $r = R$, the influence of the vacuum is no longer felt. If we express this by taking the pressure at this outer boundary to be identical to the atmospheric pressure, the second boundary condition on r is

$$P = 0 \quad \text{for} \quad r = R_o, 0 < z < H_o \quad (10.4.6)$$

Along the lower boundary $z = H_o$ we will assume that there is some impermeable medium. (This could be the water table, or a rock formation.) This assumption corresponds to a boundary condition that takes the form[4]

$$\frac{\partial P}{\partial z} = 0 \quad \text{for} \quad z = H_o, R < r < R_o \quad (10.4.7)$$

Along the upper boundary we will take the pressure to be ambient, or

$$P = 0 \quad \text{for} \quad z = 0, R < r < R_o \quad (10.4.8)$$

Although the partial differential equation and the boundary conditions are linear, this is not a simple boundary value problem to solve. Fortunately, someone else has already solved it for us (though not in the context of this application), and we can write down the solution in the form of an infinite series of modified Bessel functions:

$$-\frac{P}{P_v} = \sum_{n=1,3,5,\ldots}^{\infty} \frac{4}{n\pi} \frac{I_0\left(\frac{n\pi r}{2H_o}\right) K_0\left(\frac{n\pi R_o}{2H_o}\right) - K_0\left(\frac{n\pi r}{2H_o}\right) I_0\left(\frac{n\pi R_o}{2H_o}\right)}{I_0\left(\frac{n\pi R}{2H_o}\right) K_0\left(\frac{n\pi R_o}{2H_o}\right) - K_0\left(\frac{n\pi R}{2H_o}\right) I_0\left(\frac{n\pi R_o}{2H_o}\right)} \sin\left(\frac{n\pi z}{2H_o}\right) \quad (10.4.9)$$

The modified Bessel functions I_0 and K_0 may be found in a good set of math tables, such as that of Abramovitz and Stegun. Note that the summation is over the odd integers $n = 1,3,5,\ldots$.

To compute the characteristics of the flow field, we would go back to Eq. 10.4.1 (Darcy's law) and calculate the components of the gradient vector:

$$\nabla P = \left(\frac{\partial P}{\partial r}, \frac{\partial P}{\partial z}\right) \quad (10.4.10)$$

The components of the superficial velocity vector would then be found from

$$\mathbf{U}_o = \left(\frac{-K_D}{\mu} \frac{\partial P}{\partial r}, \frac{-K_D}{\mu} \frac{\partial P}{\partial z}\right) \quad (10.4.11)$$

The volumetric flowrate across the surface $r = R$ of the pipe would then follow from the integral of the velocity normal to the pipe surface, over the vertical extent of the pipe.

$$Q = 2\pi R \left(\frac{-K_D}{\mu}\right) \int_0^{H_o} \left[\frac{\partial P}{\partial r}\right]_{r=R} dz \quad (10.4.12)$$

Although we have an analytical solution (Eq. 10.4.9), it is not a trivial matter to compute numbers with this model because of the need to manipulate these Bessel functions. One would probably want to write a computer program to do the tedious number crunching. Obviously the program would have to include "tables" for evaluating the various Bessel functions.

Figure 10.4.3 is a graphical representation of the pressure field for a specific set of geometrical parameters. From the pressure contours, we can infer that the region in

---

[4] Since there is no flow, the pressure gradient must vanish.

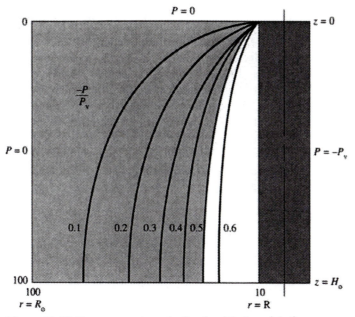

Figure 10.4.3 Pressure contours in the simplified model of a vacuum well. Outside the clear region surrounding the pipe, the pressure is less than half the vacuum pressure P_V.

which the soil is well ventilated is confined to within a few radii of the well. This implies that if a volatile contaminant were trapped in the surrounding soil, the rate of removal of the contaminant would depend on its molecular diffusion into the well-ventilated region (the clear region of the figure), after which it would be rapidly swept out of the soil, into the well.

SUMMARY

In one of the most complex flow situations we face, the fluid moves through a volume occupied in part by discrete particles such as spheres or fibers. The solid boundaries of such flows are geometrically convoluted to the point that no simple mathematical form of the boundary conditions can be expressed. Further, the "architecture" of the porous medium usually varies spatially, and some kind of averaging must be done. One approach to modeling flow through a column packed with spheres (or more generally, with odd-shaped but identical particles) is illustrated in this chapter because it leads to a useful model. The complex geometry is replaced by an "equivalent" set of capillaries, and arguments are presented for relating the actual characteristics of the packed bed (void fraction, surface-to-area ratio) to an equivalent capillary diameter. Next we use the earlier results for pipe flow friction factors in laminar and turbulent flow to predict the pressure drop–flowrate relationship for the porous medium. Surprisingly, this approach works very well and requires only minor adjustments to the coefficients that appear in the model. The final result is the Ergun equation (Eq. 10.1.15).

When the porous medium cannot be thought of in terms of *discrete* particles, it is no longer possible to relate the characteristics of the "equivalent capillary" to the (nonexistent) particulate and packing characteristics. For such "consolidated" materials we usually characterize the relationship of pressure to flowrate through the use of an empirical coefficient. Darcy's law (Eq. 10.2.1) is the most common example of this approach. The *permeability coefficient* K_D then contains all the information about the "architecture" of the consolidated porous medium. Section 10.2 ends with the presentation of several examples of the application of these modeling concepts.

Sections 10.3 and 10.4 present the development of models for two very complex fluid dynamic problems. One occurs on a microscopic scale, the penetration or "wicking" of ink into paper, driven by surface tension, and the other is a "macro scale" problem of predicting the flow of air through soil in the vacuum venting method of decontamination. In each case a particular viewpoint of the structure of the porous medium must be adopted (the equivalent capillary approach, or the application of Darcy's law), after which the information presented in this and earlier chapters is brought to bear on the development of a useful model.

PROBLEMS

10.1 The data tabulated below were obtained for airflow through a Dacron filter with 11 μm diameter fibers. The function Φ is the one defined in Eq. 10.1.16. Compare the data to the models of Blake–Kozeny and Langmuir.

α	L (cm)	$\Omega = 4\pi(1 - \alpha)\Phi$
0.0086	2.22	3.8
0.0168	1.73	4.3
0.0434	0.67	6.2
0.0823	0.36	8.5
0.151	0.36	11.7
0.299	0.18	14.3

10.2 A filter with a cross-sectional area of 1 m^2 and a thickness of $L = 2$ cm is made of the material described in Problem 10.1. Plot pressure drop, in pounds per square inch, as a function of the flowrate of air for flowrates in the range 0.05–1.0 m^3/s at standard temperature and pressure.

10.3 A reservoir holds water at 25°C. The water flows by gravity through a sand filter contained within the exit pipe. For the system sketched in Fig. P10.3, find the volume flowrate of water. The sand grains are uniform, with a diameter of 200 μm. The void fraction is $\varepsilon = 0.3$. The sand fills the 20 ft length of 1 ft diameter pipe. There is no sand in the reservoir itself.

10.4 Find the volume flowrate out of the vertical pipe in the system shown in Fig. P10.4. The pipe is packed with spherical beads of diameter 1/32 inch. The void fraction is 0.30. How would your answer change if the reservoir also were packed with the same beads?

10.5 A glass tube of inside diameter 2 cm is packed with glass beads for an axial length $L = 20$ cm. The beads have a uniform diameter of 1 mm, and the void fraction of the packing is $\varepsilon = 0.35$. Draw a curve of the anticipated flowrate–pressure drop relationship for airflow at 25°C. Do this for the pressure drop range of 0.01–1.2

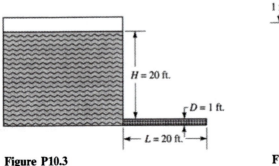

Figure P10.3

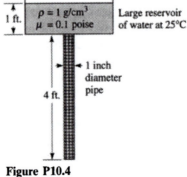

Figure P10.4

atm. The downstream (outlet) end of the tube is at atmospheric pressure. Assume ρ = constant, and justify the assumption.

10.6 A glass tube of inside diameter 2 cm is packed with glass beads for an axial length L = 20 cm. The beads have an unknown but uniform diameter, and the void fraction of the packing is ε = 0.35. At a pressure drop of 10 psi the flowrate of air is measured as 4 standard liters per minute (slm). Estimate the bead diameter.

10.7 Repeat Problem 10.6, but take the measured flowrate of air to be 40 slm.

10.8 A glass tube of inside diameter 2 cm is packed with sand for an axial length L = 20 cm. The sand particles are quite uniform in size, and they have an average diameter of 1 mm. At a pressure drop of 10 psi, the flowrate of air is measured as 4 slm. Estimate the void fraction of the bed of sand.

10.9 A porous sintered metal disk has a diameter of 1 cm and a thickness of 2 mm. When an oil that wets the metal is placed dropwise on the face of the disk, the oil wicks into the disk until the disk is saturated with oil. Measurements indicate that the disk has a capacity for 50 mg of oil. The oil has a density of 850 kg/m³. Airflow tests on the dry disk indicate that the disk will transmit 2×10^{-3} slm with a pressure differential of 10^4 N/m², and with the downstream (low pressure) side at atmospheric pressure. If the disk is regarded as a bed packed with spherical particles, what particle diameter corresponds to this performance? What is the superficial velocity upstream of the disk? What is the Reynolds number (Eq. 10.1.13) under this test condition?

Suppose the same disk were placed into an airflow system such that the downstream side were at an absolute pressure of 100 atm. What flowrate, in standard liters per minute, would correspond to a pressure differential of 10^4 N/m²? What is the superficial velocity upstream of the disk? What is the Reynolds number (Eq. 10.1.13) under this test condition?

10.10 Repeat the model development of Section 10.2.2, but for the case of a gas surrounded by a container that is a spherical shell.

10.11 Repeat the model development of Example 10.2.3, but for the case of a pipe filled with water and then pressurized to 10 atm by a piston at one end. The position of the piston is controlled to maintain the pressure constant. In par-

ticular, derive a model that yields the leakage rate in units of cubic centimeters of water per meter of axial pipe length, if the pipe material has the permeability given by Eq. 10.2.37.

10.12 A drop of a viscous oil (μ = 1 Pa·s, σ = 0.03 N/m, ρ = 800 kg/m³) is brought to one end of an open capillary tube of diameter 0.1 mm. How long will the drop take to rise through the vertical capillary to within 90% of its equilibrium rise? Give an answer for three assumed values of contact angle: θ_d = 0, 30, and 60°. What initial volume of drop is required to support this rise? Assume that the drop remains spherical at the tip of the capillary. Assume that the pressure within the drop that arises from surface tension is negligible, but evaluate the validity of that assumption.

10.13 The fluid described in Problem 10.12 is placed upon a porous substrate. The initial drop volume is 100 μL. The porous material is a sintered ceramic made from particles with a radius of 200 μm. The void volume of the porous medium is ε = 0.25. Estimate the time required for the drop to "disappear" from the exposed plane of the substrate.

10.14 Darcy's law may be written in the vector form

$$\mathbf{u} = -\frac{K_D}{\mu} \nabla p \qquad \textbf{(P10.14.1)}$$

where it is assumed that the medium is isotropic (i.e., the permeability is independent of direction in the medium). Show that for an incompressible fluid, the pressure field satisfies Laplace's equation:

$$\nabla^2 p = 0 \qquad \textbf{(P10.14.2)}$$

10.15 Repeat Problem 10.14, but let the fluid be compressible. In particular, assume that the fluid is an ideal gas and the flow isothermal. Show that now it is p^2, not p, that satisfies Laplace's equation.

10.16 The method of "soil venting," described in this problem, is used to remove volatile organic compounds from soil into which accidental leakage has occurred. Your task is to formulate (and, if you've had a good course in numerical analysis, to solve) the equations that define the airflow through this system.

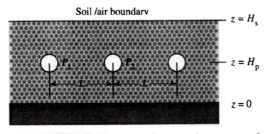

Figure P10.16 Section of soil normal to a set of parallel pipes.

A series of horizontal pipes (P_1, P_2, etc.: Fig. P10.16) is buried at some depth beneath the surface of the ground. The pipes are perforated and air is blown into the pipes from some external aboveground source. Air "leaks" out of the pipes and permeates the porous medium, the soil, that lies between the surface and some boundary (labeled $z = 0$), which we regard as impermeable. (This impermeable layer could be rock or the water table of the region.) We assume that the soil medium can be characterized by a homogeneous and isotropic permeability coefficient, as defined in Eq. P10.14.1. If there is any volatile organic material in the soil, it will evaporate into the airflow and be convected out of the soil and into the atmosphere across the plane $z = H_s$. In this example we ignore the volatile material; we want to develop a model only for the airflow through the soil.

a. Make a qualitative sketch of the streamlines of the airflow.

b. Write the partial differential equation from which the pressure field throughout the soil may be obtained. Include a set of boundary conditions on pressure.

c. What criterion would you use to decide on the spacing L between pipes? Assume that the pipe diameter is small compared to the depth of soil and that there is no significant resistance to airflow across the pipe/soil interface.

10.17 In Example 10.2.1 a value is given for the permeability of a tumor, in the form $K_D/\mu = 4.1 \times 10^{-8}$ (cm²/mmHg·s). Assuming that the fluid is an aqueous solution with a viscosity of 0.002 Pa·s, calculate a value for K_D itself, in units of meters squared.

10.18 Go back to Eq. 10.2.19:

$$\frac{\partial \rho}{\partial t} = -\nabla \cdot \rho \mathbf{u} \qquad \text{(P10.18)}$$

If we do not neglect the time derivative term, what partial differential equation must the pressure field $p(r, t)$ satisfy? Write the boundary conditions and an initial condition on $p(r, t)$. What general equation does $p_i(t)$ satisfy?

10.19 Return to Example 10.2.2, but do the case of pressurized air confined between two square porous slabs of area A and thickness H. The planes are separated by an air gap of thickness $2B \ll \sqrt{A}$. Show that the quasi-steady solution for the air pressure in the confined space is

$$\frac{p_i(t) - p_a}{p_i(t) + p_a} = \frac{p_i(0) - p_a}{p_i(0) + p_a} \exp(-2p_a\beta t) \qquad \text{(P10.19.1)}$$

where

$$\beta = \frac{K_D}{2\mu HB} \qquad \text{(P10.19.2)}$$

10.20 Equation 10.2.34 is indeterminate in the limit of the exterior pressure p_a going to zero. Show that in that case the solution for the air pressure in the confined space is

$$p_i(t) = \left[\frac{1}{p_i(0)} - \beta t \right]^{-1} \qquad \text{(P10.20)}$$

Think about the physics of this case. Does anything "interesting" happen if the absolute pressure at the outer boundary of the container is zero? What is the gas velocity u at the outer boundary?

10.21 Pressurized air is confined between two square porous slabs of area A and thickness H. The initial pressure is $p_i(0)$. The planes are separated by an air gap of thickness $2B \ll \sqrt{A}$. Gas "leaks" through the walls of the system to an ambient that is at an absolute pressure of zero. Show that the quasi-steady solution for the air pressure profile in the porous space is

$$p(r, t) = \frac{(1 - y/H)^{1/2}}{1/p_i(0) + \beta t} \qquad \text{(P10.21.1)}$$

where

$$\beta = \frac{K_D}{2\mu HB} \qquad \text{(P10.21.2)}$$

The high pressure side of the porous slab is at $y = 0$.

10.22 Consistent with our treatment of many other problems in this text, we could examine a model of flow through a vacuum venting well

(Fig. 10.4.1) based on geometrical simplifications that permit us to obtain an analytical solution to Eq. 10.4.4. Consider the following model. The pipe is so long, compared to its diameter, that we may call it infinite in the axial direction. If the entire surface of the pipe is perforated, we may assume that the flow in the surrounding soil is one-dimensional and that P is a function only of the radial position r. Give the one-dimensional solutions $P(r)$ to Eq. 10.4.4 for two cases:

a. $P = -P_v$ at $r = R$
 $P = 0$ at $r = R_{outer} > R$
b. $P = -P_v$ at $r = R$
 $P = 0$ at $r = \infty$

Explain what these solutions mean physically. Argue that neither case **a** nor case **b** is of practical use.

10.23 Equation 10.2.8 may not be a familiar differential equation. Show that the change of variable $W = sP$ converts Eq. 10.2.8 to the "hyperbolic" equation

$$\frac{d^2W}{ds^2} - \beta^2 W = 0 \qquad \textbf{(P10.23.1)}$$

whose general solution is

$$W = a \sinh \beta s + b \cosh \beta s \qquad \textbf{(P10.23.2)}$$

10.24 Derive Eq. 10.3.5.
10.25 In Eq. 10.3.6 we define h and state that it represents the *equilibrium* capillary rise. How do we conclude this?
10.26 The data presented below are from an experimental study of airflow through the porous wall of a cylindrical tube [Ramachandran et al.,

Exp. Fluids, **18,** 119 (1994)]. Are these data consistent with what we might expect from our study of this topic? Is gas compressibility a factor in these data? Is flow in the laminar regime?

The porous tube is made of sintered steel. It is a cylinder with an axial length of 14 cm and an inside diameter of 5 cm. The wall thickness is 1 cm. The porosity of the wall is $\varepsilon = 0.2838$. The circular ends of the tube are impermeable. The temperature is assumed to be 300 K. The external pressure is atmospheric.

ΔP (psi)	13.3	22.5	42.5	58.7	77
w (lb/s)	0.33	0.67	1.72	2.78	4.2

10.27 Calculate the Darcy's law permeability K_D of the tube wall for which the data of Problem 10.26 were obtained. Using Eq. 10.2.38, calculate the diameter of the spheres that would produce a packed bed of the same permeability, assuming the same porosity of that example ($\varepsilon = 0.2838$). Do you think such a calculation produces a meaningful particle size? Explain your answer.
10.28 Look at Eq. 10.2.40. Does it seem remarkable (or even incorrect!) that the mass flowrate does not depend on the size of the tube? If you pressurized two tubes with the same internal pressure, and the tubes were fabricated of the same materials, differing only in radius R, wouldn't you expect that the larger tube would permit the higher mass flow of air? Explain the result shown in Eq. 10.2.40.
10.29 Write the differential equation to which Eq. 10.3.12 is the solution.
10.30 The data shown in Fig. P10.30 are avail-

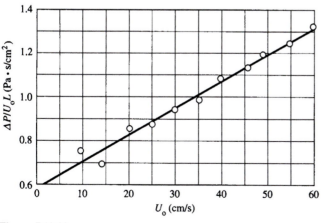

Figure P10.30

able for the flow of air through a bed of silica sand. The length of the bed is $L = 12.8$ cm, and the mean particle size is $D_p = 2.2$ mm. The void fraction is $\varepsilon = 0.366$. Test the ability of Eq. 10.1.15 to fit these data. Then fit an equation of the *form* of Eq. 10.1.15 to the data, but find the best pair of coefficients to replace 150 in the laminar term and 1.75 in the turbulent term. Assume that the downstream pressure is atmospheric and the temperature is 300 K. If the upstream pressure is much greater than atmospheric, it may be necessary to correct for compressibility effects. Examine this point: that is, evaluate the ratio of upstream to downstream pressures.

Chapter 11

Macroscopic Balances

Often we are less interested in the detailed or *microscopic* features of a flow (such as the velocity profile or the streamlines) than we are in overall behavior (such as the pressure drop–flowrate relationship). In this final chapter, therefore, we bring together a number of ideas that will permit us to solve fluid dynamics problems centering around the *macroscopic* behavior of systems. In a sense we focus our attention on the events that occur on the *boundaries* of a system, more than those taking place in its interior.

11.1 THE MACROSCOPIC MASS BALANCE

Figure 11.1.1 shows a system within which, and through which, a flow occurs. The boundaries of the system are fixed—nonmoving. Flow enters the system across an area A_1 at an average velocity U_1 and leaves across an area A_2 at an average velocity U_2. The fluid has densities ρ_1 and ρ_2 at the planes A_1 and A_2. We could allow for more entrances and exits, but a single pair will suffice to make the point.

Mass is a conserved quantity. If M is the total mass of fluid in the system at any time, then mass can change within the boundaries of the system only by virtue of a net inflow or efflux to or from the system. In other words:

$$\text{Time rate of change of mass within the system} = \text{rate of flow of mass in} - \text{rate of flow of mass out}$$

(11.1.1)

or

$$\frac{dM}{dt} = [\rho U A]_{\text{in}} - [\rho U A]_{\text{out}} = \rho_1 U_1 A_1 - \rho_2 U_2 A_2 \qquad \textbf{(11.1.2)}$$

This is such a simple idea that it seems hardly worthy of a mathematical expression. Let's look immediately at a couple of examples of applications of this principle.

EXAMPLE 11.1.1   *The Egyptian Water Clock*

An invention attributed to the ancient Egyptians is the water clock—a vessel shaped to ensure that as water flows from an orifice in the bottom, the height of the water falls *linearly* with time. Because the water level is a *linear* function of time, a uniformly

441

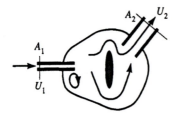

Figure 11.1.1 Flow enters and leaves a fixed system.

scaled ruler would serve as a clock "face." The vessel is shown in Fig. 11.1.2. We measure the radius of the vessel r as a function of the height H above the orifice, and we seek the $r(H)$ relationship that yields the desired linear $H(t)$ result. Water leaves the vessel under the action of the hydrostatic head $H(t)$. We need a model for the efflux velocity U_2 as a function of H. We will later derive this model, called Torricelli's law:

$$U_2 = \sqrt{2gH(t)} \qquad (11.1.3)$$

Torricelli's law

For the moment we accept the validity of this model, without discussing any restrictions on its use.

Now we apply the mass balance given above. Note that there is *no flow* across plane 1. The system volume has *fixed boundaries,* and therefore we do not choose the *moving* surface of the liquid as plane 1. Our mass balance then states that

$$\frac{dM}{dt} = -\rho_2 U_2 A_2 \qquad (11.1.4)$$

The upper surface of the liquid has area πr^2 and moves with velocity $-dH/dt$. Hence the mass flow out of the system at any instant of time is expressible in terms of H as

$$\frac{dM}{dt} = \rho \pi r^2 \frac{dH}{dt} \qquad (11.1.5)$$

Keep in mind that r is the radius of the falling surface at any time, so $r = r(t)$. We are assuming that the liquid has constant density, and we denote that density by ρ from now on, without subscript. Putting these equations together we find

$$\rho \pi r^2 \frac{dH}{dt} = \frac{dM}{dt} = -\rho U_2 A_2 = -\rho \sqrt{2gH(t)}\, A_2 \qquad (11.1.6)$$

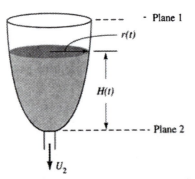

Figure 11.1.2 The Egyptian water clock.

If we design the clock to give a constant rate of change in height of

$$-\frac{dH}{dt} = \dot{H}_\text{o} \tag{11.1.7}$$

then Eq. 11.1.6 provides an expression in the form of the function $r(H)$ for the shape and size of the reservoir:

$$\frac{r}{A_2^{1/2}} = \frac{(2g)^{1/4}}{(\pi \dot{H}_\text{o})^{1/2}} H^{1/4} = \alpha H^{1/4} \tag{11.1.8}$$

Note that the coefficient α is a design parameter. We are free to select it (really, we are selecting $\dot{H}_\text{o}$) to create a fast clock for short-term events or a slow clock for long-term events.

As H goes toward zero, Eq. 11.1.8 indicates that r should approach zero. Obviously we need a nonzero r at H near zero. However, we do not expect Torricelli's law to hold near the bottom of the reservoir, for reasons we will discuss when we derive the law. Hence we take this result as valid as long as the height of the water is several times the diameter of the outlet hole; when the level gets too low, we regard the clock as inaccurate.

Note that since the principle of conservation of mass is so natural, it does not seem to have played a major role in this design problem. Nevertheless, the whole analysis rests on the combination of Eqs. 11.1.3 and 11.1.4.

EXAMPLE 11.1.2   *Pumping Rate Required for a Cleaning Jet*

A tool consisting of a jet of water that issues at high speed from a nozzle can be used in a surface cleaning operation. From an independent experimental program we know that we need a jet efflux velocity of 50 m/s from the nozzle. The nozzle orifice has an area of 10^{-5} m^2. What pumping capacity is required to maintain this jet?

The system we'll analyze lies within the vertical planes 1 and 2 of Fig. 11.1.3. In contrast to Example 11.1.1, this is a *steady state* problem. The mass of water within the system does not change in time. Hence the mass balance states that

$$0 = \rho_1 U_1 A_1 - \rho_2 U_2 A_2 \tag{11.1.9}$$

We assume that the water has a constant density, and we introduce the volume flowrate, to find

$$Q_\text{pump} \equiv U_1 A_1 = U_2 A_2 = 50 \times 10^{-5} \text{ m}^3/\text{s} = 0.5 \text{ L/s} \tag{11.1.10}$$

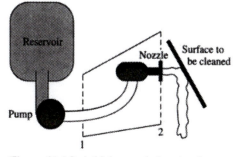

Figure 11.1.3 A high speed cleaning jet.

Again, this is a very simple problem. The mass balance just does not address important design issues such as the pressure required to produce this jet or the force required to hold the nozzle. We must move on to momentum and energy balances in order to deal with these questions. Before doing so we examine one more problem, this time under conditions of *compressible nonisothermal* flow.

EXAMPLE 11.1.3   *Effect of Temperature and Pressure on the Flowrate of a Gas*

A gas flows through several processes, undergoing changes in temperature and pressure. No chemical reactions occur, however. In the system, shown in Fig. 11.1.4, the temperature and pressure are measured at planes 2, 3, and 4. At plane 3 the volumetric flowrate is also measured. We want to know the volumetric flowrate at planes 2 and 4.

We begin by examining the system between planes 2 and 3. For an assumed steady state operation, the mass balance takes the form

$$\rho_3 U_3 A_3 = \rho_3 Q_3 = \rho_2 U_2 A_2 = \rho_2 Q_2 \tag{11.1.11}$$

It follows that

$$Q_2 = \frac{\rho_3}{\rho_2} Q_3 \tag{11.1.12}$$

Note that since the density of a gas is a strong function of temperature and pressure, we do not assume that the fluid has the same density before and after the process being carried out between these two planes. Clearly, then, we need a model for the dependence of density on temperature and pressure. This is supplied by the assumption of the applicability of the ideal gas law, from which it follows that

$$\frac{\rho_3}{\rho_2} = \frac{P_3 \, T_2}{T_3 \, P_2} \tag{11.1.13}$$

Since only the density ratio is required, we do not have to write the ideal gas law out in full, showing the gas constant and molecular weight. The molecular weight does not change because of our assumption that no chemical reactions occur. We now find

$$Q_2 = \frac{P_3 \, T_2}{P_2 \, T_3} Q_3 = \frac{10^5 \, 200}{10^6 \, 298} \times 1 = 0.067 \ \text{m}^3/\text{s} \tag{11.1.14}$$

Note how low the volume flowrate is coming across plane 2. In the process between planes 2 and 3, the gas expands because of the 10-fold drop in pressure, and it expands

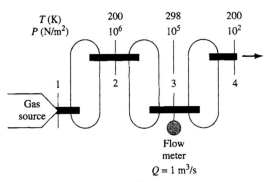

Figure 11.1.4   A gas transfer system.

slightly by virtue of the increase in temperature. Hence the *volume* flowrate at the meter is increased. Of course, the *mass* flowrate is unchanged, since mass is conserved.

By the same procedure we find that at plane 4

$$Q_4 = \frac{P_3}{P_4}\frac{T_4}{T_3}Q_3 = \frac{10^5}{10^2}\frac{200}{298} \times 1 = 670 \text{ m}^3/\text{s} \tag{11.1.15}$$

Primarily because of the letdown in pressure, there is a tremendous increase in the volume flowrate across plane 4 relative to that at plane 3.

In setting up this analysis, we assume that the temperatures and pressures entering and leaving the processes are known. We do not address the question of how these changes are brought about. We take them as given.

In stating volume flowrates in this example, we imply that each is measured at the temperature and pressure of the flowmeter. It is also possible, and in fact common, to state volume flowrates "corrected" to standard conditions, such as 273 K and one atmosphere. The actual volume flowrate would then follow from a knowledge of the actual temperature and pressure at the point of the measurement.

We move on now to a consideration of the momentum balance for a macroscopic system.

11.2 THE MACROSCOPIC MOMENTUM BALANCE

The system for analysis is the same as in Fig. 11.1.1, but we now consider forces acting on the solid boundaries, exerted by the pressure and shear stresses in the fluid at those boundaries. In addition, because momentum is a *vector* quantity, we will consider the *orientation* of the entrance and exit planes. Figure 11.2.1 will aid the analysis.

The momentum of the fluid within the system at any time arises from the local internal velocity distribution $\mathbf{u}(t)$ at every point within the volume. The *density* of momentum at any point and time is just $\rho\mathbf{u}(t)$. Hence the total momentum within the system is

$$\text{total momentum} = \int \rho\mathbf{u} \, d\text{V} \tag{11.2.1}$$

where the integration is over the volume of the system, V. [Note the use of V for volume in Sections 11.2 and 11.3, to distinguish this term from V for velocity.]

Momentum enters and leaves the system by virtue of the flows across the open boundaries. Over a small region of area dA_1 anywhere on the area A_1, the differential *rate* of flow of momentum is

$$d\mathbf{C}_1 = \rho_1 u_1^2 \, d\mathbf{A}_1 \tag{11.2.2}$$

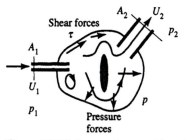

Figure 11.2.1 System for consideration of momentum conservation.

In this expression we find the *rate* of flow of momentum by multiplying the density of momentum normal to an entrance plane ($\rho_1 u_1$ for the 1 plane) by the volume flowrate across that plane ($u_1\, dA_1$). Since momentum is a vector quantity, we must somehow denote the direction of the momentum vector at each entrance plane. This is most easily done by regarding the area as a vector quantity with a magnitude A_1 and a direction given by that of the velocity vector normal to the plane. We simply write this as vector $d\mathbf{A}_1$. Because these flows are what we would call *convection* across the open boundaries, we denote the convective flow of momentum by $d\mathbf{C}$.

Across the *entire* area A_1, then, we find

$$\mathbf{C}_1 = \int \rho_1 u_1^2\, d\mathbf{A}_1 = \rho_1 \langle U_1^2 \rangle\, \mathbf{A}_1 \tag{11.2.3}$$

We have to be careful about the averaging definition here. We define the average of the squared velocity as

$$\langle U_1^2 \rangle \equiv \frac{\int u_1^2\, d\mathbf{A}_1}{\mathbf{A}_1} \tag{11.2.4}$$

This is not the same as the square of the average velocity. In other words,

$$\langle U_1 \rangle \equiv \frac{\int u_1\, d\mathbf{A}_1}{\mathbf{A}_1} \neq \sqrt{\langle U_1^2 \rangle} \tag{11.2.5}$$

The net flow of momentum is then the difference

$$\mathbf{C} = \rho_1 \langle U_1^2 \rangle\, \mathbf{A}_1 - \rho_2 \langle U_2^2 \rangle\, \mathbf{A}_2 \tag{11.2.6}$$

Pressure forces act on the open ends of the system. The net force due to these pressures is simply

$$\mathbf{F_P} = P_1 \mathbf{A}_1 - P_2 \mathbf{A}_2 \tag{11.2.7}$$

Note again that the force is a vector quantity, and the area is written as a vector to provide that information. If the system is in a gravitational field, there is a body force acting on the fluid given by

$$\mathbf{B} = M\mathbf{g} \tag{11.2.8}$$

where M is the total mass of fluid in the system at any time.

Finally, the fluid exerts forces on the interior surfaces of the system. These forces may arise from viscous shear effects as well as from the pressure distribution along these surfaces. Their calculation would require a knowledge of the microscopic flow within the system. We will simply represent them by the symbol $-\mathbf{F}$. (Our sign convention on $\mathbf{F}$ is that $+\mathbf{F}$ is the sum of forces exerted *by the fluid on the boundaries*.)

We now write the principle of conservation of momentum:

rate of change of   net rate of flow of   sum of the forces exerted
momentum of the fluid = momentum across the + *on the fluid by the boundaries*
within the system   surfaces of the system   of the system

+ body force
$$\tag{11.2.9}$$

Thus we write the momentum balance:

$$\frac{\partial}{\partial t} \int \rho \mathbf{u}\, dV = \rho_1 \langle U_1^2 \rangle\, \mathbf{A}_1 - \rho_2 \langle U_2^2 \rangle\, \mathbf{A}_2 + P_1 \mathbf{A}_1 - P_2 \mathbf{A}_2 - \mathbf{F} + M\mathbf{g} \tag{11.2.10}$$

One final comment is necessary with regard to how we write the pressure forces. The force **F**, as we have defined it, includes the force exerted on the solid boundaries by the external absolute pressure of the surrounding atmosphere. The atmospheric pressure terms cancel out on any closed surface except at the entrance and exit surfaces, which are not solid boundaries. Hence the pressures that appear in Eq. 11.2.7 must be taken to be the "gage" pressures (absolute pressure minus atmospheric pressure). If it is necessary to calculate a gas density from the pressure at an entrance or exit section, the absolute pressure must be used for that part of the calculation. We will elaborate on this point shortly.

Let's turn immediately to an example.

EXAMPLE 11.2.1 *Axial Force Exerted on the Walls of a Straight Pipe*

Air flows steadily through a portion of a straight pipe of inside diameter 0.1 m. The efflux velocity is 30 m/s into ambient air at atmospheric pressure. The flow is isothermal, at a temperature of 300 K, and the *absolute* pressure at the left-hand end of the pipe is 7×10^5 N/m^2. What is the frictional force exerted on the inside surface of the pipe? Figure 11.2.2 defines the boundaries of the system. The pressures P_1 and P_2 are the absolute pressures.

Inspection of Eq. 11.2.10 reveals that we are missing three pieces of information required in the solution of this problem: the entrance velocity U_1, and the gas densities at the entrance and exit of the pipe. If we assume that the ideal gas law is valid,[1] the gas density follows from

$$\rho = \frac{PM_w}{R_G T} \tag{11.2.11}$$

Ideal gas

where M_w is the molecular weight of the gas (29 for air). In SI units we use

$$R_G = 8314 \text{ J/kg–mol} \cdot \text{K} \tag{11.2.12}$$

for the gas constant. At the entrance to the system the density is

$$\rho_1 = \frac{7 \times 10^5 \times 29}{8314 \times 300} = 8.1 \text{ kg/m}^3 \tag{11.2.13}$$

(Note that we use absolute pressure for the density calculation.) At the exit, only the absolute pressure has changed, so the density ρ_2 is smaller by a factor of 7.

To find the entrance velocity U_1, we must apply the mass balance. From

$$\rho_1 \langle U_1 \rangle A_1 = \rho_2 \langle U_2 \rangle A_2 \tag{11.2.14}$$

it follows that

$$\langle U_1 \rangle = \frac{\rho_2}{\rho_1} \langle U_2 \rangle = \frac{1.16}{8.1} \times 30 = 4.3 \text{ m/s} \tag{11.2.15}$$

We may simplify Eq. 11.2.10 since the flow is at steady state. If the pipe is horizontal, then there is no component of gravity in the axial direction. Even if the pipe is not

---

[1] You should look up the critical temperature and pressure of air and verify, using a compressibility chart, that the ideal gas assumption is reasonable. See, for example, Reid, Prausnitz, and Poling, *The Properties of Gases and Liquids*, McGraw-Hill, NY, 1987.

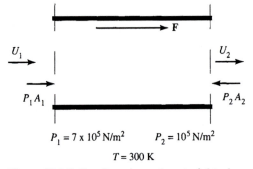

$P_1 = 7 \times 10^5 \, \text{N/m}^2 \qquad P_2 = 10^5 \, \text{N/m}^2$

$T = 300 \, \text{K}$

Figure 11.2.2 Gas flow through a straight pipe.

horizontal, the weight of air is so small that the body force would be negligible. Hence the force component in the axial direction is found from

$$\mathbf{F} = \rho_1 \langle U_1^2 \rangle \mathbf{A}_1 - \rho_2 \langle U_2^2 \rangle \mathbf{A}_2 + P_1 \mathbf{A}_1 - P_2 \mathbf{A}_2 \qquad (11.2.16)$$

We could easily confirm that this is a highly turbulent flow by calculating the Reynolds number. It is known that the velocity profile across the pipe radius is very flat, except very near the pipe wall, in turbulent flow. When that is the case, a good approximation is[2]

$$\langle U_1^2 \rangle = \langle U_1 \rangle^2 \qquad (11.2.17)$$

Finally, we must be careful about the signs of the area vectors $\mathbf{A}$. By our definition, earlier, the area vector is in the direction of the mean velocity vector across that area. In this example, both velocity vectors point in the same direction, so the area vectors have the same sign, which we take arbitrarily as positive toward the right. Thus we find $\mathbf{F}$ from

$$\mathbf{F} = 8.1(4.3)^2 \left[\frac{\pi}{4}(0.1)^2 \right] - 1.16(30)^2 \left[\frac{\pi}{4}(0.1)^2 \right] + [(7 \times 10^5) - 10^5] \left[\frac{\pi}{4}(0.1)^2 \right]$$

$$- (10^5 - 10^5) \left[\frac{\pi}{4}(0.1)^2 \right] \qquad (11.2.18)$$

$$= 1.18 - 8.2 + 4715 - 0 = 4708 \, \text{N}$$

We may offer several comments on this result. First of all, this is a large force. It is equivalent to hanging a mass of approximately 470 kg (divide newtons by $g = 9.8$) on the end of the pipe, in a gravitational field aligned with the pipe axis. (Equivalently, the force corresponds to a weight of about 1000 pounds.) We also see that the force is dominated by the pressure terms, which really correspond to (i.e., arise from) the frictional force on the pipe walls. The acceleration of the fluid makes a modest contribution to the force balance in this case. Notice, also, that the result does not depend on the length of the pipe! This is misleading, because the only effect of the pipe length is on the required pressure drop. Since the pressure drop was specified, then by implication the length of the pipe was fixed, but not stated.

Notice that the atmospheric pressure (10^5 Pa) was subtracted from the absolute pressure in this calculation. Had we neglected to do so, we would still have gotten the correct result, because the areas are identical and the pressure terms are subtractive. This is not the case in general, as we shall see when we look next at a similar problem.

---

[2] See Problem 11.12.

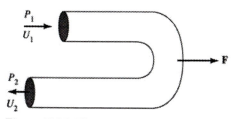

Figure 11.2.3 Flow of air through a U-bend.

EXAMPLE 11.2.2 *Force Exerted on a U-Bend in a Pipe*

Our input data will be identical to those in Example 11.1.1 except for the geometry of the pipe, which will be as shown in Fig. 11.2.3. Since the pressures and flows are assumed to be the same as in Example 11.2.1 (although the total axial length of the pipe could be different), each term in Eq. 11.2.16 has the same magnitude as before. However, the *vector* area A_2 is opposite in sign to that in Example 11.2.1 because the flows across the respective areas are opposite in direction. As a consequence, each term in Eq. 11.2.18 has the same sign, and we find

$$\mathbf{F} = 4724\,\text{N} \tag{11.2.19}$$

The force is slightly greater than that of the straight pipe, and it is positive in the direction shown by the arrow in Fig. 11.2.3.

We can knock off another problem along these lines quickly, and gain some more insight into this class of problem, by looking at the flow of an incompressible fluid.

EXAMPLE 11.2.3 *Flow of Water Through a U-Bend in a Pipe*

Figure 11.2.3 still serves for the geometry of this system. We take the following data:

$$\text{Pipe diameter } D = 0.1\,\text{m} \qquad \langle U_1 \rangle = 20\,\text{m/s}$$
$$P_1 = 2 \times 10^5\,\text{Pa (gage)} \qquad P_2 = 1.6 \times 10^5\,\text{Pa (gage)}$$

The pipe is horizontal.

Since the fluid (water) is incompressible, and the areas for flow are the same, it follows that $\langle U_1 \rangle = \langle U_2 \rangle$. Looking at Eq. 11.2.16, we are tempted to cancel the convective terms, since the densities and areas are also identical at the two ends of the pipe. While the magnitudes of the areas are the same, however, their signs differ because the area vectors have the signs of the velocity vectors, which are opposite in this example. Hence in Eq. 11.2.16 the terms will all add with positive signs. Then the force is calculated from

$$F = 2\left[1000\,\text{kg/m}^3 \times (20\,\text{m/s})^2 \times \frac{\pi}{4}(0.1\,\text{m})^2 \right]$$
$$+ 2 \times 10^5\,\text{Pa} \times \frac{\pi}{4}(0.1\,\text{m})^2 + 1.6 \times 10^5\,\text{Pa} \tag{11.2.20}$$
$$\times \frac{\pi}{4}(0.1\,\text{m})^2 = 6283 + 2827 = 9110\,\text{N}$$

Note that the pressures here are stated as "gage" pressures and so they are used directly in the calculation, since they include the required subtraction of atmospheric pressure.

In this problem we see that the convective term—the change in the rate of flow of momentum due to the velocity of the water—dominates the force on the bend. The pressure term is smaller. This distinction from earlier examples, which involved *gas* flow, arises in part because a liquid is much higher in density than a gas.

EXAMPLE 11.2.4   *Force on a Converging Nozzle*

We choose an example now that can clarify some issues related to the way we handle pressure (gage vs. absolute) acting on the boundaries of a system. In addition, we indicate that we often have options for defining the system over which the macroscopic balances are written. Figure 11.2.4 shows a converging nozzle, similar to a nozzle on a garden hose. We want to find the force acting on the nozzle, arising from the flow through the nozzle. As a control volume for the analysis, we isolate the body of the nozzle, *including the fluid within the boundaries of the nozzle.* This is shown by the **heavy** line in Fig. 11.2.4.

With this choice of control volume, the moving fluid exerts pressure forces only at sections 1 and 2. The shear forces *within* the nozzle do not act on the chosen boundaries of the control volume, because the *interior* surface of the nozzle is not a boundary of the control volume. It is the *exterior* surface of the nozzle that is the boundary. We now write the momentum equation, with "down" as the positive direction for forces. We just rearrange Eq. 11.2.16 to the form

$$0 = \rho_1 \langle U_1^2 \rangle A_1 - \rho_2 \langle U_2^2 \rangle A_2 + P_1 A_1 - P_2 A_2 + Mg + \mathbf{F}_{conn} \qquad (11.2.21)$$

where $\mathbf{F}_{conn}$ represents the force exerted *by* the connection upstream of the nozzle, *on* the nozzle. A *plus* sign is used because this is a force exerted *on* the system *by* the surroundings; it is not a force exerted on the nozzle by the fluid within the system. We expect to find that this is a negative (upward-acting) force, but that expectation will be confirmed by the analysis. We do not have to know anything about the sign of that force now, but we do have to use the sign convention that puts a positive sign in front of $\mathbf{F}_{conn}$ in the force balance. The nozzle exerts an equal and opposite force (downward, according to our expectation) on the connection. The mass M that appears here is the mass of the system, which is the sum of the mass of fluid within the nozzle and the mass of the nozzle itself.

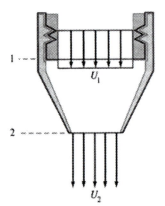

Figure 11.2.4 Analysis of flow through a nozzle.

To proceed, we next assume that the velocity profiles are flat at planes 1 and 2. Hence

$$\langle U_1^2 \rangle = \langle U_1 \rangle^2 \tag{11.2.22}$$

and the same for U_2. Then the axial component of Eq. 11.2.21 becomes

$$-F_{\text{conn}} = \rho_1 \langle U_1 \rangle^2 A_1 - \rho_2 \langle U_2 \rangle^2 A_2 + P_1 A_1 - P_2 A_2 + Mg \tag{11.2.23}$$

From the mass conservation principle we write

$$\rho_1 U_1 A_1 = \rho_2 U_2 A_2 = \dot{m} \tag{11.2.24}$$

where $\dot{m}$ is the mass flow through the nozzle. We have dropped the brackets on the velocity terms.

The result is

$$-F_{\text{conn}} = \dot{m}(U_1 - U_2) + (P_1 A_1 - P_2 A_2) + Mg \tag{11.2.25}$$

Remember that $\mathbf{F}_{\text{conn}}$ is the force exerted *by* the connection upstream of the nozzle, *on* the nozzle. It consists of three contributions: a momentum term, a pressure force, and the weight of the fluid within the nozzle plus the weight of the nozzle itself. The first term is negative (we know U_1 must be less than U_2 since A_1 is greater than A_2). Hence this is an upward-directed force. It is a reactive thrust that pushes the nozzle against the connector. The second set of terms, $(P_1 A_1 - P_2 A_2)$ is positive, as is the weight of the fluid plus the nozzle, Mg. Hence the connector must "pull" against the nozzle to hold it on, in opposition to these forces.

Now let's recall that the pressures that appear in a momentum equation (Eq. 11.2.10) are the gage pressures—absolute pressure minus atmospheric pressure. Let's look more carefully at the pressures acting on the nozzle. Figure 11.2.5 presents three pressures, acting on the three surfaces of the control volume. At the planes 1 and 2 we have forces $(P_1 + p_{\text{atm}})A_1$ and $-(P_2 + p_{\text{atm}})A_2$. Note that each force is the product of the absolute pressure—gage pressure P plus atmospheric pressure—times the area normal to the pressure. Along the exterior surface only the atmospheric pressure acts, and its component in the axial direction (see the inset in Fig. 11.2.5) is $p_{\text{atm}}(A_1 - A_2)$. Then the net pressure force is found as

$$(P_1 + p_{\text{atm}})A_1 - (P_2 + p_{\text{atm}})A_2 - p_{\text{atm}}(A_1 - A_2) = P_1 A_1 - P_2 A_2 \tag{11.2.26}$$

Hence, as stated after Eq. 11.2.10, it is the gage pressures (P_1 and P_2) that appear in the force balance.

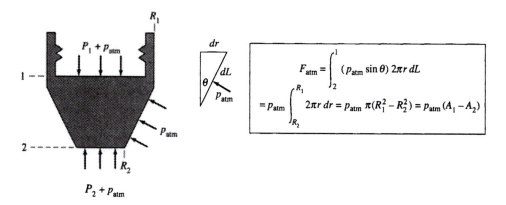

Figure 11.2.5 Pressures acting on the nozzle and the fluid within it.

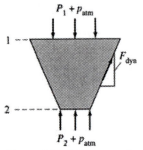

Figure 11.2.6 Forces acting on the fluid within the nozzle.

Suppose, now, that we choose to write a momentum balance just on the fluid within the nozzle, as suggested in Fig. 11.2.6. At the planes 1 and 2 we have forces $(P_1 + p_{atm})A_1$ and $-(P_2 + p_{atm})A_2$, as before. The momentum and body force terms also appear, as they do in Eq. 11.2.25. *There is no connector force,* since the connector is not in contact with this control volume. Instead, there is a force exerted by the moving fluid on the inner surface of the nozzle, which is a boundary of the control volume. It is a dynamic force, and its component in the axial direction is labeled F_{dyn}. The axial force (momentum) balance becomes

$$F_{dyn} = \dot{m}(U_1 - U_2) + [(P_1 + p_{atm})A_1 - (P_2 + p_{atm})A_2] + M_{fluid}g \quad \textbf{(11.2.27)}$$

Note that only the mass of the fluid within the nozzle appears in the body force, not the sum of the fluid plus the nozzle, as in Eq. 11.2.25. The absolute pressure appears here. The external atmospheric pressure term is omitted from this momentum balance because the exterior of the nozzle, upon which atmospheric pressure acts, is not a boundary of *this* control volume. If all the terms on the right-hand side of Eq. 11.2.27 were measured, we could calculate the dynamic force arising from the interaction of the fluid with the inner bounding surface of the nozzle. The connector force does not appear here, so this particular choice of control volume is not an aid in the calculation of that force.

Finally, let's select yet another choice for a control volume—just the nozzle body itself. Figure 11.2.7 shows the forces relevant to that choice. The forces $(P_1 + p_{atm})A_1$ and $-(P_2 + p_{atm})A_2$ do not act on the body of the nozzle. The connector exerts a force, as shown, and the fluid dynamics exerts a force $\mathbf{F}_{dyn}$ on the interior surface of the nozzle. The body force includes just the mass of the nozzle itself, M_{noz}. As a result, the force

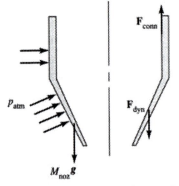

Figure 11.2.7 Forces acting on the body of the nozzle.

balance takes the form

$$0 = F_{\text{conn}} + F_{\text{dyn}} + M_{\text{noz}}g - p_{\text{atm}}(A_1 - A_2) \qquad \textbf{(11.2.28)}$$

The signs in front of both force terms are positive because these are the forces exerted *by* the surroundings *on* the boundaries of the control volume.

Now let's add Eqs. 11.2.28 and 11.2.27. The result is

$$\begin{aligned} F_{\text{dyn}} = \dot{m}(U_1 - U_2) &+ [(P_1 + p_{\text{atm}})A_1 - (P_2 + p_{\text{atm}})A_2] \\ &+ M_{\text{fluid}}g + F_{\text{conn}} + F_{\text{dyn}} + M_{\text{noz}}g - p_{\text{atm}}(A_1 - A_2) \end{aligned} \qquad \textbf{(11.2.29)}$$

which can be simplified to

$$-F_{\text{conn}} = \dot{m}(U_1 - U_2) + (P_1 A_1 - P_2 A_2) + Mg \qquad \textbf{(11.2.30)}$$

where $M = M_{\text{fluid}} + M_{\text{noz}}$. Of course this is identical to Eq. 11.2.25, as it must be.

An important observation to draw from this example is that various choices of control volumes are possible, some of which may lead to more direct results for specific features of the system. But no matter what choices we make, the physics demands and ensures that the results will be consistent.

We look next at a problem that displays a transient character. In particular, the control system is accelerating.

EXAMPLE 11.2.5   *Forces on a Reversing Bucket*

Motion can often be imparted to a body by converting the momentum of a jet of fluid to a force available to accelerate the body. This occurs commonly in turbines, for example. In the simple geometry shown in Fig. 11.2.8, the solid body, of total mass M, has a curved vane that is capable of reversing the direction of a high speed jet of water, as suggested in the sketch. The jet enters the system boundaries with speed U_1, relative to the fixed nozzle, and we assume that the exit speed U_2 relative to the fixed nozzle is unchanged in magnitude. Of course the *velocity vector* is reversed in direction. (In a real system, viscous losses would reduce the speed U_2 of the exiting jet.) The areas of the entering and exiting jets are taken as known, and of equal magnitudes A.

In this problem the *system* moves with velocity V relative to the fixed nozzle. To calculate the forces acting on the system we use Eq. 11.2.10. The volume and mass that appear in that equation refer to that of the fluid inside the system. In particular, the mass $M = \int \rho \, dV$ refers only to the water within the boundaries of the system, and not to the mass of the body, M.

If the body moves horizontally, we may drop the gravitational term in Eq. 11.2.10. The system is assumed to be exposed to the atmosphere, and so the gage pressures at the surfaces A_1 and A_2 are zero. The velocity profile across the jet is assumed to be flat, so the average of the squared velocity is the square of the averaged velocity. The

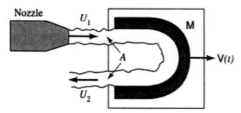

Figure 11.2.8 A water jet imparts motion to a body.

fluid inside the system boundaries has the average velocity of the system, relative to the fixed nozzle. Hence we find the force exerted *by* the fluid *on the boundaries* of the system from

$$\mathbf{F} = \rho_1 \langle U_1^2 \rangle_{\text{rel}} \mathbf{A}_1 - \rho_2 \langle U_2^2 \rangle_{\text{rel}} \mathbf{A}_2 - \frac{\partial}{\partial t} \int \rho \mathbf{u} \, d\mathsf{V} \qquad (11.2.31)$$

But the system moves with velocity V, so the velocities crossing the system boundaries are relative to that motion (hence the subscript notation) and

$$\langle U_1 \rangle_{\text{rel}} = U_1 - V \quad \text{and} \quad \langle U_2 \rangle_{\text{rel}} = -U_1 - V \qquad (11.2.32)$$

Hence Eq. 11.2.31 becomes, in scalar form,

$$F = \rho(U_1 - V)^2 A + \rho(-U_1 - V)^2 A - \frac{d}{dt}(MV) \qquad (11.2.33)$$

(Keep in mind that M is the mass of water within the system boundaries, while M is the mass of the body itself.) Under steady flow conditions, $dM/dt = 0$.

To find the dynamics of the body, we must write Newton's second law for a rigid body of mass M acted upon by the force F, and possibly by other forces. To gain an idea of what this dynamic problem would look like, let us make three assumptions:

The mass of water within the system boundaries is very much less than that of the body. Hence we drop the term $M \, dV/dt$ relative to $d(MV)/dt$.

There is a resistance to the motion of the body that is proportional to the velocity V. The speed of the water jet is constant at the value U_1.

Then Newton's second law gives us the following differential equation for the motion:

$$\frac{d}{dt}(MV) = F - K_{\text{res}}V = 2 \times \rho(U_1^2 + V^2)A - K_{\text{res}}V \qquad (11.2.34)$$

Imposition of some initial condition, such as

$$V = 0 \quad \text{at} \quad t = 0 \qquad (11.2.35)$$

then permits us to solve this differential equation and find the expected $V(t)$ behavior.

11.3 THE MACROSCOPIC ENERGY BALANCE

We examine the same generalized flow system as before, but we allow for the possibility that energy and work can cross the solid boundaries of the system. We write a *macroscopic* energy balance along the following lines.

The total energy within the boundaries of the system at any time is written as the sum of internal, kinetic, and potential (gravitational) energies, all written on a *per unit mass* basis:

$$\frac{\text{energy}}{\text{fluid mass}} \equiv \mathsf{E} = \varepsilon + \frac{U^2}{2} + gh \qquad (11.3.1)$$

Our macroscopic expression of the first law of thermodynamics for a flow system is

$$\frac{d}{dt} \int \rho \mathsf{E} \, d\mathsf{V} = \langle \rho U A \mathsf{E} \rangle_1 - \langle \rho U A \mathsf{E} \rangle_2 + Q_{\text{H}} + W_{\text{s}} + W_{\text{f}} \qquad (11.3.2)$$

where Q_{H} is the rate of heat transfer across the boundaries of the system and W_{s} is the rate at which "shaft work" crosses the boundaries of the system (see Fig. 11.3.1); W_{f}

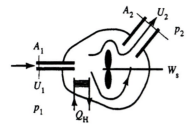

Figure 11.3.1 System for application of a macroscopic energy balance.

is the "flow work rate" associated with the flow across the surfaces at planes 1 and 2. We need to define some of these terms. (Remember, the rate at which work is performed is power.)

In the integral, E is the total energy per unit mass at any point within the volume. "Total" here means the sum of internal, kinetic, and potential energies *per unit of mass*—not the total *amount* within the volume. Angular brackets ⟨ ⟩ are used as a reminder that the quantities within have been averaged across the surfaces A_1 and A_2.

The heat transfer term Q_H accounts for any heat exchange directly from the surfaces of the system that is due to a temperature difference between the surfaces of the system and the surroundings. In addition, we may have a heat exchanger within the boundaries of the system which is connected by some plumbing arrangement to a second system in the surroundings, external to the system. In any event, we take the term Q_H to be positive if heat is transferred *into* the system.

The internal energy per unit mass of a fluid is a thermodynamic property that depends on temperature, pressure, and composition. If the fluid is incompressible—an assumption we will generally make if the fluid is a liquid—and if the flow is isothermal and no chemical reactions occur, then the internal energy per unit mass will not change at all.

For an incompressible fluid in which temperature changes but no chemical reactions occur, the internal energy density may be written as

$$\varepsilon = C_v(T - T_{ref}) \tag{11.3.3}$$

where T_{ref} is a reference temperature for a zero internal energy and C_v is the heat capacity of the fluid at constant volume.

Because the internal energy is a thermodynamic property, we will find it necessary to introduce some additional thermodynamic concepts shortly. (The relationships above suffice only if the processes that take place within the boundaries of the system are so simple that ε does not change, or changes according to Eq. 11.3.3). First, however, let's examine the "mechanical" terms in Eq. 11.3.2. We define the "shaft work" term as the work that is introduced into the system or transmitted by the system by virtue of some mechanical action across the system boundaries. For example, if a stirrer or a turbine is within the system, we can put work *into* the system by turning the shaft of the turbine (hence the name, "shaft work"). In other situations the motion of the fluid within the system may rotate a shaft from which we could *extract* work by connecting the shaft to a series of pulleys or cams, for example. We take the work to be positive if work is done *on* the system *by* the surroundings; the work term is negative if work is *extracted* from the system. (Our language is somewhat loose here. When we say "work," we mean the rate at which work is performed on, or delivered by, the system. We really mean "power," but one often sees these concepts expressed simply as "work.")

"Flow work" is a little more complex. Imagine an element of fluid passing across an entrance plane of area A at a velocity U. The absolute pressure in the plane of the surface is p. The fluid in the plane is acted upon by a force pA, which is moving that fluid at a velocity U. Hence we may think of the pressure on the entrance plane as if it were performing work on the fluid crossing the plane at a *rate* (force times velocity) given by pAU.

With these ideas in mind we rewrite Eq. 11.3.2 in the form

$$\frac{d}{dt}\int \rho E\, dV = \langle \rho UAE + pUA \rangle_1 - \langle \rho UAE + pUA \rangle_2 + Q_H + W_s \qquad \textbf{(11.3.4)}$$

In the following we will assume that the velocity profiles across the entrance and exit planes are flat. This is a good assumption for turbulent flows, but not for laminar flows. At issue is how we write average velocities, and their powers, as we noted earlier in Eqs. 11.2.4 and 11.2.5.

11.3.1   Application to Steady State Processes

In the development that follows we will assume that the problems of interest involve *steady state processes*. This restriction will simplify our analyses here, and a great many engineering processes to which we apply the energy equation are indeed at steady state. We will later do some examples of unsteady flow problems, to show how the unsteady terms can be brought back into the equation. In the steady state, Eq. 11.3.4 becomes

$$0 = \Delta \langle \rho UAE + pUA \rangle - Q_H - W_s \qquad \textbf{(11.3.5)}$$

Instead of writing out the difference between the energy and flow work terms entering and leaving the system, we use the simpler notation $\Delta \langle \ \rangle$ to denote the difference, but it is the difference between quantities leaving and entering the system, not the opposite. (Be careful here! Δ means $\langle \ \rangle_2 - \langle \ \rangle_1$, not the opposite.)

In the steady state, the term ρUA is the mass flowrate through the system, which we will denote by w, and since mass is conserved, w is unchanged across the boundaries of the system. Hence we may write Eq. 11.3.5 in the form (using Eq. 11.3.1 to replace E)

$$0 = \Delta \left\langle \varepsilon + \frac{U^2}{2} + gh + \frac{p}{\rho} \right\rangle - \hat{Q}_H - \hat{W}_s \qquad \textbf{(11.3.6)}$$

In Eq. 11.3.6 the terms Q_H and W_s of Eq. 11.3.5 have been divided by the mass flowrate w to produce the terms topped by the caret mark ˆ. The new terms represent the rate of heat flow and the rate of shaft work, respectively, *per mass flowrate through the system.*

While Eq. 11.3.6 is formally correct, it is not in a form that is very useful for solving engineering problems. We need to convert the internal energy term into a form that involves temperature, pressure, and density changes across the system. You may want to review some thermodynamics, or you may recall that a general thermodynamic relationship (that will prove useful later) is, in differential form,

$$d\varepsilon = T\, dS - pd\left(\frac{1}{\rho}\right) \qquad \textbf{(11.3.7)}$$

where dS is the differential change in entropy per unit mass. For an idealized *reversible* process in which no energy dissipation occurs, the entropy change arises from heat transfer across the boundaries:

$$T\, dS = dQ_H \qquad \textbf{(11.3.8)}$$

In any real process of engineering interest there is dissipation of energy (irreversibility), and the second law of thermodynamics states that

$$T\,dS - dQ_H \geq 0 \tag{11.3.9}$$

To use these thermodynamic relationships, we must first convert Eq. 11.3.6 to an energy balance on a *differential* section of the system. Hence Eq. 11.3.6 becomes

$$0 = d\left\langle \varepsilon + \frac{U^2}{2} + gh + \frac{p}{\rho} \right\rangle - d\hat{Q}_H - d\hat{W}_s \tag{11.3.10}$$

When Eq. 11.3.7 is used to replace the internal energy term this equation may be written[3] in the form

$$T\,dS - d\hat{Q}_H + \frac{dU^2}{2} + g\,dh + \frac{dp}{\rho} = d\hat{W}_s \tag{11.3.11}$$

The second law of thermodynamics tells us that the difference between the entropy term and the heat exchange per unit mass is positive except in the case of a thermodynamically reversible process, in which case the difference vanishes. Hence we write

$$T\,dS - dQ_H \equiv d\hat{E}_v \tag{11.3.12}$$

In this way we have defined a term that accounts for energy losses due to irreversible processes within the system. In the following we will restrict attention to the case $dQ_H = 0$; then dE_v represents the irreversible *mechanical* energy losses. The most common examples of these losses are those due to viscous friction associated with the deformation of the fluid. With the introduction of this term (i.e., dE_v), Eq. 11.3.11 takes the form

$$\frac{dU^2}{2} + g\,dh + \frac{dp}{\rho} = d\hat{W}_s - d\hat{E}_v \tag{11.3.13}$$

This is a good time to pause and review the meaning of our terms. We must also be careful about the sign of each term, in particular. The first two terms on the left-hand side of Eq. 11.3.13 represent the kinetic energy and potential (gravitational) energy changes in the system. The next term, involving pressure, arises from two sources. There is a contribution to the energy balance from the *flow work* term (this first appeared in Eq. 11.3.4), and pressure also entered when we replaced the internal energy using Eq. 11.3.7. It is the combination of these two pressure terms that ends up in the form of dp/ρ in Eq. 11.3.13.

The shaft work term W_s has been converted to its present form (with the ^ mark) by dividing the rate at which work is done *on* the system, W_s, by the mass flowrate $w = \rho U A$. This term is negative if work is done *by* the system on the surroundings. If we find a negative value for W_s when we apply the energy balance to a problem, we find that the system *delivers power* to the surroundings. A positive value means we must put power into the system.

Finally, there is a term E_v, defined by

$$E_v = w\hat{E}_v \tag{11.3.14}$$

which represents the rate at which "mechanical" energy is irreversibly lost to the system because of friction. This term is always positive.

---

[3] We have used the fact that $d(p/\rho) = (1/\rho)\,dp + p\,d(1/\rho)$.

Recall, now, that Eq. 11.3.10 is written for a differential section of the system—not for the system as a whole. Hence we now have to integrate Eq. 11.3.13 across the ends of the system, that is, between planes 1 and 2. The result is

$$\Delta \frac{U^2}{2} + g\,\Delta h + \int_{P_1}^{P_2} \frac{dp}{\rho} = \hat{W}_s - \hat{E}_v \tag{11.3.15}$$

and we remind ourselves that the Δ operator is the difference $\langle\ \rangle_2 - \langle\ \rangle_1$, not the opposite.

In the special but common case of a flow that is of constant density, Eq. 11.3.15 simplifies to the form

$$\Delta \frac{U^2}{2} + g\,\Delta h + \frac{\Delta p}{\rho} = \hat{W}_s - \hat{E}_v \tag{11.3.15'}$$

The Bernoulli equation

Although usually the term "Bernoulli equation" is reserved for the case of a frictionless flow with no shaft work, we will apply it to the energy equation for constant density flow even when the W_s and E_v terms are not zero.

This is a good point at which to consider a series of examples that will clarify the terms in the energy balance, and to illustrate the use of the macroscopic energy equation to analyze engineering systems.

EXAMPLE 11.3.1   *Turbulent Incompressible Steady Flow Through a Long Straight Pipe*

The system is defined in Fig. 11.3.2. We provide no mechanism for introducing or extracting shaft work. The pipe is horizontal, so there is no change in potential energy of the fluid in the system. There is no heat transfer across the boundaries of the system. The cross-sectional areas of the pipe inlet and outlet are identical.

Conservation of mass tells us that

$$w_1 = w_2 \tag{11.3.16}$$

Since the fluid is incompressible and the cross-sectional areas are identical, it follows that

$$U_1 = U_2 \tag{11.3.17}$$

Equation 11.3.15 then simplifies to

$$\int_{P_1}^{P_2} \frac{dp}{\rho} = \frac{\Delta p}{\rho} = -\hat{E}_v = \frac{-E_v}{\rho U A} \tag{11.3.18}$$

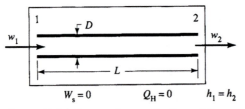

Figure 11.3.2 System for application of the macroscopic energy equation to flow through a long straight pipe.

Keeping in mind the definition of the Δ operator, we may write this equation in the form

$$p_2 - p_1 = \frac{-E_v}{UA} \tag{11.3.19}$$

There should be nothing surprising about this result, which simply states that the pressure drop along the pipe is associated with an irreversible loss of energy due to friction (viscous effects) in this flow. Note that the sign is consistent with our statement that E_v is always positive: hence $p_2 < p_1$ as expected. An imposed pressure drop from left to right is required to overcome the friction.

For flow through a straight pipe, we may make a simple connection between the energy loss term and the friction factor. The friction factor for this flow may be written as (see Eq. 5.3.30)

$$f \equiv \frac{D(p_1 - p_2)}{2L\rho U^2} \tag{11.3.20}$$

Then we can show the following two relationships that may help us in subsequent problems:

$$\frac{E_v}{\rho U A} = \hat{E}_v = \frac{2LU^2}{D}f \tag{11.3.21}$$

and

$$e_v \equiv \frac{\hat{E}_v}{U^2/2} = \frac{4L}{D}f \tag{11.3.22}$$

Evidently, e_v is a dimensionless measure of the energy loss due to friction. It is called the "friction loss factor." For a fixed geometry (fixed L/D), e_v is simply proportional to the friction factor f.

While we are on the topic of friction losses through a pipeline for turbulent flow, we mention that the definition of the friction factor given in Eq. 11.3.20, which is called the Fanning friction factor, is not universal. Another commonly defined friction factor is the Moody friction factor, which is four times the friction factor defined and used in this textbook. One often finds graphs of friction factor versus Reynolds number, and it is important to distinguish the Fanning from the Moody factor. Often, however, these charts are not labeled "Fanning" or "Moody." In such cases, the clue will lie in the laminar region. If, at the lowest Reynolds numbers, the laminar portion corresponds to $f = 16/Re$, then we have a Fanning chart with f defined as in Eq. 11.3.20. On the other hand, if the low Reynolds number region corresponds to $f = 64/Re$, we have a Moody chart.

A second point to note is that while we have assumed smooth pipes in all of our examples, real pipes are rough, and roughness can strongly increase the friction loss. Figure 11.3.3 shows the Fanning friction factor chart, including the effect of roughness. The roughness parameter is δ/D, the ratio of the mean roughness of the pipe to the pipe diameter. Commercial steel piping has roughness parameter values in the range of 10^{-2} to 10^{-4}, for pipe diameters in the range of several millimeters to a half-meter, respectively. Large concrete pipes have roughness parameter values in the range 5×10^{-2} to 5×10^{-4} for pipe diameters in the range of several inches to several feet, respectively.

We note in particular that for very rough pipes, the friction factor in the turbulent regime approaches a constant value, independent of the flowrate. We used this observation in Chapter 10 when we developed a model for turbulent flow through a packed bed.

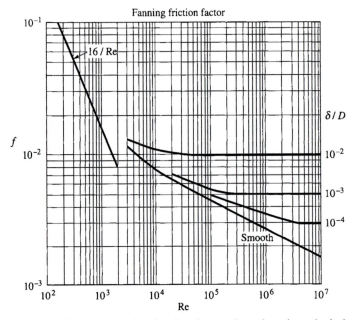

Figure 11.3.3 The Fanning friction factor chart for pipes, including roughness.

Next we look at an example of a macroscopic system that exchanges work with its surroundings.

EXAMPLE 11.3.2 **An Elevated Reservoir as a Power Source**

In the system of interest (Fig. 11.3.4), a supply of water is maintained at constant head in a large reservoir. When water is withdrawn from the reservoir, it is at such a slow rate that the head h_r does not fall measurably. The water flow from the reservoir is conducted through a pipe in such a way that it is forced to turn a turbine connected to a shaft.

With reference to Fig. 11.3.4, we see that shaft work is *removed* from the system. There is a mass flowrate w_2 out of the system. At steady state, there must be an equal mass flowrate into the system. We will come back to this shortly. If we take the elevation

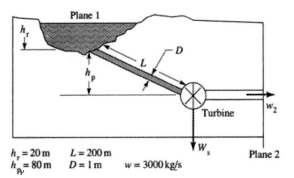

Figure 11.3.4 Flow from a reservoir through a turbine.

of the pipe at the turbine exit to be $h_2 = 0$, then the elevation of the plane 1 is $h_r + h_p$. We assume that the reservoir surface and the exit flow from the turbine are exposed to atmospheric pressure. Hence there is no pressure difference to account for. (Note that the hydrostatic pressure difference that drives the flow will be accounted for in the potential energy term.)

The Bernouilli equation (Eq. 11.3.15′) takes the form

$$\frac{U_2^2 - U_1^2}{2} + g(h_2 - h_1) = \hat{W}_s - \hat{E}_v \tag{11.3.23}$$

or

$$-\hat{W}_s = \frac{(U_1^2 - U_2^2)}{2} + gh_1 - \hat{E}_v \tag{11.3.24}$$

Normally, the rate of fall of a reservoir surface is very small. Since at steady state there must be a mass flowrate of water into the reservoir that balances the outflow downstream of the turbine, there really is a finite velocity across the plane 1. But it must be a very small velocity, in comparison to that through the turbine, because it will be spread across a wide area, and so we neglect U_1 in this calculation. This is essentially the assumption that the kinetic energy of the replacement flow to the reservoir is very small compared to that of the outlet flow from the turbine. To proceed further, we must know something about the energy losses in the turbine, and in the conduit between the reservoir and the turbine. This requires knowledge of the dynamic behavior of large rotating machines, in addition to the background we already have with respect to flow through long pipes.

In the absence of any of that information, we can at least state that this system would produce maximum power under the following conditions: *no frictional losses anywhere,* and negligible kinetic energy imparted to the fluid downstream of the turbine (the ρU_2^2 term). Under these conditions the energy balance gives us

$$-\hat{W}_s = gh_1 = 9.8(20 + 80) = 980 \, \text{m}^2/\text{s}^2 \tag{11.3.25}$$

We see that the maximum possible power output occurs if all of the gravitational head is converted to energy at the turbine. After introducing the mass flowrate, we find the maximum power output to be

$$-W_s = -w\hat{W}_s = 3000 \, \text{kg/s} \times 980 \, \text{m}^2/\text{s}^2 = 2940 \, \text{kW} = 3940 \, \text{hp} \tag{11.3.26}$$

Of course frictional losses in the pipe, and in the turbine, will reduce this value.

It is also interesting to calculate the mass flowrate through the pipeline to the turbine *in the absence of the turbine but given significant frictional losses.* We simply go back to Eq. 11.3.23 and drop the shaft work term. If the only flow resistance between the reservoir and the point where the turbine is installed is the 200 m length of 1 m diameter pipe, we may use Eq. 11.3.22 and a friction factor chart to relate the energy loss term to the flowrate.

The Bernoulli equation takes the form, in this case of no shaft work, of

$$\frac{U_2^2}{2} - gh_1 + \frac{2LU_2^2}{D}f = 0 \tag{11.3.27}$$

We may get an approximate solution by using a simple model for the f/Re relationship for a smooth pipe:

$$f = 0.079 \, \text{Re}^{-0.25} = 0.079 \left(\frac{4w}{\pi D \mu}\right)^{-0.25} \tag{11.3.28}$$

Note in particular that we may write the Reynolds number directly in terms of the mass flowrate w. When we do the same with the rest of Eq. 11.3.27, we must solve the following equation for w:

$$0.24 \frac{L\mu^{-0.25}}{\rho^2 D^{4.75}} w^{1.75} = gh_1 - \frac{0.81}{\rho^2 D^4} w^2 \qquad (11.3.29)$$

The data given in Fig. 11.3.4 must be introduced, and we take the density and viscosity of water to be $\rho = 1000$ kg/m^3 and $\mu = 0.001$ Pa·s. Then we solve Eq. 11.3.29 for w by trial and error and find

$$w = 2.6 \times 10^4 \text{ kg/s} \qquad (11.3.30)$$

The observed flow through the system, given above as 3000 kg/s, is only 10% of the maximum possible flow. This difference arises from energy losses in the system associated with the turbine, and with restricting valves between the reservoir and the turbine that provide control over the flow through the system.

We look next at an application of the Bernoulli equation that is, in a certain sense, the inverse of the example just completed.

EXAMPLE 11.3.3   Pump Requirements for a Water Supply System

A plumbing layout for a water supply system is shown in Fig. 11.3.5. We want to find the power requirement for a pump that can transfer water between the reservoirs at the specified flowrate. The following data are available:

$h_1 = 10$ m $h_2 = 3$ m $L_1 = 50$ m $L_2 = 300$ m $L_3 = 1$ m
$D_1 = 0.2$ m $D_2 = 0.5$ m $D_3 = 0.03$ m

We wish to supply water to the lower reservoir at a flowrate of 60 kg/s. A pump, labeled "P" on the figure, is available at the end of the 0.03 m pipe. We want to estimate the required power for this pump.

The first question we might ask is whether *any* pump is necessary. Is it possible that the gravitational head is sufficient to provide the required flow? To find out, we go to the Bernoulli equation. If we select the reference planes shown in Fig. 11.3.5, it takes the form

$$g(h_1 + L_1 - L_3) = \frac{U_3^2}{2} + gh_2 + \sum_i^n \hat{E}_{v,i} \qquad (11.3.31)$$

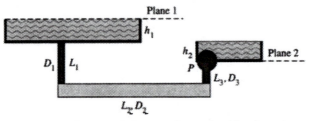

Figure 11.3.5 Pump requirement for a plumbing layout.

We neglect any kinetic energy in the reservoir, at plane 1. Hence the only energy term is that arising from potential energy (gravity). At plane 2, the velocity that appears, U_3, is the velocity in the pipe D_3 at the *bottom* of the downstream reservoir. Hence the pressure at plane 2 (the bottom of that reservoir) is the hydrostatic pressure associated with the head h_2. Thus, formally, the gh_2 term represents pressure, not gravitational potential energy, in Eq. 11.3.31. (See Problem 11.16.)

We make the assumption, in writing this equation as above, that the only friction losses are in the straight sections of piping. In particular, we neglect losses that are associated with the bends and fittings that connect the various sections together. (We will address that issue later.) Since there are three separate pipe sections, we must write the friction losses as the sum of the losses over each section, as shown in writing Eq. 11.3.31. If we introduce the friction factor for each section, we may write the friction loss term as

$$\sum_i^3 \hat{E}_{v,i} = \sum_i^3 \frac{2L_i U_i^2}{D_i} f_i \qquad (11.3.32)$$

For an incompressible liquid we may write

$$D_i^2 U_i = \text{constant} = D_3^2 U_3 \qquad (11.3.33)$$

This lets us replace U_1 and U_2 in the summation in favor of U_3, which we need in order to find the flowrate through the system. In the summation, however, we still use the L_i/D_i for each pipe section, and the friction factor f_i is based on the flowrate in each section. Putting these ideas to work, we find the following equation to solve for the velocity U_3, from which the desired flowrate w_3 follows:

$$U_3^2 = \frac{2g(h_1 + L_1 - L_3 - h_2)}{1 + \Phi} \qquad (11.3.34)$$

where

$$\Phi \equiv \frac{4L_3}{D_3} f_3 + \frac{4L_2}{D_2}\left(\frac{D_3}{D_2}\right)^4 f_2 + \frac{4L_1}{D_1}\left(\frac{D_3}{D_1}\right)^4 f_1 \qquad (11.3.35)$$

Note that the term Φ includes the friction factors, which depend on the unknown velocities. Hence Eq. 11.3.34 is not explicit in U_3. We can solve it by a simple iterative method, which has the advantage of giving us some insight into the relative roles played by the various terms in the energy balance, for this example.

We begin by calculating an approximate solution from Eq. 11.3.34 with the assumptions that the friction losses are minor and the flow is dominated by the gravitational terms. Hence we find our first approximation to U_3 as (setting $\Phi = 0$ in Eq. 11.3.34)

$$\tilde{U}_3 = [2g(h_1 + L_1 - L_3 - h_2)]^{1/2} \qquad (11.3.36)$$

where the tilde $\sim$ on U_3 reminds us that this is a rough estimate. We then calculate the friction losses from Eq. 11.3.35, using this approximate value for U_3 to calculate a Reynolds number, hence a friction factor. This gives us an approximate value of Φ with which we can "correct" our estimate of U_3, using Eq. 11.3.34. We may then repeat this procedure until the process converges on a value for U_3 that no longer changes significantly. The iterative process converges rapidly in this case, because f is such a weak function of Reynolds number for turbulent flow.

For the data of this specific problem we find

$$\bar{U}_3 = [2 \times 9.8 \times 56]^{1/2} = 33 \text{ m/s}$$

$$\bar{w}_3 = \rho \frac{\pi D_3^2}{4} \bar{U}_3 = 1000 \frac{\pi (0.03)^2}{4} 33 = 23.4 \text{ kg/s} \tag{11.3.37}$$

Next we write

$$1 + \Phi = 1 + 132f_3 + 0.03f_2 + 0.51f_1 \tag{11.3.38}$$

The Reynolds numbers in each pipe section are found to be

$$Re_3 = 10^6 \qquad Re_2 = 6 \times 10^4 \qquad Re_1 = 1.5 \times 10^5 \tag{11.3.39}$$

We will calculate the friction factors using the simple analytical approximation for smooth pipes:

$$f = 0.079 \, Re^{-0.25} \qquad \text{for } Re > 10^4 \tag{11.3.40}$$

We find

$$f_3 = 2.5 \times 10^{-3} \qquad f_2 = 5 \times 10^{-3} \qquad f_1 = 4 \times 10^{-3} \tag{11.3.41}$$

Now we use Eq. 11.3.34 to get the next approximation for U_3:

$$U_3 = \left(\frac{33^2}{1 + 0.33} \right)^{1/2} = 28.6 \text{ m/s} \qquad \text{and} \qquad w_3 = 20 \text{ kg/s} \tag{11.3.42}$$

Note that the f_i are all of the same order of magnitude. Hence, upon inspection of Eq. 11.3.38, we see that the losses in pipes 1 and 2 are negligible. This is so because the diameters of these pipes are so large, and the friction loss is a very strong function of pipe diameter, but only a linear function of pipe length.

We will be satisfied with this second approximation for w_3. Since the value calculated is only a third of the desired mass flow, it *is* necessary to introduce a pump into the system to achieve a mass flowrate of 60 kg/s.

Our procedure now is to go back to the energy equation in the form of Eq. 11.3.31, but include and solve for the shaft work of the pump:

$$\frac{U_3^2}{2} = g(h_1 + L_1 - L_3) - gh_2 - \sum_i^n \hat{E}_{v,i} + \hat{W}_s \tag{11.3.43}$$

or

$$+\hat{W}_s = -g(h_1 + L_1 - L_3) + gh_2 + \frac{U_3^2(1 + \Phi)}{2} \tag{11.3.44}$$

Since the mass flow was originally specified as $w = 60$ kg/s (corresponding to $U_3 = 85.8$ m/s), we may solve for the shaft work directly from Eq. 11.3.44, finding

$$+\hat{W}_s = 4100 \text{ m}^2/\text{s}$$
$$W_s = w\hat{W}_s = 60 \text{ kg/s} \times 4100 \text{ m}^2/\text{s} = 243 \text{ kW} = 326 \text{ hp} \tag{11.3.45}$$

The positive sign is a reminder that the work must be done *on* the system *by* the pump.

EXAMPLE 11.3.4 *Derivation and Application of Torricelli's Law*

In Chapter 7, Section 7.3, we did a problem involving the draining of a tank. We return to that type of problem to clarify some points.

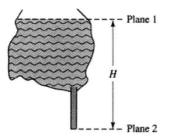

Figure 11.3.6 Liquid flows under gravity through an outlet.

We may apply the Bernoulli equation between the open planes of the system illustrated in Fig. 11.3.6. The pressure is atmospheric at the open ends. The Bernoulli equation takes the form

$$\frac{U_1^2 - U_2^2}{2} + gH = -\hat{E}_v \tag{11.3.46}$$

We will neglect U_1^2 with respect to U_2^2. This is a good approximation if the area ratios at planes 1 and 2 are as different as they appear to be in Fig. 11.3.6. Then we write

$$U_2^2 = 2gH + 2\hat{E}_v \tag{11.3.47}$$

For a frictionless flow, the energy loss term vanishes and we find

$$U_2 = \sqrt{2gH} \tag{11.3.48}$$

Torricelli's law

Note that $H = H(t)$.

This is Torricelli's law, assumed earlier in Eq. 11.1.3. It is valid only under the idealized conditions of no viscous losses in the upper vessel, the entrance to the outlet tube or orifice, or within the outlet tube itself. It gives us a reasonable approximation to the behavior of low viscosity liquids such as water flowing out of large-scale systems. It is not a good approximation for viscous liquids, or even for low viscosity liquids flowing through small diameter capillaries of any significant length, where frictional (viscous) effects would be important.

Note that nothing in the derivation is related to the shape of the vessel containing the liquid, as long as there are no narrow restrictions across which there might be viscous losses.

We now apply Torricelli's law to the problem of draining of a tank. Of particular interest is the distinction between the situation of negligible viscous effects and the case presented in Section 7.3, where viscous effects dominated. We will simplify the geometry by taking the tank to be cylindrical.

With reference to Fig. 11.3.7, we define the plane 1 as *fixed,* and not attached to the moving free surface. The area of that plane is A_1, which is also the area of the falling free surface. There is flow across the plane 2, but there is *no flow* across the plane 1.

Clearly this is an *unsteady* flow, since liquid leaves the system and is not replaced. We have derived the Bernoulli equation for *steady* flows. However, it is not difficult to return to Eq. 11.3.4 and carry out the subsequent arguments for the unsteady flow case. The result is

$$\frac{d}{dt} \int \rho E \, dV = -\Delta\left(\frac{U^2 w}{2}\right) - g\Delta(wh) - \frac{\Delta(wp)}{\rho} - w\hat{E}_v \tag{11.3.49}$$

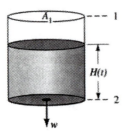

Figure 11.3.7 Draining of a cylindrical tank.

Because the liquid in the tank is in motion, it has a kinetic energy. It also has a potential energy that is changing with time, since the center of mass of the liquid is falling as the liquid level falls. Thus the transient terms (the integral on the left-hand side of Eq. 11.3.49) are found to be

$$\frac{d}{dt}\int_0^{H(t)} \rho\left(\frac{U_s^2}{2} + gz\right) A_1 \, dz = \frac{\rho U_s^2 A_1}{2}\frac{dH}{dt} + \underline{\rho g A_1 \frac{1}{2}\left[\frac{d}{dt} H^2(t)\right]} \qquad \textbf{(11.3.50)}$$

In the steady state, and in the absence of viscous losses, Torricelli's law tells us that the velocity at the outlet, U_2, will be given by Eq. 11.3.48. The velocity U_s will be less than U_2 by a factor of the ratio of the cross-sectional areas for flow. If the outlet area is smaller than A_1, as is the case when we drain a tank, then the *square* of U_s will be *much* smaller than $2gz$ (except near the end of draining, when z approaches zero). Hence we are going to neglect the kinetic energy term (the first term on the right) in Eq. 11.3.50. Only the underlined term will survive.

In considering the right-hand side of Eq. 11.3.49 we must keep in mind that nothing crosses plane 1, and there is no liquid at plane 1. In a sense *there is no plane 1,* and so there are no terms evaluated at plane 1 in the Δ terms. Only the underlined term in Eq. 11.3.49 will survive. [In Eq. 11.3.49, the $g\Delta(wh)$ term disappears because $w = 0$ at plane 1 and $h = 0$ is *defined* for plane 2.)

We may now put the energy balance together, and it takes the form

$$\rho g A_1 \frac{1}{2}\left[\frac{d}{dt} H^2(t)\right] = w\left[-\frac{U_2^2}{2}\right] \qquad \textbf{(11.3.51)}$$

Conservation of mass tells us that

$$-\frac{d}{dt} H = U_s = U_2 \frac{A_2}{A_1} = \frac{w}{\rho A_1} \qquad \textbf{(11.3.52)}$$

This lets us write Eq. 11.3.51 as

$$2gH = U_2^2 = \left(\frac{A_1}{A_2}\right)^2\left(\frac{dH}{dt}\right)^2 \qquad \textbf{(11.3.53)}$$

Note that the left-hand half of this equation is just Torricelli's law, which we have found to hold for an unsteady-state system (as long as the time rate of change of the kinetic energy in the tank is small compared to the potential energy term. We may now write Eq. 11.3.53 as a differential equation for $H(t)$:

$$\frac{dH}{dt} = -\left[2g\left(\frac{A_2}{A_1}\right)^2\right]^{1/2} H^{1/2} = -\alpha H^{1/2} \qquad \textbf{(11.3.54)}$$

The solution is

$$H^{1/2} = H_o^{1/2} - \frac{\alpha}{2}t \qquad (11.3.55)$$

when we use the initial condition that

$$H = H_o \quad \text{at} \quad t = 0 \qquad (11.3.56)$$

The time to drain the tank is found by setting $H = 0$ in this solution, and the result is

$$t_\infty = \frac{2(H_o)^{1/2}}{\alpha} = \frac{\sqrt{2}(H_o)^{1/2}}{\sqrt{g}}\frac{A_1}{A_2} \qquad (11.3.57)$$

This result is quite different from Eq. 7.3.14. In particular, the time to drain the tank is finite, and the time depends on the initial height H_o.

11.3.2   Energy Losses Through Bends and Fittings

It is useful to have information available regarding the frictional losses through the bends, fittings, contractions, and expansions that are typical of plumbing elements in a flow system. One of the simplest ways to do this is in terms of the friction loss factor e_v introduced earlier in Eq. 11.3.22:

$$e_v \equiv \frac{\hat{E}_v}{U^2/2} \qquad (11.3.58)$$

These e_v factors can be calculated for some simple geometries. For others they are determined experimentally by measuring the pressure drop through a pipe in which the particular plumbing element resides, and using the Bernoulli equation to calculate $\hat{E}_v$ as a sum of the loss in the straight section plus the loss across the element. We can illustrate the calculation of the energy loss with the example of flow through a pipe that undergoes a sudden enlargement in diameter, as shown in Fig. 11.3.8. The planes 1 and 2 are taken as shown: just at the expansion for 1, and just beyond the point downstream where the flow becomes "straight," for 2. We begin with the Bernoulli equation in the form

$$\hat{E}_v = \frac{U_1^2 - U_2^2}{2} + \frac{P_1 - P_2}{\rho} \qquad (11.3.59)$$

We are going to ignore any frictional losses along the solid bounding surfaces in this system. For that reason we will take the pressure P_1 to be the same as the pressure

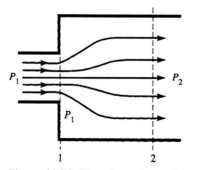

Figure 11.3.8 Flow through a sudden enlargement.

upstream from the expansion, as the figure suggests. For an incompressible fluid, mass balance gives

$$U_1 = U_2 \frac{A_2}{A_1} = \beta_{21} U_2 \qquad (11.3.60)$$

...ere β_{21} (> 1) is the ratio of the cross-sectional areas.

The pressure difference does not come from a solution to the energy equation. We must write the momentum balance (Eq. 11.2.10) next. In the axial direction we have

$$w_1 U_1 - w_2 U_2 + P_1 A_1 - P_2 A_2 = F \qquad (11.3.61)$$

What are the forces acting along the surfaces between planes 1 and 2 in the direction of flow? Since we are neglecting frictional losses on the bounding surfaces the only contribution to the force F is the product of the pressure on the annular surface of area $A_2 - A_1$, times that area, and Eq. 11.3.61 becomes

$$w_1 U_1 - w_2 U_2 + P_1 A_1 - P_2 A_2 = -(A_2 - A_1) P_1 \qquad (11.3.62)$$

Note the sign on F: it points opposite the flow.

We may solve for the pressure difference and find (noting $w_1 = w_2$)

$$(P_2 - P_1) A_2 = w(U_1 - U_2) = \rho U_2 A_2 (U_1 - U_2) \qquad (11.3.63)$$

With this result inserted into the energy equation we find, after some algebra,

$$\hat{E}_v = \frac{U_2^2}{2} (\beta_{21} - 1)^2 = \frac{(U_1 - U_2)^2}{2} \qquad (11.3.64)$$

or

$$e_v = (\beta_{21} - 1)^2 \qquad (11.3.65)$$

Sudden enlargement in a pipe

Note that e_v is defined here using the *downstream* velocity U_2, according to the usual convention.

Keep in mind that this result would not be a good model for a laminar flow, where viscous effects on the surfaces bounding the expansion region would be significant.

Energy losses are often presented in terms of a so-called head loss, which may be defined by writing the Bernoulli equation in the form (*cf.*: Eq. 11.3.59)

$$\rho g h_L = \frac{\rho(U_1^2 - U_2^2)}{2} + (P_1 - P_2) \qquad (11.3.66)$$

In this format we regard h_L as a *hypothetical hydrostatic head* that reduces the energy of the fluid to the same extent as the friction loss, as the fluid passes from plane 1 to plane 2. For flow through an expansion in a pipe, the *head loss factor* is usually defined by

$$K_L \equiv \frac{2 g h_L}{U_1^2} \qquad (11.3.67)$$

Note that the *upstream velocity* U_1 is used in this definition. Using the results of our analysis of flow through an expansion, we find, after some rearrangement,

$$K_L = \left(1 - \frac{1}{\beta_{21}}\right)^2 \qquad (11.3.68)$$

Sudden enlargement in a pipe

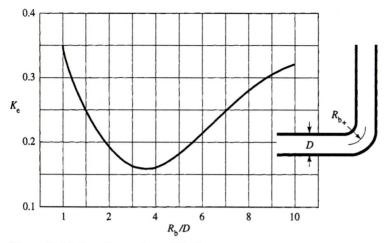

Figure 11.3.9 Loss factor for a 90° elbow.

Similar results are available in engineering handbooks for a variety of plumbing elements. For example, for flow through a 90° elbow, as shown in Fig. 11.3.9, the head loss may be written as

$$h_L = K_e \frac{U^2}{2g} \tag{11.3.69}$$

where U is the average velocity of flow through the elbow. The loss coefficient K_e, given in Fig. 11.3.9, is a function of the ratio of the elbow turning radius to the pipe diameter. We assume that there is no change in area as the fluid enters and moves through the elbow, so the velocity U that appears in Eq. 11.3.69 is simply the approach or upstream velocity to the elbow. The loss factor plotted here includes *all the friction loss along the curved axis* of the 90° elbow. Some loss factors are defined in a way that calls for the addition of the equivalent loss through a straight pipe of the same length $(\pi R_b/2)$ as the curved axis.

Another common "plumbing" element is a sudden contraction in a pipeline, as shown in Fig. 11.3.10. Experimental studies show that the friction loss factor for this case may be written in the form

$$\hat{E}_v = \frac{U_2^2 e_v}{2} \text{ with } e_v = 0.45\,(1 - \beta_{21}). \tag{11.3.70}$$

Sudden contraction in a pipe

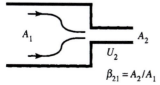

Figure 11.3.10 Contraction in a pipe.

11.3.3   Optimal Pipe Diameter to Minimize Power Costs[4]

In many situations we pump a liquid from one point to another through a long straight pipe under such conditions that the primary contribution to the pumping requirement is the frictional loss through the long section of the pipe, not through any bends or fittings across the pump, valves, or flow measuring devices in the system, or elevation changes. We would like to minimize the cost required to pump the liquid, at a specified flowrate through a system of fixed length, by selecting an optimum pipe diameter. The goal we have set assumes that there is in fact an optimum pipe diameter, and the validity of that assumption is easily argued.

To begin, we note that the "cost" should include two main terms. One is the capital cost of the pipeline, to which we must "add" in some sense the maintenance and installation costs. The other cost is that associated with running the pump—the power cost. We know that we can minimize the friction loss, and presumably therefore the pumping power costs, by using a pipe with a *large* diameter, since this will reduce the friction losses through the pipe. On the other hand, for a given distance over which we must pump the liquid, we expect to have lower capital and installation costs if we use a pipe of small diameter. Hence, we argue, there is an optimum pipe diameter.

The power cost follows from the Bernoulli equation, since the required power, in watts, is easily found in the form

$$-W_s = -w\hat{E}_v = w\frac{\Delta p}{\rho} = Q\,\Delta p \tag{11.3.71}$$

(We have used Eqs. 11.3.15' and 11.3.18.) The power cost may then be written as

$$C_{power} = \frac{Q\,\Delta p\,H(K_p/1000)}{\varepsilon_{pump}} \;(\$/yr) \tag{11.3.72}$$

where H is the number of hours of operation per year and K_p is the power cost, in dollars per kilowatt-hour. We have introduced a pump efficiency ε_{pump} to account for energy losses in the pump itself.

If we introduce the definition of the friction factor (Eq. 11.3.20), and the simple analytical representation for f (Eq. 11.3.40), we may write Eq. 11.3.72 in the form

$$C_{power} = \frac{AL\,\rho^{0.75}\mu^{0.25}Q^{2.75}}{D^{4.75}} \;(\$/yr) \tag{11.3.73}$$

where A lumps together the cost factors in Eq. 11.3.72 and the numerical coefficients from the friction factor relationship.

The capital cost of piping depends approximately, but not exactly, on the weight of material in the pipe. A simple model of piping capital cost in dollars is found to be

$$C_{pipe} = K\left(\frac{D}{D_o}\right)^n L \tag{11.3.74}$$

where n is found to be about 1.4. The coefficients K and D_o, (and n as well) in Eq. 11.3.74 are usually found by examining and then curve-fitting cost data for piping in the range of sizes of interest. We need to convert this capital cost to an annual cost, so that it can be added to the annual power cost (Eq. 11.3.73). While there are various methods for doing this, one of the simplest is to write off the cost on an annual basis

---

[4] This presentation is based on an example given in Denn, *Process Fluid Mechanics* (Prentice-Hall, Englewood Cliffs, 1980).

by assuming that the pipe has a useful life of N years, after which it has no salvage value. Then the simplest model of the "annualized capital cost" is

$$C_{\text{pipe/yr}} = \frac{K}{N}\left(\frac{D}{D_o}\right)^n L \text{ (\$/yr)} \tag{11.3.75}$$

Construction, installation, and maintenance costs will be taken to be proportional to the annualized capital cost, with the result that we write the total annualized cost of buying, installing, and maintaining the pipe in the form

$$C_{\text{pipe/yr,total}} = BD^n L \text{ (\$/yr)} \tag{11.3.76}$$

where B lumps several of the coefficients of the model. In our simple model, then, we add the power costs and the capital, installation, and maintenance costs, all on an annualized basis, to yield the total annual cost:

$$C_{\text{per year,total}} = \frac{AL\,\rho^{0.75}\mu^{0.25}Q^{2.75}}{D^{4.75}} + BD^n L \text{ (\$/yr)} \tag{11.3.77}$$

We find the optimum D by examining

$$0 = \frac{dC_{\text{per year, total}}}{dD} = \frac{-4.75AL\,\rho^{0.75}\mu^{0.25}Q^{2.75}}{D^{5.75}} + nBD^{n-1}L \tag{11.3.78}$$

from which we find

$$D_{\text{opt}} = \left(\frac{4.75\,A}{nB}\,\rho^{0.75}\mu^{0.25}Q^{2.75}\right)^{1/(4.75+n)} \tag{11.3.79}$$

It is important to note that with $n = 1.4$, the dependence of D_{opt} on the coefficients A and B (and, for that matter, on n itself) is quite weak. Hence uncertainties in these coefficients have a modest effect on the optimum D. This is an important observation— for example, we wouldn't want the optimum design to depend strongly on the prevailing power costs at different geographical locations.

11.3.4 Mammalian Circulation and Optimal Design

A form of optimal design of the circulatory system apparently occurs in mammals. The efficient delivery of blood to tissue involves a "plumbing" system in which the large blood vessels, the arteries, branch to smaller vessels, which themselves branch through several generations until the finest vessels, the capillaries, are reached. One can predict some features of the optimum geometry of branching based on the ideas of this chapter, and these predictions can be compared with physiological observations.

In the mammalian circulatory system one finds many "generations" of branching of the blood vessels from the arterial to the capillary system. A physiological principle known as Murray's law states that the cube of the radius of a parent vessel equals the sum of the cubes of the radii of the daughter vessels. This law can be derived on the hypothesis that the situation corresponds to a minimization of the work required to "operate" the blood distribution system. Refer to Fig. 11.3.11. In the derivation we make several assumptions that are not consistent with the known physiology of the circulatory system. The first is that the flow is laminar and steady, when in fact blood flow is pulsatile (i.e., unsteady) and turbulent in the larger arteries. The second assumption is that the friction losses associated with the entrance flow into the next generation of vessels, including the effects of the angles of the bifurcations, are negligible compared to losses due to viscous friction in the straight portions of the vessels. This is a questionable

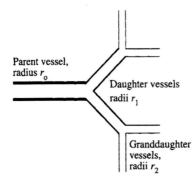

Figure 11.3.11 Two generations of branches in the arterial system.

assumption in the large vessels, but becomes more accurate after several generations as the capillaries are approached.

The pressure drop across any vessel is taken from Poiseuille's law, and the power required to pump a flow Q with that pressure drop, in watts, is given by

$$-W_s = Q\,\Delta p = \frac{8\mu L Q^2}{\pi r_o^4} = \frac{aQ^2}{r_o^4} \qquad \textbf{(11.3.80)}$$

There is also a metabolic power requirement for the maintenance of the health of the blood vessel itself. The wall of the vessel is a living tissue with a metabolic requirement m (W/m³). Hence we may write the metabolic power in the form

$$W_m = m\pi\left[\left(\frac{1}{\kappa}\right)^2 - 1\right]Lr_o^2 = br_o^2 \qquad \textbf{(11.3.81)}$$

where κ is the ratio of the inner to outer radii of the vessel and the result is in watts. Then the total power required for utilization and maintenance of a blood vessel is

$$W_{tot} = \frac{aQ^2}{r_o^4} + br_o^2 \qquad \textbf{(11.3.82)}$$

The optimum radius is found by the usual procedure (setting $\partial W_{tot}/\partial r_o = 0$) to be

$$r_o^* = \left(\frac{Q}{k}\right)^{1/3} \qquad \textbf{(11.3.83)}$$

where the constant k is defined as

$$k \equiv \left(\frac{b}{2a}\right)^{1/2} \qquad \textbf{(11.3.84)}$$

Note that the constant k depends on the metabolic rate of the vessel, the viscosity of the fluid (blood, in this case), and the vessel wall thickness parameter κ, but is independent of the length L. If we make the assumption that m and κ are constant through the generations of bifurcations of the blood vessels, then Eq. 11.3.82 is valid for each generation, when the Q appropriate to that generation is used. For example, we may write for the daughter vessels

$$r_1^* = \left(\frac{Q/2}{k}\right)^{1/3} \qquad \textbf{(11.3.85)}$$

since the flows splits in half at this branch.

Thus in the first generation (the daughter generation), we have the relationship

$$r_o^3 = 2r_1^3 \tag{11.3.86}$$

or, more generally, between any pair of generations,

$$r_n = 1.26r_{n+1} \tag{11.3.87}$$

Murray's law (which is Eq. 11.3.86), stating that the cube of the radius of a parent vessel equals the sum of the cubes of the radii of the daughter vessels, while often cited in the physiology literature, is not easy to confirm. In the first place, the mammalian vasculature is not a simple bifurcating system, as suggested in Fig. 11.3.11. Furthermore, two of our assumptions [that Poiseuille's law can serve as the $Q(\Delta P)$ model and that entrance effects can be neglected] do not hold uniformly through the generations. Vessel wall metabolism may also vary from one generation to another. Finally, the κ value may not be the same in small and large vessels. From an experimental point of view, it is extremely difficult to measure the radii of the smaller vessels with precision, keeping in mind that we are examining a relationship that involves the cube of the radius. Assessments of Murray's law are available [e.g., Sherman, *J. Gen. Physiol.*, **78**, 431 (1981)], and the best one can say is that it seems to be valid in a gross or average anatomical sense, but not in detail from one generation of bifurcation to the next. The point of the analysis displayed here is that if the details of the geometry and metabolism are known, the optimum characteristics of the branching system can be found. Thus Eq. 11.3.86 is Murray's law, and the derivation clarifies the assumptions that lead to this specific result.

11.4 FLOWRATE MEASURING DEVICES

In this section we select some examples of application of the macroscopic balances to the analysis of flows of practical importance. In particular, we see that we can design and analyze flow measuring devices based on the Bernoulli equation.

11.4.1 The Orifice Meter

If a restriction of some kind is placed in a pipeline, its presence will cause an excess pressure drop relative to the unrestricted flow. The extent of this excess pressure drop depends on the flowrate, and we may use this relationship to analyze and design a device that provides a measurement of the flowrate from the observed pressure drop.

Figure 11.4.1 shows one of the simplest flow measuring devices for a pipeline—the orifice meter. A pipe of cross-sectional area A_1 is "broken" at some point along the

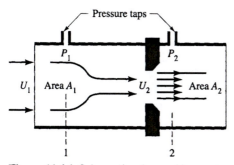

Figure 11.4.1 Schematic of an orifice meter.

axis, and a plate with a hole of smaller area A_2 is inserted. This plate is the orifice plate. Flow is forced to accelerate through this restriction, and there is a loss in pressure just downstream of the orifice. Pressure is measured by tapping into the pipe wall at two points, typically about one pipe diameter upstream, for the measurement of P_1, and a half-diameter downstream for P_2. The pressure difference $P_1 - P_2$ is measured.

The analysis of such a flowmeter begins with the Bernoulli equation, in the form given in Eq. 11.3.66. The planes 1 and 2 coincide with the pressure taps, as shown in Fig. 11.4.1. In the idealized case of no friction losses across the meter, we would set the head loss to zero and solve for the upstream velocity:

$$U_1 = \beta \left[\frac{2(P_1 - P_2)}{\rho(1 - \beta^2)} \right]^{1/2} \tag{11.4.1}$$

where

$$\beta \equiv \frac{A_2}{A_1} < 1 \tag{11.4.2}$$

is the ratio of orifice area to upstream pipe area. (Note that in the preceding analysis of the flow through a sudden *expansion,* we defined an area ratio β_{21} as the ratio of downstream to upstream areas. We do the same here, but since this is a contraction, this β is less than one.)

We may write the volume flowrate in the form

$$Q = A_2 U_2 = A_1 U_1 = A_2 \left[\frac{2(P_1 - P_2)}{\rho(1 - \beta^2)} \right]^{1/2} \tag{11.4.3}$$

This gives the desired quantity, Q, in terms of the measured pressure drop, the contraction area ratio β, and the orifice area A_2. The result also depends on the density of the fluid. Comparison of this model to the actual performance of an orifice meter shows that the predicted flowrate is larger than the actual flowrate. This error in the model arises from our neglect of frictional losses in the flow through the orifice and in the region just downstream of the orifice. These losses depend largely on the geometrical design: the area ratio β, and the placement of the pressure taps, and to some extent on the Reynolds number of the upstream flow. Commercial orifice meters of a specific design are provided with a calibration factor that converts Eq. 11.4.3 to an accurate relationship. This is usually put in the form of a so-called orifice discharge coefficient C_d, so that we write Eq. 11.4.3 in the form

$$Q = C_d A_2 \left[\frac{2(P_1 - P_2)}{\rho(1 - \beta^2)} \right]^{1/2} \tag{11.4.4}$$

Typical discharge coefficients are plotted in Fig. 11.4.2: note that the Reynolds number used is based on the diameter d *of the orifice,* not on the diameter of the pipe. Note also that curves are shown extended into the low Reynolds number region. Since the flow is through an orifice (essentially a pipe of zero axial length), it does not follow that the low Reynolds number flow through the orifice is laminar in the usual sense of one-dimensional parallel streamlines. Since, however, the Reynolds number of the upstream flow is even *smaller* than that of the flow through the orifice (because of the larger diameter—remember, this is at constant flowrate w), the flow *to* the orifice is laminar, at low Reynolds number, in the sense of being nonturbulent. Therefore the analysis based on the Bernoulli equation is suspect, since frictional effects were neglected. Nevertheless, one can write something similar to Eq. 11.4.4 on the basis of dimensional analysis, without any assumption regarding energy losses, and still expect

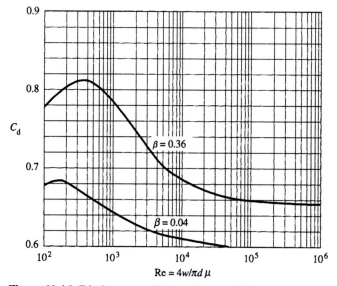

Figure 11.4.2 Discharge coefficients for an orifice meter.

C_d to be a function of Reynolds number and geometry. In any event, the curves in Fig. 11.4.2 are based on *experimental* observations over the range of Reynolds numbers shown.

When flowrates are high, there can be a substantial pressure drop across the orifice. For a gas, the ratio of absolute pressures P_2/P_1 can differ considerably from unity, and so there is a significant density change in the gas on either side of the orifice. Then a correction to the mass balance must be made in deriving Eq. 11.4.3, and the resulting corrected model for the orifice meter is of the form of Eq. 11.4.4, but it is written in terms of the *mass* flow as

$$w = Q_1 \rho_1 = C_d Y A_2 \left[\frac{2\rho_1(P_1 - P_2)}{(1 - \beta^2)} \right]^{1/2} \tag{11.4.5}$$

The correction factor Y is found experimentally to be, to a good approximation,

$$Y = 1 - \frac{1 - P_2/P_1}{C_p/C_v}(0.41 + 0.35\,\beta^2) \tag{11.4.6}$$

The heat capacity ratio C_p/C_v is 1.4 for air.

11.4.2 The Venturi Meter

The orifice meter is very inexpensive, but it has the potential disadvantage of causing a significant loss in pressure in a pipeline. A more expensive meter, which retards the flow less, is the Venturi meter shown in Fig. 11.4.3. Application of the Bernoulli equation between planes 1 and 2 leads to Eq. 11.4.4, with A_2 being the throat area and β the ratio of throat area to upstream pipe area. The experimentally determined flow coefficient for a Venturi meter having $\beta = 0.25$, and with the contraction and expansion angles shown in Fig. 11.4.3, is given in Fig. 11.4.4.

As in the case of an orifice meter, compressibility effects in the Venturi meter must be accounted for if the pressure ratio P_2/P_1 is much below unity. The correction factor Y is different from that given in Eq. 11.4.6 for an orifice, and the usually recommended

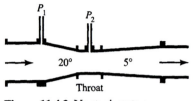

Figure 11.4.3 Venturi meter.

expression is

$$Y = \left[r^{2/\gamma} \left(\frac{\gamma}{\gamma - 1} \right) \left(\frac{1 - r^{(\gamma-1)/\gamma}}{1 - r} \right) \left(\frac{1 - \beta^2}{1 - \beta^2 r^{2/\gamma}} \right) \right]^{1/2} \tag{11.4.7}$$

where

$$r = P_2/P_1 \qquad \text{and} \qquad \gamma = C_p/C_v \tag{11.4.8}$$

Equation 11.4.7 is valid for a frictionless (isentropic) flow that is also adiabatic ($Q_H = 0$). It can be derived from simple thermodynamic relationships, along with the assumption of the ideal gas law. Figure 11.4.5 shows the function defined by Eq. 11.4.7 for the case of $\gamma = C_p/C_v = 1.4$ for air. The lowest pressure ratio shown, $P_2/P_1 = 0.53$, has the significance that at this ratio, the gas velocity in the throat of the Venturi meter reaches sonic velocity (Mach 1). For any pressure ratio below this value, the gas can flow no faster than its sonic velocity, and the pressure ratio is no longer a measure of the flow rate. Venturi (and other) flow meters are usually used at conditions so far from sonic gas velocities that we need not worry about the behavior of a gas flow near sonic conditions. In some situations, however, high speed gas flows are created, and we must understand the behavior of the gas—especially the pressure, temperature, flowrate relationships—under these conditions. Hence we turn to this topic now.

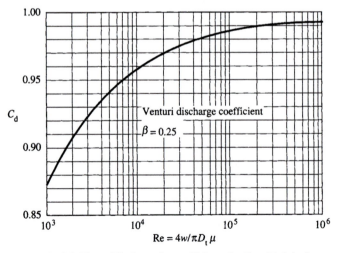

Figure 11.4.4 Flow (discharge) coefficient in Eq. 11.4.4, for a Venturi meter. D_t is the throat diameter.

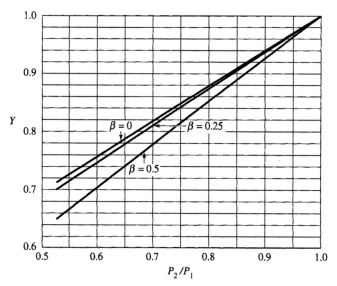

Figure 11.4.5 Correction factor from Eq. 11.4.7 for a Venturi meter.

11.5   COMPRESSIBLE FLOW IN A PIPE WITH FRICTION

To a large degree we have ignored compressibility effects in the examples of this textbook. Liquids are so incompressible that this is a very good assumption for practically any liquid flow of interest to us. Gases are compressible, of course, but often the effect of compressibility is minor as long as the gas flow velocity is small compared to the sonic velocity in the gas. Many important problems involve *subsonic* gas flow, in which we may safely simplify any compressibility effects that occur. We now carry out an analysis of compressible gas flow appropriate to the consideration of very high speed flows, where the sonic velocity is more nearly approached.

Thermodynamics plays an important role in these analyses, and it will be useful to recall some important thermodynamic relationships and catalog them here. To begin, we will assume that the ideal gas law holds for the cases of interest to us. Hence the pressure–temperature–density relationship is

$$p = \frac{\rho R_G T}{M_w} \tag{11.5.1}$$

where ρ is the mass density and M_w is the molecular weight of the gas. The Universal gas constant R_G has a value of 8314 (kg $\cdot$ m^2/s^2 $\cdot$ kg–mol $\cdot$ K) in SI units.

The enthalpy (per unit mass) is related to the internal energy per unit mass ε by

$$H = \varepsilon + \frac{p}{\rho} \tag{11.5.2}$$

Two important gas properties are the heat capacities C_v and C_p, defined by

$$C_p \equiv \left(\frac{\partial H}{\partial T}\right)_p \quad \text{and} \quad C_v \equiv \left(\frac{\partial \varepsilon}{\partial T}\right)_v \tag{11.5.3}$$

Note that these are the heat capacities *per unit mass,* since the enthalpy and energy per unit *mass* are used in these definitions. Similar definitions hold for the heat capacities *per mole,* when the enthalpy and internal energy *per mole* are used in the equations above.

We will find that the ratio of these heat capacities is important, so we define the property γ by means of a ratio:

$$C_p/C_v \equiv \gamma \tag{11.5.4}$$

As we noted before, for air γ has a value of 1.4.

For an ideal gas, it is known that C_v and C_p are related by

$$C_p - C_v = \frac{R_G}{M_w} \tag{11.5.5}$$

With these relationships at our fingertips, we may now proceed with an analysis of compressible gas flow. There are two important limiting cases to consider. One is appropriate to high speed flow through short pipes, such that hardly any heat is transferred across the walls of the pipe. In the limit of vanishing heat transfer such a flow is called "adiabatic." The second case, isothermal flow, is much more difficult to achieve for high speed flows under practical operating conditions.

11.5.1   Adiabatic Compressible Flow in a Pipe with Friction

For a pipe of constant cross-sectional area, the principle of conservation of mass ensures that

$$\rho U = \text{constant} \tag{11.5.6}$$

We may write this in differential form as

$$\frac{dU}{U} + \frac{d\rho}{\rho} = 0 \tag{11.5.7}$$

If we go back to the energy balance (Eq. 11.3.10), assume steady horizontal adiabatic flow with no shaft work, and then introduce the definition of enthalpy ($H = \varepsilon + p/\rho$), we find that for such a flow

$$H + \frac{U^2}{2} = \text{constant} \tag{11.5.8}$$

or

$$dH + U\,dU = 0 \tag{11.5.9}$$

By manipulating Eqs. 11.5.3, 11.5.4, and 11.5.1 we may write this as

$$\frac{\gamma R_G\,dT}{M_w\,(\gamma - 1)} + U\,dU = 0 \tag{11.5.10}$$

In a real fluid, friction losses will affect the pressure along the pipe axis, and we may write the momentum equation for this system in differential form as (see Eq. 11.2.10 as a starting point)

$$\rho U\,dU + dp + \frac{4\tau_w}{D}\,dz = 0 \tag{11.5.11}$$

where τ_w is the shear stress along the inside surface of the pipe. If we introduce the friction factor as

$$f \equiv \frac{2\tau_w}{\rho U^2} \tag{11.5.12}$$

then we may write Eq. 11.5.11 as

$$\rho U\, dU + dp + \frac{2f\rho U^2}{D} dz = 0 \tag{11.5.13}$$

The first thing we want to do is find the velocity variation along the axis of the pipe. Since our interest is in high speed compressible flow, we will introduce the Mach number Ma as the primary dependent variable:

$$\text{Ma} = \frac{U}{c} \tag{11.5.14}$$

where c is the speed of sound in the gas at the temperature T. For an ideal gas this is given by

$$c = \left(\frac{R_G T \gamma}{M_w}\right)^{1/2} \tag{11.5.15}$$

This lets us write Eq. 11.5.13 as

$$\gamma\, \text{Ma}^2 \frac{dU}{U} + \frac{dp}{p} + \frac{2\gamma f \text{Ma}^2}{D} dz = 0 \tag{11.5.16}$$

After some manipulation, the equation of state for the gas (Eq. 11.5.1) may be written in differential form as

$$\frac{dp}{p} = \frac{d\rho}{\rho} + \frac{dT}{T} \tag{11.5.17}$$

Equation 11.5.7 may be used to replace ρ with U, and Eq. 11.5.10 may be used to replace T with U, so that

$$\frac{dp}{p} = -\frac{dU}{U} - (\gamma - 1)\, \text{Ma}^2 \frac{dU}{U} \tag{11.5.18}$$

The pressure may now be eliminated from the momentum balance to yield

$$(\text{Ma}^2 - 1)\frac{dU}{U} + \frac{2\gamma f \text{Ma}^2}{D} dz = 0 \tag{11.5.19}$$

Since U may be written in terms of the Mach number and T, and Eq. 11.5.10 may be used to replace T with U, we may write this equation in terms of Ma only, and it takes the form, after some algebra, of a differential equation for Ma as a function of position z:

$$\frac{1 - \text{Ma}^2}{\text{Ma}^3 \{1 + [(\gamma - 1)/2]\, \text{Ma}^2\}} d\,\text{Ma} = \frac{2\gamma f}{D} dz \tag{11.5.20}$$

Equation 11.5.20 may be integrated analytically if f is constant, but we first note that if the entrance flow is subsonic (i.e., if Ma < 1), then $d\,\text{Ma}/dz > 0$ and the Mach number increases along the axis of the pipe. The reverse is true for a flow that is supersonic at the entrance—the Mach number will decrease in that case.

We may use the second law of thermodynamics to prove that the gas flow cannot change from sonic to subsonic, or *vice versa,* along the pipe axis. To do so we introduce the *entropy S* per unit mass, defined in terms of the enthalpy as

$$T \, dS = dH - \frac{dp}{\rho} \tag{11.5.21}$$

The second law tells us that in any process (such as flow of a gas along the pipe axis), the entropy of the gas can never decrease. Let us see where this statement leads us, beginning with Eq. 11.5.21. Dividing by T, introducing the definition of C_p, and using the ideal gas law, we find

$$dS = \frac{C_p \, dT}{T} - \frac{R_G}{M_w} \frac{dp}{p} = \frac{C_p \, dT}{T} - \frac{R_G}{M_w} \left(\frac{dp}{\rho} + \frac{dT}{T} \right) \tag{11.5.22}$$

or

$$\frac{dS}{dT} = \frac{C_p}{T} - \frac{R_G}{M_w} \left(\frac{d\rho}{\rho \, dT} + \frac{1}{T} \right) \tag{11.5.23}$$

From Eq. 11.5.6 we find

$$\frac{d\rho}{\rho} + \frac{dU}{U} = 0 \tag{11.5.24}$$

From Eq. 11.5.9 we find

$$\frac{dH}{dT} = -U \frac{dU}{dT} = C_p \tag{11.5.25}$$

Then Eq. 11.5.23 takes the form

$$\frac{dS}{dT} = \frac{C_p}{T} - \frac{R_G}{M_w} \left(\frac{C_p}{U^2} + \frac{1}{T} \right) \tag{11.5.26}$$

Somewhere along the pipe axis, the entropy may reach a maximum, and if it does, from setting $dS/dT = 0$ we find

$$\frac{C_p}{T^*} - \frac{R_G}{M_w T^*} = \frac{R_G C_p}{M_w U^{*2}} \tag{11.5.27}$$

where the asterisks on T and U are reminders that these are evaluated at the point of maximum entropy. We may now solve for U^*, and if we use the ideal gas law (hence, Eq. 11.5.5) and Eq. 11.5.4, we find

$$U^* = \left(\frac{R_G T^* \gamma}{M_w} \right)^{1/2} = c \tag{11.5.28}$$

or Ma $= 1$ when entropy reaches its maximum value in the pipe.

We have proven the following: *for a subsonic flow, the gas will accelerate in speed until Ma = 1.* If the Mach number were to continue to increase beyond that point in the pipe, the entropy would *decrease*—a violation of the second law of thermodynamics. It follows then that we could have a flow such that the gas speed would increase along the axis until the achievement of Ma $= 1$, and beyond that point the gas speed would remain at Ma $= 1$, even though the pressure would continue to drop toward the pipe exit. This condition is called "choked flow."

Let's return now to Eq. 11.5.20. We note first that conservation of mass implies that ρU is constant along the pipe axis. The temperature will vary along the axis in an

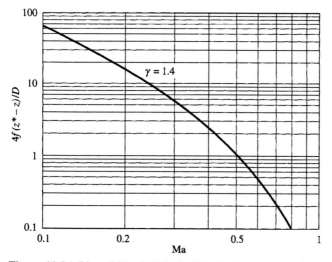

Figure 11.5.1 Plot of Eq. 11.5.30 (adiabatic flow in a pipe).

adiabatic flow, but we can show that typical temperature variations are about 20% of the absolute value of the mean temperature; hence gas viscosity will change by only 10%. Hence the Reynolds number for flow will vary by a small amount along the pipe axis. But in turbulent flow (we expect high speed gas flow to be turbulent) the friction factor is a weak function of Reynolds number. Hence a good approximation is that the friction factor is constant along the axis of the pipe. Then Eq. 11.5.20 may be integrated analytically, and the result is

$$-\frac{1}{2\,\mathrm{Ma}^2} - \frac{1+\gamma}{2}\ln\mathrm{Ma} + \frac{1+\gamma}{4}\ln\left(1 + \frac{\gamma-1}{2}\mathrm{Ma}^2\right) = \frac{2\gamma f z}{D} + \mathrm{A} \quad \textbf{(11.5.29)}$$

where A is a constant of integration. We fix a value for A by defining a position z^* along the axis where Ma = 1. This may be a *hypothetical* position, in the sense that z^* may be past the exit of the pipe. We shall see in a moment that this is just a mathematical trick that lets us solve flow problems. With this definition for A we find

$$\frac{1 - \mathrm{Ma}^2}{\gamma\,\mathrm{Ma}^2} + \frac{1+\gamma}{2\gamma}\ln\frac{(1+\gamma)\,\mathrm{Ma}^2}{2 + (\gamma-1)\,\mathrm{Ma}^2} = \frac{4f(z^* - z)}{D} \quad \textbf{(11.5.30)}$$

This expression gives us the variation of Ma, hence of speed U, along the axis of the pipe. For a given value of γ (i.e., for a specific gas), we may plot the right-hand side of Eq. 11.5.30 as a function of Ma. Such a plot is shown in Fig. 11.5.1.

Let's look at some typical problems to which we can apply this analysis.

EXAMPLE 11.5.1   *Pipe Length to Achieve Sonic Flow*

Air enters a pipe at $T = 373$ K and $p = 10^6$ N/m² (approx. 10 atm) and a speed of 100 m/s. What length of pipe (for $D = 2.5$ cm) is required to achieve sonic velocity?

First let's find the speed of sound at this temperature, for air. From Eq. 11.5.15 we calculate

$$c = \left(\frac{8314 \times 373 \times 1.4}{29}\right)^{1/2} = 387 \text{ m/s} \quad \textbf{(11.5.31)}$$

Hence the Mach number at the entrance is

$$Ma = \frac{100}{387} = 0.26 \tag{11.5.32}$$

From Fig. 11.5.1, at $Ma = 0.26$ we find

$$\frac{4f(z^* - z)}{D} = 7 \tag{11.5.33}$$

In this expression, $z^* - z$ is the distance required for the Mach number to increase from 0.26 to 1.0. (We could take $z = 0$ at the entrance, so that z^* is the required length.) We need an estimate for f, which means we must first calculate a Reynolds number for this flow. The viscosity and density of air at 373 K and 10^6 N/m^2 are found to be $\mu = 2.2 \times 10^{-5}$ Pa$\cdot$s and $\rho = 9.3$ kg/m^3. The Reynolds number is found to be Re = 6.5×10^5.

If we assume a roughness factor of $\delta/D = 0.001$, we find a friction factor of (see Fig. 11.3.3)

$$f = 0.005$$

Then from Eq. 11.5.33 we find

$$z^* - z = 17.5 \text{ m}$$

If the pipe is this long, *or longer*, the air will exit at sonic velocity.

EXAMPLE 11.5.2 *Pipe Length to Achieve a Specified Mach Number*

For the conditions specified in Example 11.5.1, find the length of pipe for which the Mach number becomes 0.5.

Equation 11.5.30 gives the distance over which the Mach number changes from $Ma = 1$ (at $z = z^*$) to some other value at z. If we want the distance over which the Mach number changes between any two values, we may calculate $z^* - z_1$ and $z^* - z_2$ for Mach numbers Ma_1 and Ma_2, and then simply find

$$z_2 - z_1 = (z^* - z_1) - (z^* - z_2) \tag{11.5.34}$$

For $Ma = 0.26$ we have

$$\frac{4f(z^* - z)}{D} = 7$$

while for $Ma = 0.5$, from Fig. 11.5.1, we find

$$\frac{4f(z^* - z)}{D} = 1 \tag{11.5.35}$$

Hence

$$\frac{4f(z_1 - z_2)}{D} = 7 - 1 = 6 \tag{11.5.36}$$

and

$$z_1 - z_2 = \frac{6D}{4f} = 7.5 \text{ m} \tag{11.5.37}$$

The Mach number rises to $Ma = 0.5$ at a distance of 7.5 m from the entrance.

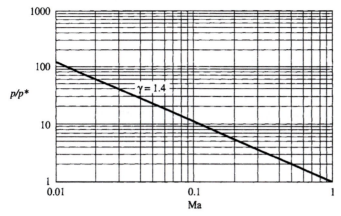

Figure 11.5.2 Plot of Eq. 11.5.42 (adiabatic flow in a pipe).

In specifying certain inlet conditions in Examples 11.5.1 and 11.5.2, we imply that we maintain a pressure drop across the ends of the pipe sufficient to drive the flow. We turn now to the means by which we find that pressure. We begin with Eq. 11.5.17. The density term is replaced with Eq. 11.5.24. The temperature term is replaced with Eq. 11.5.10. This gives us

$$\frac{dp}{p} = -\frac{dU}{U}[1 + (\gamma - 1)\,\text{Ma}^2] \tag{11.5.38}$$

From Eqs. 11.5.14 and 11.5.15, we find

$$\frac{d\,\text{Ma}}{\text{Ma}} = \frac{dU}{U} - \frac{1}{2}\frac{dT}{T} \tag{11.5.39}$$

Again, Eq. 11.5.10 is used to replace temperature with speed. The result is

$$\frac{d\,\text{Ma}}{\text{Ma}} = \frac{dU}{U}\left(1 + \frac{\gamma - 1}{2}\text{Ma}^2\right) \tag{11.5.40}$$

This lets us write Eq. 11.5.38 entirely in terms of p and Ma:

$$\frac{dp}{p} = -\frac{d\,\text{Ma}}{\text{Ma}}\left[\frac{1 + (\gamma - 1)\,\text{Ma}^2}{1 + (\gamma - 1)\,\text{Ma}^2/2}\right] \tag{11.5.41}$$

This differential equation may be integrated with the result

$$\frac{p}{p^*} = \frac{1}{\text{Ma}}\left[\frac{\gamma + 1}{2 + (\gamma - 1)\,\text{Ma}^2}\right]^{1/2} \tag{11.5.42}$$

where p^* is the pressure at which the Mach number becomes unity. This relationship is shown in Fig. 11.5.2.

We may now use this result to find the outlet pressure required to drive the flow of Example 11.5.1.

EXAMPLE 11.5.3 *Outlet Pressure to Achieve Sonic Flow*

In Example 11.5.1 the inlet Mach number is Ma = 0.26. From Fig. 11.5.2 we find

$$\frac{p}{p^*} = 4.2 \tag{11.5.43}$$

Since the inlet pressure is $p = 10^6$ N/m^2, we find

$$p^* = \frac{10^6}{4.2} = 2.38 \times 10^5 \, \text{N/m}^2 = 2.38 \times 10^5 \, \text{Pa} \qquad (11.5.44)$$

As the downstream pressure falls from 10^6 Pa, the gas accelerates toward sonic velocity. If the outlet pressure falls below the value of p^* given in Eq. 11.5.44, the gas flow remains sonic—it is choked beyond that point in pressure.

EXAMPLE 11.5.4   *Inlet Pressure to Achieve Sonic Flow*

Nitrogen flows through a capillary of diameter 1 cm and length 10 cm. The flow is adiabatic. The downstream (outlet) pressure is 2500 Pa. What upstream pressure yields sonic velocity at the outlet? What is the Reynolds number for this flow? The inlet temperature is 373 K.

This is a difficult problem: because the temperature variation along the pipe axis is unknown, the temperature at the outlet also is unknown. However, we find that there is a relatively small drop in temperature in an initially subsonic flow as it accelerates toward the exit. We will begin with the assumption that the downstream temperature is 373 K. From Eq. 11.5.15 we may find (using $M_w = 28$ and $\gamma = 1.4$ for nitrogen)

$$c = U^* = 387 \, \text{m/s} \qquad (11.5.45)$$

The viscosity of nitrogen at 373 K may be found to be

$$\mu = 2.2 \times 10^{-5} \, \text{Pa} \cdot \text{s} \qquad (11.5.46)$$

The density at the outlet (from the ideal gas law) is

$$\rho = \frac{p \, M_w}{R_G \, T} = \frac{2500 \times 28}{8314 \times 373} = 2.25 \times 10^{-2} \, \text{kg/m}^3 \qquad (11.5.47)$$

Thus we find that the Reynolds number is

$$\text{Re} = \frac{\rho D U^*}{\mu} = \frac{2.25 \times 10^{-2} \times 10^{-2} \times 387}{2.2 \times 10^{-5}} = 3975 \qquad (11.5.48)$$

From the friction factor chart (assuming a smooth tube), we find $f = 0.01$. This gives a value of

$$\frac{4f(z^* - z)}{D} = \frac{4 \times 0.01 \times 0.1}{0.01} = 0.4 \qquad (11.5.49)$$

From Fig. 11.5.1 we find, at the inlet

$$\text{Ma} = 0.62 \qquad (11.5.50)$$

From Fig. 11.5.2 we find

$$\frac{p}{p^*} = 1.7 \qquad (11.5.51)$$

or

$$p = 4250 \, \text{Pa} \qquad (11.5.52)$$

A detailed analysis of the temperature change in this flow would yield an outlet temperature of 335 K. This would have such a small effect on the calculations that we do not bother to modify Eqs. 11.5.45–11.5.47 to improve the final result.

11.5.2   Isothermal Compressible Flow in a Pipe with Friction

The analysis of isothermal flow is much simpler than that of adiabatic flow. Equation 11.5.16 is still valid as written earlier. Equation 11.5.17 simplifies to

$$\frac{dp}{p} = \frac{d\rho}{\rho} \tag{11.5.53}$$

The conservation of mass statement takes the form

$$\frac{d\rho}{\rho} = -\frac{dU}{U} \tag{11.5.54}$$

This lets us write the momentum equation (Eq. 11.5.16) with the aid of Eq. 11.5.14 as

$$\frac{d\,\mathrm{Ma}}{dz} = \frac{2f}{D}\frac{\gamma\,\mathrm{Ma}^3}{1 - \gamma\,\mathrm{Ma}^2} \tag{11.5.55}$$

Note that $d\,\mathrm{Ma}/dz$ changes sign about the critical value

$$\mathrm{Ma}^* = \left(\frac{1}{\gamma}\right)^{1/2} \tag{11.5.56}$$

Thus this constitutes the theoretical maximum value of the Mach number under isothermal flow conditions. Since γ exceeds unity, Ma^* is always less than unity—the flow is never sonic if the fluid enters the pipe as a subsonic flow.

We may integrate Eq. 11.5.55 and find

$$\frac{4f(z_T - z)}{D} = \frac{1 - \gamma\,\mathrm{Ma}^2}{\gamma\,\mathrm{Ma}^2} + \ln(\gamma\,\mathrm{Ma}^2) \tag{11.5.57}$$

where z_T is the axial position (downstream of position z) where the gas reaches Ma^*. Figure 11.5.3 shows the function defined in Eq. 11.5.57.

The pressure ratio at which the maximum gas speed is reached is easily found in the isothermal case. Since density is proportional to pressure, and Mach number is proportional to speed, under isothermal conditions we may write Eq. 11.5.54 (using Eq. 11.5.14) as

$$\frac{dp}{p} = -\frac{d\,\mathrm{Ma}}{\mathrm{Ma}} \tag{11.5.58}$$

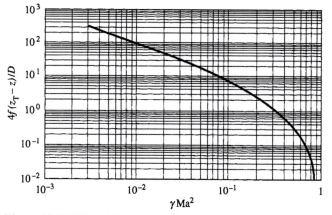

Figure 11.5.3 Plot of Eq. 11.5.57 (isothermal flow in a pipe).

We solve Eq. 11.5.58 by means of the boundary condition (really, the definition of p_T) that the maximum Mach number occurs at the pressure p_T. Then the inlet pressure at which the maximum outlet gas speed is reached is found as

$$\frac{p}{p_T} = \frac{1}{\sqrt{\gamma}\,\text{Ma}} \tag{11.5.59}$$

EXAMPLE 11.5.5   *Inlet Pressure to Achieve Maximum Speed: Isothermal Flow*

Nitrogen flows through a capillary of diameter 1 cm and length 10 cm. The flow is isothermal. The downstream (outlet) pressure is 2500 Pa. What upstream inlet pressure yields the maximum speed at the outlet? What is the Reynolds number for this flow? The inlet temperature is 373 K.

From Example 11.5.4 we found $c = 387$ m/s at this temperature. The maximum Mach number is given by Eq. 11.5.56 as

$$\text{Ma}^* = \frac{1}{\sqrt{\gamma}} = \frac{1}{\sqrt{1.4}} = 0.845 \tag{11.5.60}$$

Hence

$$U^* = 0.845 \times 387 = 327 \text{ m/s} \tag{11.5.61}$$

Using the data from Example 11.5.4, we find the Reynolds number:

$$\text{Re} = 3359 \tag{11.5.62}$$

Since the Reynolds number does not vary along the pipe axis for isothermal flow, the friction factor at the pipe entrance is also unchanged. Hence we find

$$\frac{4f(z_T - z)}{D} = \frac{4 \times 0.01 \times 0.1}{0.01} = 0.4 \tag{11.5.63}$$

From Fig. 11.5.3 we find

$$\gamma\,\text{Ma}^2 = 0.46 \tag{11.5.64}$$

From Eq. 11.5.59 we find

$$\frac{p}{p_T} = \frac{1}{\sqrt{\gamma}\,\text{Ma}} = \frac{1}{\sqrt{0.46}} = 1.47 \tag{11.5.65}$$

Hence the upstream pressure is

$$p = 1.47 \times 2500 = 3686 \text{ Pa} \tag{11.5.66}$$

EXAMPLE 11.5.6   *The Design of a Mass Flow Controller*

It is often necessary to be able to monitor and control the rate of delivery of a gas to a process. For example, the design of the system may incorporate factors that have the potential to cause variations in flowrate to the system. The analyses of this section provide a basis for the design of a controller that will maintain constant mass flow of a gas despite *downstream* variations in the system. Figure 11.5.4 shows a conceptual design for the mass flow controller (MFC) and the upstream and downstream ends of the system.

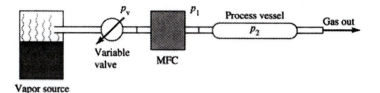

Figure 11.5.4 A mass flow controller (MFC) between a vapor source and a process vessel downstream.

The concept of this MFC is as follows. The vapor will flow through a set of parallel capillaries. The design criteria will ensure that the capillary length $(z^* - z)$ and the pressure ratio p_1/p_v permit the vapor flow to accelerate under *adiabatic* conditions to its maximum Mach number, Ma* = 1. Then *small* changes in the downstream conditions will not affect the flow from the MFC, which is "choked" at Ma = 1.

We will take the following properties for the vapor:

$$M_w = 60 \qquad \gamma = 1.4 \qquad \rho = 1.2 \text{ kg/m}^3 \text{ at 300 K and 1 atm}$$
$$\mu = 10^{-5} \text{ Pa} \cdot \text{s at 373 K}$$

The vapor enters the MFC at 373 K. Vapor flows from a liquid source by evaporation. The pressure p_v at the entrance to the MFC is essentially the vapor pressure of the liquid, and we take $p_v = 1000$ Pa at 373 K.

The process vessel is maintained at a very low operating pressure of $p_2 = 100$ Pa, but this pressure is subject to fluctuations arising in the high vacuum system downstream of the process vessel.

We must specify the length and diameter, and the number, of the capillaries that make up the MFC.

The system is to be designed to maintain a vapor flow of 600 sccm to the process vessel.

Our first task is to interpret the constraint that the vapor flow is to be measured in "sccm." This is a common notation for standard cubic centimeters per minute. The modifier "standard" means that the volume flow is measured at 273 K and 1 atm. At standard conditions, one gram-mole of any gas occupies 22,400 cm^3, so we may easily find the *mass* flowrate (in *kg/s*) to be

$$\dot{m} = \rho UA = \frac{600 \text{ sccm}}{60 \text{ s/m}} \frac{60 \text{ g/g-mol}}{22,400 \text{ cm}^3/\text{g-mol}} = 2.7 \times 10^{-5} \text{ kg/s} \qquad \textbf{(11.5.67)}$$

As a tentative design, the MFC will have a bundle of 100 capillaries of diameter $D = 0.001$ m. This gives an area for the flow of

$$A = 100 \frac{\pi}{4} D^2 = 25\pi \times 10^{-6} = 7.85 \times 10^{-5} \text{ m}^2 \qquad \textbf{(11.5.68)}$$

At the entrance to the MFC we have $p = 1000$ Pa and $T = 373$ K, so the gas density is

$$\rho = \frac{300}{373} \times \frac{1000}{10^5} \times 1.2 \text{ kg/m}^3 = 9.6 \times 10^{-3} \text{ kg/m}^3 \qquad \textbf{(11.5.69)}$$

This yields a value for the speed of the gas at the inlet to the MFC of

$$U = \frac{2.7 \times 10^{-5}}{9.6 \times 10^{-3} \times 7.85 \times 10^{-5}} = 36 \text{ m/s} \qquad \textbf{(11.5.70)}$$

The speed of sound at the system temperature is

$$c = \left(\frac{\gamma R_G T}{M_w} \right)^{1/2} = \left(\frac{1.4 \times 8314 \times 373}{60} \right)^{1/2} = 269 \text{ m/s} \qquad \textbf{(11.5.71)}$$

Hence the Mach number at the entrance is Ma = 36/269 = 0.13. Since this value is below unity, the gas will speed up toward Ma = 1.

From the value of the Mach number at the entrance to the MFC and Fig. 11.5.1, we find

$$\frac{4f(z^* - z)}{D} = 38 \qquad \textbf{(11.5.72)}$$

The Reynolds number for this flow is

$$\text{Re} = \frac{\rho U D}{\mu} = \frac{9.6 \times 10^{-3} \times 36 \times 0.001}{10^{-5}} = 34 \qquad \textbf{(11.5.73)}$$

Note that this is a laminar flow!

The friction factor for laminar tube flow is just $f = 16/\text{Re}$, so we find $f = 0.47$. Then, from Eq. 11.5.72 we find

$$z^* - z = \frac{38 \times 0.001}{4 \times 0.47} = 0.02 \text{ m} \qquad \textbf{(11.5.74)}$$

Hence a set of capillaries whose length is 2 cm will yield sonic flow at the exit of the capillaries, and thence into the process vessel, where the gas will expand and continue subsonically through the vessel. The pressure at the exit of the MFC follows from Fig. 11.5.2. At Ma = 0.13 we find

$$\frac{p_1}{p^*} = \frac{p_1}{p_2} = 8.4 \qquad \text{or} \qquad p_2 = \frac{1000}{8.4} = 119 \text{ Pa} \qquad \textbf{(11.5.75)}$$

As long as the flow from the MFC exits to a region whose pressure is below 119 Pa, the flow to the process vessel will be maintained at a constant mass flow. In short, as long as the pressure in the process vessel does not increase by more than 20% of the design value (p_2 was specified to be 100 Pa), this MFC will indeed control the flow.

A practical problem with this conceptual design is that the mass flow is not independent of changes in the *upstream* conditions. If the temperature on the liquid source varies to a small degree, there will be a significant change in the vapor pressure, and the mass flow of 600 sccm will not be maintained.

EXAMPLE 11.5.7   *Effect of Downstream Pressure on Flow into a Tube (Isothermal Flow)*

Suppose we create an airflow through a capillary tube by pulling a vacuum on the downstream end of the tube. The upstream end is open to the ambient, which is at 298 K and atmospheric pressure, which we take here as 1×10^5 Pa. By some means we maintain *isothermal* flow through the tube. Our goal is to find the flowrate through the tube as a function of downstream pressure. The tube dimensions are $L = 12.7$ mm and $D = 0.19$ mm.

We begin with a calculation for the specific case of a downstream pressure of 0.79 atm. This is a small pressure drop across a small diameter tube, and we expect a low flowrate. We will begin with Eq. 11.5.57, and assume that $\gamma \, \text{Ma}^2 \ll 1$ so that a good approximation is

$$\frac{4f(z_T - z)}{D} = \frac{1}{\gamma \, \text{Ma}^2} + \ln(\gamma \, \text{Ma}^2) \qquad \textbf{(11.5.76)}$$

We may write Eq. 11.5.76 twice, once for a position z_2 and once for z_1. For a tube of length $L = z_2 - z_1$, we then find

$$\frac{4fL}{D} = \frac{4f(z_2 - z_1)}{D} = \frac{1}{\gamma \, \text{Ma}_1^2}\left(1 - \frac{\text{Ma}_1^2}{\text{Ma}_2^2}\right) + 2\ln\frac{\text{Ma}_1}{\text{Ma}_2} \tag{11.5.77}$$

$$= \frac{1}{\gamma \, \text{Ma}_1^2}\left(1 - \frac{p_2^2}{p_1^2}\right) + 2\ln\frac{p_2}{p_1}$$

Let's assume laminar flow and solve for the Mach number Ma_1 at the tube entrance. Since the pressure ratio is given and the mass flow is constant through the tube, we can relate the downstream Mach number to that upstream through the pressure ratio at the two ends. (That in fact is how we introduced the pressure ratio into Eq. 11.5.77.) We will also find the Reynolds number in the tube, and thereby evaluate the assumption of laminar flow.

With the isothermal and laminar flow assumptions we write

$$f = \frac{16}{\text{Re}} = \frac{16\mu}{\rho_1 \, \text{Ma}_1 \, cD} \tag{11.5.78}$$

where the subscript 1 refers to the tube entrance. Combining Eqs. 11.5.77 and 11.5.78, we find

$$\text{Ma}_1 = \frac{p_1 D^2}{64\mu cL}\left(1 - \frac{p_2^2}{p_1^2}\right) + 2\left(\ln\frac{p_2}{p_1}\right)\frac{\rho_1 cD^2}{64\mu L}\text{Ma}_1^2 \tag{11.5.79}$$

which is simply a quadratic expression for Ma_1. From Fig. 3.4.3 we find $\mu = 1.85 \times 10^{-5}$ Pa·s at $T = 298$ K. From Eq. 11.5.31 we find the sound speed $c = 346$ m/s. These numbers, along with the other data of this example, yield an upstream (entrance) Mach number of

$$\text{Ma}_1 = 0.23 \tag{11.5.80}$$

This corresponds to a volume flowrate (measured at the upstream pressure of 1 atm) of

$$Q_1 = c \, \text{Ma}_1 \, A = 346 \times 0.23 \times \frac{\pi}{4}(0.19 \times 10^{-3})^2 = 2.3 \times 10^{-6}\,\text{m}^3/\text{s} \tag{11.5.81}$$

It is not difficult to confirm that the Reynolds number at the entrance is approximately 1000, so the assumption of laminar flow at the entrance is good. The flow accelerates only slightly, since the pressure ratio is only 0.79. Nevertheless, for isothermal flow, the Reynolds number remains constant along the flow direction. Hence the flow is laminar throughout the tube.

A little thought suggests that since the friction loss associated with the entrance flow into the tube has been neglected, the result just obtained is an overestimate of the flow through the tube. In effect, we have to correct the friction factor f. The easiest way to do this is to go back to the definition of the energy loss factor e_v (Eq. 11.3.22) in the form

$$e_v = \frac{4L}{D}f \tag{11.5.82}$$

For the factor e_v we use the contraction factor given by Eq. 11.3.73. Then the friction loss due to the contraction is equivalent to a straight portion of the tube of equivalent length

$$\frac{L_{eq}}{D} = \frac{e_v}{4f} = \frac{0.45}{4f} \tag{11.5.83}$$

For a calculated Re = 1000, we find that we should add a length

$$L_{eq} = \frac{0.45D}{4f} = \frac{0.45 \times 0.19}{4(16/1000)} = 1.5 \text{ mm} \tag{11.5.84}$$

to the actual length of $L = 12.7$ mm. This slightly reduces the calculated entrance Mach number to

$$Ma_1 = 0.23 \times \frac{12.7}{12.7 + 1.5} = 0.21 \tag{11.5.85}$$

At this point, since

$$\gamma \, Ma_1^2 = 1.4 \times 0.21^2 = 0.062 \tag{11.5.86}$$

we can confirm that Eq. 11.5.76 is a good approximation.

Now let's see what happens if we reduce the downstream pressure to 0.2 atm. We expect that the flowrate into the tube will increase over that of the case considered earlier. Equation 11.5.79 still applies if the flow is laminar, in which case we find

$$\begin{aligned} Ma_1 = &\frac{(1 \times 10^5)(0.19 \times 10^{-3})^2}{64(1.85 \times 10^{-5}) \, 346(12.7 \times 10^{-3})} (1 - 0.2^2) \\ &+ 2 \ln(0.2) \frac{(1.17 \times 346)(0.19 \times 10^{-3})^2}{64(1.85 \times 10^{-5})(12.7 \times 10^{-3})} \, Ma_1^2 \end{aligned} \tag{11.5.87}$$

The solution for Ma_1 is

$$Ma_1 = 0.33 \tag{11.5.88}$$

Now we calculate a Reynolds number of 1435, which is in the laminar regime. This supports our use of Eq. 11.5.78 for f. Correcting again for L_{eq} (using Eq. 11.5.84) we can find

$$Ma_1 = 0.33 \times \frac{12.7}{12.7 + 1.9} = 0.29 \tag{11.5.89}$$

In Fig. 11.5.5 we show the upstream (inlet) Mach number Ma_1 as a function of the ratio of downstream to upstream pressures. The equations we are working with yield solutions such that Ma_1 goes through a maximum at $p_2/p_1 = 0.38$. If we also plot the

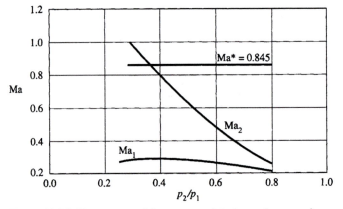

Figure 11.5.5 Upstream and downstream Mach numbers as a function of the pressure ratio.

downstream Mach number Ma_2 on the same graph, we see that as the pressure ratio falls to the level at which Ma_1 goes through its maximum, Ma_2 just reaches the critical value $Ma^* = 1/\sqrt{\gamma} = 0.845$ for air. Hence the flow is choked at a pressure ratio of approximately 0.38, and the theoretical maximum inlet Mach number is found to be $Ma_1 = 0.32$. The algebraic solutions for Ma_1 when $p_2/p_1 < 0.38$ are not physically realizable. Hence the calculation that leads to Eq. 11.5.88 is correct, but not physically meaningful.

A clue to the reason that the critical value $Ma^* = 1/\sqrt{\gamma}$ cannot be achieved in an *isothermal* flow can be obtained by considering a heat balance on the flow. We begin with Eq. 11.3.10, but we neglect the effect of gravity and assume that there is no shaftwork:

$$0 = d\left(\varepsilon + \frac{U^2}{2} + \frac{p}{\rho}\right) - d\hat{Q}_H \tag{11.5.90}$$

Next, we introduce the enthalpy $H = \varepsilon + p/\rho$ and write this equation as

$$0 = d\left(H + \frac{U^2}{2}\right) - d\hat{Q}_H \tag{11.5.91}$$

Equations 11.5.90 and 11.5.91 are valid whether the flow is isothermal or not. Although we are considering here an isothermal flow, we make use of a definition of the so-called adiabatic stagnation temperature. We begin by noting that if the flow were adiabatic ($d\hat{Q}_H = 0$), we would find

$$H + \frac{U^2}{2} = \text{constant} \tag{11.5.92}$$

If the flow were brought to a halt (stagnation), the enthalpy itself would increase to the value (setting $U = 0$)

$$H_s = H + \frac{U^2}{2} \tag{11.5.93}$$

But we may also define the enthalpy in terms of a temperature relative to some reference state:

$$H_s - H = C_p(T_s - T) \tag{11.5.94}$$

If we solve this equation for T_s, and introduce Eq. 11.5.93 and the definition of the Mach number, we find

$$T_s = \frac{U^2}{2C_p} + T = T\left(1 + \frac{\gamma - 1}{2}Ma^2\right) \tag{11.5.95}$$

This result is valid for an adiabatic flow (because of Eq. 11.5.92). *If the flow were brought to stagnation adiabatically*, T_s is the temperature we would measure. Hence T_s is called the *adiabatic stagnation temperature*. Equation 11.5.94 is valid whether the flow is adiabatic or not. We may write it in differential form as

$$dH_s = C_p dT_s \tag{11.5.96}$$

With these ideas in mind we now return to the case of interest—isothermal flow—for which Eq. 11.5.91 is valid.

$$0 = d\left(H + \frac{U^2}{2}\right) - d\hat{Q}_H = dH_s - d\hat{Q}_H \tag{11.5.97}$$

or, using Eq. 11.5.96

$$d\hat{Q}_H = C_p dT_s \tag{11.5.98}$$

Now, from Eq. 11.5.95, at constant temperature T,

$$dT_s = T(\gamma - 1)\text{Ma } d\text{Ma} \tag{11.5.99}$$

Using Eq. 11.5.55 we may write this in the form

$$d\hat{Q}_H = C_p T(\gamma - 1)\text{Ma } d\text{Ma} = \frac{2f}{D}\frac{C_p T(\gamma - 1)\gamma \text{ Ma}^4}{1 - \gamma \text{ Ma}^2} dz \tag{11.5.100}$$

This result states that as the Mach number approaches the critical value Ma* = $1/\sqrt{\gamma}$, the denominator of this expression goes to zero, and *the heat input required to maintain isothermal flow becomes unbounded.* From a practical point of view, when a compressible flow approaches the critical value isothermally, heat transfer eventually limits the approach, and the critical value is never achieved. As a consequence, the equations that describe isothermal compressible flow through a tube should not be used when the downstream Mach number approaches Ma*. The permissible degree of approach would depend on the heat transfer characteristics of the system.

11.6 ADIABATIC FRICTIONLESS COMPRESSIBLE FLOW

Often we must solve a flow problem for a gas, contained within a chamber at high pressure, that exits the chamber through a tube, orifice, or other restriction. When compressibility effects are important, this is a more complex flow than those considered in earlier chapters. Part of the difficulty lies in the possibility that the entrance pressure to the tube p_o (see Fig. 11.6.1) is not the chamber pressure p_U. We need to discuss this issue in detail now, and to develop a model that accounts for the flow field just upstream of the tube entrance.

In Fig. 11.6.1 we suggest a system in which there is a large volume of stationary gas at temperature T and pressure p_U, which leaks through a tube of length L and diameter D. Leakage of the gas does not alter the chamber pressure. At issue here is what we assume about the flow as the entrance to the tube is approached. This is a question we have already answered in our considerations of flows that were dominated by friction losses in the tube. For such flows we assumed that the pressure right at the tube entrance was the upstream pressure p_U and that there were no losses associated with the acceleration of the fluid *toward* the tube. For large Reynolds numbers, or very short tubes, we allowed for the possibility of some excess pressure loss associated with acceleration of the fluid *just beyond* (downstream of) the tube entrance, in the so-called entry region. But observations suggest that compressible flows at relatively large Mach

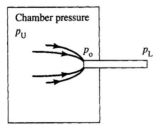

Figure 11.6.1 Adiabatic frictionless flow into a tube.

numbers lead to entrance pressures p_o that are quite different from the upstream chamber pressure p_U. Our goal here is to develop a model of this upstream flow.

In the analysis that follows, we will assume that the flow is *adiabatic* (no mechanism of heat transfer to the fluid as it approaches the entrance) and *frictionless* (viscous effects are negligible). Then we can go back to Eq. 11.5.11 and drop the shear stress term:

$$\rho U \, dU + dp = 0 \tag{11.6.1}$$

which can be rewritten in the form

$$\frac{dp}{\rho} + d\left(\frac{U^2}{2}\right) = 0 \tag{11.6.2}$$

If we assume that the gas is ideal and the flow adiabatic, it follows that

$$\frac{p}{\rho^\gamma} = \text{constant} = \frac{p_U}{\rho_U^\gamma} \tag{11.6.3}$$

where, as before, $\gamma = C_p/C_v$. When this is put into the momentum equation (Eq. 11.6.2), we find

$$\frac{p_U^{1/\gamma}}{\rho_U} \frac{dp}{p^{1/\gamma}} + d\left(\frac{U^2}{2}\right) = 0 \tag{11.6.4}$$

We now integrate this from some upstream position where we assume $U = 0$ and find

$$\frac{\gamma}{\gamma - 1}\left(\frac{p_U}{\rho_U} - \frac{p_o}{\rho_o}\right) - \left(\frac{U_o^2}{2}\right) = 0 \tag{11.6.5}$$

For an ideal gas

$$p = \rho \frac{R_G T}{M_w} \tag{11.6.6}$$

so we find

$$\frac{\gamma R_G}{M_w(\gamma - 1)}(T_U - T_o) - \frac{U_o^2}{2} = 0 \tag{11.6.7}$$

Using the definition of the Mach number (Eqs. 11.5.14 and 11.5.15), we may solve for the temperature ratio

$$\frac{T_o}{T_U} = \left(1 + \frac{\gamma - 1}{2} \text{Ma}^2\right)^{-1} \tag{11.6.8}$$

(See Eq. 11.5.95. Should we have expected this result?)

Using Eqs. 11.6.6 and 11.6.3 we can show that this leads to two additional expressions:

$$\frac{p_o}{p_U} = \left(\frac{T_o}{T_U}\right)^{\gamma/(\gamma-1)} = \left(1 + \frac{\gamma - 1}{2} \text{Ma}^2\right)^{-\gamma/(\gamma-1)} \tag{11.6.9}$$

and

$$\frac{\rho_o}{\rho_U} = \left(1 + \frac{\gamma - 1}{2} \text{Ma}^2\right)^{-1/(\gamma-1)} \tag{11.6.10}$$

With these results we may now estimate the entrance pressure p_o for compressible flow into a tube when the entrance region may be approximated as adiabatic and frictionless.

EXAMPLE 11.6.1   *Prediction of Compressible Flow Through a Tube*

Data are available [Kumagai, *J. Vac. Sci. Technol.*, **A8**, 2865 (1990)] for the type of flow we are considering here. Predict the flowrate of nitrogen through a tube of inside diameter $D = 0.015$ inch and length $L = 0.25$ inch, at a pressure $p_U = 50$ psig in a large chamber upstream from the entrance to the tube. The temperature in the upstream chamber is $T_U = 293$ K. The downstream pressure p_L is atmospheric (0 psig). Convert the estimated flowrate to units of standard liters per minute (slm): that is, volume flowrate "corrected" to a pressure of one atmosphere and a temperature of 273 K. The measured value is given as 4.2 slm.

We must begin with a guess of the friction factor f, since we do not know the flowrate. We guess the flow is turbulent, and looking at Fig. 11.3.3 we take an arbitrary value of $f = 0.003$. We will find that the trial-and-error procedure that follows is not very sensitive to the initial guess. From Fig. 11.5.1 we find an entrance Mach number, using

$$\frac{4fL}{D} = \frac{4(0.003)(0.25)}{0.015} = 0.2 \tag{11.6.11}$$

This yields an initial estimate of Ma = 0.7.

From Eq. 11.6.9 (using $\gamma = 1.4$ for nitrogen) we find

$$\frac{p_o}{p_U} = \left(\frac{T_o}{T_U}\right)^{\gamma/(\gamma-1)} = \left[1 + \frac{1.4-1}{2}(0.7)^2\right]^{-1.4/0.4} = (1.098)^{-3.5} = 0.72 \tag{11.6.12}$$

From $p_U = (14.7 + 50)/14.7 = 4.4$ atm we can now find

$$p_o = 0.72(4.4) = 3.2 \text{ atm} \tag{11.6.13}$$

and similarly,

$$T_o = 0.91(293) = 267 \text{ K} \tag{11.6.14}$$

The upstream density follows from the ideal gas law:

$$\rho_U = \frac{p_U M_w}{R_G T_U} = \frac{4.4 \times 10^5 (28)}{8314(293)} = 5.1 \text{ kg/m}^3 \tag{11.6.15}$$

Note carefully that we have converted the absolute pressure to SI units, and we use $R_G = 8314$ since we are in SI units, with $M_w = 28$ kg/kg–mol for nitrogen. From Eq. 11.6.10 this yields

$$\rho_o = 4 \text{ kg/m}^3 \tag{11.6.16}$$

We now have the gas conditions at the entrance to the tube. Note how different they are from the upstream conditions.

At the estimated entrance Mach number Ma = 0.7 we can find the velocity of the gas as

$$U = \left(\frac{R_G T_o \gamma}{M_w}\right)^{1/2} \text{Ma} = \left(\frac{8314 \times 267 \times 1.4}{28}\right)^{1/2} 0.7 = 233 \text{ m/s} \tag{11.6.17}$$

Now we can find the mass flowrate from

$$\dot{m} = \rho_o U_o A = 4(233)\frac{\pi}{4}(3.8 \times 10^{-4})^2 = 1.06 \times 10^{-4} \text{ kg/s} \tag{11.6.18}$$

Using a density for nitrogen (at STP) of $\rho_L = 1.23$ kg/m^3, we find

$$q_L = \frac{\dot{m}}{\rho_L} = \frac{1.06 \times 10^{-4}}{1.23} = 8.6 \times 10^{-5} \text{ m}^3/\text{s} = 5.2 \text{ slm} \tag{11.6.19}$$

This flowrate is in units of standard liters per minute because we have used the gas density at STP (and of course because we converted the flow to a minute basis).

With this result we may now calculate a Reynolds number and do a second iteration by correcting the f value from our initial guess of $f = 0.003$. For the Reynolds number we calculate

$$\text{Re} = \frac{4\dot{m}}{\pi D \mu} = \frac{4(1.06 \times 10^{-4})}{\pi(3.8 \times 10^{-4})1.8 \times 10^{-5}} = 2 \times 10^4 \tag{11.6.20}$$

From Fig. 11.3.3 we correct the f value to $f = 0.007$. Hence we find

$$\frac{4fL}{D} = \frac{4(0.007)(0.25)}{0.015} = 0.47 \tag{11.6.21}$$

This leads to a new estimate of the entrance Mach number (from Fig. 11.5.1) of Ma = 0.6. Note how insensitive the Ma estimate is to our poor initial guess of f. Following the procedure outlined above, we now estimate the flow conditions to be $T_o = 274$ K, $\rho_o = 4.2$ kg/m^3, $U_o = 203$ m/s, and $q_L = 4.5$ slm. The Reynolds number at this flowrate is changed only slightly from that given in Eq. 11.6.20, so the f value is accurate, and no further iteration is required.

This estimated value ($q_L = 4.5$ slm) is very close to the measured value of 4.2 slm. Problem 11.58 presents additional data, and we can prove that the method outlined here provides a very good model of this type of flow. The assumption of adiabatic frictionless flow up to the entrance to the tube leads to a model that is consistent with observations.

SUMMARY

In many areas of technology our interest in the fluid dynamics does not require a detailed (microscopic) knowledge of the velocity and pressure distributions throughout the system. Instead, it is often sufficient that we have models that relate the flows crossing the inlet and outlet boundaries of the system to the overall pressure difference across these boundaries, to the work (if any) transmitted from (or to) the interior of the system to (or from) the exterior, and to the forces exchanged across the boundaries that separate the exterior and the interior of the system.

The macroscopic statements of conservation of mass, momentum, and energy take the forms

$$\frac{dM}{dt} = \frac{d}{dt}\int \rho \, dV = \langle \rho U A \rangle_1 - \langle \rho U A \rangle_2 \tag{11.S.1}$$

$$\frac{d}{dt}\int \rho \mathbf{u} \, dV = \langle \rho U^2 \mathbf{A} \rangle_1 - \langle \rho U^2 \mathbf{A} \rangle_2 + \langle P\mathbf{A} \rangle_1 - \langle P\mathbf{A} \rangle_2 - \mathbf{F} + M\mathbf{g} \tag{11.S.2}$$

and

$$\frac{d}{dt}\int \rho \mathsf{E} \, dV = \langle \rho U A \, \mathsf{E} + pUA \rangle_1 - \langle \rho U A \, \mathsf{E} + pUA \rangle_2 + Q_H + W_s \tag{11.S.3}$$

The latter equation, in the special case of steady state constant density flows, becomes the Bernoulli equation

$$\Delta \frac{U^2}{2} + g\,\Delta h + \frac{\Delta p}{\rho} = \hat{W}_s - \hat{E}_v \qquad \text{(11.S.4)}$$

and is the foundation of much of the modeling that follows in this chapter. In particular, in Section 11.4 we apply the Bernoulli equation to the analysis of commercial flow measuring devices.

Throughout most of this text we have restricted our attention to *incompressible* fluids. In Sections 11.5 and 11.6 we apply the macroscopic energy balance, prior to the introduction of the assumption of constant density, to the modeling of *compressible* flows through pipes. We find that thermodynamic considerations enter the modeling process. Putting these ideas together, we are able to predict important features of the dynamics of compressible flows, as illustrated in Example 11.5.6 for design of a mass flow controller and Example 11.6.1 for predicting the flowrate out of a high pressure system through a small capillary.

PROBLEMS

11.1 Design a water clock (see Example 11.1.1) that will "run" for 12 hours, such that the height of the water is a linear function of time. How much water does your design require for start-up?

11.2 Design a "viscous liquid" clock that will run for 24 hours. Specify the size and shape of the reservoir, the viscosity of the working fluid, and the outlet design.

11.3. Find the force (magnitude and direction) required to hold the water nozzle shown in Fig. P11.3 in place. Assume that the pressure at the entrance to the converging section is 500 kPa (gage), and the exit velocity of the water jet is 50 m/s.

11.4 Find the force (magnitude and direction) required to hold in place the plate (the surface being cleaned) of Example 11.1.2. Assume the plate is at 45° to the jet axis, and is a square 1 m on a side. The water jet approaches the plate with a speed of 50 m/s.

11.5 A piece of flexible plastic tubing is used to deliver a liquid (water) to a process, as shown in Fig. P11.5. A short nozzle is made by squeezing a glass tube into the flexible tube. Find the axial force acting on the glass tube. The tube has an inner diameter of 1 cm and an outside diameter of 1.2 cm.

11.6 Air flows at a Reynolds number of 10^5 through a smooth circular pipe of diameter 2 cm. A 90° elbow is placed in the line at some point. The inside diameter of the elbow is 2 cm, and the radius of the elbow (see Fig. 11.3.9) is $R_b = 8$ cm. What length of straight pipe has the same pressure drop as this elbow?

11.7 Using Fig. 11.3.9, prepare a plot of the "equivalent length" L_e of a 90° elbow, as a function of R_b/D. In your plot, normalize the equivalent length to the pipe diameter. Define the equivalent length as the length of straight pipe that has the same pressure drop as the elbow, for a given pipe diameter. Plot your curve as $f L_e/D$ versus R_b/D.

11.8 A hydroelectric power plant takes in 40 m³/s of water through a supply line and into a turbine. The water is discharged at 5 m/s to atmospheric pressure. The reservoir that supplies the turbine is 100 m above the level of the turbine. Friction in the supply line, and in the turbine itself, corresponds to a "head loss" of

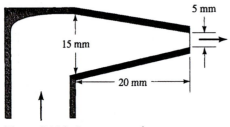

Figure P11.3 A water nozzle.

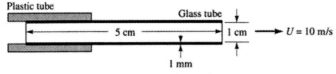

Figure P11.5 Forces acting on a glass tube.

25 m. How many megawatts of power can this turbine deliver, if all other losses are neglected?

11.9 Water flows at a rate $Q = 0.15 \text{ m}^3/\text{s}$ through a turbine, as shown in Fig. P11.9. Estimate the power generated by the turbine. Neglect all friction losses associated with the turbine itself.

11.10 With reference to Fig. P11.9, suppose the turbine is replaced by a pump. What horsepower pump is required to deliver water at $0.02 \text{ m}^3/\text{s}$ from the lower to the upper reservoir? Neglect all friction losses associated with the pump itself.

11.11 Water is being pumped from a reservoir to a site 50 m above the reservoir surface. The pump is rated at 150 kW, but it is only 65% efficient. (This means that 35% of the power is lost in overcoming frictional losses within the pump.) The total piping system is equivalent to 1000 m of 500 mm diameter smooth pipe. Find the discharge rate of the water.

11.12 In the momentum balance (Eq. 11.2.17) and in the energy balance (Eq. 11.3.10) we assumed flat velocity profiles across the entrance and exit sections. If we do not introduce that approximation, it is necessary to rewrite these equations by introducing two factors:

$$K_2 \equiv \frac{\langle U_1^2 \rangle}{\langle U_1 \rangle^2} \quad \text{and} \quad K_3 \equiv \frac{\langle U_1^3 \rangle}{\langle U_1 \rangle^3} \quad \textbf{(P11.12.1)}$$

Return to Eqs. 11.5.8 and 11.5.11 and show that they must be written in the forms

$$H + K_3 \frac{U^2}{2} = \text{constant} \quad \textbf{(P11.12.2)}$$

and

$$K_2 \rho U \, dU + dp + \frac{4\tau_w}{D} dz = 0 \quad \textbf{(P11.12.3)}$$

Carry out the analyses that yield Eqs. 11.5.30 and 11.5.42, leaving the coefficients K_2 and K_3 as constants in the resulting equations.

Show that for laminar flow in a tube of circular cross section

$$K_2 = 1.33 \quad \text{and} \quad K_3 = 2 \quad \textbf{(P11.12.4)}$$

11.13 Derive an expression for the "head loss" h_L equivalent to turbulent flow through a long smooth horizontal pipe of diameter D and length L, in terms of the friction factor. Ignore entrance and exit effects.

11.14 Prove that Eq. 11.5.57 has the following behavior:

$$\frac{4f(z_T - z)}{D} \to \frac{1}{\gamma \, \text{Ma}^2} \quad \text{for} \quad \text{Ma} \ll 1$$
$$\textbf{(P11.14.1)}$$

$$\frac{4f(z_T - z)}{D} \to \frac{1}{2}(1 - \gamma \, \text{Ma}^2)^2 \quad \text{for}$$
$$(\gamma \, \text{Ma}^2 - 1) \ll 1 \quad \textbf{(P11.14.2)}$$

11.15 A variable resistance valve is designed along the lines shown in Fig. P11.15. The "plug" is a truncated cone that moves along its axis (e.g., from position A to position B) and reduces the annular gap between the plug and the coaxial body of the valve. Hence the gas flows through an annular region of constant gap but varying radius. Develop a model of the pressure drop across this valve for a compressible gas. Design a valve suitable for use under the conditions specified in Example 11.5.7.

11.16 Go back to Example 11.3.3, and put plane 2 at the top of the *outlet* reservoir. How does Eq. 11.3.31 change? Do the individual terms have

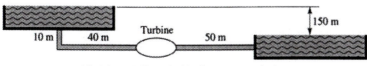

All piping is 250 mm inside diameter smooth pipe

Figure P11.9 A turbine connecting two reservoirs.

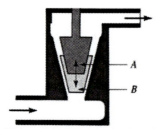

Figure P11.15 A variable resistance valve.

the same physical interpretations? In doing this problem, include the energy loss associated with the expansion of the flow from pipe 3 into the outlet reservoir. Use Eq. 11.3.64 for that energy loss.

11.17 A 90° square elbow connects two lengths of equal diameter pipe. The Reynolds number within the straight lengths is 10^5. The energy losses associated with the elbow can be considered to be equivalent to the loss due to flow through a certain length of straight pipe. Give the length, in number of pipe diameters, equivalent to a 90° square elbow. Take $R_b/D = 3.5$.

11.18 For flow through a circular orifice, as in Fig. P11.18a, we wish to create a plot of discharge coefficient versus Reynolds number. The discharge coefficient is defined in Eq. 11.4.4. We will simplify matters by considering cases for which the area ratio $S_o/S \ll 1$, where S_o is the area of the orifice, and S is that of the upstream flow. For laminar flow through an orifice, the

following theoretical result is available:

$$\Delta P = \frac{24Q\mu}{D^3} \qquad \textbf{(P11.18)}$$

The *Chemical Engineers Handbook* gives information that leads to the discharge coefficient plot shown in Fig. P11.18b. Plot C_d versus Re according to Eq. P11.18, and compare your graph to this figure. What do you infer about the value of Re at which flow through an orifice is no longer laminar?

11.19 Water at 60°F is pumped at 100 gal/min from a reservoir (1 in Fig. P11.19) through a system of piping into an open tank (2), the level of which is maintained constant at 17 ft above the level of the reservoir. Pipe lengths, diameters, and fittings also are shown in Fig. P11.19. The pump is known to be 70% efficient. Calculate the required horsepower rating of the pump.

11.20 A liquid flows through a tube of circular cross section at a Reynolds number of $UD\rho/\mu = 10$. The liquid has a density of 1200 kg/m^3 and a viscosity of 1 Pa · s; the tube is 0.1 m long and has an inside diameter of 0.001 m. What is the numerical value of the friction loss factor e_v defined in the text by Eq. 11.3.22?

11.21 The system shown in Fig. P11.21 is designed to produce a high speed water jet that will remove particulate contaminants from surfaces placed in front of the jet. For the design and operating conditions specified below, find the required gage pressure on the reservoir, ΔP, in units of newtons per meter squared.

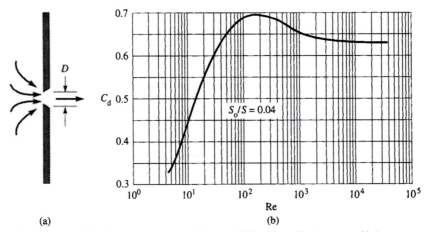

(a)   (b)

Figure P11.18 (a) Flow through an orifice and (b) orifice discharge coefficient versus Reynolds number.

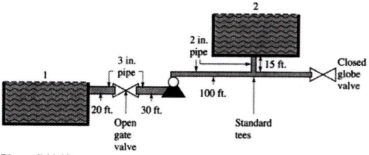

Figure P11.19

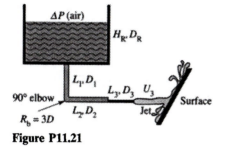

Figure P11.21

$H_R = 0.1$ m $L_1 = L_2 = 0.2$ m $L_3 = 0.06$ m
$D_R = 0.5$ m $D_1 = D_2 = 0.01$ m $D_3 = 0.003$ m
$U_3 = 30$ m/s

11.22 Water flows steadily from a tube and forms a coherent vertical cylindrical column, as in Fig. P11.22. The column *does not* break down into droplets over the distance L. Derive an analytical expression for the diameter of the jet as a function of L. State clearly any assumptions you make.

11.23 For the system sketched in Fig. P11.23, estimate the required theoretical horsepower for a pump that will produce the conditions shown. Give the pressure P_1 just downstream of the discharge of the pump. *Do not neglect friction losses in the piping.* Assume that the pump is frictionless. The supply reservoir is at atmospheric pressure, and there are no energy losses between the reservoir and the pump.

11.24 An orifice meter was calibrated in air, and later put in service to monitor the flow of helium. The user of the meter was unaware that the calibration had been done in air. What would be the error in the reported flowrate? High or low, and by what percentage?

11.25 Derive Eq. 11.5.11. Define carefully the definitions of the difference $\Delta(..)$ and the differential $d(..)$.

11.26 Derive Eq. 11.5.20.

11.27 Derive Eq. 11.5.29.

11.28 Does Fig. 11.5.1 depend strongly on the value of γ? What about Fig. 11.5.2?

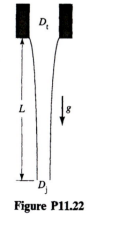

Figure P11.22

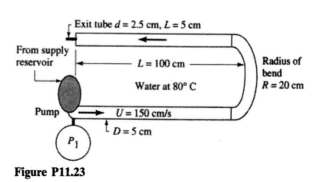

Figure P11.23

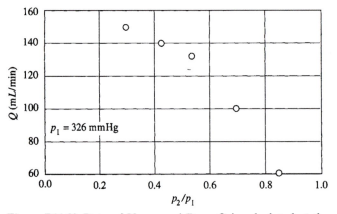

Figure P11.35 Data of Urone and Ross: Q is calculated at the upstream pressure.

11.29 Derive Eq. 11.5.59.

11.30 Using the results presented in Section 11.3.3, find the linear velocity for turbulent pipe flow in a pipe of optimum diameter. By what factor does this velocity change if the optimum pipe diameter changes by an order of magnitude?

11.31 Using the results presented in Section 11.3.3, find the optimum pipe diameter under the following conditions. The liquid is water at 25°C. The Reynolds number is to be 10^4. The system runs 80% of the year ($H = 7008$ h). The annualized construction, installation, and maintenance costs may be assumed to be 30% of the annualized capital costs.

$$K_p = 0.12 \text{ \$/kW·h} \quad \varepsilon_{pump} = 0.80$$
$$K = \$2.25/\text{m} \quad D_o = 0.1 \text{ m} \quad N = 10 \text{ years}$$

11.32 Following the method presented in Section 11.3.3, develop a model for the optimum pipe diameter under *laminar* flow conditions. Then rework Problem 11.31 for the case that $Re = 10^3$.

11.33 Go back to the comment that precedes Eq. 11.3.51. Since mass is conserved, isn't w a constant in this flow? How can w be zero at plane 1 and nonzero at plane 2?

11.34 In the calculations of Examples 11.5.6 and 11.5.7, we use the laminar flow friction factor and the actual length of the capillary. In particular, we do not account for entrance effects. Argue that it is reasonable to neglect entrance effects in Examples 11.5.6 and 11.5.7.

11.35 Urone and Ross [*Environ. Sci. Technol.,* **13**, 351 (1979)] present the data shown in Fig. P11.35. Using the method outlined in Example 11.5.7, predict the volumetric flowrate Q, measured at the upstream pressure ($p_1 = 326$ mmHg), for each downstream pressure p_2 and compare your predictions to the data. Assume *isothermal* flow at $T = 298$ K. The tube dimensions are $L = 12.7$ mm and $D = 0.19$ mm.

11.36 Repeat Problem 11.35, but assume that the flow is *adiabatic* throughout the tube.

11.37 Show that we may obtain Eq. 11.5.77 by starting with Eq. 11.5.55 and integrating from $z = 0$ to $z = L$, where the Mach numbers are denoted Ma_1 and Ma_2, respectively. Is the approximation $\gamma Ma^2 \ll 1$ necessary?

11.38 Find the volumetric flowrate Q through the piping system shown in Fig. P11.38. The fluid is water at 25°C. State clearly any approximations that led you to your result.

11.39 For the system shown in Fig. P11.39, the observed flowrate is found to be $Q = 4$ ft³/s. Is there really a pump inside the pump house? State clearly the basis for your answer.

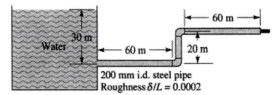

Figure P11.38

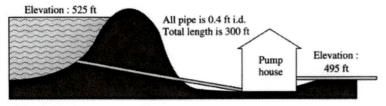

Elevation : 525 ft

All pipe is 0.4 ft i.d.
Total length is 300 ft

Pump
house

Elevation :
495 ft

Figure P11.39

11.40 Repeat Problem 4.66 for turbulent flow.
11.41 Does Eq. 11.3.79 give the same D_{opt} that we find in Problem 9.4 of Chapter 9?
11.42 Find the volumetric flowrate from each pipe depicted in Fig. P11.42.
11.43 Plot volumetric flowrate Q (mL/s) versus pipe length L, for L in the range 0.5–12 inches, for the system shown in Fig. P11.43. Account for entrance losses. For what pipe length does the flow become turbulent?

Using the relationship

$$\Delta P_{21} = -\rho g L + \frac{E_{v23}}{\langle U_2 \rangle S_2} \quad \textbf{(P11.43)}$$

find ΔP_{21} as a function of Q, using the results of the first part of this problem.

11.44 An agricultural water sprinkler is designed as suggested in Fig. P11.44. The water pressure is 50 psig just before the entrance to the rotating arms of the sprinkler. What is the magnitude of the reactive thrust exerted on the tip of the sprinkler, in units of pounds force?

11.45 Water flows under the "sluice gate" shown in Fig. P11.45. Find the force exerted by the water on the gate, in units of newtons per meter of gate width.

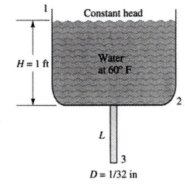

Constant head

$H = 1$ ft

Water
at 60° F

2

L

3

$D = 1/32$ in

Figure P11.43

11.46 A long pipe ends in a short converging nozzle bolted onto the end of the pipe, as shown in Fig. P11.46. Find the force that must be provided by the flange bolts, and note the direction of the force. Neglect friction losses in the nozzle itself.

11.47 It is necessary to transfer water, at 20°C, from the upper to the lower tank shown in Fig. P11.47, taking advantage of the available hydrostatic head. If the required flowrate is $Q = 0.17$

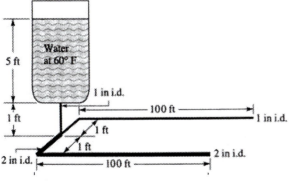

5 ft

Water
at 60° F

1 in i.d.

1 ft

2 in i.d.

100 ft

1 ft

1 ft

100 ft

1 in i.d.

2 in i.d.

Figure P11.42

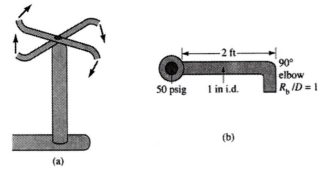

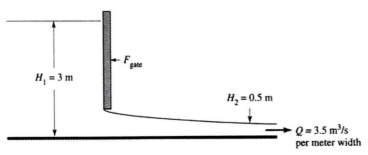

(a)

(b)

Figure P11.44 (a) A sprinkler system and (b) detail of one sprinkler arm.

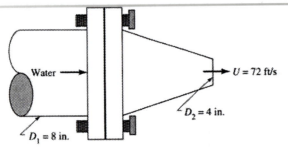

Figure P11.45

Figure P11.46

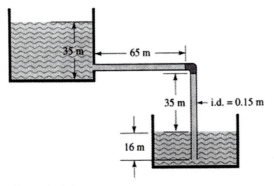

Figure P11.47

m^3/s, what is the required inside diameter of the horizontal pipe? Assume that all pipe is smooth. The water levels in both reservoirs are maintained constant.

11.48 The four-armed water sprinkler described in Fig. P11.44 is fed by a delivery line as sketched in Fig. P11.48.

a. If P_1 is 50 psig, as in Problem 11.44, what is the required pressure P_o?

b. If we add two more "arms" and require that P_1 still be 50 psig, what pressure is now required at P_o?

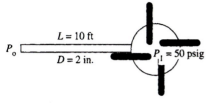

Figure P11.48

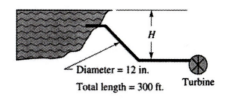

Figure P11.50

c. If, with the addition of the two "arms," we must maintain the pressure P_0 that is found in part a, what diameter delivery pipe can we use between points 0 and 1?

11.49 A turbine is driven by a flow of 0.28 m³/s from a reservoir, as shown in Fig. P11.49. What is the ideal power (in kilowatts) that we can expect to derive from this system?

11.50 A turbine is driven by flow from a reservoir as indicated in Fig. P11.50, and it is observed that the turbine delivers 100 hp when the flowrate is 20 ft³/s. The elevation H of the surface of the reservoir above the turbine is not known. The "plumbing" details are not known either, except that the 300 ft of piping is 12 inch i.d. We want to estimate the flowrate that can be achieved if the turbine is removed from the system.

11.51 We want to determine the energy loss factor for our product, a globe valve, as a function of the "percent closure" of the valve. Current practice is to put the valve into a plumbing system like that shown in Fig. P11.51 and measure the flowrate. From the data given, calculate values for the loss factor K_v defined as

$$K_v = \frac{h_v}{U^2/2g} \qquad \text{(P11.51)}$$

where h_v is the equivalent head loss due to the presence of the valve.

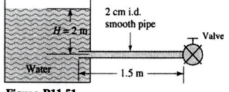

Figure P11.51

VALVE CLOSURE (%)	OBSERVED FLOW (ft³/min)
0	2
25	1.3
40	1.1

What percentage of reduction in flow is due to the fully opened valve? That is, what would the flow be in the absence of the valve?

11.52 Derive a model for the time required for a low viscosity liquid to drain from a tank of conical shape, as shown in Fig. P11.52. The lower opening through which the liquid drains has a radius R_0. The tank is initially filled to the height H_L above the opening. Note that $h(t)$ is mea-

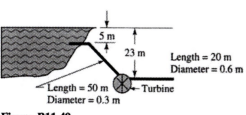

Figure P11.49

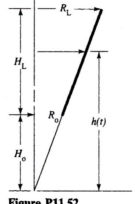

Figure P11.52

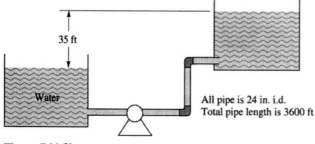

Figure P11.53

sured above the apex of the cone—not above the level of the opening.

11.53 Find the power required to pump water at a flowrate of 0.65 m³/s from the lower to the upper reservoir of the system shown in Fig. P11.53. The pump efficiency is 75%. Assume that the piping is smooth.

What additional power is required after the pipe corrodes to the point that the roughness factor (see Fig. 11.3.3) is $\delta/D = 0.001$?

11.54 A 150 kW pump with an efficiency of 70% is available to pump water to a level 39 m above the surface of the reservoir shown in Fig. P11.54. What flowrate can we expect? Give the relative contributions of potential energy, kinetic energy, and friction to the pumping requirement.

11.55 An oil of viscosity $\mu = 20$ mPa·s and density $\rho = 900$ kg/m³ is transferred by gravity flow between two tanks, as shown in Fig. P11.55. Find the flowrate.

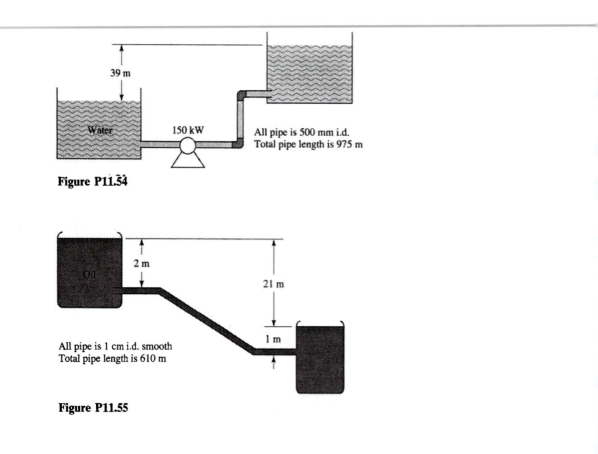

Figure P11.54

Figure P11.55

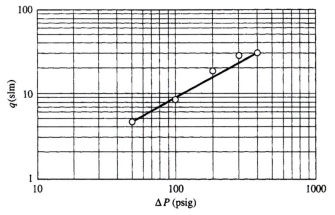

Figure P11.58 Data for q versus ΔP: The line is the solution to the equations for adiabatic flow.

11.56 Explain why K_e for an elbow (see Fig. 11.3.9) goes through a minimum. Wouldn't you expect the energy loss associated with the bend to become very small as the bend "straightens out" (i.e., as R_b/D gets large)?

11.57 Begin with Eq. 11.5.13, and derive a model for the isothermal compressible turbulent flow of an ideal gas through a long pipe of circular cross section. Do the same for adiabatic flow. In both cases, give an expression for the mass flowrate through the pipe in terms of the pressures p_o and p_L at two axial positions $z = 0$ and $z = L$ within the pipe.

11.58 Predict the flowrate of nitrogen through a tube of inside diameter $D = 0.015$ inch and length $L = 0.25$ inch, as a function of the pressure p_U in a large chamber upstream from the entrance to the pipe. The temperature in the upstream chamber is 293 K. The downstream pressure is atmospheric in all cases. Convert the flowrates at each upstream pressure to units of standard liters per minute (slm): that is, volume flowrate "corrected" to a pressure of one atmosphere and a temperature of 273 K. Compare your predictions to the data of Kumagai [*J. Vac. Sci. Technol.*, **A8**, 2865 (1990)], shown in Fig. P11.58. Assume that the flow into the pipe from the upstream chamber is adiabatic and frictionless.

ΔP (psig)	50	100	200	300	400
q (slm)	4.2	8.7	18.2	25.6	30.1

11.59 For the flow restriction and conditions described in Problem 11.58, a chamber pressure of 30 psig is measured for the flow of *silane*. What is the flowrate, in standard liters per minute?

11.60 Figure P11.60 shows a selection of data presented by Harvey and Kubie [*Proc. Inst. Mech. Eng.*, **209**, 363 (1995)], who injected water into a cylindrical tank of height 370 mm and internal radius 120 mm. The tank had an orifice of radius R centered on the bottom of the tank. The injection method caused the water to swirl circumferentially inside the tank, but a set of baffles placed on the bottom of the tank prevented vortex formation without impeding the outflow. For a given flowrate Q the investigators measured the steady state height of the water in the tank, h. Use Torricelli's law to predict the data shown, and explain any deviations you observe in terms of the conservation of energy principle that leads to Torricelli's law.

11.61 Show that the equivalent length of a 90° elbow is

$$\frac{L_e}{D} = \frac{K_e}{4f} \qquad \textbf{(P11.61)}$$

where K_e is found from Fig. 11.3.9.

11.62 Air is flowing through the two-dimensional, wedge-shaped region shown in Fig. P11.62. The pressure difference $p_A - p_B$ is being measured with a liquid manometer, as shown. The air flowrate (per unit width normal to the

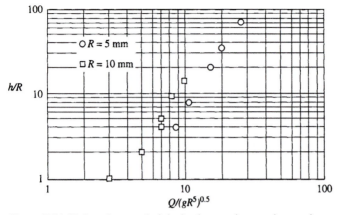

Figure P11.60 Steady state height h of water in a tank, as a function of flowrate Q, for two values of the outlet orifice radius R.

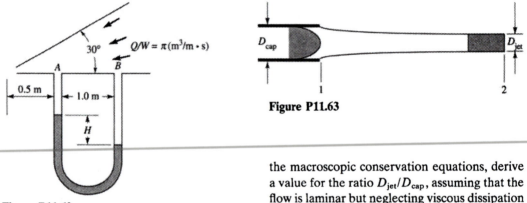

Figure P11.62

Figure P11.63

page) is π m³/s · m. Find the pressure difference $p_A - p_B$ as well as the elevation difference H. The manometer fluid has a density of 1000 kg/m³.

11.63 As suggested in Fig. P11.63, a liquid in laminar flow issues from a long capillary of circular cross section as a horizontal jet into still air. At sufficiently high speed, the jet remains nearly horizontal for a distance of many diameters downstream of the capillary exit. Hence gravity is unimportant. The jet diameter far downstream (but still in the horizontal region) is observed to be different from the capillary diameter. Using the macroscopic conservation equations, derive a value for the ratio D_{jet}/D_{cap}, assuming that the flow is laminar but neglecting viscous dissipation as the velocity profile changes from parabolic at the capillary exit to flat far downstream. The experimentally observed value is $D_{jet}/D_{cap} = 0.867$.

11.64 Police arrive at the scene of a murder and find a body and a leaking barrel of wine. The barrel has two 6 mm bullet holes 20 cm above the floor level. Wine is still leaking, and the wine level at the time of arrival of the police is noted to be 36 cm above the level of the bullet holes. The total height of the wine barrel is 1.2 m, and its diameter is 0.6 m. Assuming that the barrel was full at the time of the shooting, how much time has elapsed between the shooting and the arrival of the police?

Appendix

Unit Conversions

While the International System (SI) of units is being adopted by most journals and texts, much of the existing and important literature uses other units. It is necessary to be able to convert from one system to another. This is especially true for the case of physical property data, which may be available in several units depending on the sources used.

The fundamental mechanical units of the SI system are the meter (m) for length, the kilogram (kg) for mass, and the second (s) for time. From these basic "measures" we can derive units for commonly measured mechanical quantities. These units, in turn, are often given special names in honor of pioneers in the fields of physics, mechanics, and thermodynamics. For example, we have the following units that appear commonly throughout this text:

Force	$kg \cdot m/s^2$	newton (N)
Pressure	$kg/m \cdot s^2 = N/m^2$	pascal (Pa)
Energy, work	$kg \cdot m^2/s^2 = N \cdot m$	joule (J)
Power	$kg \cdot m^2/s^3 = J/s$	watt (W)

In fluid dynamics we commonly require values for viscosity and surface tension of fluids. In SI units we find

Viscosity	$kg/m \cdot s = Pa \cdot s$
Surface tension	$kg/s^2 = N/m$

Of course the mass density is simply

Density	kg/m^3

The dominant system of units of the second half of the twentieth century has been the metric, or cgs, system. In that system we commonly find

Force	$g \cdot cm/s^2$	dyne
Pressure	$g/cm \cdot s^2$	dyn/cm^2
Energy, work	$g \cdot cm^2/s^2$	erg

The British system of units (ft, lb, s) is largely being replaced in modern technical literature, but one must still be able to convert to SI units because of the numerous commonplace uses of the British system, such as gallons and cubic feet for volume measure, and pounds force for weight.

Unit Conversion Table

To Convert from:	To	Multiply by
	Acceleration	
ft/s^2	m/s^2	0.3048
	Area	
in.^{2}	m^2	6.45×10^{-4}
ft^2	m^2	0.0929
cm^2	m^2	10^{-4}
	Density	
g/cm^3	kg/m^3	1000
lb/ft^3	kg/m^3	16.02
	Energy	
Btu (thermochemical)	kg $\cdot$ m^2/s^2 = N $\cdot$ m = joule (J)	1054
calorie (thermochemical)	joule (J)	4.184
kilocalorie (thermochemical)	joule (J)	4184
erg	joule (J)	10^{-7}
ft $\cdot$ lb$_f$	joule (J)	1.356
kW $\cdot$ h	joule (J)	3.6×10^6
	Force	
dyne	kg $\cdot$ m/s^2 = N	10^{-5}
lb$_f$	N	4.448
poundal	N	0.138
	Heat Transfer Coefficients	
Btu/ft$^2 \cdot$ h $\cdot$ °F	kg/s$^3 \cdot$ K = W/m$^2 \cdot$ K	5.678
lb$_f$/ft $\cdot$ s $\cdot$ °F	W/m$^2 \cdot$ K	26.27
lb$_m$/ $\cdot$ s$^3 \cdot$ °F	W/m$^2 \cdot$ K	0.816
g/s$^3 \cdot$ K	W/m$^2 \cdot$ K	0.001
cal/cm$^2 \cdot$ s $\cdot$ K	W/m$^2 \cdot$ K	41,840
	Length	
angstrom (Å)	m	10^{-10}
ft	m	0.3048
inch	m	0.0254
micrometer (μm)	m	10^{-6}
	Mass	
pound (lb)	kg	0.4536
ton (2000 lb)	kg	907
	Power	
Btu/s	kg $\cdot$ m^2/s^3 = J/s = watt (W)	1054
cal/s	W	4.184
ft $\cdot$ lb$_f$/s	W	1.356
horsepower	W	745.7
kcal/s	W	4184

Pressure

atmosphere (atm)	$kg/m \cdot s^2 = N/m^2 =$ pascal (Pa)	1.013×10^5
bar	Pa	10^5
cmHg	Pa	1333
cmH_2O	Pa	98.06
dyn/cm^2	Pa	0.1
$in.H_2O$	Pa	249
lb_f/ft^2	Pa	47.88
$lb_f/in.^2$ (psi)	Pa	6895
mmHg	Pa	133.3
torr	Pa	133.3

Thermal Conductivity

$Btu/h \cdot ft \cdot °F$	$kg \cdot m/s^3 \cdot K = W/m \cdot K$	1.7307
$lb_f/s \cdot °F$	$W/m \cdot K$	8.007
$cal/s \cdot cm \cdot K$	$W/m \cdot K$	418.4
$g \cdot cm/s^3 \cdot K$	$W/m \cdot K$	10^{-5}
$lb_m \cdot ft/s^{-3} \cdot °F$	$W/m \cdot K$	0.2489

Thermal Diffusivity

ft^2/h	m^2/s	2.58×10^{-5}
cm^2/s	m^2/s	10^{-4}

Viscosity

centipoise (cP)	$kg/m \cdot s = Pa \cdot s$	0.001
poise	$Pa \cdot s$	0.1
$lb_f \cdot s/ft^2$	$Pa \cdot s$	47.88

Volume

ft^3	m^3	0.0283
gallon (U.S.)	m^3	3.785×10^{-3}
$in.^3$	m^3	1.639×10^{-5}
liter (L)	m^3	0.001

Index

CPSIA information can be obtained at www.ICGtesting.com
Printed in the USA
LVOW051604250911

247793LV00003B/24/A